GRAPHENE SIMULATION

Edited by **Jian Ru Gong**

Graphene Simulation
Edited by Jian Ru Gong

Published by InTech
Janeza Trdine 9, 51000 Rijeka, Croatia

Publishing Process Manager Iva Simcic

Technical Editor Teodora Smiljanic

Cover Designer InTech Design Team

Additional hard copies can be obtained from orders@intechopen.com

Graphene Simulation, Edited by Jian Ru Gong
p. cm.
ISBN 978-953-307-556-3

For EU product safety concerns:
IN TECH d.o.o., Prolaz Marije Krucifikse Kozulić 3, 51000 Rijeka, Croatia,
info@intechopen.com or visit our website at intechopen.com.

Meet the editor

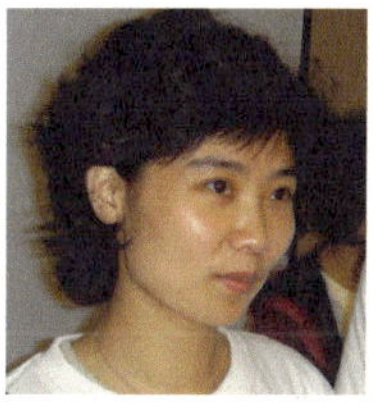

Professor Jian Ru Gong received her Ph.D. degree in Physical Chemistry from Institute of Chemistry, Chinese Academy of Sciences under the direction of Prof. Chunli Bai in 2005. She was a Postdoctoral fellow with Professor Charles M. Lieber at Harvard University, Chemistry & Chemical Biology Department during 2005-2008, and has joined National Center for Nanoscience and Technology, China in 2009. She won Chinese National Natural Science Prize, among other prizes and honors; published high-impact papers in Proc. Natl. Acad. Sci. USA, J. Am. Chem. Soc., Nano Lett., Small, J. Phys. Chem. B, Chem. Mater., etc. She is a member of the editorial board for Journal of Bioterrorism & Biodefense, and a reviewer for Journal of the American Chemistry Society and Wiley-VCH. Graphene nanomaterial is one of her current research interest, including synthesis, characterization, application in nanoelectronics, catalysis, biology, and many others similar fields.

Contents

Preface

Graphene, a single layer of graphite formed by a repetitive hexagonal lattice, has recently raised extensive interest of the world-wide scientific community since it has been obtained successfully in 2004. The graphene is of significant value for the fundamental studies of condensed matter and quantum physics. The main techniques of graphene simulation based on density functional theory (DFT) and non-equilibrium Green's function techniques provides the accurate results of electronic structure and transport. In this section, semi-empirical simulation as a necessary implement facilitates the larger-scale system simulation. However, an important aspect to address real world problems is the limitation to rather small system sizes of typically a few hundred atoms, which makes it necessary to combine them with the analysis of simulation results based on molecular dynamics techniques. Recent developments of quantum theory on graphene can assist in the exploration of Dirac fermion behavior, external field interaction etc.

Dr. Jian Ru Gong,
National Center for Nanoscience and Technology
China (NCNST)

Part 1

Electronic Structure and Transport Simulation

1

DFT Calculation for Adatom Adsorption on Graphene

Kengo Nakada and Akira Ishii
Department of Applied Mathematics and Physics
Tottori University, Tottori
JST-CREST, 5 Sanbancho
Chiyoda-ku, Tokyo
Japan

1. Introduction

Graphene is well-known to be two-dimensional material made of carbon atoms. Graphene is the basic material to form nanotube, fullerene and graphite. Graphene is a substance that attracts attention not only as parts of the nanocarbons but also for its own interesting electronic and mechanic properties (T. Ando, A. K. Geim et.al, K.S. Novoselov et.al). In the last decade, the most significant problem is how to make wide and high quality graphene itself. Nowadays, good quality graphene can be made in labolatories, for example, using SiC(0001) surface. Thus, one of the next step of the research related to graphene is how to make nano structures on graphene plane (V. M. Karphan et.al).

Recently, H.Fujioka et al. has suceeded for the growth of GaN on graphite using PLD (Pulsed Laser Deposition) method and they success to make light emission diode using GaN on graphite (K. Ueno et.al, A. Kobayashi et.al 2006, A. Kobayashi et.al 2007, G. Li et.al, M.-H. Kim et.al, H.Fujioka 2009). Since the graphite is made of the stacking of many graphene plane, similar growth will be possible for graphene, if the graphene plane is supported mechanically with the other certain material. Such growth is not limited to GaN, but many other possibilities to form nano structures, nano devices or thin films on graphite plane. In order to apply graphene for such purpose, however, we should first investigate deeply for the interaction between adatoms and graphene plane (A. Ishii. et.al 2008, K.Nakada et.al). In experiments, the adatom adsroption on graphene is reported for some atomic species (I. Zanella et.a, K. Kong et.al, K. Okazaki et.al, A. Lugo-Solis et.al, M. Wu et.al, H.Gao et.al, J. Dai et.al) , but not for all atomic species.

In this chapter, we introduce the adsorption mechanism, atomic structures, stabilty, migration barriere energies and electronic properties of the adatom adsorption system on graphene plane for most all atomic species using the density functional theory. First, we review briefly the basic properties of graphene used in the following sections as remarks. The band structure of graphene is shown in figure 1. Because of the in-plane hexagonal symmetry, the p_x and p_y orbitals are degenerated. The s-orbital and the degenerated p_x and p_y orbitals make sp^2 hybrid orbitals in the graphene plane as σ bonds. The p_z orbitals are out of the sp^2 hybrid orbitals and they form the pi bonds normal to the graphene plane. These p_z orbitals forms bonding orbital (π) and anti-bonding orbital (π^*) below and above the Fermi energy level. These two band has no bandgap and they contact at one point in the momentum space just at the Fermi energy.

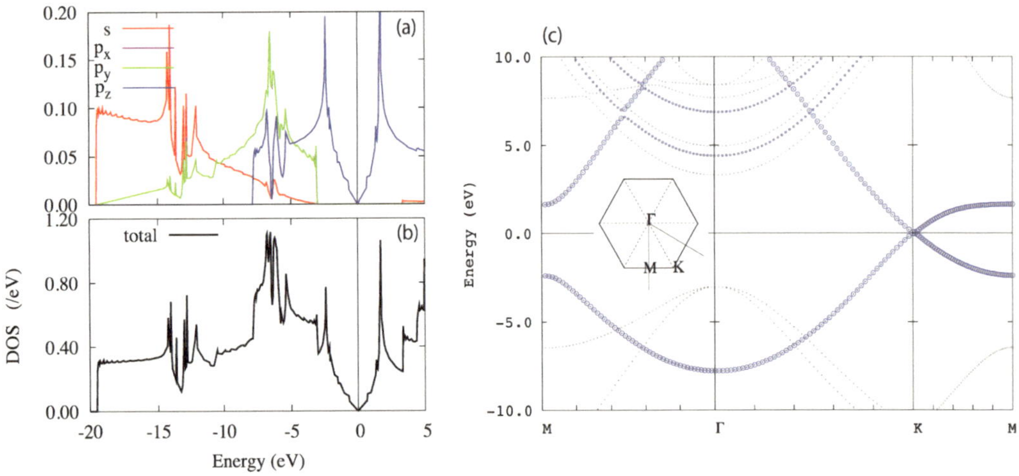

Fig. 1. (a) Local density of states projected to s, p_x, p_y, p_z orbitals for graphene. (b) Total local density of states for graphene. The LDOS projected to p_x and p_y are degenerated. (c) Band structure of 1×1 graphene. The blue lines in the figure correspond to wave functions of p_z orbitals.

Near the point, the band is linear so that the effective mass of the electron of the bands are zero. This point is known to be "Dirac point", because the band structure around the point is similar to the massless Dirac particle as the solution of the Dirac equation of the relativistic quantum mechanics. The Dirac point is very important in the physics of graphene. Because of this feature, the mobility of electron in graphene is very large. The theoretical prediction of the mobility is 1000 times larger than silicon and experimentally observed mobility is more than 100 times, at least. This large electron mobility is one of the significant reason that the graphene is expected to be the material of the future nano device. Using the large mobility, we can expect a lot of application for graphene for small gate voltage for electrons and holes of the device. Moreover, interesting features of graphene are large heat conductivity, large Young's modulus and light weight because of carbon atoms. Because of the two-dimensionality of graphene, adsorption of atoms or molecules on graphene affects the electronic properties of graphene itself dominantly through the p_z-orbitals. It means that the doping effect for graphene is very interesting.

2. Calculation method

In this work, we used a first-principles band calculation technique based on density functional theory. We used VASP (G.Kresse et.al 1993, G.Kresse 1993, G. Kresse et.al 1996, G. Kresse et.al 1996) which is a first-principles calculation code with high precision using the PAW method (G. Kresse and D. Joubert). We adopted LDA (P. Hohenberg and W. Kohn, W. Kohn and L. J. Sham) as the term exchange correlation with a cutoff energy of 500 eV and all calculations performed nonmagnetically. The unit cell for the graphene sheet adopted a 3×3 structure. The lattice constant of the grapheme used the value optimized by

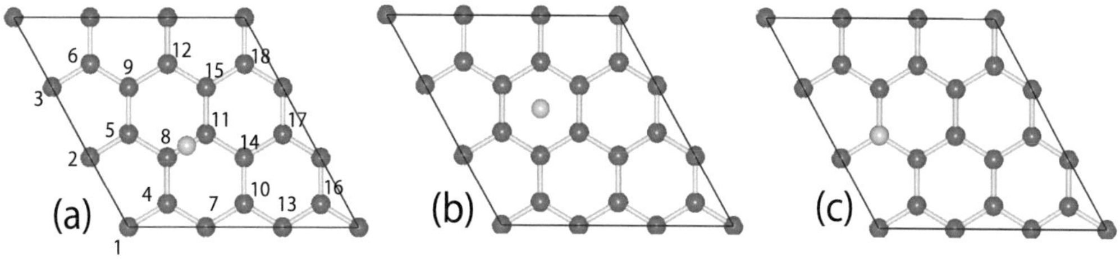

Fig. 2. Adsorption sites of graphene of 3×3 supercell. (a) Bridge site (B-site) positioned at the center of C-C bond. Number in the figure shows us the numbering of each carbon atoms of the 3×3 supercell. (b) Hexagonal site (H-site) positioned at the center of the six membered ring. (c) Ontop site or tetrahedoral site (T-site)

calculation. The distance between graphene sheets is about 14.7Å and the distance between adatoms is about 7.3Å. The final potential is constructed self-consistently from eigenstates at 24 sampling k-points in the irreducible Brillouin zone (IBZ). For the calculation of adatoms at certain sites, the position coordinate of the adatom parallel to the surface is fixed and the coordinate normal to the surface is fully relaxed. One atom of the edge of the 3×3 structure of the graphene sheet is fixed during the relaxation of the other carbon atoms of the sheet. Using the coordinate which converged potential, we performed convergent calculation using 240 k-sampling point in the IBZ. To obtain a final potential, we calculated a 3×3 graphene. Fig.3 is band structure and BZ of 3×3 graphene. Owing to the supercell used in the calculations, the K point of the 1×1 grapehen BZ in Fig.1(c) is folded into the Γ point of the supercell BZ. Similarly, the M point of the 1×1 graphene BZ is foled into the Γ of the supercell BZ. We calculated supercell BZ with M-Γ-K-M line to follow the dispersion of the Dirac point. The calculation was carried out at three adsorption sites, H, B and T shown in Fig. 1. We calculated the adsorption energy from the formula

$$E_{bond} = (E_{graphene} + E_{adatom} - E_{total}). \quad (1)$$

E_{bond} is the binding energy of the adsorbed atom to the graphene sheet. $E_{graphene}$ is the total energy of one sheet of the graphene and E_{adatom} is the total energy as an isolated atom of the adatom. We treated almost all the elements of the periodic table except the lanthanoids and noble gases as adatoms and carried out the calculation from H to Bi.

3. Result and discussion

3.1 Adsorption energy

The 3d transition metal is spin polarized at low temperature. In some groups (K. T. Chan et.al, P. A. Khomyakov et.al), calculations considering spin polarization are performed for a few adatoms. However, if the nonmagnetic state is one of the ground state, its calculation is as important as the spin polarization calculation. In other words, the nonmagnetic state is in the condition to take an average of the spin, and it is the starting point for discussion of the ferromagnetic state and the discussion of high temperature. As the first step in the discussion of the growth of a compound semiconductor on graphene, we discuss the electronic state in the nonmagnetic state. Fig. 4 is the calculation result for the nonmagnetic state. This figure shows the most stable adsorption site and bond energy when various adatoms are adsorbed at three adsorption sites. In the figure, the most stable sites for each adatom are indicated by colors. The green, red and yellow boxes mean that the most stable

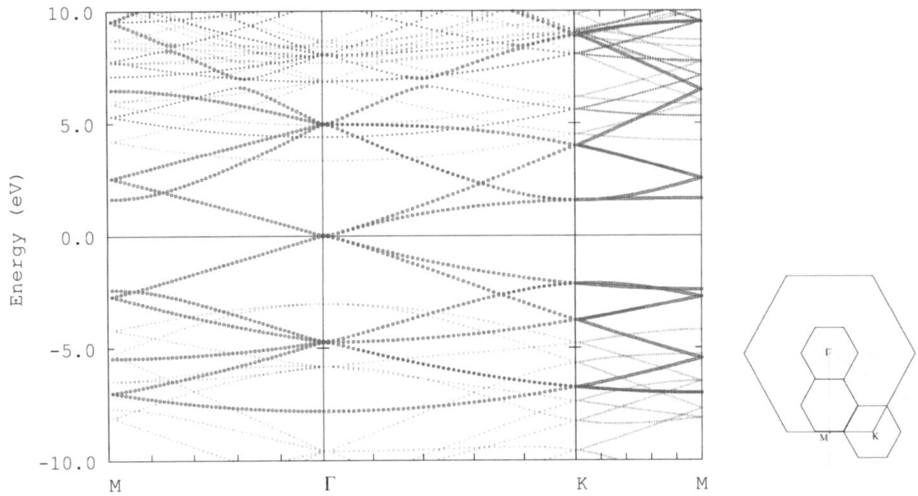

Fig. 3. The band structure and BZ of 3×3 graphene. The blue lines in the figure correspond to wave functions of p_z-orbitals.

site is the B-site, H-site or T-site respectively. The value in Fig. 4 shows the magnitude of the adsorption energy when each adatom is adsorbed to the most stable adsorption site. This result shows that for the transition metal elements the most commonly stable is the H-site. For the nonmetallic elements, the B-site is most stable. For H, F, Cl, Br and I, where the valence electron number of the adatom is 1, the T-site is the most stable adsorption site. In addition, for transition metal elements, the magnitude of the adsorption energy of each adatom is very large. The largest adsorption energy is shown for the adatoms of the nonmetallic elements C, N, O. For the transition metal elements, the bond energy shows an increasing tendency with an increasing number of d-electrons. Furthermore, it shows a tendency for the bond energy to decrease when the number of d-electrons increases to more than half occupancy, because the d-orbitals are shifted down. Therefore we find very large bond energy for Mn, where the d-orbital is half occupied. The bond energy is a very large on metal element adsorption, but this state is unstable because it is constructed from a localized non-bonding band at the Fermi level and the number density of states at the Fermi level (DOS(E_F)) is very large. There is low bond energy for Cu, Ag, Au because all d-orbitals are occupied. Furthermore, the bond energy becomes very small for Zn, Cd, and Hg, where the s-orbital is close to the d-orbital. In addition, there is almost no difference in the adsorption energy of the three adsorption sites when the adsorption energy of the adatom is small. In contrast, the adsorption energy is large for the nonmetallic elements C, N, O and the difference of adsorption energy between sites is also large (higher than 3.0 eV). This shows it to be easy to adhere strongly to the B-site. However, there are a few differences between the adsorption energy of the T-site and the B-site when the adatom is C. Fig. 5 is a table of the bond distance between graphene and the adatom. ('The bond distance' means the distance between the average of the position of the graphene sheet and the adatom.) In the case that the bond distance between the adatom and graphene is large, the binding energy tends to reduce. In the case of a large bond distance, the adatom shows physical adsorption-like bonding. When the bond distance is short, the bond energy tends to increase. In this case the bonding feature is like chemical adsorption.

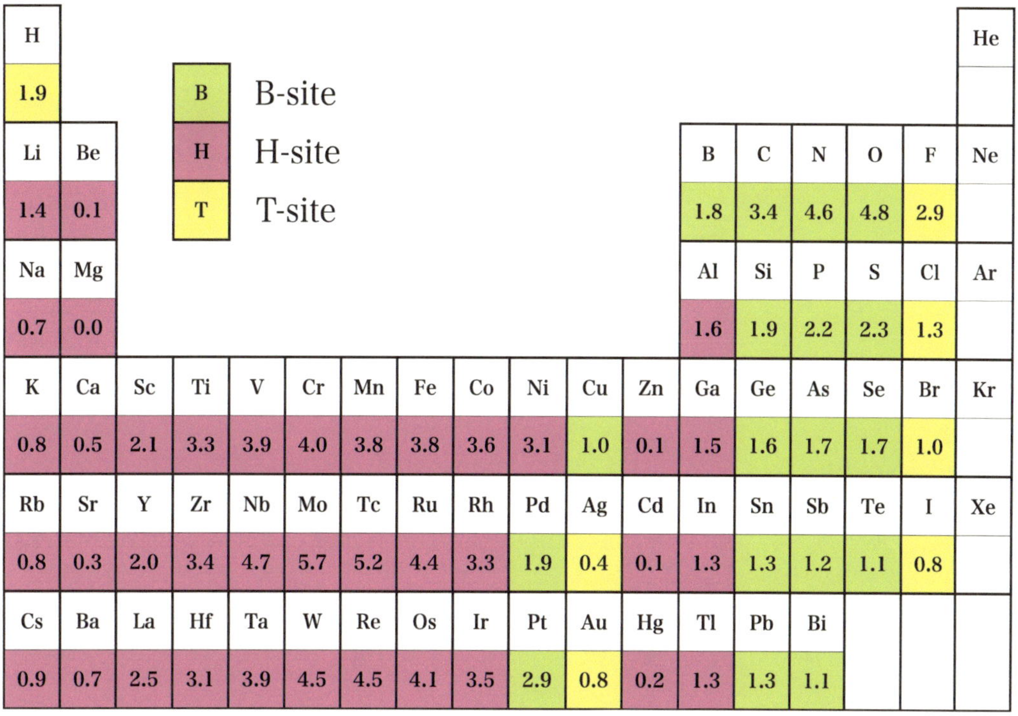

Fig. 4. The most stable sites and bond energy when an adatom was adsorbed: green is B-site, red is H-site and yellow is T-site.

In figure 6, typical examples for physical and chemical adsorption are shown. We show the band structure, the amount of carbon p_z character is indicated by a red of the bands. The Fermi level is at zero energy. The adatom of model metal element is Fig.6(a) Na, (b) K, (c) Ru and (d) Cs. The adatom of nonmetallic element is (e) C, (f) N (g) O and (h) F. The Dirac point corresponds to the crossing of bands at Γ with predominantly p_z character, as is clearly visibel for physisorbed Na, K, Ru and Cs on graphene. For chemisorbed C, N, O and F on graphene, the Dirac point disappear and the bands have a mixed character. If the binding energy is small, (if adsorbed Na,K,Ru or Cs), the characteristic conical points of band structure at Γ can still be clearly identified. In contrast, if the binding energy is large, (if adsorbe C,N,O or F), the graphene bands are strongly perturbed. In particular, the characteristic Dirac point of graphene at Γ are destroyed because the graphene p_z states hybridize strongly with the adatom. However, a very wide hybridized orbital is constructed between graphene and adatom. Furthermore, Fermi level shifts to upper on the C,N,O and F atomic adsorption. In other words, it corresponds to an electronic dope. The tendency of the adsorption is classifiable to two widely as had shown in typical example. The physical adsorption and the chemical adsorption can discuss from the bond distance. We discuss on tendency of the stability from adsorption energy and an adsorption distance. The long bond distance between graphene and adatom shows physical adsorption, and the short bond distance shows chemical adsorption.

Figure 7 shows local density of state and density of states when we adsorbed element of the fourth period (K,Ca,Sc,Ti,V,Cr,Mn,Fe,Co,Ni,Cu,Zn,Ga,Ge,As and Se). We arranged model metal, transition metal and the calculation which adsorbed the nonmetallic element. The (a)-(p) in fig.7 corresponds to K,Ca,Se,Ti,V,Cr,Mn,Fe,Co,Ni,Cu,Zn,Ga,Ge,As and Se adsorption.

less than 2.0Å (red)
more than 2.0Å (white)

H																	He
1.49																	
Li	Be											B	C	N	O	F	Ne
1.62	2.93											1.72	1.65	1.62	1.59	1.87	
Na	Mg											Al	Si	P	S	Cl	Ar
2.22	3.21											2.04	2.03	2.09	2.08	2.56	
K	Ca	Sc	Ti	V	Cr	Mn	Fe	Co	Ni	Cu	Zn	Ga	Ge	As	Se	Br	Kr
2.58	2.14	1.76	1.56	1.47	1.42	1.38	1.38	1.42	1.47	2.03	3.02	2.11	2.16	2.22	2.25	2.78	
Rb	Sr	Y	Zr	Nb	Mo	Tc	Ru	Rh	Pd	Ag	Cd	In	Sn	Sb	Te	I	Xe
2.74	2.37	2.00	1.74	1.66	1.59	1.57	1.62	1.71	2.08	2.42	3.18	2.35	2.42	2.46	2.49	3.26	
Cs	Ba	La	Hf	Ta	W	Re	Os	Ir	Pt	Au	Hg	Tl	Pb	Bi			
2.84	2.49	2.07	1.85	1.65	1.60	1.58	1.61	1.70	2.12	2.41	3.13	2.48	2.53	2.57			

Fig. 5. Distance between the adatom and grapehen in the most stable adsorption site: a red substrat shows distances less than 2Å, a white substrate shows distances more than 2Å.

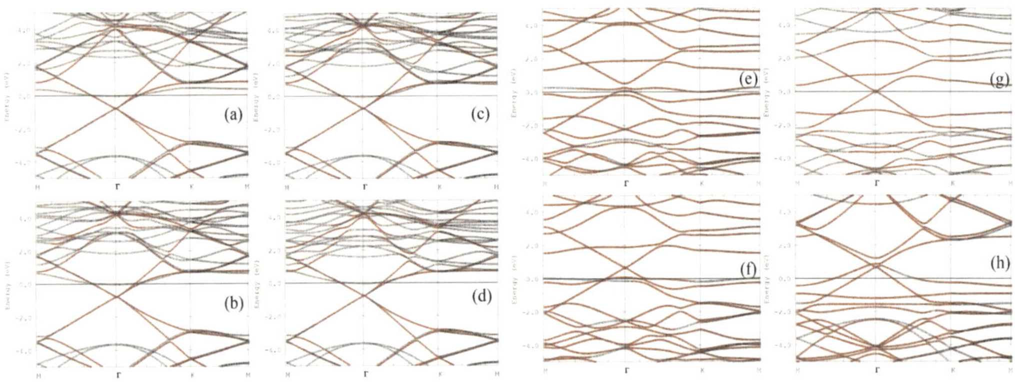

Fig. 6. Typical example for physical and chemical adsorption. A red circles correspond to the amount of adatom of p_z charactor. The adatom is (a) Na, (b) K, (c) Ru, (d) Cs, (e) C, (f) N (g) O and (h) F.

The left axis of fig. 6 shows density of states (/eV), and the right axis shows local density of state of adatom (/eV). It is performed a projection of a wave function of the adatom to s,p,d orbital. The wave function character is calculated, either by projecting the orbitals onto spherical harmonics that are non-zero within spheres of a radius. Each calculation is result of the H-site adsorption. In a model metallic element and the transition metal element, H site adsorption is most stable. B-site adsorption is most stable in the metalloids such as Ge,As and Br, but shows calculation result of the H-site adsorption for comparison. 18 C atoms constituting graphene is included in the 3×3 unit cell. We change an axial contraction scale of LDOS to compare LDOS of the adatom with the DOS of the total. As for the K adatom adsorption on the graphene in figure 7(a), peak structure of the p-orbital appears around 12eV. The 3p-orbital of the K atom is a closed shell, but it is necessary to treat 3p-orbital as a valence electron because there is the orbital in a shallow rank. In other words, we prepared for pseudopotential to treat 3p-orbital as a valence electron to treat it as shallow core. Almost of 3p bands are lone status, but hybridization with the graphene is slightly shown. The 3d-orbital is not occupied with an electron. Conduction band constructed from 3d is fermi level upper 5eV. In the K and the Ca atomic adsorption, the 3d-orbital is located in the conduction electron band of the Fermi surface upper part. However, the valence band around the Fermi level is occupied by d-orbital of the adatom when we adsorbed adatom from Sc to Cu. One of the reasons of strong energy when transition metal was adsorbed is a hybridized orbital between 3d-orbital of adatom and p orbital of graphene. The peak structure is located around the Fermi level. The peak structure around the Fermi level is made from the wave function of adatom which does not hybridize orbital. In figure 8, for typical example, we show a band structure of the Cr atomic adsorption on the graphene with high adsorption energy. Fig. 8 shows a band structure close to the Fermi level when a Cr atom adsorbed to the graphene B-site, H-site and T-site. The wave function projected to p_z orbital in the 8-site of C atom is plotted in blue color circle. The most stable adsorption site is H-site. The π bonding bands made from the p_z orbital of the graphene is broken by B-site adsorption and the T-site adsorption. However, the π bonding band is kept when a Cr atom adsorbs on the H-site. The wave function projected to pz orbital in the 8-site of C atom is plotted in blue color circle. The adatom was adsorbed each in (a) the B-site, (b) the H-site and (c) the B site. We can understand from a surface structure (Fig. 9) of the graphene after the Cr atomic adsorption. The Figure 9(a) is a side view of the graphene structure when a Cr atom adsorbed on the B-site. When adatom adsorbed H-site, we showed it in figure 9(b) and showed it to figure 9(c) when we adsorbed T-site. When adatom adsorbs on the B-site and the T-site, a structure of the graphene after the adsorption is warped. However, a structure of the graphene does not change when adatom adsorbs on the H site.In other words a π bonding band of the p_z orbital of the graphene functions enough because symmetry of the graphene does not collapse when adatom adsorbs in the H-site. Almost all H-site adsorption is most stable in the adsorption of the transition metal in many cases. When adatom adsorbs on the H-site (the center of the six-membered ring), a lot of adjacent bond between adatom and the C atom is made. However, when adatom adsorbs on the B site and the T-site, bonding between the specific C atom and adatom becomes strong, but bonding with the other C atom becomes weak. Besides, π bonding is made weak when adatom adsorbs on the H-site and the T-site because a surface structure of the graphene is broken. Figure 10 drew the wave function (charge density map) of the band of around the Fermi level when a Cr atom adsorbed to the graphene. In fig. 10, A charge density map when a Cr atom adsorbed to the graphene. (a-1) and (a-2) show a charge density map made from around fermi the 39th band from the bottom. A shape of typical $d_{3z^2-r^2}$ **-orbital** is shown in the 39th band. We can understand that 39th band is very localized by comparing fig. 7(f) with fig.10(a). The p-d orbital hybridization of C atom and the adatom begins with a band than 37th band below.

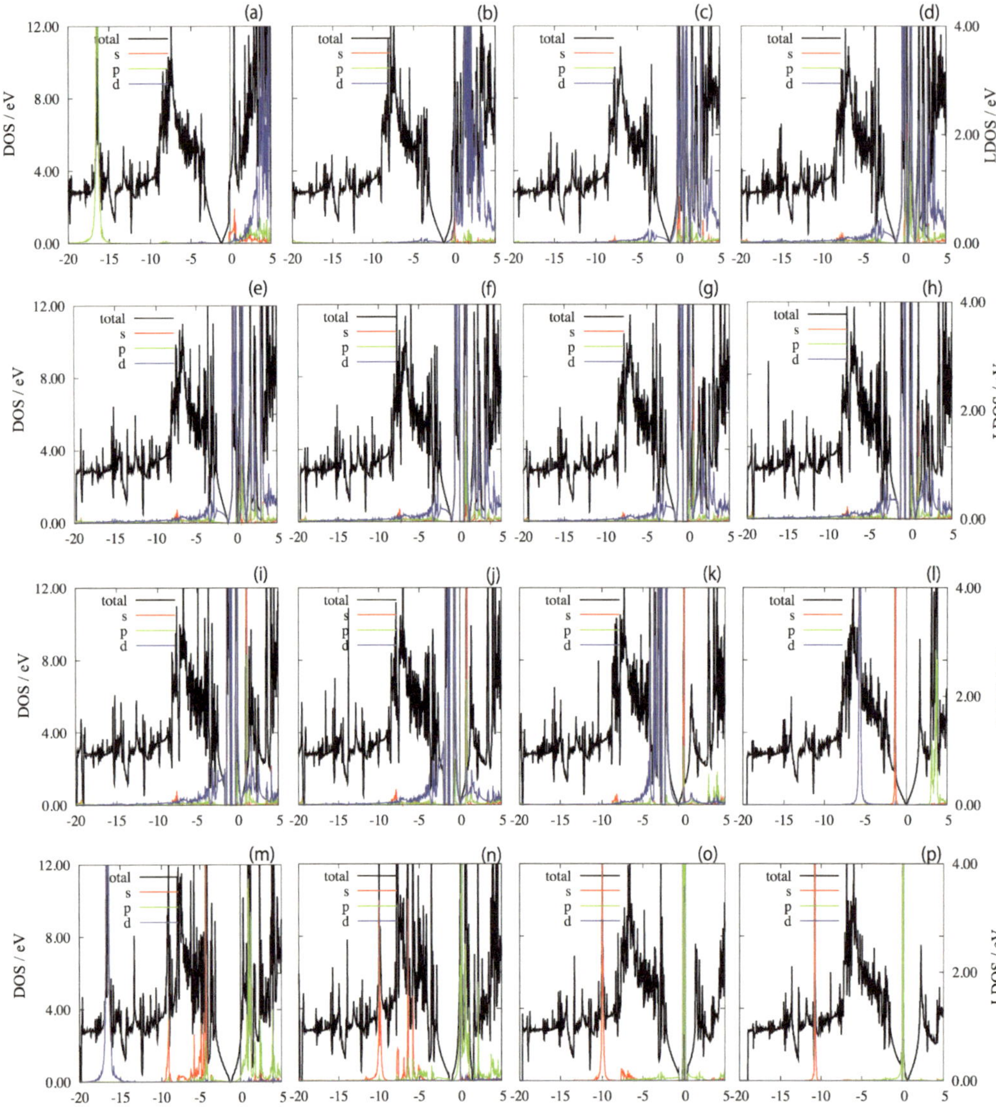

Fig. 7. Density of states of the fourth period and local density of state.There is no spin polarization. A black line is density of states, and the value is shown in a left axis. The local density of state of the adatom is shown with axis of the right side. Local density of states projected to s, p, d orbitals for graphene. A red line shows s orbital, a green line shows p orbital, and blue line shows d-orbital component. (a)-(p) is the calculation result which adsorbed K,Ca,Sc,Ti,V,Cr,Mn,Fe,Co,Ni,Cu,Zn,Ga,Ge,As and Se atom on graphene.

In Fig.7, we follow a band located very much just under the fermi level. In fig. 7, a band structure of the Cr,Mn,Fe,Co and Ni atomic adsorption which is 3d transition metal element is illustrated. These very localized around fermi level and the number of the density of status is very large. In addition, the π band of the graphene is divided by localized d-orbital. The number of the very large status just under the fermi level shows that status is instability.

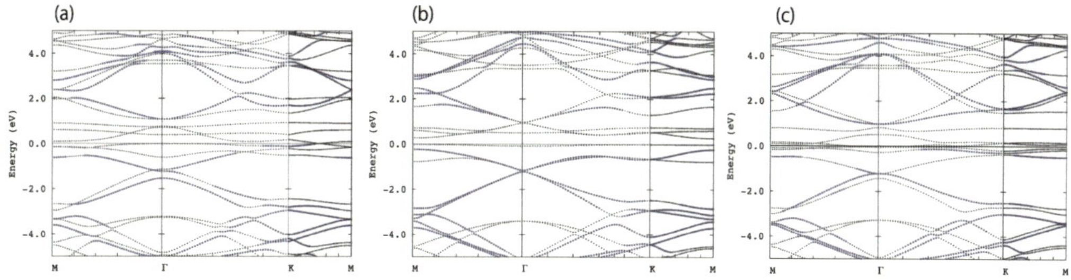

Fig. 8. A figure of band when Cr adatom adsorbed in each adsorption site in the graphene. The wave function projected to p_z-orbital in the 8-site of C atom is plotted in blue color circle. The adatom was adsorbed each in (a) the B-site, () the H-site and (c) T-site.

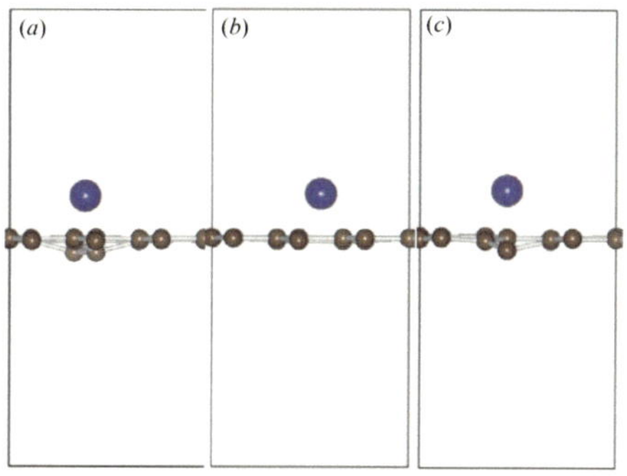

Fig. 9. A surface structure when a Cr atom adsorbed in each adsorption site in the graphene. (a) B-site, (b) H-site, (c) T-site.

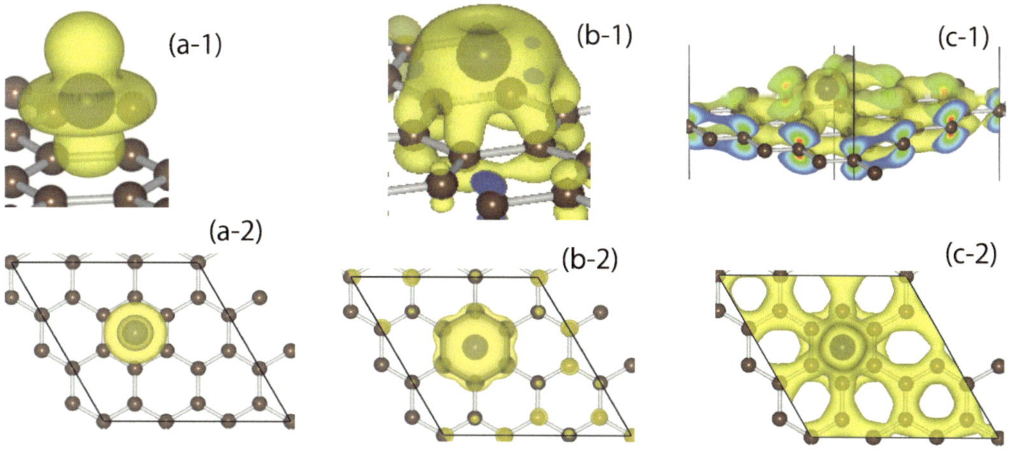

Fig. 10. A charge density map when a Cr atom adsorbed to the graphene. (a-1) and (a-2) show a charge density map made from the 39th band from the bottom. (b-1) and (b-2) is made from 37th band. (c-1) and (c-2) is made from 35th band.

We performed all nonmagnetic (we do not consider spin polarization) calculation. In other words, this shows that status is stabilized by spin polarization. In figure 11, we show calculation result in consideration of spin polarization of the 3d transition metal Ti,V,Cr,Mn,Fe,Co and Ni. A left axis shows total density of states (/eV) in the figure, and the right axis shows local density of state (/eV) of the adatom. The local density of state of the graphene does not almost influence polarization. Only 3d electron of the adatom and C atom around the Fermi surface cause polarization. We plotted only 5eV from -5eV around the fermi level in fig. 11. In fig. 12, a calculation in consideration of spin polarization. A black line is density of states, and the value is shown in a left axis. The local density of state of the adatom is shown with axis of the right side. Local density of states projected to s, p, d orbitals for graphene. A red line shows s orbital, a green line shows p orbital, and blue line shows d-orbital component. (a)-(g) is the calculation result which adsorbed Ti,V,Cr,Mn,Fe,Co and Ni atom on graphene.

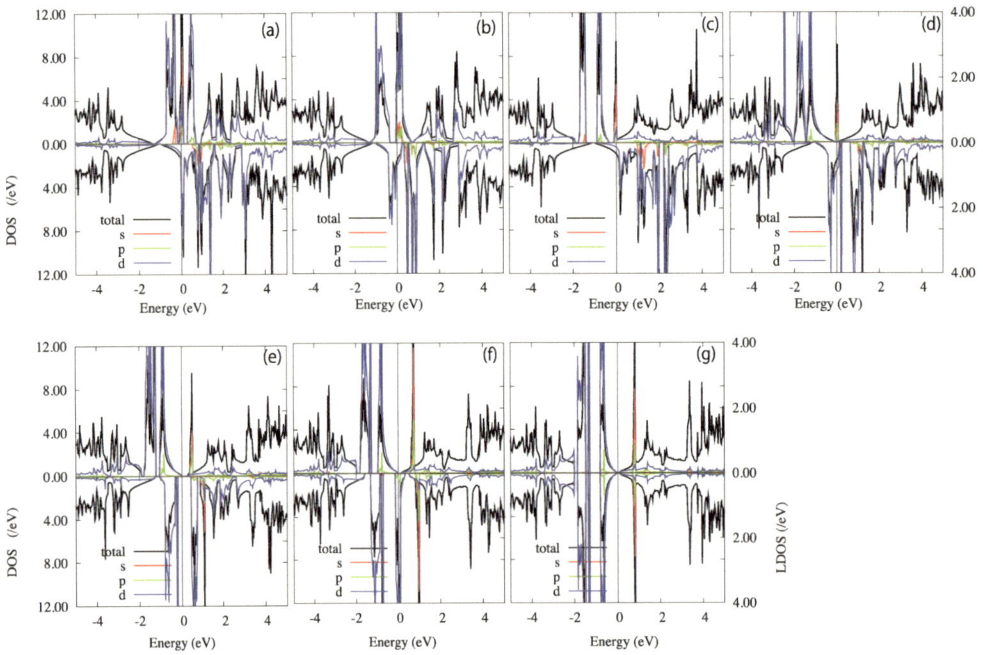

Fig. 11. Calculation in consideration of spin polarization. A black line is density of states, and the value is shown in a left axis. The local density of state of the adatom is shown with axis of the right side. Local density of states projected to s, p, d orbitals for graphene. A red line shows s orbital, a green line shows p orbital, and blue line shows d orbital component. (a)-(g) is the calculation result which adsorbed Ti,V,Cr,Mn,Fe,Co and Ni atom on graphene.

For the calculation in consideration of spin polarization, the case which the adsorption energy shows most high value when a Ti atom adsorbed, Fe atom adsorbed shows the second largest value. Migration energy is maximum the case which a Fe atom adsorbs, the case which a Ti atom adsorbs shows the second largest migration energy. The stable adsorption site is not different from spin polarization in calculation without the spin polarization. The adsorption energy decreases in comparison with the result that does not consider spin polarization widely generally. A tendency to decrease is seen in the adsorption energy in comparison with the result that does not consider spin polarization. Most of the difference of the adsorption energy are because total energy (formula 1) of the isolated atom is different by spin polarization. By the total energy of the 3d transition metal isolated atom by spin polarization of number eV make a great difference. In other words, originally the adsorption energy showed a slight overestimate for the calculation that did not consider spin polarization. The large difference by the spin polarization is a change of the migration energy. In other words, for the calculation in consideration of spin polarization, the energy difference between adsorption sites decreases. When specially spin polarization is large, we appear conspicuously. When spin polarization is large, it is remarkable. When a Cr atom adsorbs, the exchange splitting with the spin is the largest (4 μ_B / adatom). In this case the energy seems to profit by exchange splitting, but the adsorption energy is small because most of the orbital hybridization with the graphene does not exist. In the graphene adsorption of the Mn atom, similar discussion is possible. However, except such a case, magnitude and the tendency of the adsorption energy in the most stable adsorption site are about the same as a result of nonmagnetism. The difference of the migration energy by the spin polarization needs attention in the 3d transition metal.

adatom	migration	B-site	H-site	T-site	moment
Ti	0.78	1.76	2.55	1.77	1.65
V	0.45	1.46	1.91	1.41	1.36
Cr	0.12	0.65	0.77	0.65	4.16
Mn	0.14	-0.01	0.26	0.12	0.78
Fe	1.06	1.20	2.31	1.25	1.86
Co	0.73	1.88	2.61	1.83	0.92
Ni	0.43	2.22	2.65	2.17	0.00

Table 1. A calculation result when the 3d transition metal adatom which considered spin polarization adsorbed to the graphene. We show migration energy when we adsorbed to the H site of the most stable adsorption site.When adatom adsorbed each to the B,H,T sites, we show adsorption energy and magnetic moment. The magnetic moment shows magnetic moment per adatom (/μB).

3.2 Migration energy

In the above section, we discuss the adsorption energy for adatoms on graphene plane. The adsorption energy is the energy to remove the adatom from the graphene. Here, we discuss the other important energy, migration energy. Migration energy or migration barrier energy is the required energy for adatom on graphene to move from a site to other site. For the case of large migration energy, adatom does not move at room temperature. For the case of small migration energy, adatom can move easily on the graphene plane even at room temperature.

In general, for making nano structures on a surface, the migration barrier energy for the adatom on the surface is very significant to discuss the temperature dependence of the nano structure. For adatom having small migration energy, the nano structure on the surface can easily disappear because of the movement of the adatoms on the surface. For the growth of thin films on surface using epitaxy technique, the choice of the growth temperature is very important, because adatoms to form the thin film should move on the surface. Thus, the growth temperature is a function of migration energy of the adatom on the surface.

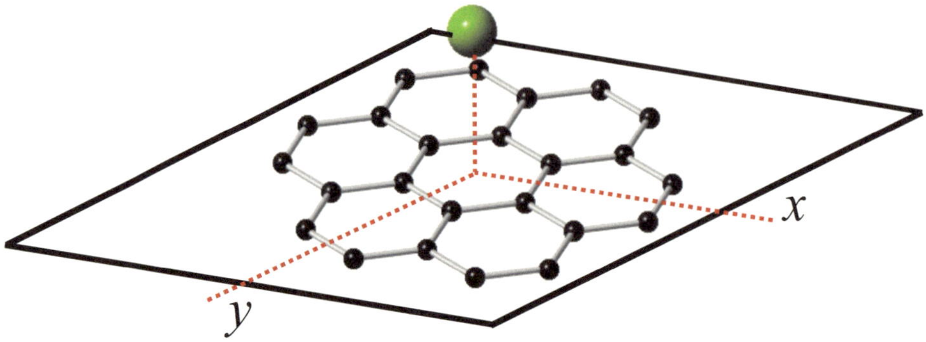

Fig. 12. A sketch of the migration energy. Migration energy or migration barrier energy is the required energy for adatom on graphene to move from a site to other site.

In figure 13, we show the table of the calculated migration energies for various adatoms on graphene. From the result, we found that adatom can be distinguished into two groups; adatoms with fixed adsorption site and mobile adatoms. Roughly speaking, the two groups are separeted at the migration energy value of 0.5 eV. Above 0.5eV, adatom does not move at room temperature. Below 0.5eV, adatom is very mobile even at room temperature.

$$R = R_0 \exp(-E / k_B T)$$

The rate of hopping of adatom per second, R can be expressed with the above formula, where E is the migration energy, T is temperature and kB is the Boltzmann constant. R_0 is a prefactor. The prefactor for Si adatom on Si(001) surface, the value is determined to be $1.25\text{x}10^{10}$ (T.Kawamura et.al). If we consider the grapnene plane having more than 1 million atoms, the migration barrier energy 0.5eV corresponds to the hopping of at least one atom within one second. For example, Ito and Shiraishi have reported a kMC simulation of MBE which takes into account the electron counting model (T. Ito et.al, K.Shiraishi et.al). They have used parameters such as the hopping barrier energies obtained by first-principles calculation. The migration barrier energy can be obtained by using the density functional method where the barrier energy can be easily found by using the contour map of the total energy of the adatom-graphene system as a function of the position of the adatom parallel to the graphene plane. As a biginning Ito and Shiraishi, migration barrier energies of an adatom on surface has been calculated by many researchers using the density functional theory.

According to our calculation, the migration barrier energies of adatom on graphene can be distinguished into two types; fixed adatom and mobile adatom. For some transition metals, Ti, V, Cr, Mn, Fe, Co, Nb, Mo, Tc, Ru, Ta, W, Re, Os, the migration barrier energy at the most stable site is very large so that these atoms adsorb on graphene strongly and do not move on graphene plane. On the contrary, for many atoms, Be, B, Na, Mg, Si, Cl, K, Ca, Cu, Zn, Ga, Ge, Br, Rb, Sr, Y, Pd, Ag, Cd, In, Sn, Sb, Te, I, Cs, Ba, Ir, Pt, Au, Hg, Tl, Pb, Bi, the migration barrier energies are very small so that these adatoms can move even at room temperature.

The 3d transition metal is polarized with low temperature. Our calculation was carried out without magnetism. However, the nonmagnetic calculation is the calculation that becomes average about a spin. It is the starting point for the discussion of the ferromagnetic state and the discussion of high temperature. For a typical example, by calculating the Ti adatom adsorption on graphene, a discussion of the metal-graphene junction is carried out [18-21], because the Ti atom has a very large adsorption energy and large migration energy. Furthermore, an important fact is that the Ti atom does not break the structure of graphene on adsorption. Fig. 15 shows a structure of the graphene adsorption of the Ti atom using a nonmagnetic calculation. The adatom disturbs the structure of graphene on adsorption at the B-site and the T-site. However, there is no change of the graphene structure from H-site adsorption. The H-site is the most stable adsorption site for Ti adsorption. It is important for the growth of a compound semiconductor on graphene that the adatom does not disturb the surface structure. In talbe 1, we performed a calculation including spin polarization for a adatom of 3d transition metal. As a result, the structure does not change with the H-site adsorption, just as for the nonmagnetic calculation. The adsorption energy decreases, but it is energy very larger than physical adsorption. For our calculation, the adsorption energy is 2.55 eV when including the spin polarization. The migration energy is very high then with 0.78eV. When a Ti atom adsorbs on graphene, it is result almost same as nonmagnetism. However, the migration energy is lower than nonmagnetic result in the 3d transition metal adsorption. The adatom which is over threshold sill level 0.5eV of the migration energy at the room temperature is only Ti,V,Fe and Co in the 3d transition metal. One of the reasons which Ti atom is used for in metal-graphene junction is caused by migration energy and adsorption energy.

more than 0.50 eV

under 0.50 eV more than 0.30 eV

less than 0.30 eV

H																	He
0.60																	
Li	Be											B	C	N	O	F	Ne
0.30	0.02											0.12	0.25	1.00	1.02	0.45	
Na	Mg											Al	Si	P	S	Cl	Ar
0.13	0.02											0.05	0.05	0.45	0.46	0.02	
K	Ca	Sc	Ti	V	Cr	Mn	Fe	Co	Ni	Cu	Zn	Ga	Ge	As	Se	Br	Kr
0.12	0.07	0.34	0.61	1.05	1.45	1.26	0.97	0.77	0.40	0.03	0.02	0.03	0.07	0.20	0.23	0.00	
Rb	Sr	Y	Zr	Nb	Mo	Tc	Ru	Rh	Pd	Ag	Cd	In	Sn	Sb	Te	I	Xe
0.09	0.04	0.12	0.39	0.83	1.47	1.40	0.96	0.39	0.06	0.01	0.01	0.02	0.03	0.03	0.09	0.00	
Cs	Ba	La	Hf	Ta	W	Re	Os	Ir	Pt	Au	Hg	Tl	Pb	Bi			
0.10	0.05	0.18	0.23	0.60	1.17	1.23	0.75	0.15	0.19	0.03	0.01	0.00	0.01	0.00			

Fig. 13. The most stable site of migration energy. A dark green color is more than 0.5 eV and a light green color is under 0.5 eV but more than 0.3 eV.

3.3 Charge transfer

Charge density analysis are done using AIM method or Bader method (R. F. W. Bader, W. Tang et.al, E. Sanville et.al, G. Henkelman et.al). In contrast to the Mulliken method using local basis, we use only spacial gradient of charge density to analyze charge density of each atoms using Bader method. Since requiered data is only charge density, we can apply the Bader method in the density functional calculation with the plane wave expansion like

VASP. In fig.15, we show the results the adatom adsorption at the most stable site. The values in the table are the difference of the number of valence electron and the calculated results with Bader method for each adatoms. The value obtained with the Bader method and that obitained with the Mulliken method is usually not equal. No one know which is correct. Neverthless, we can discuss the charge transfer between adoms using the Bader method calculation where we obtain elecron charge density in a certain volume around each atoms. The boundary of each atoms are determined simly as the minimum point of the electron charge density between each two atoms. Therefore, positive values in the table mean that electron transfer from the adatom to the graphene. On the other hand, negative values in the table mean that the electron transfer from the graphene to the adatom.

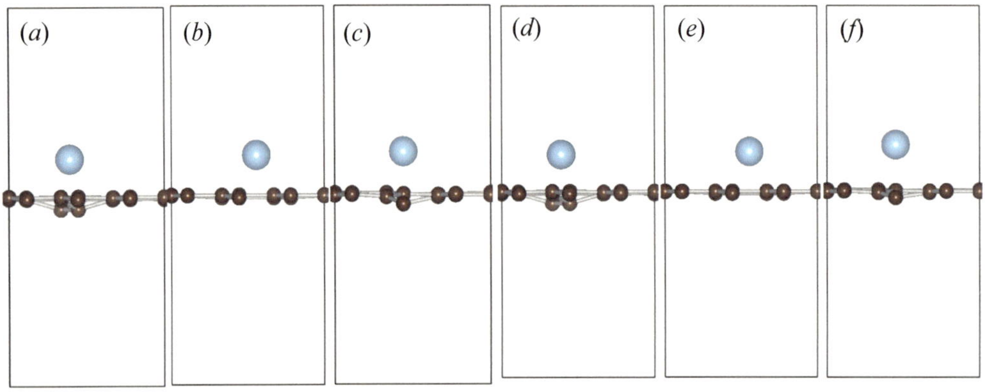

Fig. 14. We showed the final structure for the calculation in the Ti adatom adsorbed the graphene. The calculation that (a)-(b) does not consider spin polarization. The calculation that (d)-(f) considered spin polarization. We show B-site, H-site, T-site adsorption each. The adatom disturbs the structure of graphene with B-site and T-site adsorp tion. The structure of graphene does not change with H-site adsorption.

The present calculation is the calculation of the charge transfer for the system of one adatom on a graphene plane. Thus, the calculation is almost equal to the study of electronegativity for each adatom on graphene. The feature of our calculation is similar to the general discussion of the electronegativity for each atomic species. For example, non metal atomic species collect electrons from graphene. On the contrary, metallic atom leaves electron to graphene. For some metals, Cu, Ag, Au, or Z, Cd, Hg, we found no electron transfer. Similarly, we found no electron transfer for Pt on graphene, also. These results can be applied for doping to graphene. We show the calculated density of states and the local density of states for non metal species, B, C, N and O adsorbed on graphene in fig.16 (a), (b), (c) and (d). The most stable adsorption sites for the four adatoms are B-site. For Boron, the adsorption energy is 1.8eV and the electrons transfers from B atom to graphene. For Oxygen, the adsorption energy is 4.8eV and the electrons transfers from graphene to the Oxygen adatom. We analyze the local density of states when adsorbed O and B on graphene. We represent typical example of low bond energy at nonmetallic element, in fig. 16(a), adsorbed B. The value of the total density of state corresponds to the left axis. The value of the local density of states of adatom. We plotted the total density of states and local density of states when adsorbed adatom B. This result shows decrease of the local density of state of the adatom in fig 16(a-1) and increase of the local density of states of the graphene in fig. 16(a-2). This result means that charge is transferred to graphene

Electron transfer from graphene to adatom (pink)

Electron transfer from adatom to graphene: greater than 0.5 (green); less than 0.5 (white)

H																	He
0.15																	
Li	Be											B	C	N	O	F	Ne
0.86	0.05											0.43	0.02	-0.68	-0.84	-0.59	
Na	Mg											Al	Si	P	S	Cl	Ar
0.62	0.10											0.81	0.72	0.38	-0.04	-0.41	
K	Ca	Sc	Ti	V	Cr	Mn	Fe	Co	Ni	Cu	Zn	Ga	Ge	As	Se	Br	Kr
0.63	0.87	1.1	1.1	0.98	0.84	0.70	0.58	0.48	0.45	0.19	0.03	0.55	0.43	0.22	-0.01	-0.34	
Rb	Sr	Y	Zr	Nb	Mo	Tc	Ru	Rh	Pd	Ag	Cd	In	Sn	Sb	Te	I	Xe
0.60	0.82	1.04	1.16	0.97	0.72	0.53	0.37	0.24	0.17	0.10	0.03	0.57	0.43	0.27	0.09	-0.28	
Cs	Ba	La	Hf	Ta	W	Re	Os	Ir	Pt	Au	Hg	Tl	Pb	Bi			
0.61	0.86	1.23	0.82	0.83	0.72	0.48	0.35	0.22	-0.00	-0.11	0.01	0.48	0.37	0.24			

Fig. 15. Charge density analysis of Bader. The inserted figure represents the three adsorption sites of graphene sheet, T,B and H. The pinkness signifies electron transfer from graphene to adatom. The green and white signifies electron transfer from adatom to graphene. As for the green, an electron number is larger than 0.5. As for the white, an electron number is less then 0.4.

from adatom when adsorbed B. As a result, the bonding orbitals between adatom and the graphene decrease, because valence electron is transferred from adatom to graphene. Furthermore, we show typical example of high bond energy at nonmetallic element, in fig 16(d), adsorbed O. The value of the total density of state corresponds to the left axis. The value of the local density of states of adatom. This result shows increase of the local density of states of the adatom in fig. 16(d-1) and decreasing of the local density of states of the graphene, in fig16(d-2). The electron of the graphene is transferred to adatom by adsorbing O, as a result, it is made a very wide hybridized orbital between things of graphene and adatom. The cause of strong bond energy when adsorbed O on graphene is a wide hybridized orbital. By the result of the Bader analysis in figure 15, 0.84 electrons per unit cell are transferred to adatom from graphene in the O atomic adsorption, and 0.4 electrons transferred from adatom to graphene in the B atomic adsorption. As well as result from analysis of the density of states, tendency of the charge transfer when we adsorbed O and B was shown from result of the charge density analysis of Bader. We performed Bader analysis about all adatom and the density of states analysis about some atom. The transference of the charge is important to discussion of the bonding of adatom and the graphene. For example, the quantity of charge transfer is important as not only the discussion of the mechanism but also an indicator when it is decided experimentally of the adatom. The results will be very helpful to plan the construction of nano structures on graphene.

4. Application

4.1 Electrode

The metal-graphene junction attracts much attention for designs such as nano devices and transistors using graphene. When the metal is used as an electrode on graphene, high understand from these results that Ti and Zr bond well with graphene and could be useful as

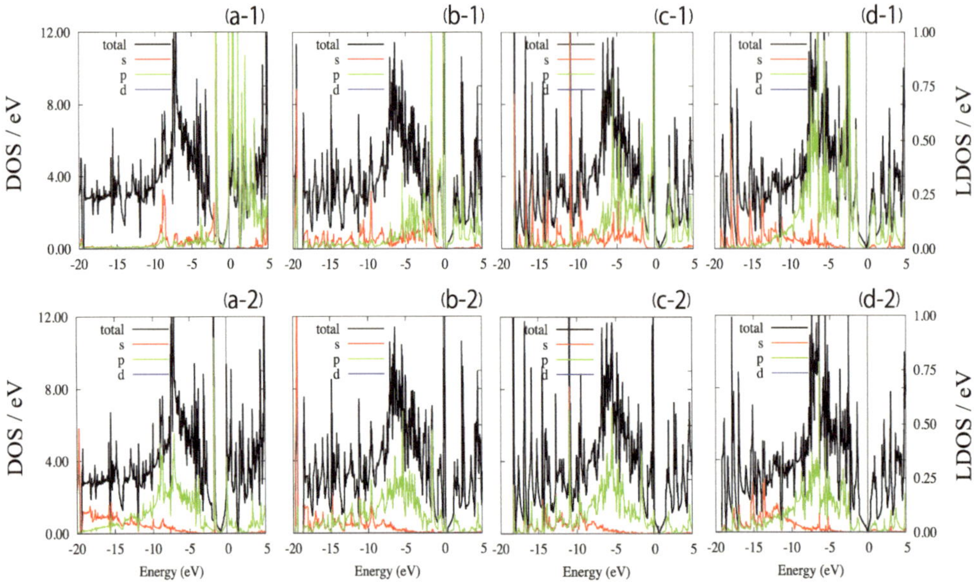

Fig. 16. The density of states and local density of state which adsorbed B,C,N and O. The value of the total density of state corresponds to the left axis. The value of the local density of states of adatom. : As for the black line, total density of states when adatom adsorbed to graphene. A red line shows s-orbital, a green line shows p-orbital, and blue line shows d-orbital component. Local density of states projected to s, p, d orbitals for graphene. (a-1) - (d-1) show local density of state of the adatom. (a-2) - (d-2) shows local density of states of the graphene (11th-site).

adsorption energy and a high migration energy are required, because an adatom is transferred if the migration energy is low at room temperature. Therefore, high migration energy is necessary for the electrode material. Before discussing the electric conduction properties or the work function for varous graphene-metal junction to design the electrode material, however, we should discuss the possibilities of electrode materials from the point of view of the migration and adsorption energy as a first step towards that purpose in these studies. In previous studies (K. T. Chan et.al, P.A. Khomyakov et.al, H. Sevincli et.al, Y. Scanchez-Paisal et.al), the bonding of Ti and Zr on graphene has been discussed. We can see why Ti and Zr attract attention as electrode materials from our calculation, because both the adsorption energy and migration energy are high and they do not disturb the surface structure on adsorption. For our calculation, we found many adatoms with high adsorption energy and high migration energy, similar to Ti and Zr. For example, Mn and Cr atom adsorption have very high adsorption energy and migration energy. However, with the calculation including spin polarization, the adsorption energy decreases in these atom. In Mn adatom adsorption, the adsorption energy is 3.3 eV for the nonmagnetic calculation, but decreases to 0.3 eV when considering spin polarization, because the state of the isolated atom is greatly stabilized by including spin polarization. As regards the state of an isolated atom of a 3d metal, the approximation of an average about spin is not appropriate at low temperature. However, for the adsorption of Ti and Zr adatoms, the adsorption energy and migration energy are almost the same as the nonmagnetic calculation. We found that Fe, Co and Ni adatom adsorption has high adsorption energy and migration energy without depending on the spin polarization. Adatoms such as Fe, Co or Ni may make good electrode materials. From this study, we can electrode materials. As a result of a similar calculation concerning transition metal element adsorption such as Fe, Co, Ni, the possibility to make an electrode material was suggested. High adsorption energy, high migration energy and the graphene surface structure are important physical quantities for electrode materials with graphene.

4.2 Epitaxy

For the epitaxial growth, the migration barrier energy for supplied atom plays an important role. Small migration barrier energy is significant for growing large-area graphene. However, in the case of large migration energy, the epitaxially grown layers on graphene have small grain size. Typically, the small migration energy of 0.03 eV for Cu makes the grain of the large size.
Using pulsed laser deposition technique, the growth of GaN on graphite has been succeeded recently. The structure of the grown GaN on graphite is calculated using the density functional theory(Ishii-Tatani-Asano-Nakada and Ishii-Tatani-Nakada) and the obtained structure agrees well with experiments. For the investigation, the adsorption site of nitrogen adatom on graphite is very important and the adsorption is very similar to the adsorption of N on graphene. Thus, the epitaxial growth on graphene or graphite is possible and will be interesting project in near future.

5. Conclusion

In this research, we calculated the adsorption and migration energy systematically for each adatom adsorbed using the band calculation with the PAW method at three adsorption sites on 3 × 3 graphene. In the case of model metal and transition metal elements, the adatom almost always adsorbed to the H-site and when it was a nonmetallic element we showed that it was mainly adsorbed to the B-site. We showed a tendency to adsorb to the T-site when the number of valence electrons of the adatom is 1, as is the case for H, F, Cl, Br, and I. The stable site for atomic species of transition metals having a very large migration barrier energy (Ti,V,Cr,Mn,Fe,Co,Nb,Mo,Tc,Ru,Ta,W,Re and Os) is H. In transition metal elements, we showed the largest bond energy when the d-orbital is half occupied. The adsorption energy showed a tendency to decrease when the d-orbital occupation exceeded a half. In addition, the adsorption of nonmetallic elements such as C, N, O shows a very large bond energy and adsorption at the B-site. When the d-orbital and s-orbital are occupied, as in Be, Mg, Ca, Sr, Ba, Zn, Cd, Hg, Cu, Ag and Au, we show a very small adsorption energy. Furthermore, we estimated the minimum limit of the migration energy of the adatom. The tendency for the magnitude of the migration energy is similar to the tendency of the adsorption energy. In the most stable adsorption site, the adatom does not break the structure of graphene. However, in an adsorption site other than the most stable site, we showed that in many cases the adatom breaks the surface structure of graphene. In the growth of a compound semiconductor on graphene, or metal-graphene junctions, we show the importance of the adsorption energy and migration energy. Because Ti and Zr showed good bonding on graphene, we showed that these were useful as electrode materials. Our calculation will be very helpful for experimental groups that are considering the use of atoms and molecules as building blocks, or graphene for making new nano devices, such as nano wires and nano switches.

6. Acknowledgment

This research was supported by the CREST project of the Japan Science and Technology Agency (JST). Figs. 2, 9, 10 and 14 were created using the VESTA package of Momma and Izumi (K. Momma et.al).

7. References

Ando T., NGP Asia Mater 1(1), 17-21 (2009)
Bader W. F. R., Atoms in Molecules -A Quantum Theory, Oxford University Press, Oxford, 1990.
Chan T. K., Neaton B. J., Cohen L. M., Phys. Rev. B 77, 235430 (2008).
Dai J., Yuan J., Phys. Rev. B 81, 165414 (2010)
Fujioka H, US Patent App. 12/735,826, 2009
Ga H., Zhou J., Lu M., Fa W., Chen Y., J. Appl. Phys. 107, 114311 (2010)
Geim K. A., Novoselov S. K., Nature Materials 6 (3), 183-191 (2007).
Henkelman G., Arnaldsson A., Jónsson H., Comput. Mater. Sci. 36, 254-360 (2006).
Hohenberg P., Kohn W., Phys. Rev. 136, B864 (1964)
Ishii A., Tatani T., Asano H. and Nakada K., phys.stat.sol. (c) 7 347-350 (2010)
Ishii A., Tatani T. and Nakada T., phys.stat.sol. in press
Ishii A., Yamamoto M., Asano H. and Fujiwara K., J.Phys.:Conf.Ser. 100 (2008) 052087
Ito T., Shiraishi K., Jpn. J. Appl. Phys. 35, 1949 (1996).
Karpan M. V., Giovannetti G., Khomyakov A. P, Talanana M., Starikov A. A., Zwierzycki M., van den Brink J., Brocks G., Kelly J. P., Phys. Rev. B 78, 195419 (2008)
Kawamura T., Toyoshima S., Ichimiya A., Surf.Sci. 493, 561 (2001).
Khomyakov A. P., Giovannetti G., Rusu C. P., Brocks G., van den Brink J., Kelly J. P., Phys. Rev.B 79, 195425 (2009).
Kim. -H.M., Ohta J., Kobayashi A., Fujioka H., Oshima M., Appl. Phys. Lett. 91, 15903 (2007)
Kobayashi A., Kawaguchi, Y., Ohta J., Fujioka H., Appl. Phys. Lett. 89, 111918 (2006).
Kobayashi A., Kawano S., Kawaguchi Y., Ohta J., Fujioka H., Appl. Phys. Lett. 90, 041908 (2007).
Kohn W., Sham J. L, Phys. Rev. 140, A1133 (1965)
Kong K., Choi Y., Ryu B., Lee J-O., Chang H., Mat. Sci, Eng. C26, 1207 (2006)
Kresse G., Furthomuller J., Comput. Math. Sci. 6, 15-50 (1996).
Kresse G., Futrthomuller J. , Phys. Rev. B 54, 1169 (1996).
Kresse G., Hafner J., Phys. Rev. B 47, RC558 (1993).
Kresse G., Joubert D., Phys. Rev. B 59, 1758 (1999)
Kresse G., Thesis, Technische Universita Wien, 1993.
Li G., Ohta J., Okamoto K., Kobayashi A., Fujioka H., Appl. Phys. Lett. 89, 182104 (2006).
Lugo-Solis A. , Vasiliev I., Phys. Rev. B 76, 235431 (2007)
Momma K., Izumi F., J. Appl. Crystallogr. 41, 635-658 (2008).
Nakada K. and Ishii A., Solid State Commun. 151 (2010) 13-16
Novoselov S. K., Geim K. A., Morozov V. S., Jiang D., Katsnelson I. M., Grigorieva V. I., Dubonos V. S., Firsov A. A., Nature 483, 04233 (2005)
Okazaki-Maeda K., Yamakawa S., Morikawa Y., Akita T., Tanaka S., Hyodo S., Kohyama M., J. phys.: conf. 10, 072044 (2008).
Sanchez-Paisal Y., Scanchez-Portal D., Ayuela A., Phys. Rev. B 80, 045428 (2009)
Sanville E., Kenny D. S., Smith R., Henkelman, G., J. Comp. Chem. 28, 899-908 (2007).
Sevincli H., Topsakal M., Durgun E., Ciraci S., Phys. Rev. B 77, 195434 (2008).
Shiraishi K., Ito T., Phys. Rev. B 57, 6301 (1998).
Tang W., Sanville E., Henkelman G., J. Phys.: Condens. Matter 21, 084204 (2009).
Ueno K., Kobayashi A., Ohta J., Fujioka H., Jpn. J. of Appl. Phys. Express Letters, 45, L1139 (2006).
Wu M., Liu En-Zuo, Jiang Z. J., Appl. Phys. Lett. 93, 082504 (2008)
Zanella I., Fagan B. S., Mota R., Fazzio A., J. Phys. Chem. C 112, 9163 (2008).

2

Structural and Electronic Properties of Graphene upon Molecular Adsorption: DFT Comparative Analysis

Ali Zain Alzahrani
Physics Department, Faculty of Science,
King Abdulaziz University
Saudi Arabia

1. Introduction

Since its discovery in 2004 (Nobel prize in 2010), graphene -a single sheet of carbon atoms forming the thinnest free standing material to date- has attracted enormous interest due to its potentially tunable and exotic structural and electronic properties (Castro Neto et al., 2009; Geim & Novolselov, 2007; Katsnelson et al., 2006, 2007; Novoselov et al., 2004, 2007; Ohta et al., 2006; Y. Zhang et al., 2005). The pristine graphene is characterized as a zero-gap semiconductor with bonding π and antibonding π^* bands touch in a single point at the Fermi level (E_F) at the corner of the Brillouin zone, and close to this so-called Dirac point the bands display a linear dispersion, leading to extremely high charge carriers mobility at room temperature of approximately 15,000 $cm^2/V.s$ (Geim & Novolselov, 2007) which is significantly higher than that of the widely-used semiconductor, *namely* silicon (Si), of approximately 1400 $cm^2/V.s$. Like carbon nanotubes, measurements (Lee et al., 2008) have shown that graphene is extremely strong and rigid compared to Si-based materials. These incredibly fascinating properties alongside the high thermal conductivity suggest that graphene is an excellent candidate for the applications in the circuits beyond the conventional complementary metal-oxide semiconductor technology and many other potential applications. Moreover and recently, the possibility of using graphene as a highly-sensitive gas sensor has been reported as the good sensor properties of carbon nanotubes are already known. It was shown that the increase of the concentration of graphene charge carrier induced by adsorbed gas molecules can be used to make highly sensitive sensors. These highly-sensitive properties of graphene can be attributed to the fact that graphene is a low-dimensional structure with only a surface but no volume which increase the chemical reaction of adsorbates and the surface atoms. Additionally, the high conductivity of graphene even in low charge density is another reason for being a highly-sensitive sensor.

Having established the importance of pristine graphene in many potential applications, the adsorption of single atoms (Chan et al., 2008; Farjam & Rafii-Tabar, 2009; Han et al., 2007; Hao et al., 2006; Li et al., 2008; Mao et al., 2008; Medeiros, 2010; Yang, 2009) and molecules (Duplock et al., 2004; Elias et al., 2009; Giannozzi et al., 2003; Ito et al., 2008; Leenaerts et al., 2008, 2009; Nakamura et al., 2008; Novoselov et al., 2004; Pinto et al., 2009; Sanyal et al.,

2009; Schedin et al., 2007; Wehling et al., 2008; Y.-H. Zhang, 2010) on the bare graphene surface has been the subject for different theoretical and experimental investigations due to their promising applications in nanoscale electronics, bioelectronics, gas sensors, and hydrogen storage devices. Among these adsorbates, hydrogen has been considered as one of the most interesting and fantastic candidates. Recently, it has been experimentally reported, using the transmission electron microscopy, that a graphene sheet can be chemically converted into graphane through a hydrogenation process by reacting with atomic hydrogen (Elias et al., 2009). This process, however, transforms the zero-gap semiconductor graphene into a wide-gap semiconductor (insulator) graphane. Theoretically reported studies (Boukhvalov et al., 2008; Sofo et al., 2007) using the density functional scheme, have revealed that the chairlike configuration, with hydrogen atoms attached to the carbon atoms in alternative manner, is the energetically most preferable structure for graphane. Sofo *et al.* (Sofo et al., 2007) have found that the chairlike and boatlike conformers are semiconducting with 3.5 eV and 3.7 eV, respectively. As has been claimed in many literatures (H. Ohno, 1998; Y. Ohno et al., 1999; Savchenko, 2009), future hydrogen-fuel technologies should make use of graphane as hydrogen storage due to its very high hydrogen density. Moreover, this extremely thin material with a finite band gap is also likely to find its use in many technological and industrial applications. Overall, graphene surface could be successfully used as a base for creating new promising and useful materials, and it would be of quite interest to theoretically investigate the effects of incorporating various molecules into its structure for different technological and industrial applications.

The adsorption of various molecules onto graphene has also been investigated. The structural and electronic properties of oxygen-adsorbed graphene have been theoretically studied by Nakamura *et al.* (Nakamura et al., 2008) and Ito *et al* (Ito et al., 2008). Their results have indicated that the adsorption of oxygen molecules onto graphene produces epoxy and ether group phases which are almost bistable. Moreover, they have concluded that the ether structure is the most energetically preferable for adsorption involving both sides of the sheet, while the one-side adsorption structure appears only as a meta-stable phase, with a finite energy gap at the K point emerges and its value increases as the number of oxygen increases with respect to the number of carbon atoms. The key charge transfer mechanisms upon adsorption of NH_3, NO, and NO_2 onto graphene have been reported by Leenaerts *et al.* (Leenaerts et al., 2008, 2009). Their theoretical results indicate that the NO_2 adsorbates induce a relatively strong doping comparing to the NO molecule. Within the framework of the local density approximation of the density functional theory, Pinto *et al.* (Pinto et al., 2009) have investigated the chemisorption of tetrafluoro-tetracyanoquinodimethane (F4-TCNQ) molecule on pristine graphene by means of the electronic properties. It was reported that the F4-TCNQ molecule acts like a p-type dopant for graphene with an approximately charge of 0.3 e/molecule being transferred from the highest occupied molecular orbital (HOMO) of graphene to the lowest unoccupied molecular orbital (LUMO) of the molecule. Zhang *et al.* (Y.-H. Zhang et al., 2010) have recently investigated the binding of organic electron donor and acceptor molecules on graphene sheets within the framework of the density functional theory. They found that the adsorption of 2,3-dichloro-5,6-dicyano-1,4-benzoquinone (DDQ) and tetrathiafulvalene (TTF) cause hybridizations between the molecular levels and the graphene valence bands. These hybridizations transform the zero-gap semiconductor graphene into a metallic graphene. Despite the available studies, there are no enough theoretical comparative study on the physics and chemistry of the adsorption of small molecules onto the surface of pristine graphene.

In the present chapter, we aim to theoretically investigate the fundamental changes of the structural and electronic properties of graphene upon the incorporation of hydrogen, benzene, and naphthalene molecules. The first-principles calculations will be performed using the density functional theory in its local density approximation scheme and the pseudopotential method.

2. Calculation methodology

The present *ab initio* calculations have been performed using the density functional theory (DFT) (Hohenberg & Kohn, 1964) with a plane wave basis set as implemented in the QUANTUM ESPRESSO simulation package (Giannozzi et al., 2009). The electron–electron interactions were expressed within the local density approximation (LDA) as parameterized by Perdew and Zunger (Perdew & Zunger, 1981). The electron–ion interaction was treated by using the ultrasoft pseudopotential for carbon and hydrogen (Vanderbilt, 1990). We expanded the single-particle Kohn–Sham (Kohn & Sham, 1965) wave functions using a linear combination of plane-wave basis sets with a kinetic energy cutoff of 45 Ry. The Kohn–Sham equations were Self consistently solved by employing a 14×14×1 **k** points Monkhorst–Pack set (Monkhorst & Pack, 1976) within the hexagonal Brillouin zone. The repeated supercell technique was used to model the studied graphene-based structures. In each surface structure of pristine and molecule-adsorbed graphene we considered a 6×6×1 unit cell containing 72 carbon atoms. We have used our calculated in-plane lattice parameter for graphene of 2.45 Å which is in good agreement with the previously reported theoretical (Ito et al., 2008; Schabel & Martins, 1992; Yin & Cohen, 1984) and experimental values for bulk graphite (D. Mckie & C. Mckie, 1986). To minimize the interactions between the graphene sheet and its periodic image, we considered a vertical separation of 14.65 Å (six times the lattice parameter) along the surface normal direction. These parameters have been carefully chosen after several calculations to obtain well-converged results. Relaxed atomic positions for carbon and hydrogen atoms were obtained by using the total-energy and force minimization methods following the Hellmann–Feynman approach. The equilibrium atomic positions were determined by relaxing all atoms in the cell except the carbon atom at the origin which was kept in its bulk position.

3. Results and discussion

In the following subsections we will present, based on *ab initio* calculations, a comparative study of the structural and electronic properties of pristine graphene, hydrogen-adsorbed graphene (graphane), benzene-adsorbed graphene, and naphthalene-adsorbed graphene. To establish well-defined comparative study we have performed the calculations using unit cells of similar sizes and parameters.

3.1 Pristine graphene

It is rather important for our present comparative study to start with the structural and electronic properties of the pristine graphene. Figure 1 shows a schematic view of the fully-relaxed structure of the pristine graphene, indicating the basic structural parameters. It is well-known that each carbon atom has two $2s$ and two $2p$ electrons in its valence

state. These four electrons lead to various *sp*-hybridized orbitals. For graphene, each carbon atom is bonded to three other carbon atoms according to an sp^2 hybridization. In the present calculations the C-C bonds are found to be 1.41 Å which are smaller than the C-C bond lengths of diamond of 1.52 Å. The C-C-C angle is measured to be 120° which is slightly larger than the prospective value of 109.5° in its diamond structure. These values suggest that, unlike the ideal sp^3 diamond structural phase, graphene has a significant sp^2 nature as stated above. This feature, therefore, leads to the considerable rigidity of graphene materials comparing with the normal semiconducting materials, such as Si.
The electronic band structure of the clean graphene sheet is plotted in Fig. 2 along the principal directions of the hexagonal Brillouin zone. It is clearly shown that the band structure of pristine graphene has a zero-gap semiconducting nature. It is important to note the folding of the bands due to the used supercell. In this plot, the top of the valence state and the bottom of the conduction state degenerate at the Γ point (Dirac point) instead of the K point of the hexagonal Brillouin zone. These two bands obey a linear in-plane dispersion relation near the Fermi energy at the Γ point of the Brillouin zone resulting in zero effective mass for electrons and holes and high mobility of charge carriers.

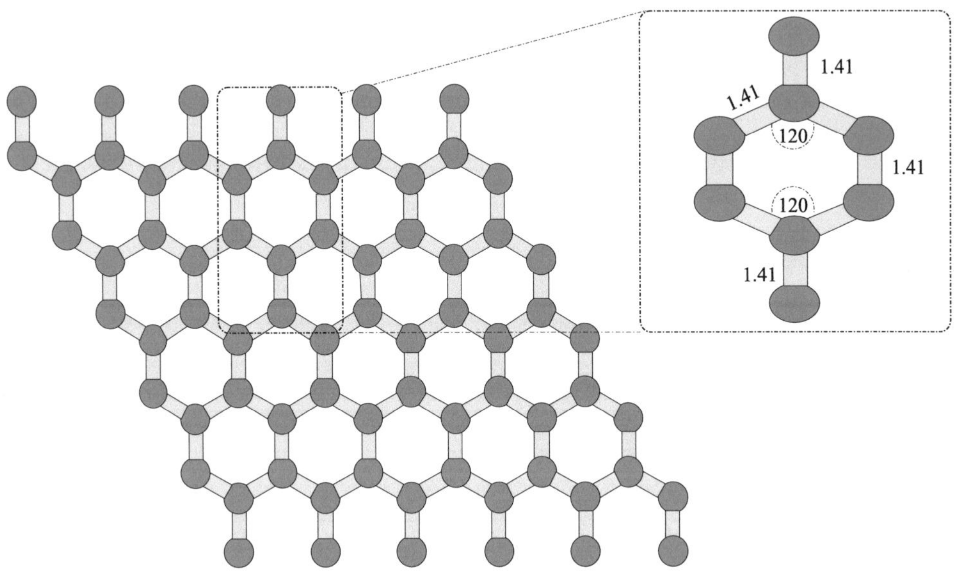

Fig. 1. Schematic top view of the optimized structure of pristine graphene. The inset shows the structural parameters of the hexagonal ring. The bond lengths are measured in angstrom (Å) and angles are measured in angles (°).

In a previous report (AlZahrani & Srivastava, 2009) we have studied the in-plane dispersion curves, at the Γ point, slightly above and slightly below the Fermi energy to extract the velocities of electron and hole carriers. These velocities were estimated to be 1.11×106 m/s and 1.04×106 m/s. The partial charge density plots of these two states at the K point confirm the bonding π and antibonding π* orbital nature of the HOMO and LUMO states of pristine graphene, as clearly shown in Fig. 3.

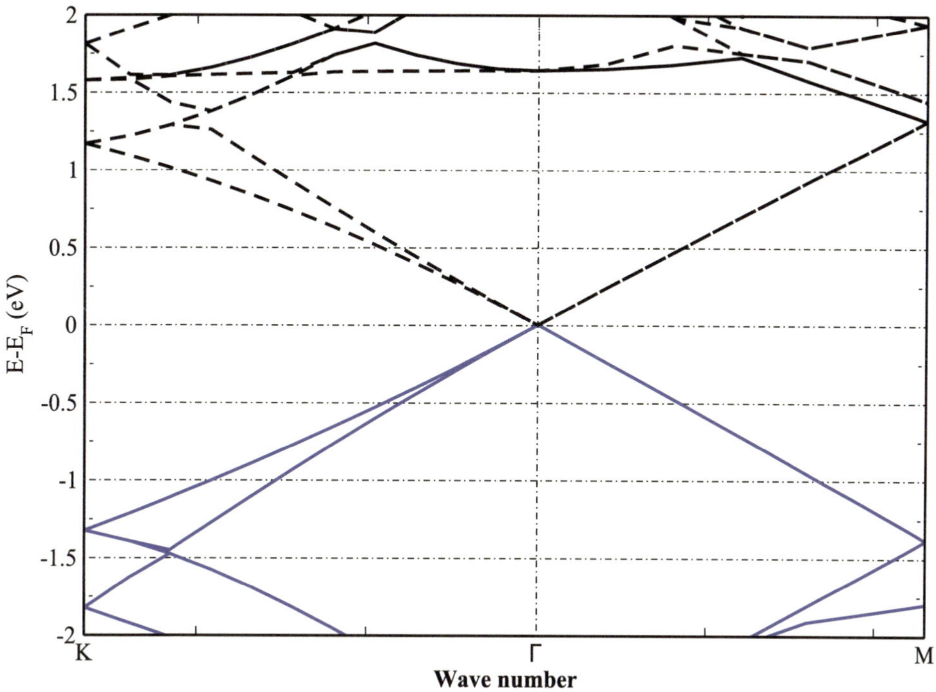

Fig. 2. Electronic band structure of the pristine graphene along the principal directions of the hexagonal Brillouin zone. The Fermi level is set at the zero.

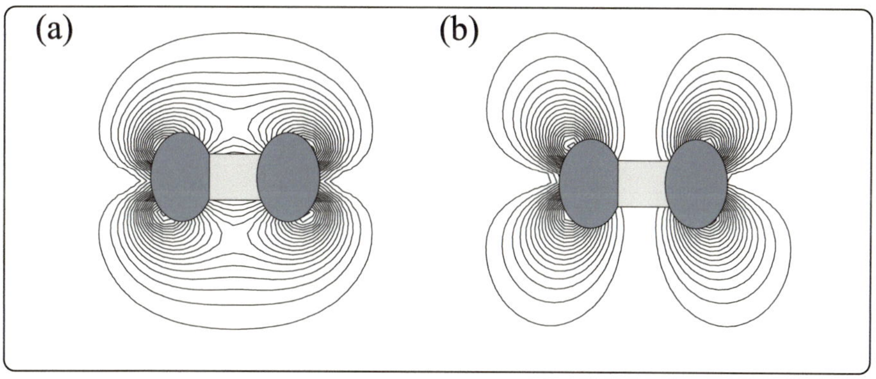

Fig. 3. Partial charge density plot, at the K point, of the (a) highest occupied state and (b) lowest unoccupied state.

3.2 Hydrogen-adsorbed graphene

The chemical adsorption of hydrogen atoms on pristine graphene has gained great interest due to the immense changes in the electronic properties of graphene. These changes lead to a new wide-gap semiconducting material which has the name of *graphane*. Subsequently, graphane has been experimentally synthesized and reported that it obeys a reversible hydrogenation-dehydrogenation process (Elias et al., 2009). This material, therefore, could open the gate for enormous technological and industrial applications, such as hydrogenstorage and two-dimensional nanoelectronics. Our purpose in this section is to find the energetically most stable geometry of graphane and then to compare its structural and electronic properties with those of pristine graphene. To model such a material, we have initially considered four different preliminary configurations depending on the adsorption sites of hydrogen atoms above and/or below the graphene sheet. These structures are chairlike, boatlike (Sofo et al., 2007), tablelike, and stirrup configurations as schematically shown in Fig. 4. The key building block of these structures is the number and orientation (up or down) of the attached hydrogen atoms in each hexagonal cell of graphene. The chairlike conformer consists of hydrogen atoms which are alternatively attached to the carbon atoms on both sides of the sheet. The hydrogen atoms in the boatlike conformer are alternatively attached in pairs to the carbon atoms on both sides. In the tablelike configuration the hydrogen atoms are attached to every carbon atom from one side of the sheet. Finally, the stirrup structure has three hydrogen atoms attached to the carbon atoms from the upper side of the sheet and also three others attached to the carbon atoms from the bottom side. Our self-consistent calculations indicate that the chairlike configuration is the energetically most stable structure (minimum energy structure) with an energy gain of approximately 0.129 eV, 0.131 eV, and 0.655 eV comparing with the boatlike, stirrup, and tablelike configurations,

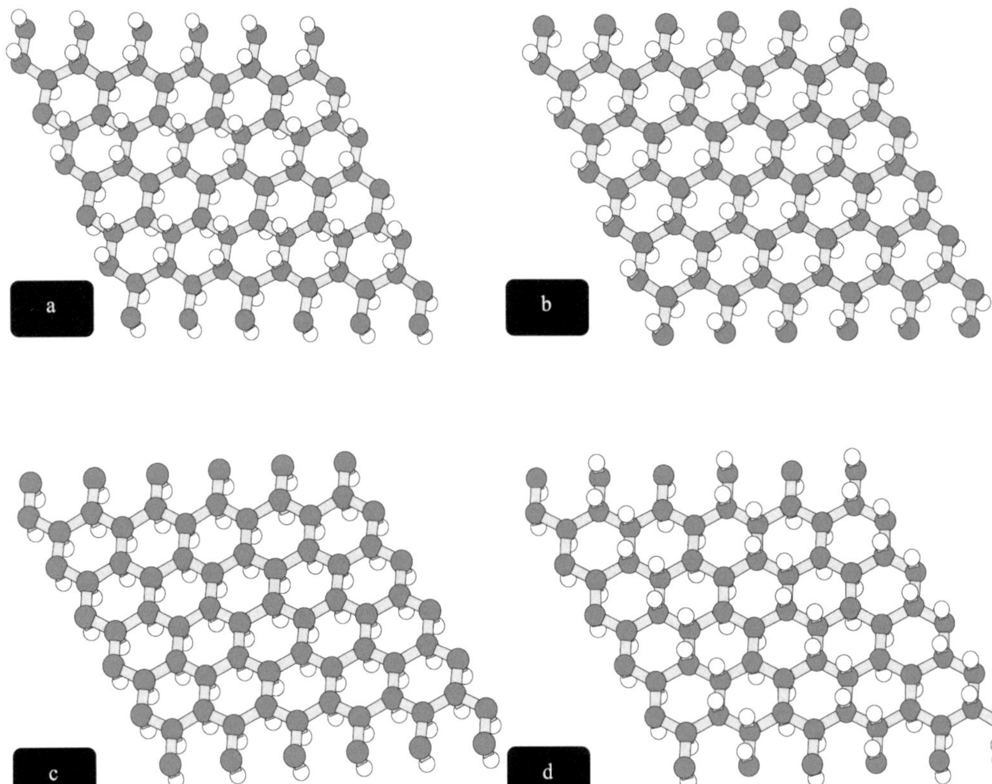

Fig. 4. Schematic view of the optimized structures of the possible structures of hydrogen-adsorbed graphene (graphane) with (a) boatlike, (b) chairlike, (c) tablelike, and (d) stirrup.

respectively. These findings for chair and boatlike structures are very comparable with the previously reported results obtained by Sofo *et al.* (Sofo et al., 2007). We note that boatlike and stirrup configurations are almost meta-stable structures and can be found in H-rich environment. In the following lines, we will focus our discussion on the structural and electronic properties of the ground state structure of graphane (chairlike conformer).

We have started our calculations for chairlike geometry of graphane with a flat sheet of graphene and hydrogen atoms at 1.0 Å above carbon species. Minimization of this structure leads to a fully-relaxed configuration as schematically depicted in Fig. 5. From this figure we have found that the C–C bond length is approximately 1.49 Å, which is larger than the C–C bond length in the ideal graphene (1.42 Å). However, this value is almost comparable with the C–C bond length obtained for graphite (1.47 Å) and diamond (1.52 Å) using similar computational parameters. Moreover, the calculated graphane C–C bond length is in excellent agreement with the bond length of 1.48 Å obtained by Igami *et al.* (Igami et al. 2001). Upon the H adsorption, the basis carbon atoms in the cell are found to experience a vertical buckling (perpendicular distance between the two carbon sublattices) of approximately 0.46 Å, which is in excellent agreement with the theoretical values obtained by Boukhvalov *et al.* (Boukhvalov et al., 2008) and Sahin *et al.* (Sahin et al., 2010) and the experimental value extracted by Elias *et al* (Elias et al., 2009). Having this amount of buckling, the lattice constant of graphene increases from 2.45 Å to 2.50 Å. This amount of buckling leads not only to a structural variation but also to a significant change in the electronic properties of graphene.

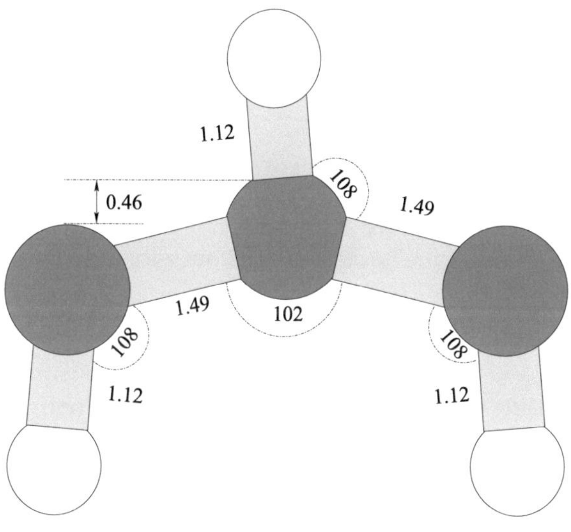

Fig. 5. The optimized atomic structure of the chairlike configuration along with the key-structural parameters. The bond lengths and angles are measured angstrom and degrees, respectively.

Consistent with previously reported results (Boukhvalov et al., 2008; Sahin et al., 2010; Sofo et al., 2007), the C–H bond length is measured to be 1.12 Å, which is identical to the typical bond length of the hydrocarbon compounds. While the angle between two adjacent C-C bonds (C-C-C angle) is found to be 102°, the angle between C-H and C-C bonds (C-C-H angle) is determined to be 108°. The average value of these angles is slightly smaller than the tetrahedral angle of diamond of 109.5°. These values are in mild agreement with the previous theoretical results (Boukhvalov et al., 2008; Sahin et al., 2010). However, this suggests that the nature of C-C and C-H bonds is not entirely sp^3 but sp^3-like. Overall, these findings for calculated bond lengths and angles clearly indicate that the bonding in graphane is sp^3-like.

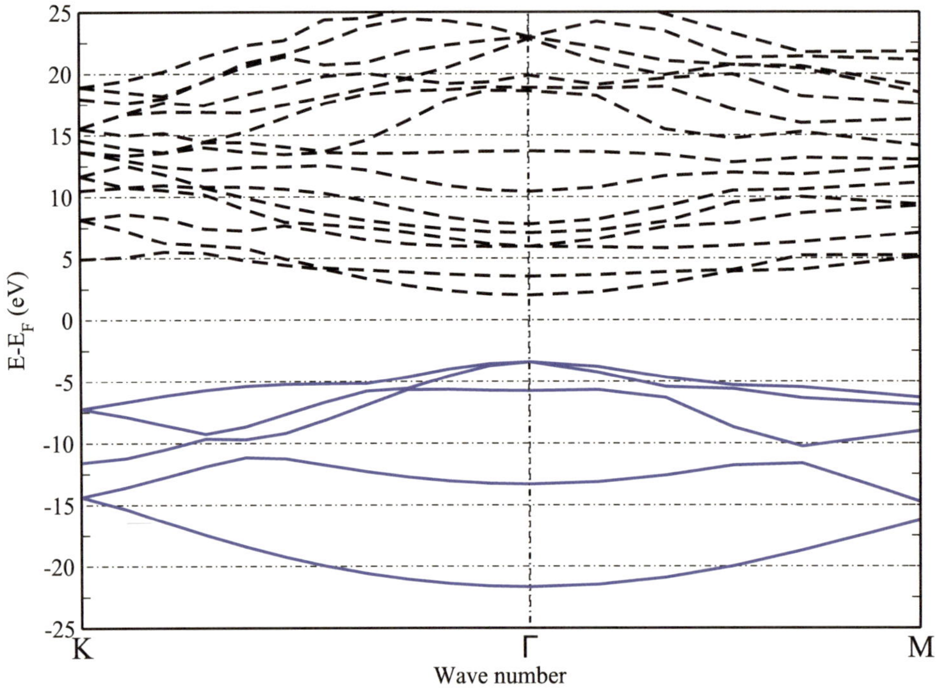

Fig. 6. Surface electronic band structure of the most stable structure of graphane (chairlike) along the principal directions of the hexagonal Brillouin zone. The Fermi level is set at the zero.

The electronic band structure of chairlike graphane is calculated along the principal directions of the hexagonal Brillouin zone as shown in Fig. 6. It is clearly noted that the bonding π and antibonding π^* states of clean graphene are now removed. Since graphane is an sp^3-like saturated structure with every C atoms being bounded to three adjacent C atoms and a single H atom, the system is found to be non-magnetic semiconducting with a direct LDA band gap of 3.9 eV, with HOMO at E_F -3.4 eV and LUMO at E_F +2.5 eV.

This value of band gap is slightly larger than the reported value of 3.5 eV (Sofo et al., 2007). It is rather important to indicate that due to the well-known deficiency of the LDA in dealing with semiconducting systems, the underestimated band gap of 3.9 eV is corrected by GW_0 approximation to become 5.97 eV (Lebègue et al., 2009). From Fig. 6, we clearly note that the uppermost occupied band is doubly degenerate at approximately 7 eV below the Fermi level at the zone edge, *namely* K point. This degeneracy has also been observed for pristine graphene but with energetic shift due to the charge transfer from H atoms towards the graphene. We also find a double degeneracy of the top of valence band at about 3 eV below the Fermi level. Such degeneracy suggests that these bands have a symmetrical orbital nature but with different effective masses. Above the Fermi level, we can identify band degeneracy at the zone edges K and M with eigenvalues of 8 eV and 5 eV, respectively. These features can be directly related to the graphene sheet as we have noted in Fig. 2 but with significant change in their energies. Inspection of the partial charge density, at the K point, of both HOMO and LUMO states reveals that the π bonding orbital in C-C bonds of graphene is broken and a new sp_z orbital is created between H and the upper C atom (i.e. the carbon atom that was tilted upwards) upon hydrogenation process. Moreover, the antibonding π^* orbital in graphene is removed and replaced by the antibonding state which is a hybridization of the H s and C p orbitals. These plots are shown in Fig. 7.

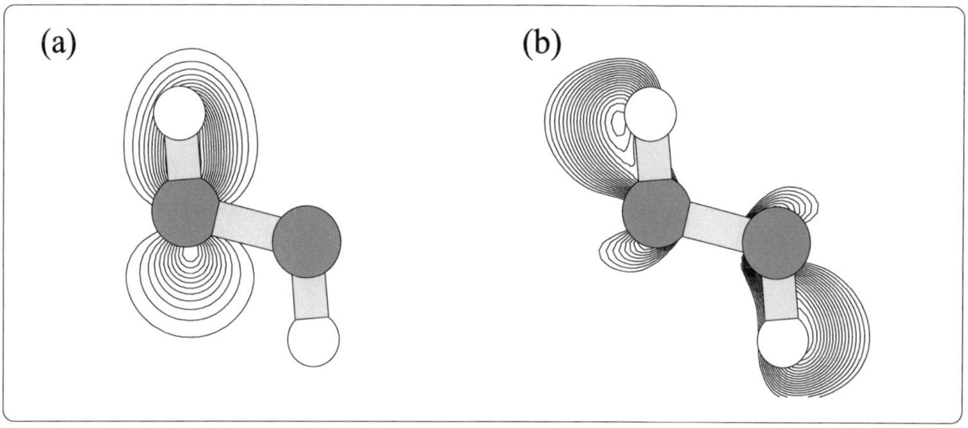

Fig. 7. Partial charge density plot, at the K point, of the (a) highest occupied state and (b) lowest unoccupied state of chairlike graphane.

It is quite important for the device engineering and manufacturing to figure out the bonding nature of C-C and C-H bonds. To perform such an examination, we performed total charge density calculations in a plane and along the C–C and C–H bonds. Figure 8(a) shows a contour map of the total charge density in [010]/[001] plane. It clearly indicates that the charge distribution around the C-C bond is supportive of that in tetrahedrally coordinated diamond as shown in Fig. 8(b). Our results indicate that the C–C and C–H bonds have a noticeable degree of covalency, as shown in panels (c) and (d) of Fig. 8. A considerable amount of charge is uniformly localized around the carbon atoms. It is interesting to note that the double-hump feature of the charge density along the neighboring C atoms (Fig. 8(c)) is typical of the diamond structure, which is not an artifact of the pseudopotential method. Moreover, we have clearly observed that a little amount of charge being transferred from the hydrogen towards the carbon atoms. Quantitatively, we have used the Löwdin population analysis scheme (Lowdin, 1950) to obtain numeral information about the atomic charges. Employing this scheme, the wave functions are projected onto linear combinations of atomic orbitals; we find that a charge of 0.2e has been transferred from the hydrogen atoms to the carbon atoms for each unit cell. Our calculated value is in good match with the result obtained by Sofo et al (Sofo et al., 2007).

3.3 Benzene-adsorbed graphene

Rather than hydrogen, it has been reported that the adsorption of organic molecules on graphene leads to significant changes in the fundamental atomic and electronic properties of the substrate. To examine the reliability of these changes we will study the mechanism of the chemisorption of small molecules (benzene and naphthalene) on a clean sheet of graphene. This subsection will be designed to study the benzene-adsorbed graphene structure whereas the next subsection will detail the naphthalene-adsorbed graphene system. To investigate the basic properties of graphene upon the adsorption of benzene, we firstly check different possible adsorption sites of the molecule onto the substrate.

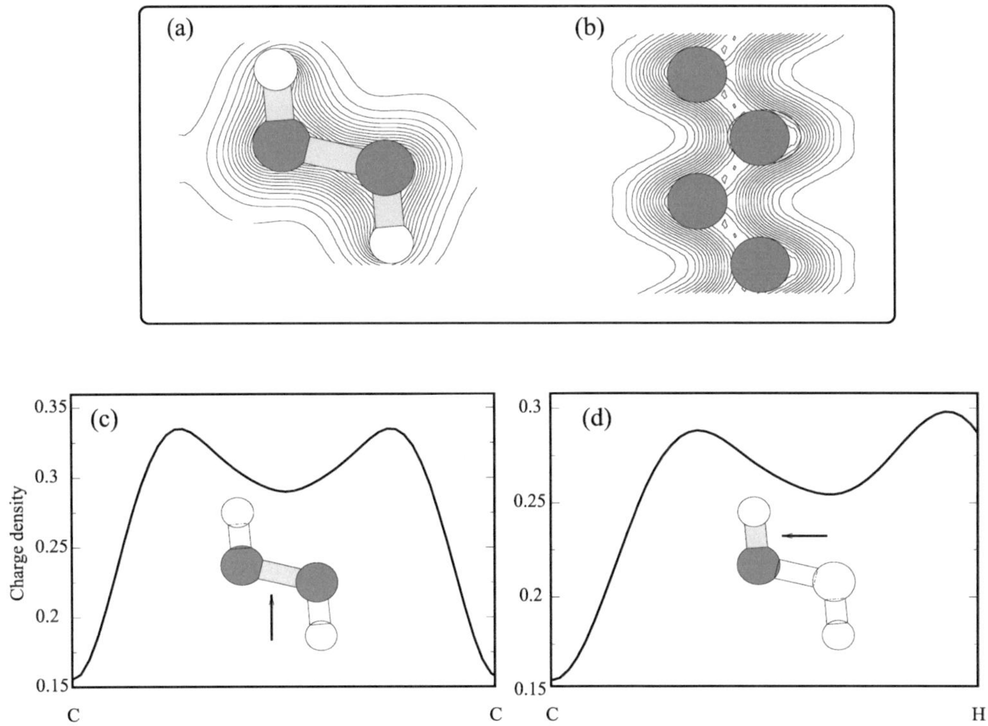

Fig. 8. Total charge density contour plots of (a) graphane and (b) diamond in a plane passing though H-C-C-H and C-C-C-C lines, respectively. Total charge density plot along the (c) C-C and (d) C-H bonds in graphane. The charge density is measured in $e/\text{a.u}^3$.

Neglecting the unfavorable substitutional sites, we have considered two configurations for the adsorption of benzene on pristine graphene. In these we have attempted a hollow and a stack adsorption sites, as schematically shown in Fig. 9. Therefore, to evaluate the energetically most preferable configuration between them, we compare their surface formation energies according to the formula:

$$\Delta E = E_{BG} - E_G - E_B. \qquad (1)$$

The symbols E_{BG}, E_G and E_B indicate the total energies of the optimized structures of the benzene-adsorbed graphene, pristine graphene, and isolated benzene molecule, respectively. It is important to state that these total energies are calculated within similar unit cells and computational parameters. Using the above equation we find that the adsorption energies of hollow and stack structures are approximately -0.25 eV and -0.30 eV, respectively. These values suggest that the stack configuration represent the ground-state structure of the benzene-graphene system with energy gain of about 0.05 eV comparing with the hollow phase. Therefore we will focus our theoretical investigation on the stack configuration to determine the structural and electronic properties of benzene/graphene structure. The starting point towards the structural optimization of the present system is the initial position of benzene molecule above the graphene sheet. The molecular atoms (benzene molecule) are initially placed at 1.5 Å above the carbon atoms of the graphene. After several iterations of relaxation process, the benzene molecule adopts a planar geometry at 3.52 Å above the graphene sheet. While this value is consistent with the experimental value of 3.6 Å that is estimated from binding energy curves performed by Chakarova-Kack *et al.* (Chakarova-Kack et al., 2006), it is slightly larger than the value of 3.17 Å obtained by Zhang *et al* (Y.-H. Zhang et al., 2010). This inconsistency is due to the quite low energy cutoff used in their calculations. Subsequently we have found that the C-C and C-H bond lengths of benzene on the top of graphene are measured to be 1.39 Å and 1.10 Å, respectively. These values are identical to those of the hydrocarbons compounds of 1.40 Å and 1.10 Å for C-C and C-H bonds, respectively. Since the calculated C-C-C and C-C-H angles of the molecule are identical and equal to 120°, we conclude that benzene reorients itself in a planar manner above the graphene. This orientation has also been noted for F4-TCNQ (Pinto et al., 2009) which indicates similar adsorption mechanism for organic molecules on graphene. Adsorbed-graphene sheet, on the other side, has been found to have a C-C bond length of 1.41 Å and a C-C-C angle of 120°. These suggest that, even though benzene is being adsorbed, graphene preserves its basic structural behaviour.

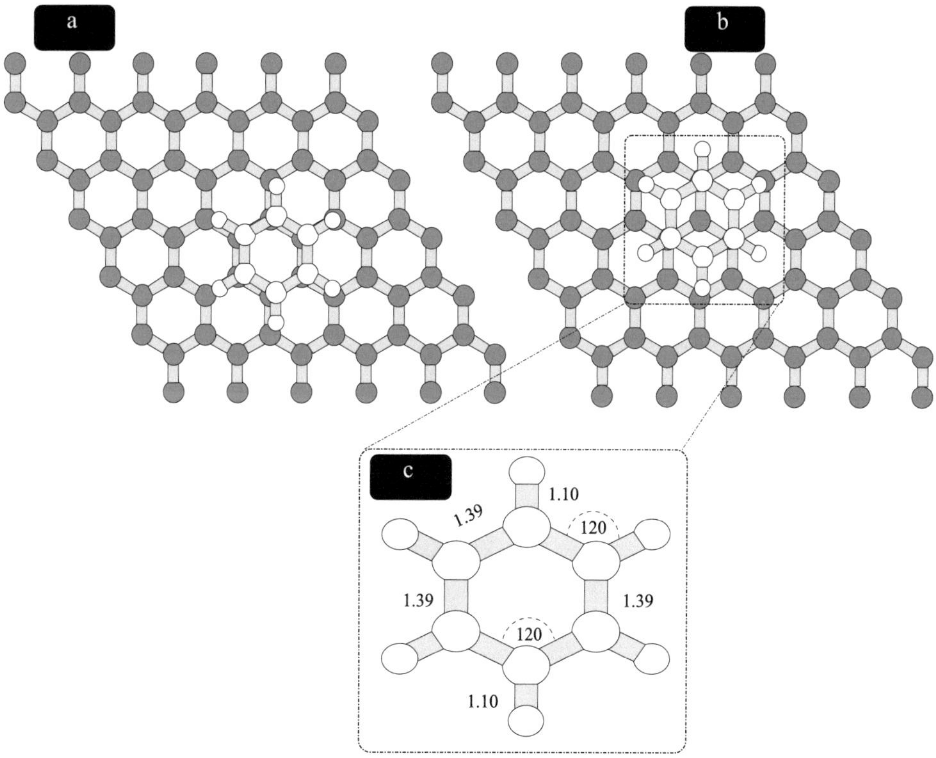

Fig. 9. Schematic view of the minimum-energy structures of the (a) hollow and (b) stack configurations of benzene-adsorbed graphene. (c) Top view of the relaxed benzene molecule with its structural parameters. The grey solid spheres represent C atoms from the graphene while dashed spheres indicate C atoms from the molecule.

Further, we have performed surface electronic band structure for the benzene-adsorbed graphene configuration along the high-symmetry directions, K→Γ and Γ→M, as shown in Fig. 10. Setting the Fermi level at the zero-energetic position, we clearly note that the Dirac point of the systems is coincided with the Fermi level, indicating a zero-gap nature. This indicates that, for low-energy states, the adsorption of benzene leads to unchanged electronic structure regarding to pristine graphene. Accordingly, this suggests that charge transfer is not expected to occur between the graphene and the molecule. Such an observation can be understood if we believe that only the states very far below/above the Dirac point of graphene are perturbed by the molecular adsorption. However, this conclusion is supportive of the result obtained by Zhang *et al.* (Y.-H. Zhang et al., 2010). In their study they found that the adsorption of benzene on pristine graphene results in insignificant amount of electronic charge being transferred from the molecule to the graphene sheet.

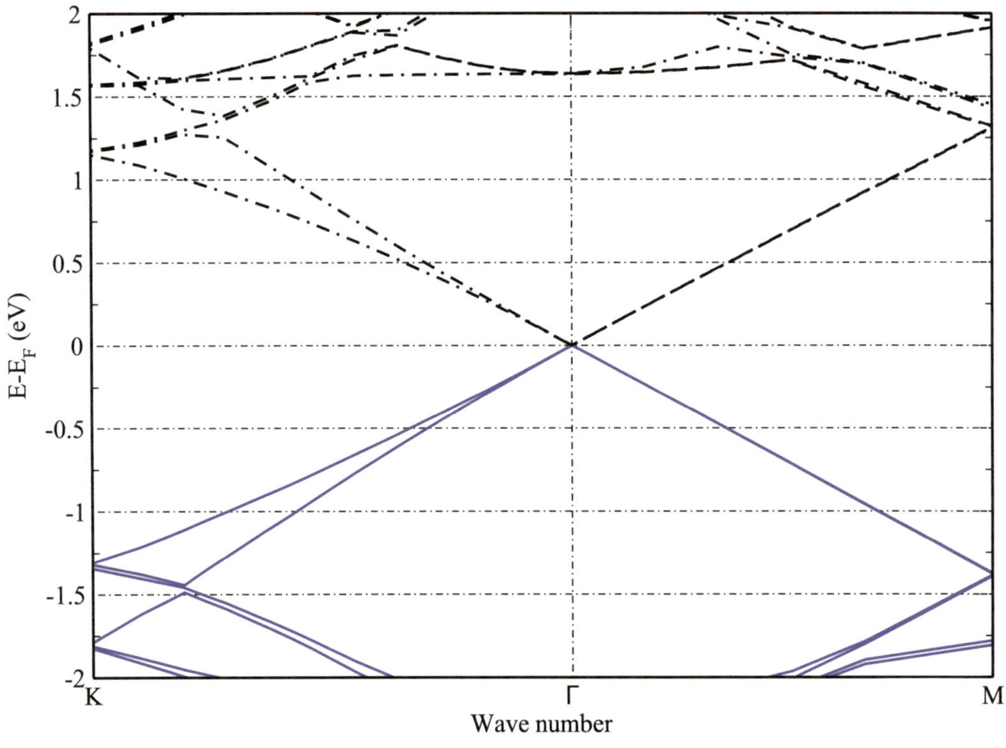

Fig. 10. Electronic band structure of the benzene-adsorbed graphene system with the molecule on a stack adsorption site. The zero-energy position indicates the Fermi level.

3.4 Naphthalene-adsorbed graphene

As has been performed for benzene-adsorbed graphene structure, we have tested at least two adsorption sites for naphthalene molecule onto graphene. Between hollow and stack configurations we have found that the latter represents the minimum-energy structure of naphthalene-adsorbed graphene, as shown in Fig. 11. Our calculations indicates that the adsorption energy of stack and hollow phases are approximately −0.47 eV and −0.39 eV, respectively. The molecule is found to be relaxed 3.15 Å above the sheet suggesting no bond formation between the molecule fragments and the carbon atoms of the graphene. Looking at the C–H bond length of the molecule we have identified no appreciable change and its typical value of 1.10 Å. Comparing the naphthalene-adsorbed system with the benzene-adsorbed system, we have clearly noted considerable alterations in the C–C bond lengths of the molecule. These bond lengths are categorized into three groups: 1.37 Å, 1.40 Å, and 1.43 Å. These values are in the acceptable range of the typical bond lengths of an isolated naphthalene molecule (1.36–1.42 Å). The C–C–C and C–C–H angles vary in the interval 121–122° and 118–120°, respectively. These results suggest a very tiny amount of vertical tilt in the carbon planes. However, the substrate keeps its original structure as also seen for benzene-adsorbed structure.

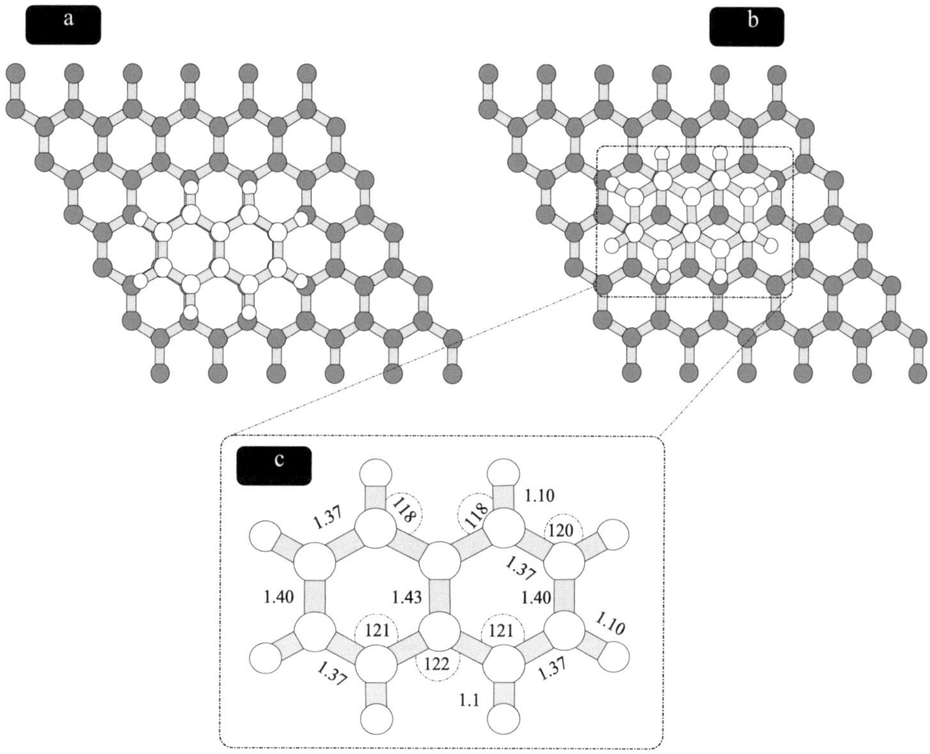

Fig. 11. The minimum-energy structures of the (a) hollow and (b) stack configurations of naphthalene-adsorbed graphene. (c) Top view of the fully relaxed molecule.

In Fig. 12, we have depicted the electronic band structure of the naphthalene/graphene system. Despite that the band structure for the benzene/graphene system looks very similar to the pristine graphene in the low-energy region (±2.0 eV with respect to Fermi level E_F), the energy bands for the naphthalene/graphene system performs little changes below the Fermi level. From the figure we clearly identify a new flat (non-dispersive) band at energy of E_F −1.3 eV. This band is believed to be originated from the molecule states. Overall, the system has an entire zero-gap behaviour with indication that no charge being transferred from/to the graphene substrate.

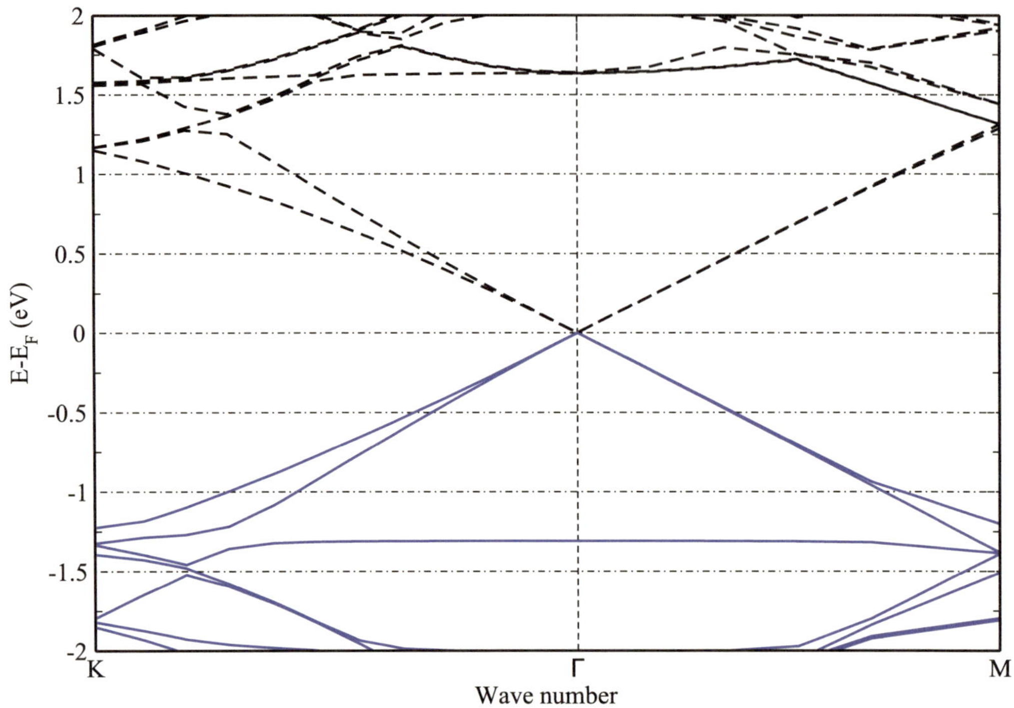

Fig. 12. Electronic band structure of the naphthalene/graphene system with the molecule on a stack adsorption site. The Fermi level is located at the zero-energy position.

4. Conclusion

Within the framework of local density approximation of the density functional theory and pseudopotential theory we have presented a comparative *ab initio* study for the adsorption of molecules on a pristine graphene. The *sp*^{2} structure and zero-gap behaviour are found to the fundamental characteristics of clean graphene with degenerate bonding π and antibonding π* states at the K point. Upon the adsorption of hydrogen atoms on pristine graphene, a chairlike configuration is found to be the energetically most stable structure for the system (graphane). As the four valence electrons of carbon atoms participate in the formation of the covalent bonzds with hydrogen atoms the π bands are removed from the band structure of graphane. The absence of these bands leads graphane to be a semiconducting with wide direct gap at the Γ point. Moreover, the structural transformation of carbon bonds from *sp*^{2} to *sp*^{3}-like hybridization results in an increase in the bond length from 1.41 Å to 1.49 Å. Unlike the hydrogen-adsorbed graphene, benzene and naphthalene-adsorbed structures are found to only stabilize the graphene sheet with no significant change in its low-energy electronic properties.

5. References

AlZahrani, A. Z. & Srivastava, G. P (2009). Gradual changes in electronic properties from graphene to graphite: first-principles calculations. *Journal of Physics: Condensed Matter* Vol. 21, No. 49, (November 2009), pp. 495503- 495509, ISSN 0953-8984

Boukhvalov, D. W.; Katsnelson, M. I. & Lichtenstein, A. I. (2008). Hydrogen on graphene: Electronic structure, total energy, structural distortions and magnetism from first-principles calculations. *Physical Review B* Vol. 77, No. 3, (January 2008), pp. 035427-035433, ISSN 1098-0121

Castro Neto, A.H.; Guinea, F.; Peres, N.M.R.; Novoselov, K.S. & Geim, A.K. (2009). The electronic properties of graphene. *Reviews of Modern Physics* Vol. 81, No. 1, (January 2009), pp. 109-162, ISSN 0034-6861

Chakarova-Kack, S.D.; Schroder, E.; Lundqvist, B. & Langreth, D. (2006). Application of van der Waals Density Functional to an Extended System: Adsorption of Benzene and Naphthalene on Graphite. *Physical Review Letters* Vol. 96, No. 14, (April 2006), pp. 146107-146110, ISSN 0031-9007

Chan, K. T.; Neaton, J. B. & Cohen, M. L. (2008). First-principles study of metal adatom adsorption on graphene. *Physical Review B* Vol. 77, No. 23, (June 2008), pp. 235430-235441, ISSN 1098-0121

Duplock, E.J.; Scheffler, M. & Lindan, P.J.D. (2004). Hallmark of Perfect Graphene. *Physical Review Letters* Vol. 92, No. 22, (June 2004), pp. 225502-225505, ISSN 0031-9007

Elias, D.C.; Nair, R. R.; Mohiuddin, T. M. G.; Morozov, S. V.; Blake, P.; Halsall, M. P.; Ferrari, A. C.; Boukhvalov, D. W.; Katsnelson, M. I.; Geim, A. K. & Novoselov, K. S. (2009). Control of Graphene's Properties by Reversible Hydrogenation: Evidence for Graphane. *Science* Vol. 323, No. 5914, (January 2009), pp. 610-613, ISSN 0036-8075

Farjam, M. & Rafii-Tabar, H. (2009). Energy gap opening in submonolayer lithium on graphene: Local density functional and tight-binding calculations. *Physical Review B* Vol. 79, No. 4, (January 2009), pp. 045417-045423, ISSN 1098-0121

Geim, A. K. & Novoselov, K. S. (2007). The rise of graphene. *Nature Materials* Vol. 6, (March 2007), pp. 183 - 191 ISSN 1476-1122

Giannozzi, P.; Car, R. & Scoles, G. (2003). Oxygen adsorption on graphite and nanotubes. *The Journal of Chemical Physics* Vol. 118, No. 3, (November 2002), pp. 1003-1006, ISSN 0021-9606

Giannozzi, P.; Baroni, S.; Bonini, N.; Calandra, M.; Car, R.; Cavazzoni, C.; Ceresoli, D.; Chiarotti, G.L.; Cococcioni, M.; Dabo, I.; Dal Corso, A.; de Gironcoli, S.; Fabris, S.; Fratesi, G.; Gebauer, R.; Gerstmann, U.; Gougoussis, C.; Kokalj, A.; Lazzeri, M.; Martin-Samos, L.; Marzari, N.; Mauri, F.; Mazzarello, R.; Paolini, S.; Pasquarello, A.; Paulatto, L.; Sbraccia, C.; Scandolo, S.; Sclauzero, G.; Seitsonen, A.P.; Smogunov, A.; Umari, P. & Wentzcovitch, R.M. (2009). *Journal of Physics: Condensed Matter* Vol. 21, No. 39, (September 2009), pp. 395502- 395520, ISSN 0953-8984

Han, M. Y.; Özylmaz, B.; Zhang, Y. & Kim, P. (2007). Energy Band-Gap Engineering of Graphene Nanoribbons. *Physical Review Letters* Vol. 98, No. 20, (May 2007), pp. 206805-206808, ISSN 0031-9007

Hao, S. G.; Zhou, G.; Duan, W. H.; Wu, J. & Gu, B. L. (2006). Tremendous Spin-Splitting Effects in Open Boron Nitride Nanotubes: Application to Nanoscale Spintronic Devices. *Journal of the American Chemical Society* Vol. 128, No. 26, (June 2006), pp. 8453-8458, ISSN 0002-786

Hohenberg, P. & Kohn, W. (1964). Inhomogeneous Electron Gas. *Physical Review* Vol. 136, No. 3B, (November 1964), pp. B864-B871, ISSN 0031-899X

Igami, M.; Okada, S. & Nakada, K. (2001). Electronic and geometric structures of fluorine adsorbed graphene. Synthetic Metals Vol. 121, No. 1-3, (March 2001), pp. 1233-1234, ISSN 0379-6779

Ito, J.; Nakamura, J. & Natori, A. (2008). Semiconducting nature of the oxygen-adsorbed graphene sheet. *Journal of Applied Physics* Vol. 103, No. 11, (June 2008), pp. 113712-113716, ISSN 0021-8979

Katsnelson, M. I.; Novoselov, K. S. & Geim, A. K. (2006). Chiral tunnelling and the Klein paradox in graphene. *Nature Physics* Vol. 2, (September 2006), pp. 620-625, ISSN 1745-2473

Katsnelson, M. I. (2007). Graphene: carbon in two dimensions. *Materials today* Vol. 10, No. 1-2, (January-February 2007), pp. 20-27, ISSN 1369-7021

Kohn, W. & Sham, L. (1065). Self-Consistent Equations Including Exchange and Correlation Effects. *Physical Review* Vol. 140, No. 4A, (November 1965), pp. A1133–A1138, ISSN 0031-899X

Lebègue, S.; Klintenberg, M.; Eriksson, O. & Katsnelson, M. I. (2009). Accurate electronic band gap of pure and functionalized graphane from GW calculations. *Physical Review B* Vol. 79, No. 24, (June 2009), pp. 245117 -245121, ISSN 1098-0121

Lee, C.; Wei, X.; Kysar, J. W. & Hone, J. (2008). Measurement of the Elastic Properties and Intrinsic Strength of Monolayer Graphene. *Science* Vol. 321, No. 5887, (July 2008), pp. 385-388, ISSN 0036-8075

Leenaerts, O.; Partoens, B. & Peeters, F.M. (2008). Adsorption of H_2O, NH_3, CO, NO_2, and NO on graphene: A first-principles study. *Physical Review B* Vol. 77, No. 12, (March 2008), pp. 125416-125421, ISSN 1098-0121

Leenaerts, O.; Partoens, B. & Peeters, F.M. (2009). Adsorption of small molecules on graphene. *Microelectronics Journal* Vol. 40, No. 4-5, (April-May 2009), pp. 860-862, ISSN 0026-2692

Li, X.; Wang, X.; Zhang, L.; Lee, S. & Dai, H. (2008). Chemically Derived, Ultrasmooth Graphene Nanoribbon Semiconductors, *Science* Vol. 319, No. 5867, (February 2008), pp. 1229-1232, ISSN 0036-8075

Löwdin, P. O. (1950). On the Non-Orthogonality Problem Connected with the Use of Atomic Wave Functions in the Theory of Molecules and Crystals. Journal of Chemical Physics Vol. 18, No. 3, (March 1950), pp. 365-375, ISSN 0021-9606

Mao, Y.; Yuan, J. & Zhong, J. (2008). Density functional calculation of transition metal adatom adsorption on graphene. *Journal of Physics: Condensed Matter* Vol. 20, No. 11, (February 2008), pp. 115209-115214, ISSN 0953-8984

McKie, D. & McKie, C. (1986). *Essentials of Crystallography*, Blackwell Science, 978-0632015740 Oxford, United Kingdom

Medeiros, P.V.C.; Mota, F.; Mascarenhas, A.J.S. & Castilho, C.M. (2010). A DFT study of halogen atoms adsorbed on graphene layers. *Nanotechnology* Vol. 21, No. 48, (November 2010), pp. 115701-115708, ISSN 0957-4484

Monkhorst, H.J. & Pack, J.D. (1976). Special points for Brillouin-zone integrations. *Physical Review B* Vol. 13, No. 12, (June 1976), pp. 5188–5192, ISSN 1098-0121

Nakamura, J.; Ito, J. & Natori, A. (2008). Structural bistability of the oxygen-adsorbed graphene sheet. *Journal of Physics: Conference Series* Vol. 100, Part 5, (May 2008), pp. 052019-052022, ISSN 1742-6588

Novoselov, K. S.; Geim, A. K.; Morozov, S. V.; Jiang, D.; Zhang, Y.; Dubonos, S. V. & Firsov, A. A. (2004). Electric Field Effect in Atomically Thin Carbon Films. *Science,* Vol. 306, No. 5696, (October 2004), pp. 666-669, ISSN 0036-8075

Novoselov, K. S.; Jiang, Z.; Zhang, Y.; Morozov, S. V.; Stormer, H. L.; Zeitler, U.; Maan, J. C.; Boebinger, G. S.; Kim, P. & Geim, A. K. (2007). Room-Temperature Quantum Hall Effect in Graphene. *Science* Vol. 315, No. 5817, (March 2007), pp. 1379, ISSN 0036-8075

Ohno, H. (1998). Making Nonmagnetic Semiconductors Ferromagnetic. *Science* Vol. 281, No. 5379, (August 1998), pp. 951-956, ISSN 0036-8075

Ohno, Y.; Young, D. K.; Beschoten, B.; Matsukura, F.; Ohno, H. & Awschalom, D. D. (1999). Electrical spin injection in a ferromagnetic semiconductor heterostructure. *Nature* Vol. 402, (December 1999), pp. 790-792, ISSN: 0028-0836

Ohta, T.; Bostwick, A.; Seyller, T.; Horn, K. & Rotenberg, E. (2006). Controlling the Electronic Structure of Bilayer Graphene. *Science* Vol. 313, No. 5789, (August 2006), pp. 951-954, ISSN 0036-8075

Pinto, H.; Jones, R.; Goss, J.P. & Briddon, P.R. (2009). p-type doping of graphene with F4-TCNQ. *Journal of Physics: Condensed Matter* Vol. 21, No. 40, (September 2009), pp. 402001-402003, ISSN 0953-8984

Perdew, J. P. & Zunger, A. (1981). Self-interaction correction to density-functional approximations for many-electron systems. *Physical Review B* Vol. 23, No. 10, (May 1981), pp. 5048–5079, ISSN 1098-0121

Şahin, H.; Ataca, C. & Ciraci, S. (2010). Electronic and magnetic properties of graphane nanoribbons. *Physical Review B* Vol. 81, No. 20, (May 2010), pp. 205417-205424, ISSN 1098-0121

Sanyal, B.; Eriksson, O.; Jansson, U. & Grennberg, H. (2009). Molecular adsorption in graphene with divacancy defects. *Physical Review B* Vol. 79, No. 11, (March 2009), pp. 113409-113412, ISSN 1098-0121

Savchenko, A. (2009). Tranforming Graphene. *Science* Vol. 323, No. 5914, (January 2008), pp. 589-590, ISSN 0036-8075

Schabel, M.C. & Martins, J.L. (1992). Energetics of interplanar binding in graphite. *Physical Review B* Vol. 29, No. 11, (September 1992), pp. 7185–7188, ISSN 1098-0121

Schedin, F.; Geim, A.K.; Morozov, S.V.; Hill, E.W.; Blake, P.; Katsnelson, M.I. & Novoselov, K.S. (2007). Detection of individual gas molecules adsorbed on graphene. Nature Materials *Vol. 6,* (July 2007), pp. *652-655,* ISSN 1476-1122

Sofo, J. O.; Chaudhari, A. S. & Barber, G. B. (2007). Graphane: A two-dimensional hydrocarbon. *Physical Review B* Vol. 75, No. 15, (April 2007), pp. 153401-153404, ISSN 1098-0121

Vanderbilt, D. (1990). Soft self-consistent pseudopotentials in a generalized eigenvalue formalism. *Physical Review B* Vol. 41, No. 11, (April 1990), pp. 7892–7895, ISSN 1098-0121

Wehling, T.O.; Lichtenstein, A.I. & Katsnelson, M.I. (2008). First-principles studies of water adsorption on graphene: The role of the substrate. *Applied Physics Letters* Vol. 93, No. 20, (November 2008), pp. 202110-12, ISSN 0003-6951

Yang, C-K. (2009). A metallic graphene layer adsorbed with lithium. *Applied Physics Letters* Vol. 94, No. 16, (April 2009), pp. 163115-163117, ISSN 0003-6951

Yin, M.T. & Cohen, M. L. (1984). Structural theory of graphite and graphitic silicon. *Physical Review B* Vol. 29, No. 12, (June 1984), pp. 6996–6998, ISSN 1098-0121

Zhang, Y.; Tan, Y.; Stormer, H. & Kim, P. (2005). Experimental observation of the quantum Hall effect and Berry's phase in graphene. *Nature* Vol. 438, (November 2005), pp. 201-204 ISSN, 0028-0836

Zhang, Y.-H.; Zhou, K.-G.; Xie, K.-F.; Zeng, J.; Zhang, H.-L. & Peng, Y. (2010). Tuning the electronic structure and transport properties of graphene by noncovalent functionalization: effects of organic donor, acceptor and metal atoms. *Nanotechnology* Vol. 21, No. 6, (January 2010), pp. 065201-065207, ISSN 0957-4484

3

Computer Simulation of Radiation Defects in Graphene and Relative Structures

Arkady M.Ilyin
Kazakh National University, Physical Department
Kazakhstan

1. Introduction

Graphene is a single layer of carbon atoms arranged in a chicken-wire-like hexagonal lattice and in spite of its recent availability for experimental investigations (Novoselov et al., 2004) it is an object of great interest for many researchers (Elias et al., 2009; Sofo et al., 2007; Luo et al., 2009; Teweldebrhan and Balandin, 2009; Ilyin et al., 2009) because of amazingly wide field of its potential applicability: electronics, sensors, materials science, biology etc. In particular, graphene and few layer graphene fragments as well as carbon nanotubes can be used in production of composites, based on metal, ceramic, polymer matrices, filled with graphene's or few layer graphene fragments as elements of reinforcement. Obviously, the main goal of using graphene or nanotubes in making composites is using their extremely high mechanical characteristics in combination with low weight. It should be noticed, that many difficulties concerning graphene's applications originate from its surface chemical inertness. In other words, the sp^2 electron structure of ideal graphene often results in very low binding energy between graphene's surface and atoms of many elements. It is one of the obstacles for modifying and applications of graphene in production of electronic devices, when controllable electronic properties are needed. Besides, it results in poor interfacial bonding of the graphene fragments with matrices in composite materials and with sliding between few layer graphene sheets under stressed state. The noticeable success in this direction has been recently achieved in the work (Elias et al., 2009), in which the hydrogenation of graphene was performed by using an ion-plasma technique. It should be noticed, that composite of graphene and hydrogen, associated with graphene's surface was theoretically predicted few years ago in the paper (Sofo et al., 2007) and named as graphane. Unfortunately, this success up to now can be surely referred only to graphene-hydrogen composition. But today's technologies especially in the field of materials science need much more wide area of possible compositions and special materials. It was reasonable to suppose, that radiation defects may essentially improve binding ability of graphene with atoms of other elements due to production of additional chemical bonds (Ilyin, 2010; Ilyin & Beall, 2010). It is especially important for application of graphene species in R&D of composites. Moreover defects in such structures may improve mechanical properties by linking reinforcing carbon nanoelements to each other and by increasing the strength and stiffness of the composite (Ilyin et al.,2010). Unfortunately, it is not yet well understood which kinds of stable radiation defects and their complexes can exist in graphene and its derivatives. Obviously, it is not easy to create definite types of radiation defects and perform direct studies of them in nanoobjects in direct laboratory experiments. In this situation computer simulation of radiation defects in graphene- and relative nanostructures becomes of great importance (Ilyin et al., 2011). In this paper we have submitted some results of computer simulation and calculations of some possible kinds of stable radiation defects in graphene, and more complex configurations, involving atoms of light metals: Be and Al, which are linking with radiation defects. In other words, we consider possible effect of radiation defects on production and modification of graphene - metal composites and calculate energetic and structural properties for some mostly possible configurations. Computer simulation and calculations were performed by the use well known molecular dynamics, extended Hückel technique and density functional theory.

2. Simulation and calculations

The first model which we consider in this chapter is a simple configuration, which consists of a graphene sheet and a carbon atom adsorbed with it. We imply, that in this case no chemical bond between graphene and C atom exists. This kind of interaction refers to well known van der Waals forces. These interactions are weak and insignificant in our common life, but their role increases dramatically among the nano-scaled objects. One of the important features of them - the additivity i.e. every particle of the system makes its contribution into the total interaction energy. Therefore, they can be relatively large for nanosystems, involving 10^2 - 10^3 atoms or molecules. But van der Waals interactions are not accounted by the use widespread computer techniques like molecular orbital linear combination of atomic orbitals by self consistent field (MO LCAO SCF) or density functional theory (DFT). Therefore, in this problem the method of molecular dynamics (MD), which is recognized as the effective tool for similar systems, was chosen. A graphene sheet for MD simulation of single- and diatomic defects on the undamaged structure was built of 78 atoms. Van der Waals interaction between adsorbed atoms and graphene was described by well known Lennard-Johns potential in the usual form.

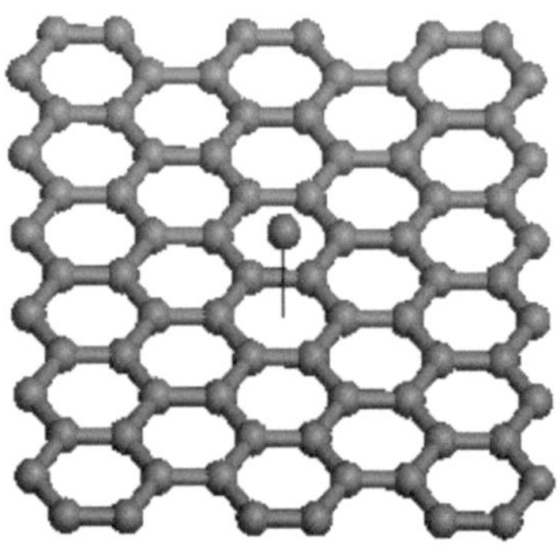

Fig. 1. Configuration of a single adsorbed atom on a graphene structure

Presented in Figure 1 is one of investigated defect configurations which is actually a single carbon atom, adsorbed on the surface of the undamaged graphene sheet. The maximum value of the binding energy for the single adsorbed atom E = – 0.18 eV at a distance Z = 0.25 nm from the graphene plane. One important characteristic is also the energy of carbon atom in the center point of graphene's hexagon (Z = 0). This position was found to be very unstable with the positive energy equals to E_0 =11.4 eV. It means also, that the graphene sheet is impermeable for displaced carbon atoms with energies lower than E_0.

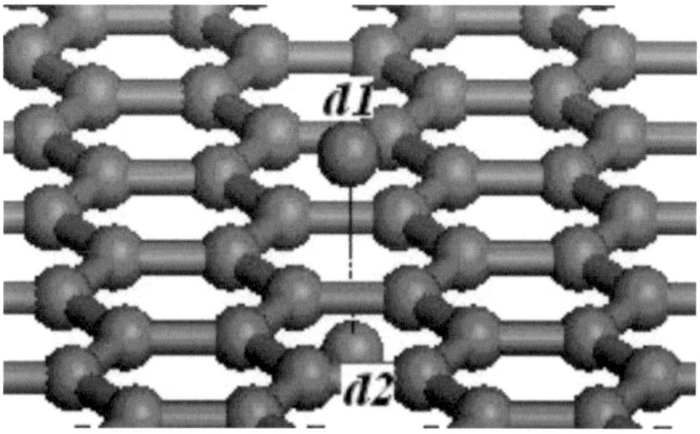

Fig. 2. Configuration of a dumbbell defect on a graphene (N=78) structure

Figure 2 presents a configuration of a more complicated, two-atom defect which is like a dumbbell with a symmetrical configuration of the atoms d1 and d2 normal to the graphene sheet. Calculations of this defect were performed by MD, using the LJ potential.

To begin, in all cases the minimum energy lateral position of the atoms adsorbed, had been found out at the normal axis of symmetry of hexagon (Z-axis). Further we performed calculations with movement of atoms along the Z-axis. Results of calculations of the binding energies for these defects as a function of a distance Z over the center of hexagon are presented in Figure 3. It can be seen, that there is an interval between approximately 2 and 3 angstroms that exhibits a trough with a negative energies, which is evidence of the existence of stable binding states. Low values of bonding energies testify of the vdW nature of the interaction.

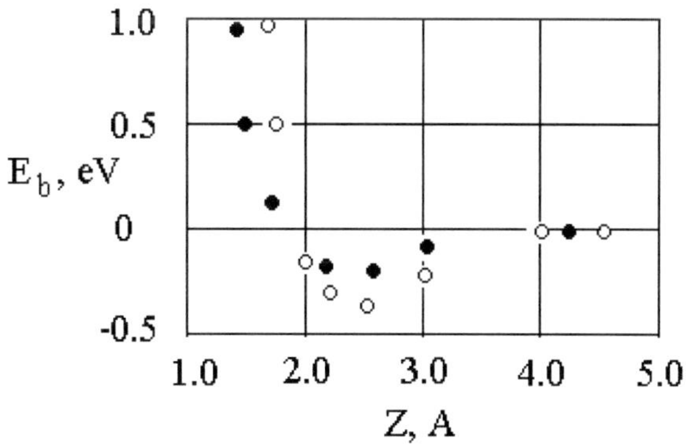

Fig. 3. The binding energy for the single atom and the dumbbell defect as a function of a distance Z over the center of hexagon. The black marks - the single atom configuration (see Figure 1), the light marks - the dumbbell configuration (see Figure 2)

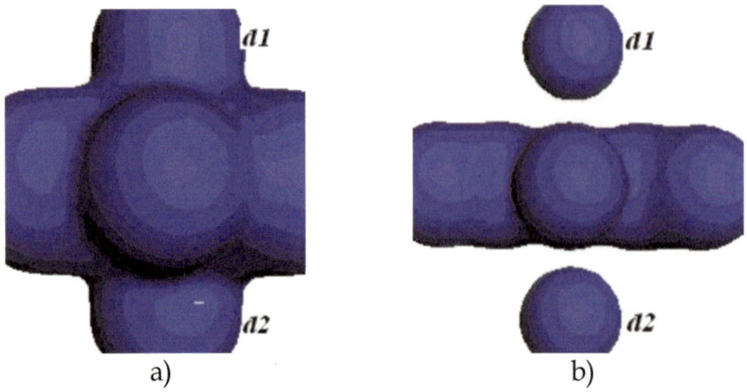

Fig. 4. The electron charge distribution for the dumbbell presented in Fig.2. Density of electron charge equals: a) 0.02 el / Å^3 , b) 0.5 el / Å^3 .

Figure 4 illustrates the results of calculations of electron charge distribution for the dumbbell presented in Figure 2, performed by MO LCAO method. These calculations were performed in order to check our assumption about vdW interaction between atoms adsorbed and graphene. One can see some overlapping of electron charge only by very low level of electron density (Figure 4,a). And, obviously, there are no signs of overlapping of electron charge at high level of electron density, which could be responsible for some kind of bonding between graphene and atoms adsorbed. One can see well distinguished electron charge clouds, obviously closed on graphene and d1 and d2 atoms with a gap between them. At the same time graphene's structure is linked with dense electron clouds, which provided strong bonding. It proves that weak bonding interaction for the defects presented above is controlled by vdW forces. It is unlikely, that such defects can be useful for essential modifying of mechanical properties of composite materials.

2.1 Vacancies in graphene

Irradiation of graphene-based electronic devices by fast electrons or ions will be always accompanied by creation of atom vacancies. Therefore, it is very important to know about changes in electronic properties of graphene fragments which should be expected under irradiation and about how they depend on defect concentration. For such estimation we have used large enough graphene's fragment (N=208).

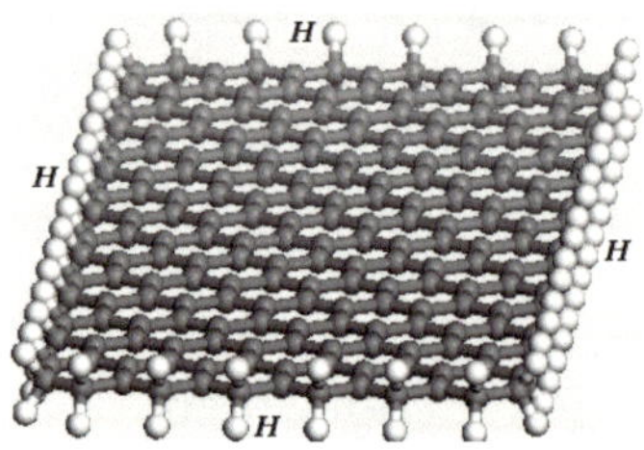

Fig. 5. Graphene for calculations with edge-bonds shut by H-atoms

In order to avoid end-effects by calculations, free end-bonds of carbon atoms were shut with hydrogen atoms. Afterwards, in order to take into account the possible effect of a larger size of a real graphene sheet, which can restrict the atoms neighboring to the vacancy, all edge atoms of graphene were fixed at their initial positions. After that we simulated and calculated one-, two- and three -vacancy configurations with using in all cases a procedure of energy minimization.

It was revealed, that in all cases, after energy optimization the vacancy zone increased so that all the three two-coordinated atoms, neighboring the vacancy, were shifted nearly symmetrically: all three distances between surrounding atoms 1-2 , 2-3, 3-1 (figure 6) become as large as 2.76 Å instead of 2.46 Å in the initial state.

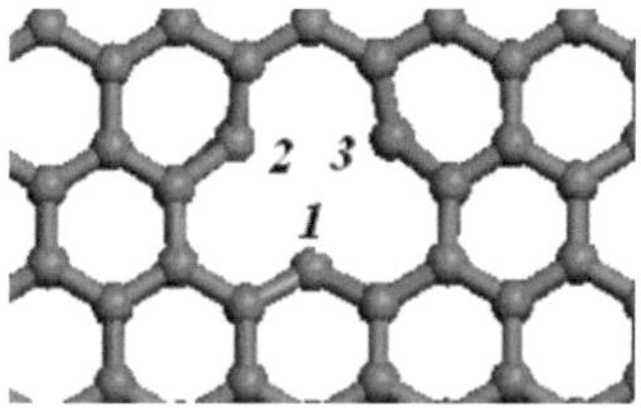

Fig. 6. Vacancy zone in graphene after the optimization procedure

Figure 7 presents a configuration of a graphene with 3 single vacancies displayed by the electron charge distribution.

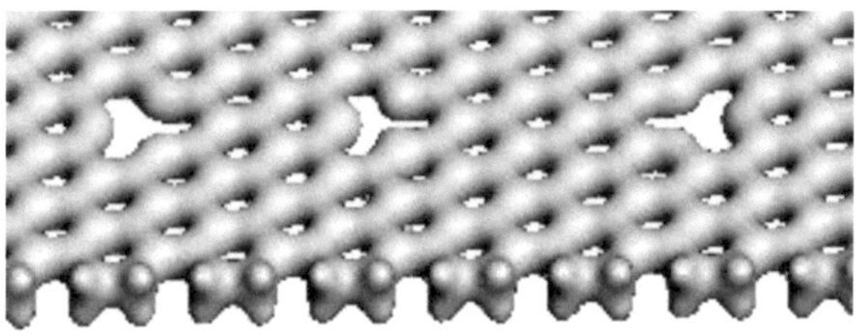

Fig. 7. Graphene with 3 single separated vacancies in the structure

Calculations of HOMO and LUMO were also performed in all cases. Figure 8 presents dependence of Eg = HOMO - LUMO for graphene -208 with different numbers of vacancies. There were one-, two-, and three vacancies in aligned configuration.

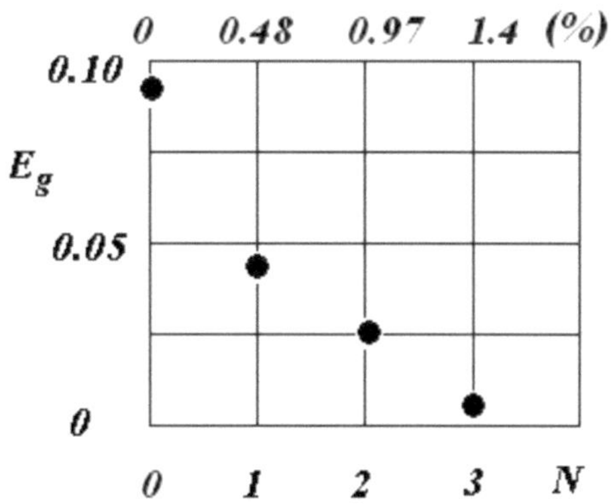

Fig. 8. The value of Eg = HOMO - LUMO as a function of concentration of vacancies in a graphene-208 fragment.

Obviously, data for E_g presented in Figure 8 illustrate the effect of limited size of graphene fragment (so called size-effect), because the infinite graphene (the initial state with N =0) is intrinsically semi-metal with Eg =0. But these and similar effects should be taken into account when physicists and technologists will design devices based on real graphene fragments of limited sizes.

2.2 Radiation defects with strong bonding

As the next step we simulated and calculated energetic and structural characteristics of 3D defect configuration presented in Figure 9. This type of radiation defect, which involves two carbon atoms arranged symmetrically over a vacancy can be named "dumbbell", like to configuration presented in Figure 2.

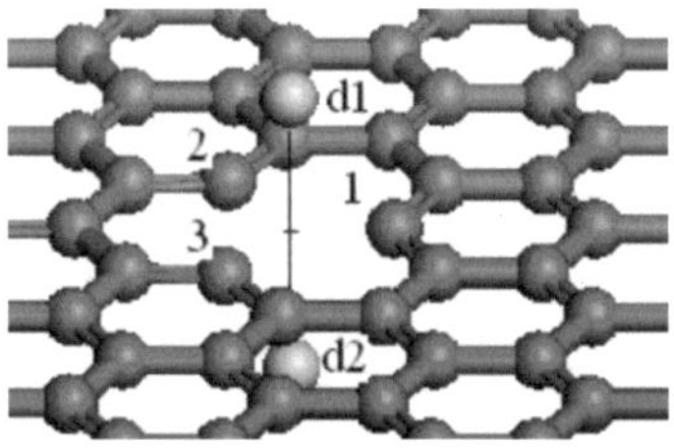

Fig. 9. The complex radiation defect, involving a vacancy and a dumbbell configuration.

But in this case the two carbon atoms (d1,d2) of the dumbbell are chemically bonded with free bonds of atoms, neighboring at the vacancy. One can see from the graph in Figure 10, that there is a strong bonding.

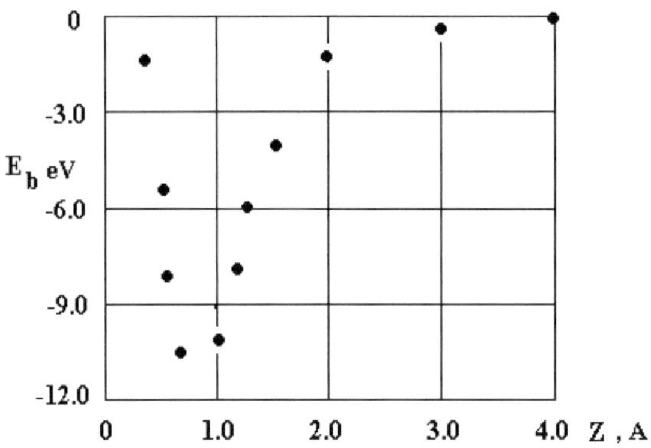

Fig. 10. The binding energy for the dumbbell placed over a vacancy as a function of the distance Z over the graphene sheet

Obviously, the elastic properties of composite materials and their uniformity are of great importance in using composite materials. The E_b - Z curve in Figure 10 can be used for estimation of elastic characteristics of the C-C dumbbell defect along the graphene sheet (under shear stress). One can see, that the maximum slope of the curve is near the point Z = 1.3 Å. The numerical estimation by using $\Delta E/\Delta Z$ at this point with small segments gives the value 0.5 TPa. The maximum binding energy of the dumbbell over the relaxed vacancy was obtained as large as -10.0 eV and the corresponding distance between the graphene's plate and atoms equals 0.7 A. The electron charge distribution presented in Figures 11 and 12 proves that there is fast chemical bond between dumbbell and graphene sheet.

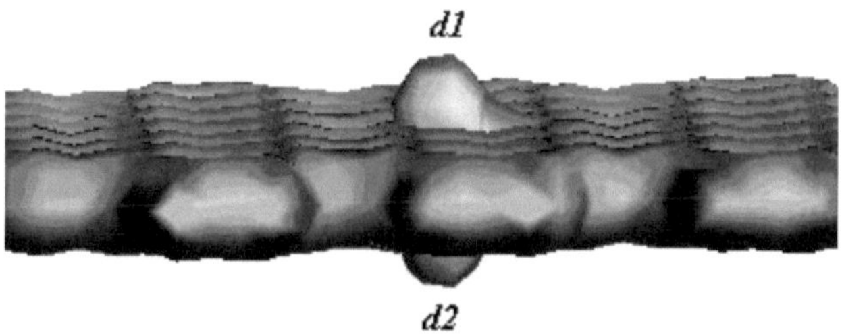

Fig. 11. The electron charge distribution for the graphene-dumbbell configuration at density of the charge equal to 1.0 el/Å^3 .

Fig. 12. The electron distribution for a graphene with two separated dumbbells

Calculation of electronic properties of dumbbell configurations presented in Figures 11 and 12 for graphene-208 were also performed. Eg = HOMO – LUMO was as large as 0.05 eV for a case with one- and two dumbbels. Figure 12 displayed no signs of noticed non-uniformity between two dumbbells in the graphene structure. So that, effect of radiation defects like dumbbells can be considered as insignificant by their concentrations about 1%. The large value of the binding energy for the dumbbell defects and the electron charge distribution calculated for these defect configurations obviously demonstrate, that there is significant interaction between the dumbbell and a graphene, as well as between atoms of the dumbbell itself.

The ability of graphene's vacancy to bind atoms of other elements by the chemical way were firstly checked by simulation and calculation of the vacancy - hydrogen complex defect, which was performed in the hydrogenated graphene ("chair"- graphane) structure (Figure 13). Our calculations proved that a vacancy zone can serve as a site with a high concentration of hydrogen. In this case the vacancy zone has non- symmetrical mode of deformation: distances between atoms: C1-C2 = C2-C3 = 2.72 Å, C1-C3 = 2.55 Å

The results of calculations of binding energy of H atoms, bonding at the vacancy are presented in Fig. 14. These data witness, that the value of binding energy depends on the total number of H atoms, placed at the vacancy. One can see, that the binding energy has well defined minimum at N = 3.

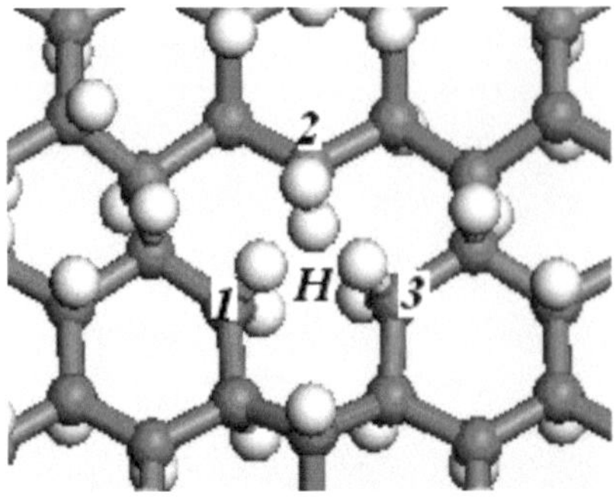

Fig. 13. The atomic structure of vacancy with 6 hydrogen atoms bonding

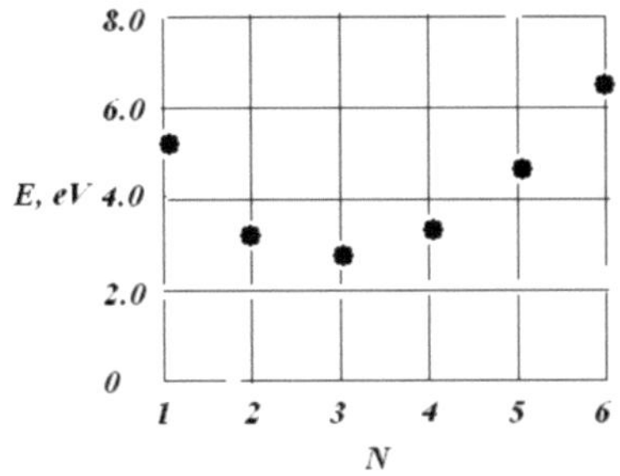

Fig. 14. The dependence of hydrogen - vacancy binding energy on the total number N of hydrogen atoms linking with the vacancy

2.3 Bridge-like radiation defects in graphene

It is known very well that two layer graphene is being one of main components by production of graphene's materials by many technologies. Therefore, it can play a significant role in some applications, using graphene materials. At least, it is reasonable to study properties of two layer graphene, keeping in mind its future applications. In particular, two layer graphene fragments can be used in production of lightweight composite materials with high stiffness and strength. Therefore, calculations were made also for bilayer graphene fragment. The usual AB graphite-like configuration, which is the most common for graphite -like materials, was chosen for calculations (see Figure 15). The Figure 16 illustrates that interaction between undamaged graphene sheets has van der Waals nature, without any signs of electron charge overlapping. The coupled atom pair which were removed by creating the vacancy pair is marked by black. The interstitial C-atom, knocked from the structure was placed between graphene layers. After that relaxation procedure was used to obtain a minimum of the total energy of the defect volume. The edge atoms of graphenes were fixed in order accounting the size effect.

In Figure 16 one can see the typical picture of the electron charge distribution for undamaged bilayer graphene, controlled by van-der-Waals interactions. There is no electron charge overlapping, between different graphene sheets.

In Figure 17 one can see the much more complex defect configuration with vacancies, faced each other, which were produced in both of graphene sheets and interstitial carbon atom (*i*), caught between them. This type of radiation defects can be called as a bridge-like defect. The essential feature of the defect is that the two graphene sheets are linked with fast covalent bond based on the interstitial atom and as one can see, neighboring atoms 1-2 and 3-4 facilitating the rising two additional bonds, because of pulling in the gap between graphenes. The distribution of the electron charge presented in Figure 18 proves existing of three covalent bonds, originated between graphene layers.

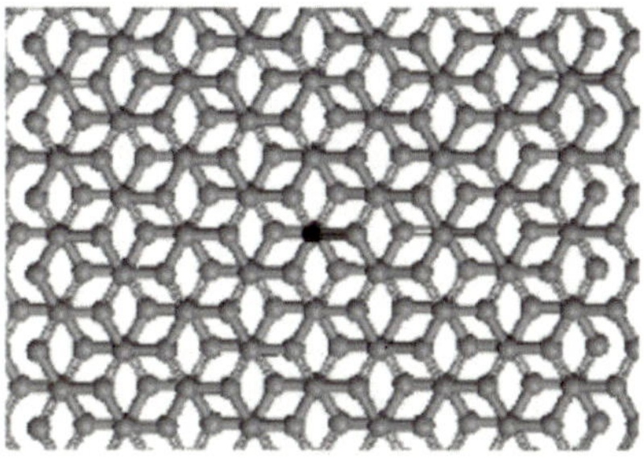

Fig. 15. Graphene AB two layer used for building complex interior defect

Fig. 16. The distribution of the electron charge for undamaged bilayer graphene (the density equals 0.4 el / A^3).

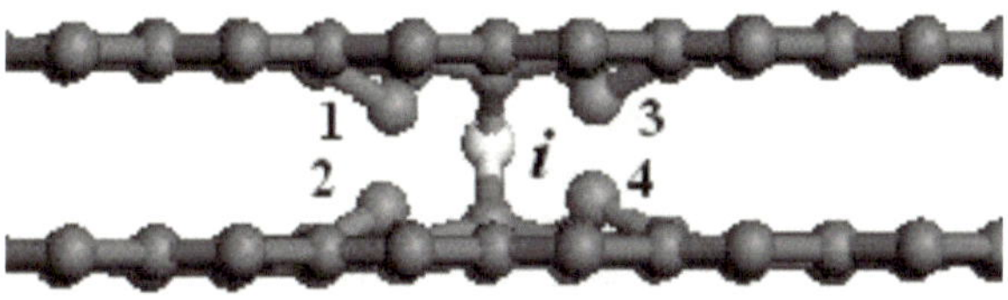

Fig. 17. A complex bridge-like radiation defect bonding together sheets of bilayer graphene after relaxation

The total binding energy for this complex defect configuration was calculated as large as -11.3 eV. We have supposed, that ends of two- or few layer graphene fragments may also serve as sites of bridge-like defects, linking graphene's sheets together. A typical configuration of end-bridge-like bonding of a carbon atom is presented in Figure 19.

Fig. 18. The distribution of the electron charge (the density equals 1.4 el / A^3)

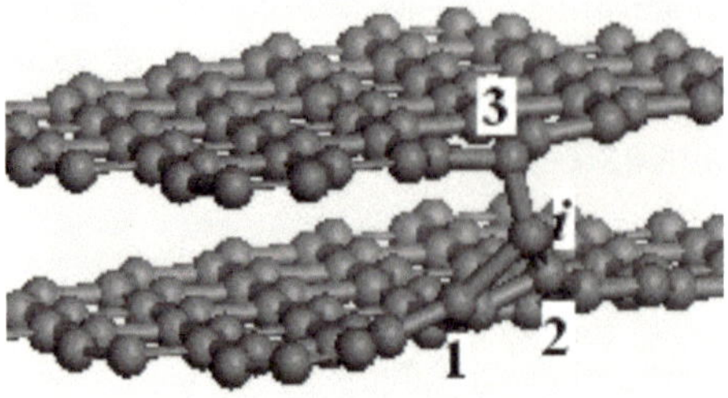

Fig. 19. The bridge-like defect based on an interstitial atom *i* at the edge of bilayer graphene.

2.4 Graphene-metal composition induced by radiation

It is known very well, that graphene and few layer graphene fragments, like to carbon nanotubes can be used as elements of reinforcement in production of composites, based on various matrices. Moreover, many important physical properties of material (particularly, electric and heat conduction, magnetic characteristics and so on) can be modified and improved by using graphene and few layer graphene fragments as filler. As was mentioned above, many difficulties concerning graphene's applications originate from its sp^2 electron structure. In other words, the electron structure of ideal graphene often results in very low binding energy between graphene's surface and atoms of many metals. It is one of the obstacles for modifying and applications of graphene in production of composite materials. In our recent papers we suggested using of radiation modification of composite materials with carbon nanostructures due irradiation by fast electrons or ions. Production of special kind of bridge-like defects may be considered as an effective technological tool of essential modification of physical-mechanical properties of composite materials, filled with carbon nanostructures (Ilyin & Beall, 2010).

Beryllium, aluminum and their alloys are being very important materials for designing new composites, especially for fields where combination of light weight with high strength is needed, for example - transportation systems and air-space technologies. Therefore, in this paper we focused on study of possible production composite materials based on Be and Al matrices, with using graphene fragments as reinforcement elements. We suppose, that radiation defects may essentially improve binding ability of graphene with atoms of light metals due to production of additional chemical bonds. Unfortunately, direct experimental study of such nanosystems with atomic scale defects and operations with them can be hardly performed today even in laboratories with high level equipment. Figure 20 presents some typical possible positions of metallic atoms (Be, Al) arranged on graphene surface in high symmetry sites. Calculations performed for all of these positions gave values of the binding energy of metal atoms on the graphene surface nearly zero.

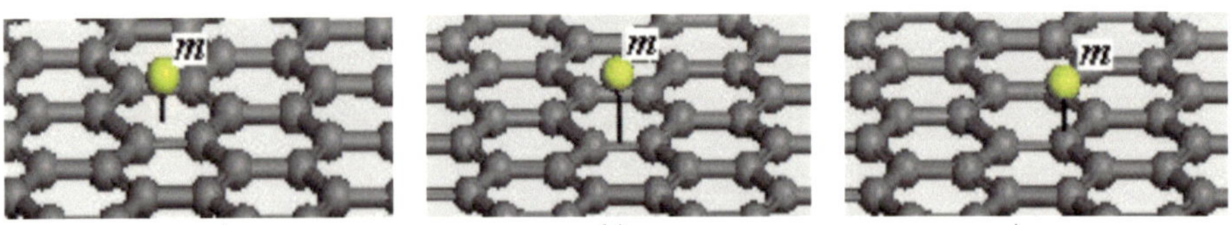

a) b) c)

Fig. 20. Some of calculated symmetrical configurations of metal atoms (Be and Al) on graphene surface with nearly zero binding energies: a) over the center of a hexagon; b) over the center of the C-C bond; c) over a C atom.

2.4.1 Be – graphene composition

Figure 22 presents a scheme of estimation elastic characteristics of defect involving Be dumbbell at vacancy by techniques like above for C-C dumbbell. The value of elastic modulus for direction "to right" in the Figure 21 was calculated as large as 0.05 TPa and 0.02 TPa in the perpendicular direction.

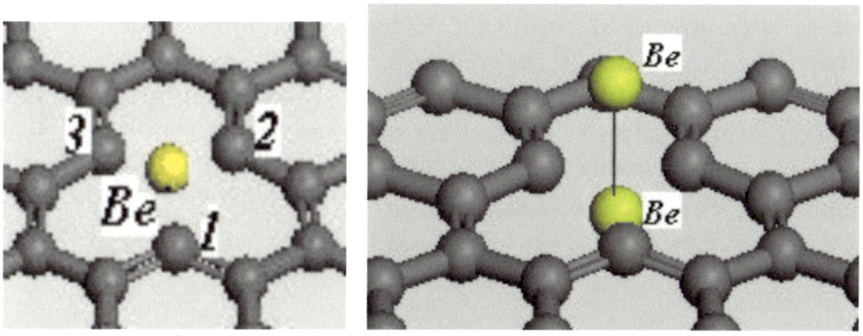

a) b)

Fig. 21. Configurations of Be atoms bonded with a vacancy: a) a stable position of the single Be atom in the graphene sheet. The binding energy E_b equals 2.6 eV; b) Configuration of "Be - dumbbell" over a vacancy with binding energy 4.1 eV.

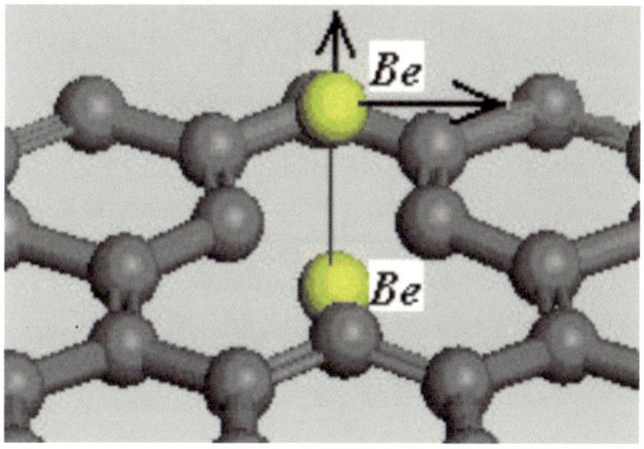

Fig. 22. Scheme of the elastic modulus calculation for the Be dumbbell at vacancy.

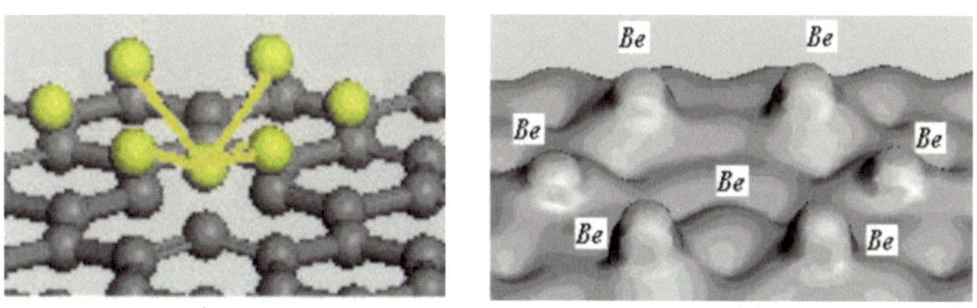

a) b)

Fig. 23. a) A configuration of an initially flat Be-cluster over vacancy after optimization. The binding energy of the cluster with graphene was obtained as large as 7.2 eV; b) the electron charge distribution in the area of the metal cluster - vacancy with a density of charge 0.2 el / $Å^3$.

2.4.2 Al – graphene composition

A stable configuration with a minimum of total energy was provided by Al atom placed in the plate of graphene sheet within the vacancy. One can see in figure 24 that the vacancy zone essentially and symmetrically increased with all three lengths of bonds equal 1.7 Å.

Figure 25 presents configuration of Al - dumbbell placed at the vacancy. The equilibrium distance between Al atoms equals 2.5 Å, the binding energy is as large as 2.9 eV.

Very interesting and important result for technological applications by production of composite materials based on Al or Al -alloys matrix with graphene filler shows computational model in Figure 27. One can see, that Al atoms can be chemically attached to a bridge-like defect in vacancy's zones. One also can see a significant deformation of graphene sheets around the defect zone. The binding energy of the interstitial carbon atom *i* at this configuration was equal to -9.3 eV.

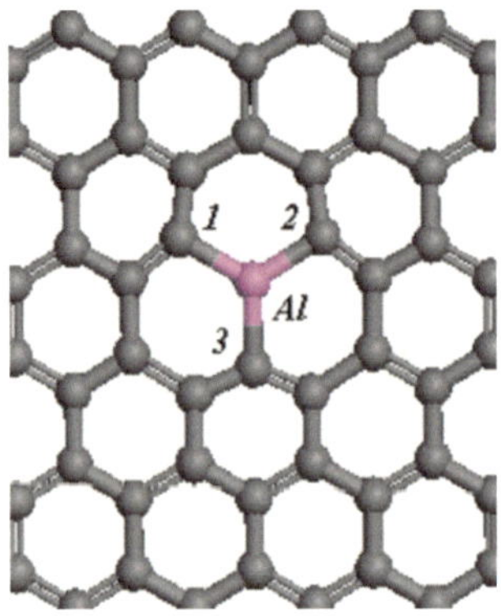

Fig. 24. Configuration of a single Al-atom bonded with a vacancy.

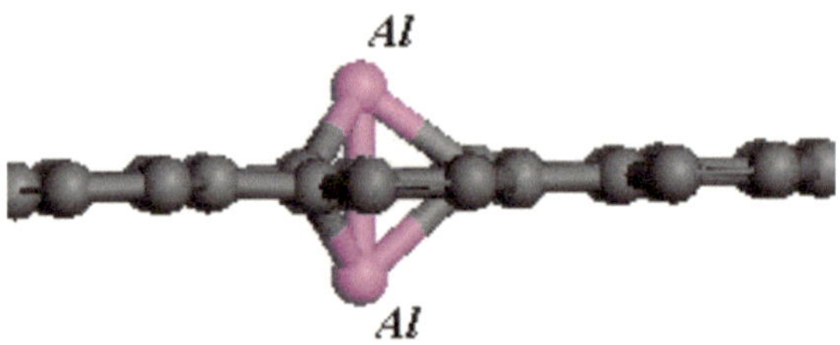

Fig. 25. Configuration of Al - dumbbell bonded with a vacancy.

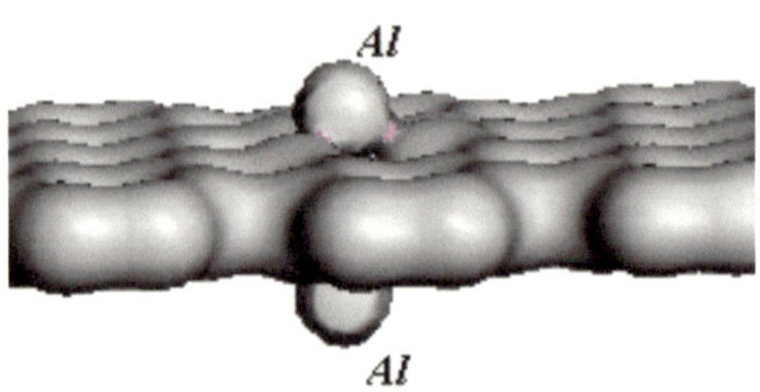

Fig. 26. The distribution of the electron charge (the density equals 1.4 el / A^3) for the Al - dumbbell bonded with a vacancy in graphene

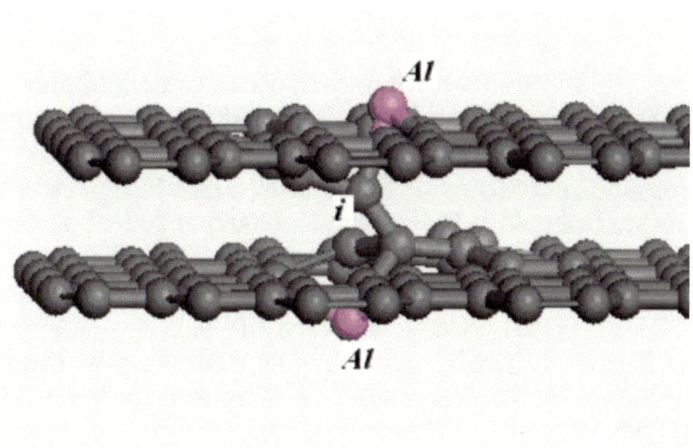

Fig. 27. Complex defect: a bridge-like defect in two- layer graphene with Al atoms attached on both graphene sheets.

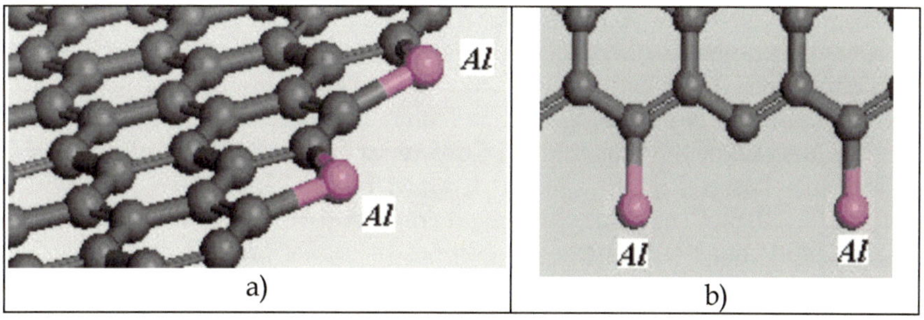

Fig. 28. Edge bonds of graphene fragments with Al atoms caught with binding energy E_b as large as 4.2 eV (a) and 3.5 eV (b).

Figure 28 illustrates one more possible way of creation additional chemical bonds between Al atoms and graphene fragments. Free end bonds can serve as additional sites of linking Al atoms providing better adhesion between metal matrix and graphene filler with binding energy as large as 4.2 eV in the case of "arm-chair" edge and 3.5 eV for "zig-zag" edge.

3.Conclusion

Some stable radiation defect configurations, involving single adsorbed carbon atom, 3D C-dumbbell defect and a vacancy-like defect in graphene sheet, associated with a dumbbell of adsorbed atoms as well as complex interior defects : bridge-like radiation defects, which can originate under fast electron or ions irradiation and attach metal atoms, were simulated. The binding energy, structure characteristics and some electron characteristics of defects have been determined by using molecular dynamics, MO LCAO and DFT. Our calculations show, that electronic properties of graphene fragments are rather stable to radiation damage in the form of vacancies and dumbbells. Results of simulations and calculations also show, that special kind of radiation defects, namely bridge-like radiation defects produced by irradiation with fast electrons or ions can become an effective technological tool by production of composite materials, based on light metal matrices with carbon nanostructures, in particular, graphene's fragments, as reinforcement elements.

4. References

Novoselov, K.; Geim A.; Morozov, S.; Jiang, D; Zhang, Y.; Dubonos, S.; Grigorieva, I.& Firsov,A.(2004). Electric field effect in atomically thin carbon films. *Science,* Vol. 306, pp.666-669, ISSN 0036-8075.

Elias,D.; Nair,R; Mohiuddin,T; Morozov, S.; Blake,P.; Halsall,M.; Ferrari,A.; Boukhvalov,D.; Katsnelson, M.; Geim,A. & Novoselov,K. (2009) . Control of Graphene's Properties by Reversible Hydrogenation: Evidence for Graphane. *Science* , Vol.*323*, pp.610–613.

Sofo,J.; Chaudhari A. & Barber,G. (2007). Graphane: a two-dimensional hydrocarbon. *Phys.Rev.*,B75, pp.153401-153404, ISSN 0974-0546 .

Luo,Zh.; Yu,T.;Kim,K.; Ni,Zh.;You,Y.; Lim,S.; Shen,Z.;Wang ,S.& Lin,J. (2009). Thickness-Dependent Reversible Hydrogenation of Graphene Layers. *ACS NANO,* Vol.3, No 7, pp.1781-1788.

Teweldebrhan D & Balandin A. (2009). Modification of graphene properties due to electron-beam irradiation. *Appl.Phys.Lett.*, Vol. 94, 013101

Ilyin,A.; Daineko,E.& Beall,G. (2009). Computer simulation and study of radiation defects in graphene. *Physica E,* Vol.42, No pp. 67-69, ISSN 1386-9477.

Ilyin,A.; Guseinov,N.;.Nikitin, A & Tsyganov,I. (2010). Characterization of thin graphite layers and graphene with energy dispersive X-ray analysis. *Physica E,* Vol.42, No 8 , pp. 2078-2080.

Ilyin,A. (2010). Simulation of end-bridge-like radiation defects in carbon multi-wall nanotubes. *Book of Abstracts* of 10th International Conference on Computer Simulations of Radiation Effects in Solids, p.123, ISBN, Poland, Krakov, July 19-23, 2010.

Ilyin,A &.Beall,G. (2010). Computer simulation and study of bridge-like radiation defects in the carbon nano-structures in composite materials. *Proceedings of NanoTech Conference,* pp.312-315, ISBN 978-1-4398-3401-5, Annaheim, California, USA, June, 21-25, 2010.

Ilyin,A; Beall, G & Tsyganov,I. (2010). Simulation and Study of Bridge-Like Radiation Defects in the Carbon Nano-Structures. *Journal of Computational and Theoretical Nanoscience* , Vol. 7, No. 10 (Oct.,2010) ,pp. 2004-2007, ISSN 1546-1955.

Ilyin,A;Guseinov, N;Tsyganov, I & Nemkaeva,R. (2011). Computer Simulation and Experimental Study of Graphan-Like Structures Formed by Electrolytic Hydrogenation. *Physica E,*Vol.43 ,pp. 1262-1265,ISSN 1386-9477.

4

Hydrogenation of Graphene and Hydrogen Diffusion Behavior on Graphene/Graphane Interface

Zhimin Ao and Sean Li
School of Materials Science and Engineering
The University of New South Wales, Sydney, NSW 2052
Australia

1. Introduction

Hydrogenation of carbon materials has been attracting a wide range of interests as an application of hydrogen storage materials in hydrogen-powered automobile as well as a methodology to manipulate the electric properties of carbon materials. Graphene with unique electronic, thermal and mechanical properties has been investigated as one of the most promising candidates for the next generation of electronic materials (Geim, 2009). However, several major challenges have to be tackled before the widespread application of graphene. For example, the absence of a band gap in the electronic spectrum of intrinsic graphene and the Klein paradox as a consequence of the Dirac-type nature of the charge carriers (Novoselov et al., 2004; Rao et al., 2009). The most efficient way to overcome these problems is hydrogenation of graphene (Luo et al., 2009). The graphane (Sofo et al., 2007), a fully hydrogenated single layer of graphene, was suggested to possess the promising semiconductor properties with a band gap of around 3.5 eV theoretically. Very recently, Elias *et al.* (Elias et al, 2009) experimentally demonstrated the formation of graphane through the exposure of a graphene membrane to hydrogen plasma. Subsequently, the rate of hydrogenation process of multilayer graphene was found to strongly depend on the number of layers (Luo et al., 2009; Ryu et al., 2008). These discoveries open important perspectives for the application of graphene-based devices because the electronic gap in those graphanes could be controlled by the degree of hydrogenation (Elias et al., 2009; Zhou et al., 2009).

The hydrogenation process of graphene in above experiments occurs through the exposure of graphene to the hydrogen plasma, which contains H^+, H_3^+, H atoms and H_2 molecules (Elias et al., 2009; Luo et al., 2009). One generally assumes that H atoms form covalent bonds with the carbon atoms in the graphene (Luo et al., 2009). In this chapter, alternative approaches to hydrogenate graphene will be proposed theoretically based on density functional theory (DFT) calculations. It suggests that the hydrogenation process can be realized through the exposure of graphene to molecular hydrogen gas in the presence of a strong perpendicular electric field. It has been demonstrated that an electric field can modify the chemical activity of materials (Liu et al., 2009). For example, the dissociation activation energy of molecular oxygen on Pt (111) is tunable by an applied electric field (Hyman et al.,

2005). To understand the mechanism of graphene hydrogenation with molecular hydrogen, the pathway of the dissociation of a H_2 molecule and the subsequent atomic adsorption on graphene in the presence of an electric field will be investigated.

On the other hand, it is believed that graphene nanoribbons (GNRs) offer the possibility to achieve tunable electronic properties. This is because their properties are highly dependent on their width and also the orientation of edges, for example, the GNRs can be turned from semiconducting to metallic by manipulating the structural parameters (Han et al., 2010; Jiao et al., 2010). Unfortunately, to manipulate the edge structure and width of freestanding GNRs is a very challenging experimental task (Han et al., 2010; Jiao et al., 2010). Both experimental data and the corresponding *ab initio* calculations demonstrated that the zigzag edge is metastable in vacuum due to a planar reconstruction to lower the energy of the system (Koskinen et al., 2009). Alternatively the high quality GNRs can be fabricated by selectively hydrogenating the graphene or by carving GNRs on a graphane sheet (Balog et al., 2010; Sessi et al., 2009; Singh et al., 2009). A bandgap opening in graphene, induced by the patterned absorption of atomic hydrogen, was recently found experimentally (Koskinen et al., 2009). Meanwhile, the hybrid graphene/graphane nanoribbons (GGNRs) were also studied by *ab initio* calculations (Hernández-Nieves et al., 2010; Lu et al., 2009; Singh et al., 2010). It was shown that the bandgap of GGNRs is dominated by the graphene rather than the graphane (Hernández-Nieves et al., 2010; Lu et al., 2009; Singh et al., 2009). Its electronic and magnetic properties strongly depend on the degree of hydrogenation of the interface (Hernández-Nieves et al., 2010). However, the hydrogen diffusion associated with high mobility of the isolated H atoms on graphene has a strong influence on the stability of the graphene/graphane interface.

In this chapter, in order to enhance the hydrogen storage capacity in graphene and also to manipulate the electronic properties of graphene, the hydrogenation of graphene with and without an applied electric field, as well as the stability of the graphene/graphane interface in the hybrid nanoribbons are studied through DFT calculations. The energy barriers for the hydrogenation reaction and the diffusion of H atoms located at the graphene/graphane interface are simulated using DFT. All the possible reaction and diffusion pathways are analyzed to find the minimum reaction and diffusion barriers.

2. Hydrogenation of graphene

2.1 Hydrogenation of pristine graphene

In the simulation, all DFT calculations are implemented in the DMOL3 code (Delley, 2000). The local density approximation (LDA) with the PWC functional is employed as the exchange-correlation functional (Perdew & Wang, 1992). A double numerical plus polarization (DNP) is employed as the basis set. The convergence tolerance of energy is taken 10^{-5} Ha (1 Ha = 27.21 eV), and the maximum allowed force and displacement are 0.002 Ha and 0.005 Å, respectively. To investigate the minimum energy pathway for the hydrogen dissociative adsorption on graphene, linear synchronous transition/quadratic synchronous transit (LST/QST) (Halgren & Lipscomb, 1977) and nudged elastic band (NBE) (Henkelman & Jonsson, 2000) tools in DMOL3 code are used, which have been well validated to find the structure of the transition state (TS) and the minimum energy pathway. In the simulation, three-dimensional periodic boundary conditions are taken. The computer simulation cell consists of a 2×2 graphene supercell with a vacuum width of 18 Å to minimize the interlayer interaction. The *k*-point is set to 20×20×1, and all atoms are allowed to relax.

The adsorption energy of a molecular H_2 on a pristine graphene layer $E_{b\text{-}H2}$ is defined as,

$$E_{b\text{-}H2} = E_{H2\text{-}graphene} - (E_{graphene} + E_{H2}) \tag{1a}$$

where $E_{H2\text{-}graphene}$, $E_{graphene}$, E_{H2} denote the energy of the system with a H_2 molecule adsorbed on graphene, the energy of the graphene layer and a H_2 molecule, respectively. For the case of atomic H chemically bonded on pristine graphene, the binding energy $E_{b\text{-}H}$ is defined as,

$$E_{b\text{-}H} = [E_{2H+graphene} - (E_{graphene} + 2E_H)]/2 \tag{1b}$$

where $E_{2H+graphene}$ is the energy of the system with 2 H atoms bound on the graphene layer, $E_{graphene}$ is the energy of the graphene layer, and E_H is the energy of a free H atom in the same slab lattice.

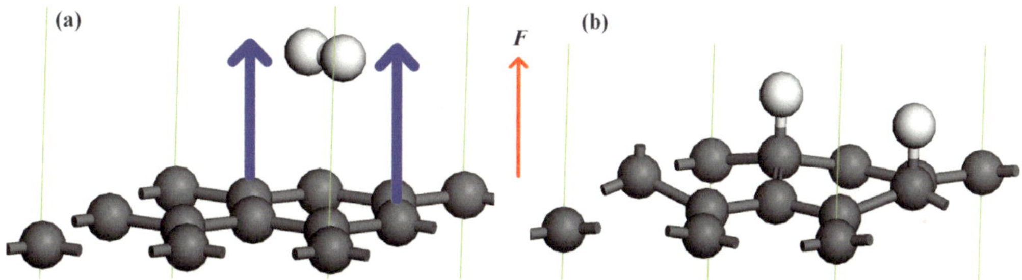

Fig. 1. The initial (panel a) and final (panel b) structures for a H_2 molecule dissociative adsorption on graphene. In this figure, a 2×2 supercell is shown where the gray and white balls are carbon and hydrogen atoms, respectively. The direction of the electric field is indicated by the arrow. (Reproduced with permission from Ref. (Ao & Peeters, 2010a). Copyright 2010, AIP)

For the case of molecular hydrogen adsorbed on graphene, previous DFT reports showed that the H_2 molecule is weakly physisorbed at the hollow site of graphene as shown in Fig. 1(a) (Ao et al., 2009, Arellano et al., 2000). The distance between the H_2 molecule and the graphene layer $d_{H2\text{-}graphene}$ is 2.612 Å with adsorption energy $E_{b\text{-}H2}$ = -0.153 eV, which are consistent with other simulation results of $d_{H2\text{-}graphene}$ = 2.635 Å and $E_{b\text{-}H2}$ = -0.159 eV (Ao et al., 2009), and $d_{H2\text{-}graphene} \approx$ 2.8 Å and $E_{b\text{-}H2}$ = -0.133 eV (Okamoto et al., 2001). For the case of atomic hydrogen adsorption on graphene, the favorable configuration is two H atoms adsorbed on two face-by-face carbon atoms in the same hexagon as shown in Fig. 1(b), which is consistent with the reported DFT result (Miura et al., 2003). The C-H bond length $l_{C\text{-}H}$ is 1.125 Å with binding energy $E_{b\text{-}H}$ = -2.184 eV, which agrees with another DFT result $l_{C\text{-}H}$ = 1.13 Å (Miura et al., 2003). In addition, the C atoms bonded with the two H atoms move up towards the H atoms with about 0.32 Å and the C-C bond length $l_{C\text{-}C}$ is 1.493 Å. This is similar to the sp^3 bond length of 1.53 Å in diamond while it is much longer than 1.420 Å for the sp^2 carbon length. The reconstruction of the graphene layer was also reported by the others, in which the C atoms bonded with the H atoms move out of the graphene plane by 0.35 Å (Miura et al., 2003). In this case, the carbon hexagon becomes more chemically active as it turns from sp^2 bonding to sp^3 like bonding (Miura et al., 2003). Based on the calculation of the band structure for the structure shown in Fig. 1(b), this hydrogenated graphene has a band gap of 3.4 eV, which is consistent with that of 3.5 eV for fully hydrogenated graphene – graphane (Sofo et al., 2007).

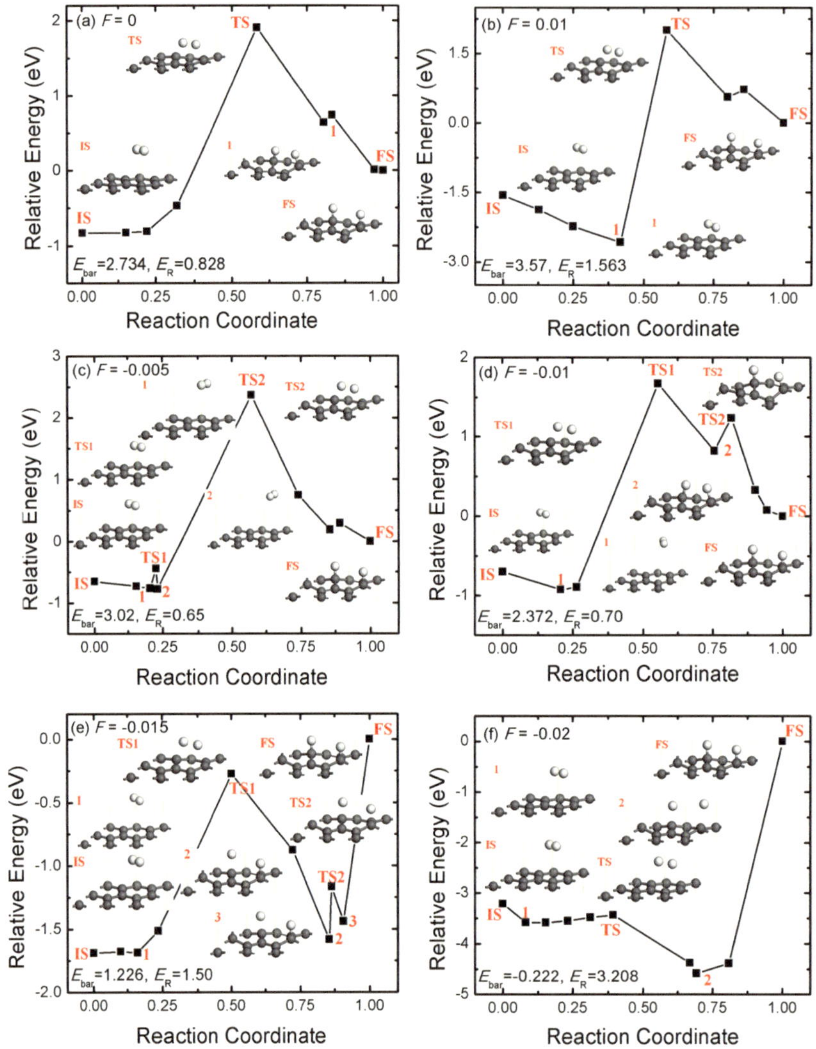

Fig. 2. The reaction pathway of a H_2 molecule that undergoes a dissociative adsorption on graphene for different electric field. IS, TS, FS, 1 and 2 represent initial structure, transition structure, final structure, energy minimum state 1 and 2, respectively. Their atomic structures are given by the inserts. The energy of FS is taken to be zero. The unit of F, E_{bar} and E_R are a.u. and eV, respectively, where E_{bar} is the energy barrier and E_R is the reaction energy. (Reproduced with permission from Ref. (Ao & Peeters, 2010a). Copyright 2010, AIP)

The energy minimum pathway for a H_2 molecule dissociative adsorption on graphene is from the structure depicted in Fig. 1(a) to the structure in Fig. 1(b), which agrees with the reported minimum pathway (Miura et al., 2003). Here, this dissociative adsorption pathway is shown in Fig. 2(a). From Fig. 2(a), the dissociative adsorption reaction barrier E_{bar} is 2.734 eV, which is smaller than 3.3 eV as reported (Miura et al., 2003). Regarding the transition state (TS), the H_2 molecule is dissociated into 2 free H atoms without any binding with the C atoms. After TS, these 2 H atoms bind with the C atoms and move to the exact top sites of the C atoms shown as the final structure (FS) in Fig. 2(a). Thus, this is a two-step reaction. Step one is that the H_2 molecule is dissociated into 2 free H atoms. Subsequently, the 2 H atoms are bound to the two C atoms. Step one needs an energy of 2.7 eV to overcome the potential barrier and the second step releases an energy of 1.9 eV. So totally the reaction energy is about 0.8 eV for the H_2 dissociative adsorption on graphene. The dissociation of H_2 is the rate-limiting step because a large energy is required.

Next the effect of an electric field F on this dissociative adsorption process is investigated, our numerical results are shown in Fig. 2. From the figure, in general E_{bar} increases with increasing F, and the reaction energy E_R increases as the absolute value of F increases. However, there is an abnormal energy barrier value E_{bar} = 3.020 eV at F = -0.005 a.u., which is larger than those for F = -0.01 a.u. and F = 0 (1 a.u. = 5.14×10^{11} V/m). The physical interpretation of this abnormal E_{bar} value for F = -0.005 a.u. will be given later. Thus, such a negative electric field can act as a catalyst to significantly facilitate the hydrogenation process of graphene. However, E_R increases as E_{bar} decreases with increasing absolute value of the negative electric field. In order to confirm the occurrence of hydrogenation reaction, the change of the Gibbs free energy $\Delta G = \Delta H - T\Delta S$ is considered where ΔH, T and ΔS are the change of enthalpy, the reaction temperature and the change of entropy, respectively. From Fig. 2, one can see that the energies of the product are higher than those of reactant. So this hydrogenation process is an endothermic reaction, i.e. $\Delta H > 0$. On the other hand, molecular H_2 is dissociated into two H atoms, this step leads to an increase of entropy, e.g. $\Delta S > 0$. Thus, if sufficient energy is available, which is the case at high temperature, ΔG can become negative and the reaction will go smoothly and efficiently.

The pathways for this reaction under different electric fields are given in Fig. 2. The energy minimum atomic structures of both reactant and product change in the presence of an electric field (Ao et al., 2008b). As shown in Fig. 2(b) at F = 0.01 a.u., the reactant was geometry optimized from the initial structure (IS) to the energy minimum state - State 1, where the physisorbed H_2 molecule moves towards the carbon layer while a negative electric field pushes the H_2 molecules upwards as shown in Figs. 2(c)-2(e). Under F = -0.02 a.u. as shown in Fig. 2(f), the H_2 molecule in the reactant is dissociated into two free H atoms as indicated by State 1. On the other hand, the energies of FS in Figs. 2(e) and 2(f) are not minimal after the TS. Thus the reaction will be terminated at State 2 in both Figs. 2(e) and 2(f) if there is not sufficient energy available. At State 2 in Figs. 2(e) and 2(f), the H_2 molecule is already dissociated. The free H atoms are considered to automatically bind with the C atoms once the applied electric field is removed since there is no potential barrier after H_2 dissociation at TS as shown in Fig. 2(a). In addition, experiments have indicated that the graphene layer can be automatically hydrogenated by free H atoms (Elias et al., 2009; Luo et al., 2009). Notice that a negative E_{bar} -0.222 eV is found when F = -0.02 a.u.. In the other words, the reaction from IS to State 2 can automatically progress when F = -0.02 a.u., and subsequently we remove the electric field in order that the H atoms bind to the C atoms. In addition, in case of Fig. 2(c) there is an extra energy barrier before reaching the transition state of the whole reaction TS2, which may be the reason behind the appearance of abnormal E_{bar} value.

The origin of the changes in E_{bar} with F can be understood through the analysis of the partial density of states (PDOS). Fig. 3 shows the PDOS of hydrogen and the p orbital of the carbon atoms at which the H_2 molecule is dissociated and the H atoms are adsorbed. It is noted that the PDOS for both C and H atoms are changed. In general, the weight of the bands corresponding to the H atoms below the Fermi level increases as $-F$ increases, while those of the C atoms decreases. These changes are induced by the electron transfer between the H and the C atoms. In the presence of a negative electric field, electrons flow from the C atoms at the bottom to H atoms at the top, and more charge is transferred as $-F$ increases. The strength of the interaction and the trend in E_{bar} can be explained by considering the positions and weights of the interaction between the bands of carbon and hydrogen. Based on the position of the bands, it is possible to identify the character of the C-H bands. The lowest C-H band around -8 ~ -10 eV are due to the interaction of the H σ_g state and the C P_z state (Zhang & Cho, et al., 2007). The second lowest C-H band around -3 ~ -5 eV is most likely related to the bonding interaction of H σ_u and C P_z states as they are above the H σ_g state. The H σ_u and the antibonding C P_z states above the Fermi level result in the highest C-H band, which determines the strength of the interaction and thus determines the energy barrier heights of the transition states (Arellano et al., 2000). As one can see from Figs. 3(e) and 3(f), the PDOS of the highest C-H bands are depressed and the lowest and the second lowest C-H bands are strongly enhanced. This agrees with the result in Fig. 2 that the barrier decreases to 1.226 eV at F = -0.015 a.u. and a negative E_{bar} -0.222 eV is found at F = -0.02 a.u.

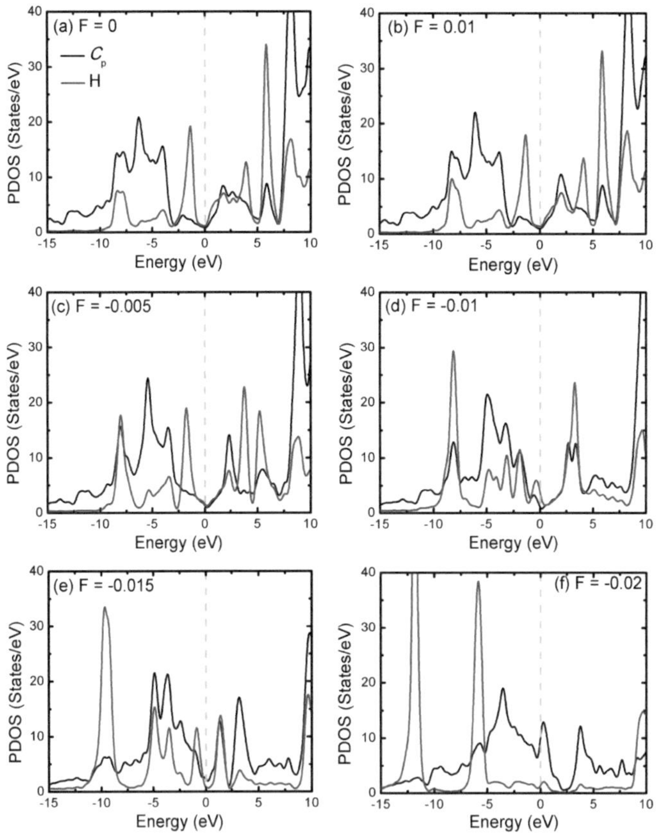

Fig. 3. Partial density of states (PDOS) of the hydrogen and the carbon atoms which are bond with hydrogen in the presences of different electric fields. The black and blue curves are the PDOS of the two H atoms and p orbit of the two C atoms at the transition states. The vertical lines indicate the Fermi level. (Reproduced with permission from Ref. (Ao & Peeters, 2010a). Copyright 2010, AIP)

2.2 Hydrogenation of N-doped graphene

As shown in the last section, the process of hydrogenation in pristine graphene is divided into two steps: molecular hydrogen dissociation under the electric field and the formation of covalent bonds between the dissociated H atoms and C atoms after removing the electric field. This two-step hydrogenation process prevents the efficiency of graphene hydrogenation, and it is desirable to propose a direct hydrogenation approach, i.e. to find a way to reduce both the reaction barrier and reaction energy.

Doping of graphene is usually used to functionalize graphene (Rao et al., 2009). For example, Al-doped graphene was theoretically found to significantly enhance CO adsorption, where CO is found to be chemically adsorbed on the doped Al atom (Ao et al., 2008a). The gas adsorption properties of graphene with other elements, such as N, B and S, were systemically investigated using density functional theory (Dai et al., 2009). Recently, an *ab initio* study of hydrogen interaction with N-doped carbon nanotubes (CNTs) showed that CNTs with nitrogen reduced the energy barrier of hydrogen dissociation from 1.3 eV to 0.9 eV (Zhang & Cho, 2007). In addition, N-doped graphene has been prepared by carrying out arc discharge in the presence of H_2 and pyridine or H_2 and ammonia. Transformation of nanodiamond in the presence of pyridine also yields N-doped graphene (Panchakarla et al., 2009). N-doped graphene can also be synthesized by the chemical vapor deposition method (Wei et al., 2009) and through electrothermal reaction with ammonia (Wang et al., 2009). Therefore, in this section, the pathway for H_2 molecular dissociative adsorption on N-doped graphene under electric field through DFT calculations is investigated.

All DFT calculations are performed with the DMOL3 code (Delley, 1990; 2000), and have the same settings as in section 2.1. The binding energy of an N atom on the graphene layer $E_{b\text{-}N}$ is defined as,

$$E_{b\text{-}N} = E_{N\text{-}graphene} - (E_{v\text{-}graphene} + E_N) \tag{1c}$$

where $E_{N\text{-}graphene}$, $E_{v\text{-}graphene}$ and E_N are the energy of the system with an N atom doped into the graphene layer, the energy of a pristine graphene layer with a vacancy defect and the energy of a free N atom in the slab, respectively.

2.2.1 The adsorption of hydrogen on N-doped graphene

As reported, the H_2 molecule is weakly adsorbed on pristine graphene, which is the reason why many people paid a lot of attention to modify graphene through doping or using the other methods to enhance the interaction between hydrogen and graphene (Ao et al., 2009; 2010a; Ataca et al., 2009). The energetic favorable adsorption configurations of one H_2 molecule and two H atoms on pristine graphene surface are given in Fig. 1, and the corresponding discussion on the structure changes before and after hydrogenation are also shown in section 2.1. The band structures of pristine graphene, the system of a H_2 molecule physi-adsorbed on graphene, and the system of graphene with 2 H atoms chemically adsorbed are shown in Figs. 4(a), 4(b) and 4(c), respectively. One can clearly see that pristine graphene has a zero band gap as expected (Elias et al., 2009; Rao et al., 2009; Shevlin & Guo, 2009; Sofo et al., 2007). After H_2 molecular adsorption, there is almost no change in the band structure and the zero band gap remains. In fact, it is a simple combination of the band structure of pristine graphene and the H_2 molecule. However, for the band structure of 2 H atoms chemically adsorbed on graphene, the significant changes are found near the Fermi level. It has an indirect band gap of 3.4 eV where the top of the valence band is located at the

M point and the bottom of the conductive band is at the Γ point. The band gap is consistent with that of 3.5 eV for fully hydrogenated graphene—graphane (Sofo et al., 2007). Note that all C atoms are fully bonded and there are no unpaired electrons. Therefore, no net spin extists in this system. For half-hydrogenated graphene as shown in Fig. 3(a) of the Ref. (Zhou et al., 2009b), strong σ-bonds are formed between C and H atoms and the π-bonding network formed by *p* orbitals of the carbon ring is broken, leaving the electrons in the unhydrogenated C atoms localized and unpaired where spins come out. These spins decrease the band gap energy to 0.43 eV significantly (Zhou et al., 2009a, 2009b). In our case, the two H atoms are dissociative adsorbed on two face-by-face C atoms where all electrons in this system are paired without spins, similar to the status of fully hydrogenated graphene-graphane. The accuracy of such a LDA of the system used here has been verified by many other studies (Yao et al., 2007).

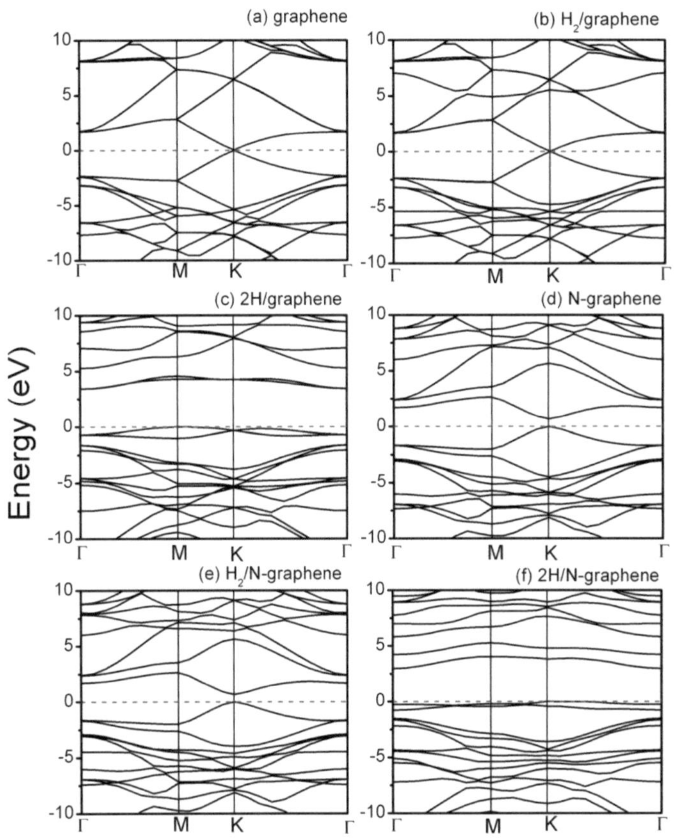

Fig. 4. The band structure of pristine graphene (a), the system with a H_2 molecule adsorbed on pristine graphene (b), the system with 2 H atoms chemically adsorbed on pristine graphene (c), N-doped graphene (d), the system with a H_2 molecule adsorbed on the N-doped graphene (e), and the system with 2 H atoms chemically adsorbed on the N-doped graphene (f). In the figure, a 2×2 supercell of graphene is taken, and the dash lines are the position of the Fermi level. (Reproduced with permission from Ref. (Ao & Peeters, 2010b). Copyright 2010, ACS)

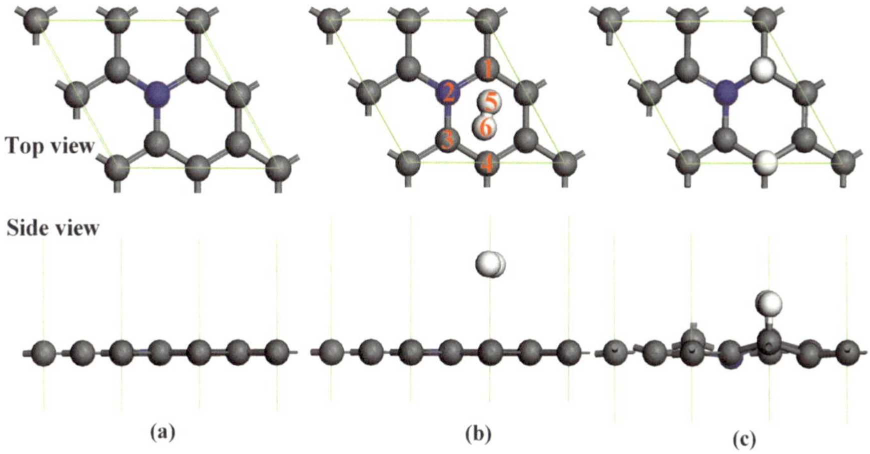

Fig. 5. The atomic structure of the N-doped graphene (a), a H_2 molecule adsorbed on the N-doped graphene (b) and 2 H atoms dissociative adsorbed on N-doped graphene (c). The blue sphere is an N atom. (Reproduced with permission from Ref. (Ao & Peeters, 2010b). Copyright 2010, ACS)

Doping graphene by both electrons and holes changes its electronic structure significantly (Rao et al., 2009). Recently, the nitrogen doped graphene was prepared by arc discharge of graphite electrodes in the presence of H_2, He, and NH_3 (Panchakarala et al., 2009), chemical vapor deposition (Wei et al., 2009), or through the electrothermal reaction with ammonia (Wang et al., 2009). Furthermore, using *ab initio* calculations, N-doped CNTs was reported to reduce the energy barrier of hydrogen molecule dissociative adsorption (Zhang & Cho, 2007). Therefore, the effect of N doping on the dissociative adsorption of a H_2 molecule is considered. In the experiment, N atoms substitute the C atoms in graphene. The doping concentration can be controlled by adjusting the experimental parameters, such as the ratio among H_2, He_2 and NH_3 (Panchakarla et al., 2009). In this simulation, a 2×2 supercell for graphene is taken, one of the C atoms is replaced by a N atom, which leads to the ratio of N:C = 1:7. After geometry optimization, the N-doped graphene still keeps the planar feature with a small shrinkage of C-N bond length $l_{C\text{-}N}$ = 1.41 Å, which $l_{C\text{-}C}$ = 1.42 Å in pristine graphene. This is consistent with the other simulation results (Dai et al., 2009). The structure with N doping is shown in Fig. 5(a).

In the arc discharge doping process, the vacancy defects in graphene were induced, thus allowing the N atoms from NH_3 to adsorb at the vacancies and forming the covalent bonds with C atoms (Panchakarla et al., 2009). In this case, the binding energy $E_{b\text{-}N}$ between N and graphene layer can be determined by Eq. (1c). It was found that $E_{b\text{-}N}$ is as strong as -12.65 eV/atom. For graphite the C-C binding energy was reported to be -9.55 eV/atom (Sofo et al., 2007). Therefore, it is favorable to form N-C bonds in the presence of a N atom, although there is an energy barrier for the reaction, which can be overcome through decalescence or by arc discharge technique in the experiment.

The favorable atomic structure of a H_2 molecule physi-adsorbed on the N-doped graphene is shown in Fig. 5(b). The result shows that the planar structure remains and the adsorbed H_2 molecule is located on the hollow site of the carbon hexagon. The distance $d_{H2\text{-}graphene}$ and adsorption energy $E_{b\text{-}H2}$ are respectively 2.615 Å and -0.159 eV, which are similar to the results above for a H_2 molecule physi-adsorption on pristine graphene. In the other words, doping N into graphene has only little effect on H_2 physisorption.

Different from the aforementioned physic-adsorbed H_2, the favorable atomic structure of 2 H atoms chemically bonded on the C atoms is given in Fig. 5(c). Similar results for the case of pristine graphene are obtained, where the average binding energy for the C-H bonds $E_{b\text{-}H}$ is -2.028 eV and the C atoms binding with the H atoms move upwards by about 0.35 Å. However, the bond lengths for the two C-H bonds $l_{C1\text{-}H5}$ and $l_{C4\text{-}H6}$ are different; they are 1.122 and 1.132 Å, respectively. Meanwhile, the doped N atom moves downwards with about 0.1 Å. Note that the 2 H atoms were adsorbed at two C atoms that are asymmetric. H5 was adsorbed onto C1, but H6 was adsorbed on C4 not C3 which is symmetric with C1 as shown in Fig. 5(c). To investigate the effect of cell size on the adsorption sites of H atoms on graphene, the cell size is increased to a 4 × 4 supercell. It was found that the two H atoms still prefer to take the sites of the C atoms near to the N atom.

Fig. 6 shows a H_2 molecule dissociative adsorption pathway on the N-doped graphene. Following the reaction coordinate, the reaction energy barrier is 2.522 eV, which is a little smaller than 2.734 eV for the H_2 molecule dissociative adsorption on pristine graphene. The result is consistent with the report that N doping into CNTs would reduce the energy barrier of molecular hydrogen dissociative adsorption (Zhang & Cho, 2007). Besides the initial and final structures of the reaction, the atomic structure with the minimum energy state, State 1, and the transition state are also given in Fig. 6. For State 1, the H_2 molecule diffuses from the hollow site of the carbon hexagon to a site near the doped N atom before the transition state. For the transition state, the H_2 molecule is already dissociated into 2 H atoms but only H5 is adsorbed on C1 and H6 is a free atom near C4. Subsequently, H6 is also adsorbed as shown in the final structure. Such a reaction pathway explains why the two H atoms are not adsorbed onto two C atoms with symmetry, since the adsorption of the two H atoms does not occur at the same time. The adsorption of the first H atom would change the electronic distribution and also the chemical reactivity of each C atom.

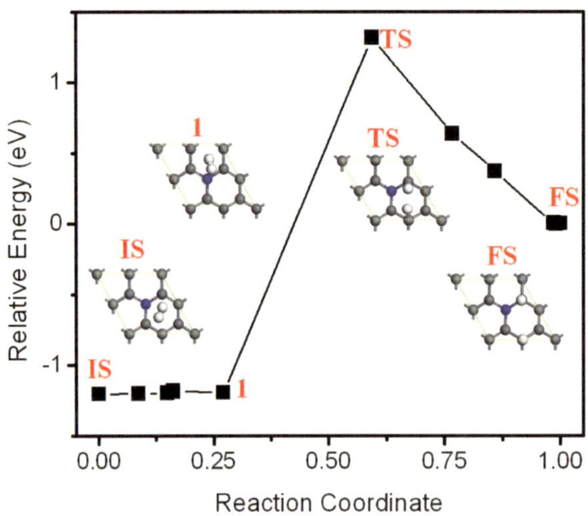

Fig. 6. The reaction pathway of a H_2 molecule dissociative adsorbed on a N-doped graphene layer. (Reproduced with permission from Ref. (Ao & Peeters, 2010b). Copyright 2010, ACS)

The band structures of N-doped graphene, the system with a H_2 molecule physi-adsorbed on N-doped graphene, and the system with 2 H atoms chemically adsorbed on N-doped graphene are given in Figs. 4(d), 4(e) and 4(f), respectively. It is clearly shown in Fig. 4(d) that there is an energy gap between the valence and conduction bands at the K point with N doping. The direct band gap is about 0.7 eV. As clarified by the others (Martine et al., 2007), the doping atoms enter into the lattice of graphene, form covalent bonds with the C atoms and change the atomic structure of graphene. This would largely modify the electronic structure of graphene and suppress the density of states near the Fermi level, thus a gap is opened between the valence and conduction bands. After a H_2 molecule is adsorbed, the band structure of the system is not fundamentally changed. However, for the system of 2 H atoms adsorbed on N-doped graphene, the band structure changes a lot and an indirect band gap of around 3 eV is found. The top of the valence band is located at the K point and the bottom of the conduction band is at the Γ point. Therefore, doping N into graphene has no significant effect for the H_2 molecule physisorption and only slightly reduces the dissociation energy barrier. However, for the band structure of graphene, doping N would induce a band gap energy of ~0.7 eV. Thus, similar to the N-doped CNTs (Zhou et al., 2006), N-doped graphene exhibits semiconductor behavior, which leads to a decreased conductivity, an improved on/off ratio and a Schottky barrier when electrodes are added.

2.2.2 The effect of an applied electric field on the dissociative adsorption

As a perpendicular electric field F will lead to a polarization of the charge density, it will have an effect on the dissociative adsorption of hydrogen. To investigate the effect of the electric field on the hydrogen dissociation desorption on N-doped graphene, a perpendicular electric field is applied on the N-doped graphene system. The energies of the initial structures, transition structures, final structures, reaction barriers and reaction energies of a H_2 molecule dissociative adsorbed on N-doped graphene under different electric fields are given in Table 1. It is found that the positive electric field reduces the barrier significantly in the N-doped graphene system. When increasing the electric field to $F = 0.009$ a.u. (1 a.u. = 5.14×10^{11} V/m), no barrier is found, the reaction occurs automatically. Note that for $F \geq 0.005$ a.u., the reaction energy is negative in the N-doped system and it decreases as F increases. A negative electric field is also applied in this system, the negative electric field leads to an increase of the energy barrier. It is understandable that reversing the electric field has an opposite effect on the energy barrier for hydrogen dissociative adsorption due to the polarization effect of the electric field. Therefore, Table 2 gives the atomic charges of the two H atoms and the two C atoms which are bound to the two H atoms, at the transition state in the presence of different electric fields in N-doped graphene. As shown in the table, electrons move towards the H atoms with increasing F, while the opposite behavior is found for the two C atoms and the N atom. Thus a positive electric field leads to a transfer of electrons from the bottom atoms to the top atoms and a negative electric field has the opposite effect.

It is interesting to notice that the reaction barrier is negative, i.e. -0.895 eV, for a H_2 molecule dissociative adsorbed on N-doped graphene under a 0.009 a.u. electric field. Therefore, this reaction pathway is shown in Fig. 7. As shown in this figure, there are two transition states TS1 and TS2, and two minimum energy states: State 1 and State 2. For State 1 and State 2, the two dissociative H atoms bond on C1 and C4, which are closest to the doped N atom. For TS1 and TS2, the H_2 molecule is dissociated and only one H atom is bonded with C1, which is closest to the doped N atom. Another H atom is free. However, the positions of the free H atom are different for the two transition states. The configurations of the two transition states are similar to that in the N-doped graphene system without electric field, where there is also one H atom bonded on C1 and another is free.

	F (a.u.)	Energy of reactant (Ha)	Energy of production (Ha)	Transition state (Ha)	Dissociation barrier (eV)	Reaction energy (eV)
Pristine graphene	0	-303.482	-303.452	-303.382	2.734	0.828
N-doped graphene	0	-320.010	-319.966	-319.918	2.522	1.205
	0.005	-319.924	-319.941	-319.891	0.878	-0.480
	0.009	-319.940	-320.022	-319.973	-0.895	-2.241
	-0.005	-319.903	-319.828	-319.791	3.033	2.046

Table 1. The energies of reactants, productions, transition state, dissociation barriers and reaction energies for the reaction of a H_2 molecule with pristine graphene or with N-doped graphene and 2 H atoms bond on pristine graphene or on N-doped graphene in the presence of different electric fields.

	Atom	*F*=-0.005	*F*=0	*F*=0.005	*F*=0.01
N-doped graphene	H5	0.287	0.126	-0.024	-0.115
	H6	0.226	-0.031	-0.291	-0.402
	C1	-0.293	-0.206	-0.113	-0.071
	C4	-0.264	-0.104	0.008	0.102
	N	-0.414	-0.378	-0.368	-0.346

Table 2. Atomic charge of the two H atoms and the two C atoms which are bonded with the two H atoms, at the transition state under different electric field in N-doped graphene. The different atoms are numbered which are given in Fig. 5. The units of the charge and electric filed are $|e|$ and a.u., respectively.

The reaction barrier -0.895 eV is the energy difference between IS and TS2. In this reaction process, there are two barriers that should be overcome, State 1 to TS1 with a barrier of 0.966 eV, and State 2 to TS2 with a barrier of 0.801 eV. However, the energy from the exothermic process of IS to State 1 is sufficient to support the endothermic process from State 1 to TS2. In the other words, the dissociative adsorption reaction can occur automatically if the energy created from IS to State 1 can be used to provide the energy needed from State 1 to TS2. Further increasing the electric field beyond 0.009 a.u., no transition state is found and the final structure can be obtained directly even by the geometry optimization method. Therefore, the electric field can induce molecular hydrogen dissociative adsorption in the N-doped graphene system, and the process changes from an endothermic to an exothermic reaction. It means that the electric field and N doping act as the catalyses of H_2 dissociation.
The origin of the differences in the energy barrier of hydrogen dissociative adsorption on graphene can be understood through the analysis of the partial density of states (PDOS). Fig. 8 shows the PDOS of the hydrogen and carbon atoms which the H_2 molecule dissociates and is adsorbed on for different electric field. The interaction between the hydrogen and graphene atoms is mainly determined by the bonding and antibonding interactions between

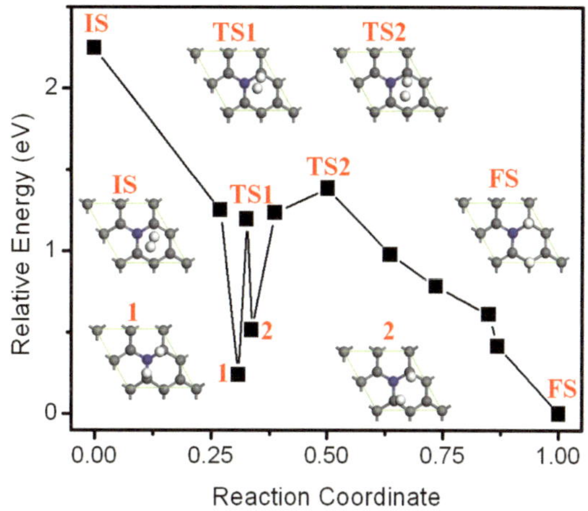

Fig. 7. The reaction pathway of a H_2 molecule dissociative adsorption on N-doped graphene in the presence of an electric field F = 0.009 a.u. (1 a.u. = 5.14×10^{11} V/m). In this figure, there are 2 transition states TS1 and TS2. (Reproduced with permission from Ref. (Ao & Peeters, 2010b). Copyright 2010, ACS)

the σ_g and σ_u states of hydrogen and the p state of the carbon atoms directed perpendicular to graphene (p_z) which bond with the hydrogen atoms. The C p_z-band, which participates strongly in bonding with hydrogen, can be identified between the Fermi level and ~-2 eV in the PDOS of C atoms of isolated graphene. These bands almost disappear in the plots of the PDOS of the C atoms with interaction with H atoms in the 2H/graphene system. Therefore, the strength of the interaction and the trends in the dissociative adsorption energy barriers can be explained by considering the positions and weights of the bonding and antibonding states between the C p_z and hydrogen bands. At the position of the C-H bands, there are distinctive differences in the PDOS of the C atoms with and without H interaction. At the other positions, their PDOS are almost identical. Based on the positions of the bands of C and H atoms, the character of the C-H bands can be easily assigned. The lowest C-H band around -8 eV are obviously due to the interaction of the H σ_g state and C p_z state. The second lowest C-H band around -4 eV is mostly due to the bonding interaction of H σ_u and C p_z states since they are above the H σ_g state. The highest C-H bands are due to a combination of the interaction between the C p_z and H σ_u states, and between the H σ_u and the antibonding C p_z states above the Fermi level. The position and weight of the highest C-H bands determine the strength of the interaction and thus determine the energy barrier heights of the transition states (Zhang & Cho, 2007). As one can see from Figs. 8(a) and 8(b), the third C-H bands correspond to the high energy barrier as shown in Table 1. In Fig. 8(d), the weight of the third C-H band is smallest and this case has the lowest energy barrier while the weight of the third C-H band is the largest in Fig. 8(a) and it has the highest barrier, which agrees with the result in Table 1. When comparing Figs. 8(a) and 8(b), N doping induces a weight increase of the two lowest C-H bands and a weight decrease of the highest C-H band slightly. On the other hand, the presence of an electric field leads to a weight increase of the two lowest C-H bands and a weight decrease of the highest C-H band for the hydrogen dissociative adsorption in the N-doped graphene system. These confirm that N doping and a positive electric field have catalytic effects on hydrogen dissociative adsorption.

The hydrogenated graphene, i.e. graphane, was first synthesized in 2009 by exposing graphene to a hydrogen plasma (Elias et al., 2009). Here, a new promising approach is proposed to hydrogenate graphene. With this technique, the hydrogenation can be realized automatically. The hydrogenated N-doped graphene has two very attractive properties for applications. It has a very high volumetric and gravimetric hydrogen density if fully hydrogenated, and has a band gap around 3 eV. The gravimetric capacity of 7.5 wt% hydrogen is higher than the 6 wt% hydrogen target indicated by the Department of Energy U.S. (DOE), and the volumetric hydrogen capacity of 0.12 kg H_2/L is higher than the DOE target of 0.081 kg H_2/L. Alternatively, the band gap of N-doped graphene is increased to 3 eV after 2 H atoms are adsorbed, i.e. in the range of UV light. This is comparable to TiO_2, graphitic C_3N_4 materials and porous graphene, which have shown the potential applications in photocatalyzed splitting of water into hydrogen (Du et al., 2010; Sum et al., 2002; Wang et al., 2008). These results suggest that the hydrogenated N-doped graphene may solve the band gap problem of graphene for nanoelectronic applications and display photocatalytic activity.

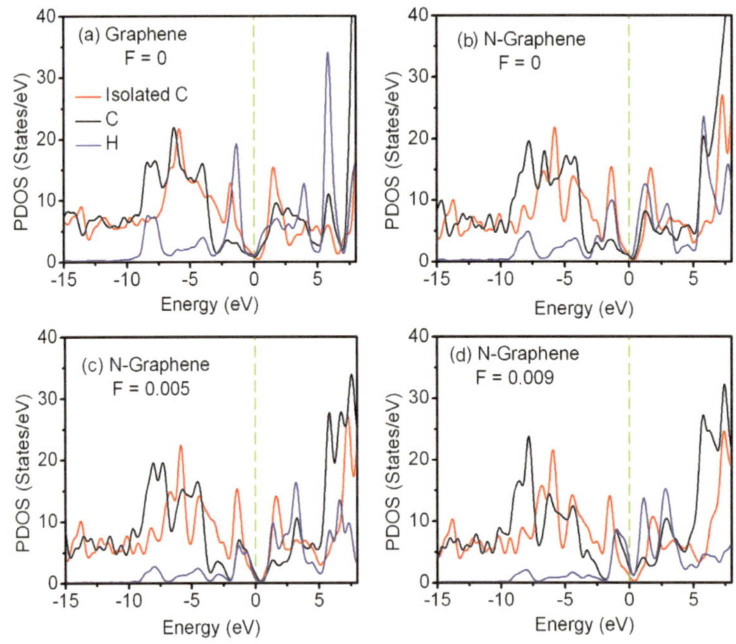

Fig. 8. Partial density of states (PDOS) of the hydrogen and the carbon atoms which are bonded with hydrogen without and with N doping of graphene in the presence of different electric fields. (a) H_2 dissociation over pristine graphene without electrical field. (b) H_2 dissociation over N-doped graphene without electrical field. (c) H_2 dissociation over N-doped graphene with $F = 0.005$ a.u.. (d) H_2 dissociation over N-doped graphene with $F = 0.009$ a.u.. The black and blue curves are the PDOS of the two H atoms and the two C atoms at the transition states, the red curves are the PDOS of the same C atoms in graphene without H, with graphene fixed at the same positions as in the transition states. The vertical lines indicate the Fermi level. (Reproduced with permission from Ref. (Ao & Peeters, 2010b). Copyright 2010, ACS)

3. Diffusion of hydrogen on graphene/graphane interface

As aforementioned, manipulating the edge structure and width of GNR is a way to tuning the electronic properties of graphene, such as opening its band gap. High quality GNRs can be fabricated by selectively hydrogenating graphene or by carving GNRs on a graphane sheet. However, the hydrogen diffusion has a strong influence on the stability of the graphene/graphane interface and such a phenomenon needs to be clarified.

Similar to the studies above, the hydrogen diffusion behavior on the graphene/graphane interface is investigated with DFT calculation and all the DFT calculations were performed using the DMOL3 code (Delley, 2000). The generalized gradient approximation (GGA) with revised Perdew-Burke-Ernzerhof (RPBE) functional was employed as the exchange-correlation functional (Hammer et al., 1999). A double numerical plus polarization (DNP) was used as the basis set, while the DFT semicore pseudopotentials (DSPP) core treatment was employed for relativistic effects that replaces core electrons by a single effective potential. Spin polarization was included in all our calculations. The convergence tolerance of energy was set to 10^{-5} Ha (1 Ha = 27.21 eV), and the maximum allowed force and displacement were 0.02 Ha and 0.005 Å, respectively. To investigate the diffusion pathways of hydrogen atoms at the graphene/graphane interface, linear synchronous transition/quadratic synchronous transit (LST/QST)(Halgren & Lipscomb, 2000) and nudged elastic band (NEB) (Henkelman & Jonsson, 2000) tools in DMOL3 code were used, which have been well validated in order to search for the structure of the transition state (TS) and the minimum energy pathway. In the simulations, three-dimensional periodic boundary conditions were imposed, and all the atoms are allowed to relax.

The supercells used for the zigzag and armchair graphene/graphane nanoribbons are shown in Figs. 9(a) and 9(b), respectively. The interlayer interaction was minimized by allowing a vacuum width of 12 Å normal to the layer. For both type of nanoribbons, the C atoms are displaced from the C plane by about 0.29 Å due to the bonded H atoms. This value is similar to the shift of 0.32 Å that C atoms experience when a H_2 molecule is dissociative adsorption on graphene (Ao & Peeters, 2010a). In both cases, this is a consequence of the change in the hybridization of the C atoms from sp^2 in graphene to sp^3 in graphane. In addition, for the zigzag GGNR both the graphene and the graphane nanoribbons are flat [see Fig. 9(a)]. However, the graphene and graphane layers are not in the same plane, they are connected with an angle of about 162° at the interface, which is consistent with previous reports (Hernández-Nieves et al., 2010; Lu & Feng, 2009). For the armchair GGNR [Fig. 9(b)], the graphene and graphane regions are almost in the same plane, while there is little curvature in the graphene nanoribbon.

Now the stability of the two types of interfaces is analysed by calculating the diffusion barriers for hydrogen atoms. For the case of a zigzag interface, there are two different types of C and H atoms, which are indicated in Fig. 9(a) as sites A and B. For the diffusion of the H atom bonded to the C atom at site A, there are two possible diffusion paths labeled as 1 and 2 in Fig. 9(a). At the site B, there are three possible diffusion pathways for the H atom that we label as 3, 4 and 5. In the case of an armchair interface, all the C atoms at the interface are equivalent from a diffusion point of view. So there are five different diffusions pathways that are labeled as 6-10 in Fig. 9(b). When analysing the diffusion paths, it is found that all the diffusions are along linear pathways, and also that the H atom is free without directly binding to any C atom at the transition state.

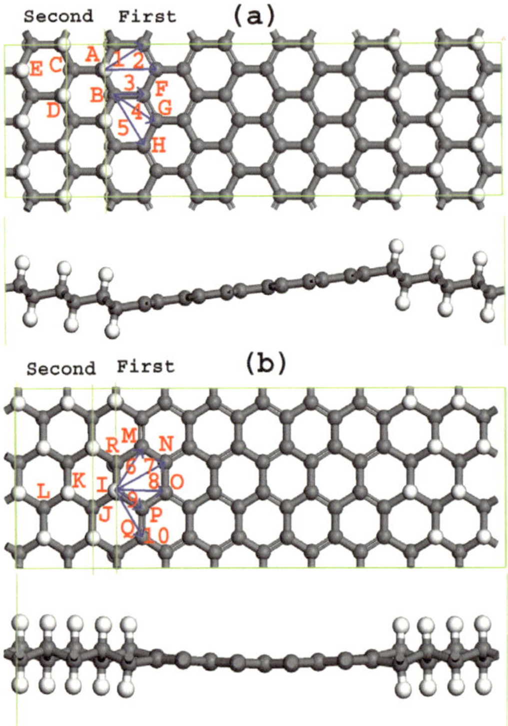

Fig. 9. Atomic structure of graphene/graphane nanoribbons with (a) zigzag and (b) armchair interfaces after relaxation. The arrows indicate the different diffusion pathways considered. The gray and white spheres are C and H atoms, respectively. (Reproduced with permission from Ref. (Ao et al, 2010c). Copyright 2010, AIP)

The diffusion barriers for the different paths and for both types of graphene/graphane interfaces are summarized in Table 3. For the zigzag interface, it is found that the barriers are 3.83, 4.48, 2.86, 3.64 and 3.86 eV for the pathways 1–5, respectively. Thus, the minimum diffusion barrier for the zigzag GGNRs involves H diffusion from the carbon atom at site B along the C–C bond to its nearest carbon atom with an energy barrier of 2.86 eV. For the armchair interface, energy barriers for pathways 6–8 and 10 are 3.17, 4.07, 4.20 and 4.05 eV, respectively. The pathway number 9 involves H diffusion to the nearest C atom at site P. However, it is found that this diffusion cannot occur because during the geometry optimization, the H atom at site P diffuses back to the C atom at site I. Thus, the energy barrier for H diffusion in armchair interfaces can be minimized to 3.17 eV to the second nearest C atom along Path 6.

Recently, it was reported that the diffusion barrier for a single hydrogen atom on pristine graphene layer is about 0.3 eV, which was obtained by DFT calculation using a similar method to this work (Boukhvalov, 2010). Furthermore, the diffusion barriers of transition-metal (TM) atoms on graphene were reported in the range of 0.2–0.8 eV (Krasheninnikov et al., 2009). If the TM adatoms are coupled to a vacancy, the diffusion barrier would increase substantially, reaching to the range of 2.1–3.1 eV. Thus, it was claimed that adatoms with the barrier in such a magnitude are stable at room temperature (Krasheninnikov et al., 2009), supporting the notion that the stability of H atoms at GGNRs interfaces are enhanced greatly and rather stable at room temperature.

	Diffusion pathway	E_F-E_I (eV)	Diffusion barrier (eV)
Zigzag interface	1	1.77	3.83
	2	3.92	4.48
	3	1.90	2.86
	4	1.47	3.64
	5	1.17	3.86
Armchair interface	6	1.39	3.17
	7	2.48	4.07
	8	1.58	4.20
	9[a]		
	10	1.09	4.05

[a] It is found that this diffusion path cannot occur.

Table 3. Diffusion barriers for several diffusion paths and energy differences between the states after and before the diffusion (E_F-E_I) in graphene/graphane nanoribbons.

From above analysis, one can see that the minimum diffusion barriers for both of armchair and zigzag interfaces are about one order of magnitude larger than the energy barrier for H diffusion on pristine graphene. From Table 3, one can also see that all the aforementioned H diffusion processes imply increases of several electronic Volts in the energy of the system. At the same time, this indicates that after the diffusion the energy needed for recovering the system back to the initial perfect thermodynamic state is always lower than the energy needed for distorting the interfaces. The barriers for backward diffusion are defined as the difference of energy between the final and the transition states (E_F-E_T), and can be obtained from Table 3 as the difference between the values of E_F-E_I and the diffusion barrier (E_T-E_I). These demonstrate that the graphene/graphane interfaces are rather stable in both types of hybrid nanoribbons.

Such stability enhancement can be understood by calculating the binding energy of the H atoms in the different conditions, which is proportional to the strength of the C-H bonds. The binding energies (E_b) were calculated by E_b=E_i-(E_f+E_H), where E_i is the initial energy of the system, E_f is the energy of the system after removing the H atom, and E_H is the energy of an isolated H atom. For the zigzag interface, we found that the binding energy of the C-H bond at sites A and B are -4.59 and -2.80 eV, respectively. While for an H atom at site I of the armchair interface, the binding energy is -3.35 eV. All these values are larger than the binding energy of an isolated H atom on a graphene supercell containing 32 C atoms that is equal to -0.88 eV. This indicates the stability enhancement of the H atoms at graphene/graphane interfaces. The results of the binding energies also explains why for the zigzag interface it is easier to move the H atoms from site B (E_b=-2.80 eV) than from site A (E_b=-4.59 eV). This explanation is also applicable for us to understand why moving the atoms at site B (E_b=-2.80eV) in the zigzag interface is easier than moving the H atoms at site I (E_b=-3.35eV) in the armchair interface. In addition, the C-H bond length at site A (1.108 Å) is smaller than that at site B (1.112 Å). It is believed that if one bond breaks, the remaining coordinated ones would become shorter and stronger (Sun et al., 2009; Zhang et al., 2009). As shown in Fig. 9(a), the C atom at site F binds with other three C atoms, while the C atoms at both sites A and B are bonded with three C atoms and one H atom. Therefore, E_b of C-C bond between sites B and F is greater than that between sites A and B. Such a strong C-C bond weakens the others bonding at site B including the C-H bond (Aizawa et al., 1990).

Hence, the C-H bond at site B is weaker than that at site A. The same explanation can be applicable to the case of armchair GGNR. On the other hand, C-C bond between sites R and I in Fig. 9(b) is weaker than that between sites A and B due to the effect from both C-C bonds between R and M as well as I and P. Thus, the E_b of C-H bond at site I is between those at sites A and B, which is consistent with the DFT result above. Therefore, the H atom at site B can diffuse easier and the GGNR with the armchair interface is more stable than the one with the zigzag interface.

To further understand the mechanism behind of the higher stability of the H atom at site A, the atomic charges are analysed through the Mulliken method. Table 4 gives the atomic charges of atoms near the interfaces. One can see that atoms at both interfaces (i.e. at sites A, B, and I) are more charged than other atoms. At the interface, C atoms are more negative and the corresponding H atoms are more positive. Furthermore, it also shows that the both interfaces mainly affect the charge distribution of the first row of atoms at interfaces, while there is slight effect on the atoms of the second row at the armchair interface. This result agrees with the fact that an interface influences mainly the atoms of the first two rows (Sun, 2007). It is known that the atomic charge is mostly affected by the atoms belonging to the same carbon ring, especially the nearest atoms. For the carbon and hydrogen atoms at the site A, they have similar nearest atoms as the sites in graphane region far apart from the interface, where the three nearest C atoms are bonded by sp^3 orbitals. For the C and H atoms at site B, only two nearest C atoms are bonded by sp^3 orbitals, the other one on its right hand side at site F is bonded by sp^2 orbitals. Therefore, the effect of the interface on site B is

	Atom Site	C atom	H atom
Zigzag interface	A	-0.045	0.045
	B	-0.086	0.057
	C	-0.031	0.033
	D	-0.030	0.033
	E	-0.030	0.031
	F	0.009	
	G	0.019	
	H	0.009	
Armchair interface	I	-0.087	0.063
	J	-0.042	0.038
	K	-0.028	0.033
	L	-0.029	0.031
	M	0.021	
	N	0.005	
	O	0.004	
	P	0.021	
	Q	0.021	

Table 4. Charges of C and H atoms at different sites on the graphene-graphane nanoribbons with different interfaces. The location of the sites is shown in Fig. 9, and the unit of charge is $|e|$.

stronger than that on the site A. On the other hand, for both sites A and B, there are three C atoms bonded by sp^2 orbitals in the carbon ring. Thus, the charge distribution of the atoms on the both sites is affected by the interface. A similar explanation can be applied to the charge difference on the atoms at sites I and J at the armchair interface. Therefore, the C atom at site B (-0.086 $|e|$) is more chemically active than the one at site A (-0.045 $|e|$) because it has more electrons.

4. Conclusion

In summary, the reaction pathways of a H_2 molecule dissociative adsorption on the pristine graphene and the N-doped graphene layer were investigated under a perpendicular applied electric field using DFT calculations. The corresponding atomic and electronic structures of the reactants, productions and transition states were analysed. The results show that the N doping and electric field facilitate the hydrogen dissociative adsorption. The energy barrier for hydrogen dissociative adsorption on N-doped graphene is suppressed by a positive electric field normal to the graphene surface, and the reaction can automatically occur when the electric field is increased above 0.005 a.u.. The origin of the differences in the energy barrier can be understood through the analysis of PDOS. N doping and the applied electric field are considered to diminish the third interaction band between C and H, which mainly determines the barrier height. Based these results, a promising approach to hydrogenate graphene is proposed. It can be used to enhance the hydrogen storage capacity, and also to manipulate the physical properties of graphene for the applications of nanoelectronic devices. Such an exotic phenomenon is associated with its photocatalytic activity due to the band gap opening caused by the applied electric field and doping effects.

On the other hand, the stability of graphene/graphane nanoribbons with both zigzag and armchair interfaces is studied by calculating the diffusion barriers of H atoms using DFT method. It is found that the H atoms can be firmly stabilized by the graphene/graphane interfaces effects. This is resulted by the increase of the C-H bond strength at the graphene/graphane interfaces. The results demonstrate that both zigzag and armchair graphene/graphane interfaces in hybrid nanoribbons are rather stable, thus increasing the feasibility for future technological applications of these systems.

5. Acknowledgement

This study was financially supported by the Vice-Chancellor's Postdoctoral Research Fellowship Program of the University of New South Wales (SIR50/PS19184), the ECR grant of the University of New South Wales (SIR30/PS24201) and ARC Discovery project of DP1096769.

6. References

Aizawa, T.; Souda, R.; Otani, S.; Ishizawa, Y. & Oshima, C. (1990). Anomalous bond of monolayer graphite on transition-metal carbide surface. *Phys. Rev. Lett.* 64: 768-771.

Ao, Z. M.; Jiang, Q.; Zhang, R. Q.; Tan, T. T. & Li, S. (2009). Al doped graphene: A promising material for hydrogen storage at room temperature. *J. Appl. Phys.* 105: 074307-074312.

Ao, Z. M. & Peeters, F. M. (2010a). Electric field: A catalyst for hydrogenation of graphene. *Appl. Phys. Lett.* 96: 253106.

Ao, Z. M. & Peeters, F. M. (2010b). Electric field activated hydrogen dissociative adsorption to nitrogen-doped graphene. *J. Phys. Chem. C* 114: 14503-14509.

Ao, Z. M.; Hernández-Nieves, A. D.; Peeters, F. M. & Li S. (2010c). Enhanced stability of hydrogen atoms at the graphene/graphane interface of nanoribbons. *Appl. Phys. Lett.* 97: 233109.

Ao, Z. M.; Yang, J.; Li, S. & Jiang, Q. (2008a). Enhancement of CO detection in Al doped graphene. *Chem. Phys. Lett.* 461: 276-279.

Ao, Z. M.; Zheng, W. T. & Jiang, Q. (2008b). The effects of electronic field on the atomic structure of the graphene/α-SiO_2 interface. *Nanotechnology* 19: 275710.

Ataca, C.; Aktürk, E. & Ciraci, S. (2009). Hydrogen storage of calcium atoms adsorbed on graphene: First-principles plane wave calculations. *Phys. Rev. B* 79: 041406(R).

Arellano, J. S.; Molina, L. M.; Rubio, A. & Alonso, J. A. (2000). Density functional study of adsorption of molecular hydrogen on graphene layers. *J. Chem. Phys.* 112: 8114.

Balog, R.; Jφrgensen, B.; Nilsson, L.; Andersen, M.; Rienks, E.; Bianchi, M.; Fanetti, M.; Lægsgaard, E.; Baraldi, A.; Lizzit, S.; Sljivancanin, Z.; Besenbacher, F.; Hammer, B.; Pedersen, T. G.; Hofmann, P. & Hornekær, L. (2010). Bandgap opening in graphene induced by patterned hydrogen adsorption. *Nat. Mater.* 9: 315-319.

Boukhvalov, D. W. (2010). Modeling of hydrogen and hydroxyl group migration on graphene. *Phys. Chem. Chem. Phys.* 12: 15367-153671.

Dai, J.; Yuan, J. & Giannozzi, P. (2009). Gas adsorption on graphene doped with B, N, Al, and S: A theoretical study. *Appl. Phys. Lett.* 95: 232105.

Delley, B. (1990). An all-electron numerical method for solving the local density functional for polyatomic molecules. *J. Chem. Phys.* 92: 508.

Delley, B. (2000). From molecules to solids with the DMol3 approach. *J. Chem. Phys.* 113: 7756

Du, A.; Zhu, Z. & Smith, S. C. (2010). Multifunctional porous graphene for nanoelectronics and hydrogen storage: new properties revealed by first principle calculations. *J. Am. Chem. Soc.* 132: 2876-2877.

Elias, D. C.; Nair, R. R.; Mohiuddin, T. M. G.; Morozov, S. V.; Blake, P.; Halsall, M. P.; Ferrari, A. C.; Boukhvalov, D. W.; Katsnelson, M. I.; Geim, A. K. & Novoselov, K. S. (2009). Control of graphene's properties by reversible hydrogenation: Evidence for graphane. *Science* 232: 610-613.

Geim, A. K. (2009). Graphene: Status and prospects. *Science* 324: 1530.

Halgren, T. A. & Lipscomb, W. N. (1977). The synchronous-transit method for determining reaction pathways and locating molecular transition states. *Chem. Phys. Lett.* 49: 225-232.

Hammer, B.; Hanse, L. B. & Nφrskov, J. K. (1999). Improved adsorption energetic within density-functional theory using revised Perdew-Burke-Ernzerhof functional. *Phys. Rev. B* 59: 7413.

Han, M. Y.; Brant, J. C. & Kim, P. (2010). Electron transport in disordered graphene nanoribbons. *Phys. Rev. Lett.* 104: 056801.

Henkelman G. & Jonsson, H. (2000). Improved tangent estimate in the nudged elastic band method for finding minimum energy paths and saddle points. *J. Chem. Phys.* 113: 9978.

Hernández-Nieves, A. D.; Partoens, B. & Peeters, F. M. (2010). Electronic and magnetic properties of superlattices of graphene/graphane nanoribbons with different edge hydrogenation. *Phys. Rev. B* 82: 165412.

Hyman M. P. & Medlin, J. W. (2005). Theoretical study of the adsorption and dissociation of oxygen on Pt(111) in the presence of homogeneous electric fields. J. *Phys. Chem. B* 109: 6304-6310.

Jiao, L.; Wang, X.; Diankov, G.; Wang, H. & Dai, H. (2010). Facile synthesis of high-quality graphene nanoribbons. *Nat. Nanotec.* 5: 321-325.

Koskinen, P.; Malola, S. & Hakkinen, H. (2009). Evidence for graphene edges beyond zigzag and aimchair. *Phys. Rev. B* 80: 073401

Krasheninnikov, A. V.; Lehtinen, P. O.; Foster, A. S.; Pyykkö, P. & Nieminen, R. M. (2009). Embedding transition-metal atoms in graphene: structure, bonding, and megneticsm. *Phys. Rev. Lett.* 102: 126807

Liu, W.; Zhao, Y. H.; Nguyen, J.; Li, Y.; Jiang, Q. & Lavernia, E. J. (2009). Electric field induced reversible switch in hydrogen storage based on single-layer and bilayer graphenes. *Carbon* 47: 3452-3460.

Lu Y. H. & Feng, Y. P. (2009). Band-gap engineering with hybrid graphane-graphene nanoribbons. *J. Phys. Chem. C* 113: 20841-20844.

Luo, Z.; Yu, T.; Kim, K.; Ni, Z.; Yu, Y.; You, Y.; Lim, S.; Shen, Z.; Wang, S. & Lin, J. (2009). Thickness-dependent reversible hydrogenation of graphene layers. *ACS Nano* 3: 1781-1788.

Martins, T. B.; Miwa, R. H.; da Silva, A. J. R. & Fazzio, A. (2007). Electronic and transport properties of Boron-doped graphene nanoribbons. *Phys. Rev. Lett.* 98: 196803.

Miura, Y.; Kasai, H.; Diño, W.; Nakanishi, H. & Sugimoto, T. (2003). First principle studies for the dissociative adsorption of H_2 on graphene. *J. Appl. Phys.* 93: 3395-3400.

Novoselov, K. S.; Geim, A. K.; Morozov, S. K.; Jiang, D.; Zhang, Y.; Dubonos, S. V.; Grigorieva, I. V. & Firsov, A. A. (2004). Electric field effect in atomically thin carbon films. *Science* 306: 666-669.

Okamoto Y. & Miyamoto, Y. (2001). Ab initio investigation of physisorption of molecular hydrogen on planar and curved graphenes. *J. Phys. Chem. B* 105: 3470-3474.

Panchakarla, L. S.; Subrahmanyam, K. S.; Saha, S. K.; Govindaraj, A.; Krishnamurthy, H. R.; Waghmare, U. V. & Rao C. N. R. (2009). Synthesis, structure, and properties of boron- and nitrogen-doped graphene. *Adv. Mater.* 21: 4726-4730.

Perdew J. P. & Wang, Y. (1992). Accurate and simple analytic representation of the electron-gas correlation energy. *Phys. Rev. B* 45: 13244.

Rao, C. N. R.; Sood, A. K.; Subrahmanyam, K. S. & Govindaraj, A. (2009). Graphene: The new two-dimensional nanomaterial. *Angew. Chem. Int. Ed.* 48: 7752-7778.

Ryu, S.; Han, M. Y.; Maultzsch, J.; Heinz, T. F.; Kim, P.; Steigerwald, M. L. & Brus, L. E. (2008). Reversible basal plane hydrogenation of graphene. *Nano Lett.* 8: 4597-4602.

Sessi, P.; Guest, J. R.; Bode, M. & Guisinger, N. P. (2009). Patterning graphene at the nanometer scale via hydrogen desorption. *Nano Lett.* 9: 4343-4347.

Shevlin, S. A. & Guo, Z. X. (2009). Density functional theory simulations of complex hydride and carbon-based hydrogen storage materials. *Chem. Soc. Rev.* 38: 211-225.

Singh, A. K. & Yakobson, B. I. (2009). Electronic and magnetism of patterned graphene nanoroads. *Nano Lett.* 9: 1540-1543.

Singh, A. K.; Penev, E. S. & Yakobson, B. I. (2010). Vacancy clusters in graphene as quantum dots. *ACS Nano* 4: 3510-3514.

Sofo, J. O.; Chaudhari, A. S. & Barber, G. D. (2007). Graphane: A two-dimensional hydrocarbon. *Phys. Rev. B* 75: 153401.

Sum, K.; Ai-Shahry, M. & Ingler, W. B. (2002). Efficient photochemical water splitting by a chemically modified n-TiO_2. *Science* 297: 2243-2245.

Sun, C. Q. (2007). Size dependence of nanostructures: Impact of bond order deficiency. *Prog. Solid State Chem.* 35: 1-159.

Sun, C. Q.; Sun, Y.; Nie, Y. G.; Wang, Y.; Pan, J. S.; Ouyang, G.; Pan, L. K. & Sun, Z. (2009). Coordination-resolved C-C bond length and the C 1s binding energy of carbon allotropes and the effective atomic coordination of the few-layer graphene. *J. Phys. Chem. C* 113: 16464-16467.

Wang, X.; Li, X.; Zhang, L.; Yoon, Y.; Weber, P. K.; Wang, H.; Guo, J. & Dai, H. (2009). N-doping of graphene through electrothermal reactions with ammonia. *Science* 324: 768-771.

Wang, X. C.; Maeda, K.; Thomas, A.; Takanabe, K.; Xin, G.; Carlsson, J. M.; Domen, K. & Antonietti, M. (2008). A metal-free polymeric photocatalyst for hydrogen production from water under visible light. *Nat. Mater.* 8: 76-80.

Wei, D.; Liu, Y.; Wang, Y.; Zhang, H.; Huang, L. & Yu, G. (2009). Synthesis of N-doped graphene by chemical vapor deposition and its electrical properties. *Nano Lett.* 9: 1752-1758.

Yao, Y.; Ye, F.; Qi, X. L.; Zhang, S. C. & Fang, Z. (2007). Spin-orbit gap of graphene: First-principles calculations. *Phys. Rev. B 75*: 041401.

Zhang, X.; Kuo, J.; Gu, M.; Bai, P. & Sun, C. Q. (2010). Graphene nanoribbon band-gap expansion: Broken-bond-induced edge strain and quantum entrapment. *Nanoscale* 2: 2160-2163.

Zhang, Z. & Cho, K. (2007). Ab initio study of hydrogen interaction with pure and nitrogen-doped carbon nanotubes. *Phys. Rev. B* 75: 075420

Zhou, J.; Wang, Q.; Sun, Q.; Chen, X. S.; Kawazoe, Y. & Jena P. (2009a). Ferromagnetism in semihydrogenated graphene sheet. *Nano Lett.* 9: 3867-3870.

Zhou, J.; Wu, M. M.; Zhou, X. & Sun, Q. (2009b). Tuning electronic and magnetic properties of graphene by surface modification. *Appl. Phys. Lett.* 95: 103108.

Zhou, Z.; Gao, X.; Yan, J. & Song, D. (2006). Doping effects of B and N on hydrogen adsorption in single-walled carbon nanotubes through density functional calculations. Carbon 44: 939-947.

5

Description of Adsorbed Phases on Carbon Surfaces: A Comparative Study of Several Graphene Models

José L. Vicente and Alberto G. Albesa
Instituto de Investigaciones Fisicoquímicas Teóricas y Aplicadas INIFTA, Depart. de Química Facultad de Ciencias Exactas, UNLP CC 16, Suc. 4 (1900) La Plata Argentina

1. Introduction

Since it was first identified (Scheele, 1777), gas adsorption process had been investigated for more than two hundred years. Among the large number of phenomena nowadays recognized involved in adsorption, the attention of the scientific community was focused on two main issues. First, from the applied point of view, characterization of solid surfaces through the observed behavior of the adsorbed phase was pursued. Second, and from the more basic science perspective, elucidation of the nature and magnitude of the interaction forces for different experimental conditions and systems.

Taking into account that the behavior of adsorbed molecules depends on the properties of both the solid surface and the adsorbate itself, and for somehow exploring basic aspect of the problem, we focused our attention only in a set of basic carbonaceous substrates and a small groups of simple adsorbates.

We have organized this review in several sections, starting in Section 2 with a brief discussion of some carbonaceous surfaces associated to graphitic structures when they are considered at different scales. This section also deals with different carbon materials and their properties as well as how description of several basic structural units and orientation degrees emerge (Bandosz et al., 2003). In Section 3.1 the adsorption of different gases such as nitrogen, carbon dioxide, argon and methane are described, with special emphasis in the last one due to its environmental relevance (Beaver & Sircar, 2010; Wagner, 1996; Wuebbles, 2000). Three approach to study gas adsorption on graphite surfaces that tackle the problem from different points of view where discussed (Albesa & Vicente, 2008): i) semi empirical quantum mechanical (SQM) model for gas - graphite interaction; ii) Grand Canonical Monte Carlo (GCMC) simulations (D. Do & H. Do, 2005; Sabzyan & Babajani, 2005); and Mean-Field Approximation (MFA) of the lattice model. To emulate by GCMC simulations the interaction gas - surface several models are compared (Albesa et al., 2008), and then a unified description is utilized for the adsorption mechanism that therefore was not only merely based on the three above three mentioned theoretical tools, but also on experimental data. The results reveal changes occurred the adsorbed species, and also can be used to evaluate interaction energy between the gas and the graphite surface as the coverage increases; these allow elucidating somehow the role played by the substrate in adsorption process at different pressures and temperatures (Llanos et al., 2003; Steele, 1974). Section 3.2 is devoted to study the adsorption of simple gases using graphitic curved surfaces, for this end we used abs initio density functional theory (DFT) calculations for graphene sheets whose surfaces had different curvature, and discuss the relation between these model systems and experimental data for adsorption on single wall carbon nanotubes (Albesa et al., 2009; Albesa et al. 2010). In section 4, Gas Separation, we present simulated adsorption isotherms obtained for two characteristic carbonaceous structures, graphite and single wall carbon nanotubes.

2. Relation between different carbon and graphitic surfaces

The great interest in graphitic surfaces stems from their high surface activity, associated with strong adsorbate-surface forces and large surface area. From the commercial point of view, such structures are generally cheap to produce, and can be prepared with different special characteristics. As a result of the strong covalent sp^2 bonding in graphite, the distance C - C in graphite (*0.142 nm*) becomes shorter than the van der Waals radius of carbon (*0.335 nm*) giving a surface density of 38.2 atoms/nm^2.

The need to improve this activity, in order to have greater adsorption capacity and selectivity in mixture separation, has motivated the development of new carbon structures, related direct or indirectly to the graphitic ones. In few years the studies that began changing simple graphitic surfaces, rapidly grow beyond the broad word of fullerenes (Harris, 2003) and reach new and more exotic structures like nanotubes and nanohorns (Dresselhaus et al., 1996).

The two most common allotropes of carbon that occur naturally are graphite and diamond. Graphite is assembled from parallel sheets of hexagonal carbon atoms arrays, each atom being linked to three other sp^2 hybridized carbon atoms by sigma bonds or length *0.1415 nm*. The value of the in-plane lattice constant is *0.2456 nm*. The parallel sheets, or graphene layers, or basal planes, are held in place by the delocalized p electrons and are *0.3354 nm* apart. Diamond is assembled forming a tree - dimensional tetragonal network where each carbon atom is linked to four other sp^3 hybridized atoms.

The C_{an} fullerenes are shell-like molecules consisting of n three-coordinated carbon atoms arranged as 12 pentagons and ½ (n - 20) hexagons where n is even, and greater than 20 except 22. In spite of great a number of these molecules have been detected experimentally, only a few high-purity fullerenes are produced in macroscopic quantities. Fullerene molecules are essentially polyhedral rather than spherical and, increasing the number of carbon atoms, each fullerene can exist as one of several different isomers.

Just like in order to obtain a fullerene one has to lose the original sheet constituted by hexagonal carbon rings (incorporating pentagonal rings to the atoms arrange), the single walled nanotube (SWNT) is other kind of structure where this not happen and graphene configuration is preserved. In other words, the unit cell of a SWNT is constructed from a portion of graphene sheet defined by the chiral vector C_h and the translation vector $\boldsymbol{T}$. If $\boldsymbol{a}$ and b are the basis vectors, the chiral vector, $\boldsymbol{C_h} = n\boldsymbol{a} + m\boldsymbol{b}$, connects two equivalent sites on the graphene sheet at an angle defined by the so-called Hamada indices (n,m). The unit cell of a carbon nanotube is obtained by rolling up the graphene sheet $(T,\ C_h)$ and joining the edges perpendicular to $\boldsymbol{C_h}$ to form a cylinder. There are three classes of nanotubes, i.e., zig –

zag, armchair and, chiral, corresponding to $n \neq m = 0$, $n = m \neq 0$ and, $n \neq m \neq 0$ respectively. From the Hamada indices one can define the diameter $d = C/\pi$ and the length $l = \sqrt{3}\ C/d_R$ of the nanotube unit cell, where $C = a\ (m^2 + mn + n^2)^{1/2}$, $a = 0.246\ nm.$, and d_R the greatest common divisor of $2m + n$ and $2n+m$. Multi wall nanotubes (MWNT) are constructed from a number of concentric SWNT. Carbon nanotubes are easier to manufacture than fullerenes and present higher stability; these properties together with their interesting physicochemical behavior, have prompted a great deal of interest reflected in the extensive experimental and theoretical studies published in the last years.

Like in the case of fullerenes, the graphenic membranes can take a conical shape only adopting arranges that are not purely formed by hexagonal carbon rings. The apex has to be defined by one or more ring atom removed in order to reach a conical hat that lead to the form of fivefold (or smaller) rings (Yudasaka et al., 2008). One class of such conical structures with a particularly sharp apical angle are the single-wall carbon nanohorns (SWNH). Conical graphenic structures with wider opening angles (corresponding to fewer pentagons at the apex) sometimes form multilayer structures). The diameter of an individual SWNH ranges from 2 to 4 nm, and the length is 40 to 50 nm. About 2000 of them assemble to form a spherical aggregate with a diameter of about 100 nm. Three types of spherical aggregates are known and, are called dahlia, bud and seed because of their appearances. The dahlia aggregate has long cone - shaped tips sticking through its surface, while the bud aggregate does not. The seed aggregates have lower graphitization that the dahlia and bud aggregates, and their tubules are corrugates (Azami et al., 2008).

In so far as increase the complexity of the carbonaceous structure the long range order of the graphene sheet is lost, because of the curvature, the presence of apexes, etc. Considering, according to the vast amount of experimental data, the hierarchical structure of the activated carbons (Bandosz et al., 2003), they would constitute perhaps a greater example of the previous mentioned loss. The first level of the hierarchy are the so-called basic structural units (BSU) that consist of a few roughly aligned polyaromatic-like molecules or "layers" that generally not preserve the graphite order between them, and due to the presence of functional groups, interlayer spacing are generally greater than that of graphite. The BSUs are assembled to form regions of local molecular orientation (the second level of the hierarchy) which are in turn assembled in space to yield the complex structure of activated carbons.

3. Adsorption on carbon surfaces

3.1 Simple gas adsorption on open graphite surfaces

The literature offers numerous experimental and theoretical studies of the gas adsorption on open graphite surfaces, considering the case of spherical molecules, e.g., rare gases (Cheng & Steele, 1990; Nicholson & Parsonage, 1982), simple molecules, e.g., nitrogen, oxygen, carbon dioxide and, methane (Bottani & Bakaev, 1994; Kowalczyk et al., 2005), and more complexes like ethylene or phenol (Bertoncini et al., 2000; Bottani, 1999). However, from a technological point of view, most of these studies restrict their descriptions to special conditions, i.e., low pressures or temperatures, soft surfaces, weak interactions, etc. In spite the powerful theoretical tools developed in the last decades, from semi empirical and DFT calculations to canonical (CMC) and grand canonical (GCMC) Monte Carlo simulations, nowadays two significant constrains limit computer performances to deal with the adsorption on graphite. One is the number of molecules considered in the calculations, and this frontier moves as fast as new computer facilities are developed. The other problem comes from the approximations adopted in each model. The structure of different adsorbates on the basal plane of graphite has been extensively studied using Monte Carlo and molecular dynamics, however, the validity of these results depend on the interaction potentials adopted. The interaction potentials used in almost all computer simulations are composed of two parts: fluid-fluid and fluid-surface interactions. Some years ago the proposal (Steele, 1974) of an approximate interaction potential between a spherical molecule and a graphite surface (averaged over carbon atoms) reduced and became accessible the expensive computation time. On the other hand many systems and/or conditions require the knowledge of the real potential beyond how well the adsorbate-adsorbate interactions are described. As in example, the variation of the entropy of adsorption with the coverage in the basal plane of graphite determines the shape of the adsorption isotherm (variation of the chemical potential), because in the submonolayer region of this homogeneous surface the adsorption energy varies slightly due to the adsorbate-adsorbate interaction and the main contribution to the change on the chemical potential comes from the entropy of adsorption. Different corrections were proposed to improve the approximate gas-surface interaction potential, considering two dimensional Fourier series expansions (Kim & Steele, 1992) or simply introducing adjusting prefactors.[32] In spite it was argue, according to different reasons (Bottani & Bakaev, 1994; Kowalczyk et al., 2005), that neglecting the energetic inhomogeneity of the surface along the graphite plane is not expected to affect the results significantly. However, considering all kind of results that are possible to be obtained from homogeneous surface potentials, one can hardly agree with this affirmation, at least at low temperatures. It's because energetic homogeneity assumption means all sites on the surface are equivalent then, during the simulations, the probability that a molecule remains in one site or another is the same. On the contrary, considering energetic inhomogeneity, a molecule adsorbed on a site of the surface can reach other more favorable site later, during the simulation steps, in order to reach equilibrium that means inhomogeneity requires more trials to obtain each simulated point, even at low coverage. Of course, while the differences can be considered negligible at very low coverage, as the pressure increase they become significant, i.e. all sites are not equivalent on the surface and there will be fewer molecules adsorbed in the monolayer, and so on.

Among the simple gases that can adsorb on graphite, now we concentrate our study on the case of methane. Different reasons have leading the interest in the adsorption of methane on carbonaceous surfaces in general and, on graphite in particular. Practical reasons come from the fact that natural gas, which is composed mostly of methane, provides an alternative to traditional liquid petroleum fuels because their environmental advantages and natural abundance. From the theoretical point of view, next to the numerous and complete descriptions about rare gases adsorption process, the interaction between methane molecules and graphite surfaces remain with many unknown subjects to study.

According to the previous arguments, all description of the adsorption process, even on a simple graphite surface, requires a previous microscopic knowledge of the interactions between gas molecules and the substrate and between adsorbed molecules, as well as the different configurations of these molecules on the surface. But describing the interaction potential between methane molecules and graphene surface by using classical quantum mechanical methods, for real systems (i.e. many molecules involved), becomes a task beyond regular computational costs. On the other hand we are interested more than absolute values of the adsorption energies on each site of the surface, about the relative differences between one site and another and, most of all how the presence of neighboring adsorbed molecules modified the adsorbed system geometrically and energetically. To tackle these problems we performed PM3 semi-empirical and density functional theory (DFT) calculations of increasing numbers of adsorbed methane molecules on a graphite surface. To mimic the graphene plane we have chosen a planar polyaromatic molecule which has chemical characteristic similar to graphite sheet in many aspects, as is shown in Fig. 1. The interaction energy, ΔE, of n $(n = 1,\ldots, 4)$ adsorbed methane molecules, has been calculated by employing PM3 semiempirical methods as:

$$\Delta E = E(G - n\,CH_4) - \left[\,E(G) + E_a(n\,CH_4)\,\right] \qquad (1)$$

where $E(G\text{–}n\,CH_4)$, $E(G)$ and $E_a\,(n\,CH_4)$ are energies of the graphite-methane complex, the isolated graphite sheet and the methane molecules corresponding to the same configuration but without the graphite surface respectively (Albesa & Vicente, 2008). The results obtained at PM3 level were subjected to further geometry optimizations using DFT calculations (Parr & Yang, 1989; Wang & Perdew, 1991) keeping the graphene structure fixed. In order to consider the interaction energy when different numbers of methane molecules approach the surface, we begin optimizing the geometry of the first methane molecule from various relative orientations with respect to the reference plane of graphite.

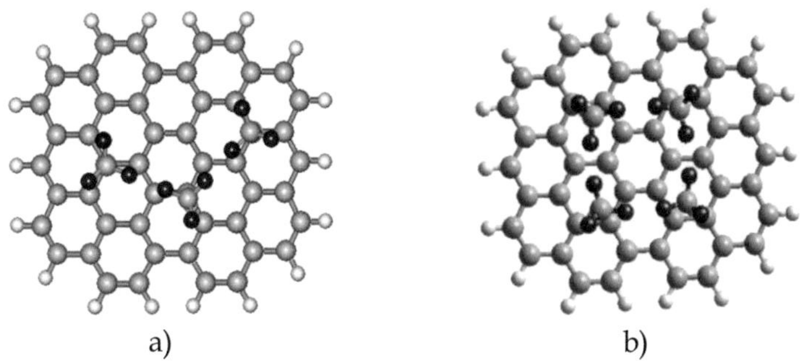

Fig. 1. Optimized geometry of three and four methane molecules on a planar polyaromatic molecule that mimics a graphene surface (top view) (Adapted from Albesa & Vicente, 2008) Copyright 2008 Argentine Chemical Society

The interaction energy between methane and the graphite surface, at low coverage, is almost the same from one adsorption site to another. Although it seems to indicate that graphene structure does not play an important role in methane adsorption, one has to take into account that this result tell us only what happened at very low coverage, but says nothing about the consequences of neighboring adsorbed molecules, in other words, when the coverage increases how near the first adsorbed molecule next molecules can be placed?.

The interaction energy, corresponding to the most favorable configuration, between one, two, and three methane molecules adsorbed on a graphene surface, is almost the same ($\Delta E \cong -\,2.9\ KJ/mol$). The distance of the adsorbed molecules to the surface diminishes from 4.2 Å for one molecule to 3.8 Å in the case of four from DFT results and 3.89 Å from PM3. Furthermore, the first three molecules form arrangements compatible with substrate hexagonal configuration of 4.5 Å sides, at almost the same distance of 3.95 Å from the

surface, this means that, with low coverage, the adsorbed phase follow graphene symmetry(see Fig 1a).
By increasing the coverage (i.e. four methane molecules), the structure is modified, the distance between methane molecules changes to 3.95 Å and between methane and graphene 3.80 Å, the interaction energy becomes ΔE (- 3.77 KJ/mol, and the hexagonal configuration move to a square one with the same average side of 4.5 Å (see Fig 1b). According to the previous quantum mechanical description of the interactions between gas molecules and the substrate and between adsorbed molecules, we conclude that, when equilibrium is reached, although at very low coverage a molecule has the same chance to adsorb on each site of the surface, letter as the coverage increase and some sites are already occupy, the probability that a molecule adsorbs in one site or another is not the same. In other words adsorption process can be approximated correctly considering energetic homogeneity of the surface at very low coverage, but when this condition is not fulfilled energetic inhomogeneity must be taken into account.
Computer Monte Carlo simulations (MCS) with the grand canonical ensemble (Albesa et al., 2008) is one of the more useful techniques that gives microscopic descriptions without expensive computational cost and allowing to over comes the quantum mechanical limitation of have to consider only a few adsorbed molecules. MCS offers not only the possibility of studying the adsorption phenomena at molecular scales but also making direct comparisons with experimental results. From the theoretical point of view MCS allow to compare how homogenous or non homogenous surface potential can be affect the results. We performed this comparison for the basal plane of graphite by taking the fluid–surface interaction (φ^{sf}) first, for the inhomogeneous case, as considering a two parameters 12-6 Lennard–Jones intermolecular potential between each atom of the adsorbate molecule (for methane we consider the spherical molecule approximation) as a site that interacts with each carbon atom of the solid, then the pair wise summation is carried over all atoms belonging to the solid that are located at a distance less than the cut off apart from the adsorbate mass center, and call this assumption atomistic model. Secondly, as an example of homogenous case we take the classical uniform Steele's 10 - 4 - 3 analytical potential (Steele, 1973)

$$\varphi^{sf}(z) = 3\varphi_w \left[\frac{1}{5}\left(\frac{\sigma^{sf}}{z}\right)^{10} - \frac{1}{2}\left(\frac{\sigma^{sf}}{z}\right)^{4} - \frac{\left(\sigma^{sf}\right)^{4}}{6\Delta(z+0.61\Delta)^{3}} \right] \quad (2)$$

where z is the distance between a fluid molecule and the substrate surface and Δ is the separation between lattice planes, the energy parameter φ_w si given by

$$\varphi_w = 4\,\pi\,\rho_s\,\varepsilon^{sf}\,(\sigma^{sf})^2 \quad (3)$$

being ρ_s = 0.382 Å^{-2} the carbon atoms density on the graphite slab surface, and the interaction parameters σ^{sf} and ε^{sf} are calculated using Lorentz - Berthelot combination rules. This approximation will be called analytical model. Then we compare the calculation by MCS with the grand canonical ensemble obtained from both models and with experimental results at different temperatures and pressures ranges.
At low pressures there is a good agreement between experimental and simulated isotherms for both models. In this pressure ranges, when the temperature changes, the curves exhibit different shapes. At temperatures lower than 113 K isotherms shows two clear horizontal

steps located at relative pressures of 13/40 and 29/40, respectively (see Fig. 2a), due to the completion of the first and second layers, that confirms an adsorption mechanism layer by layer (Hamilton & Goodstein, 1983). As the temperature increases, this kind of ordered adsorption is less significant, the steps are absent (see Fig. 2b) and all layers are available to be filled.

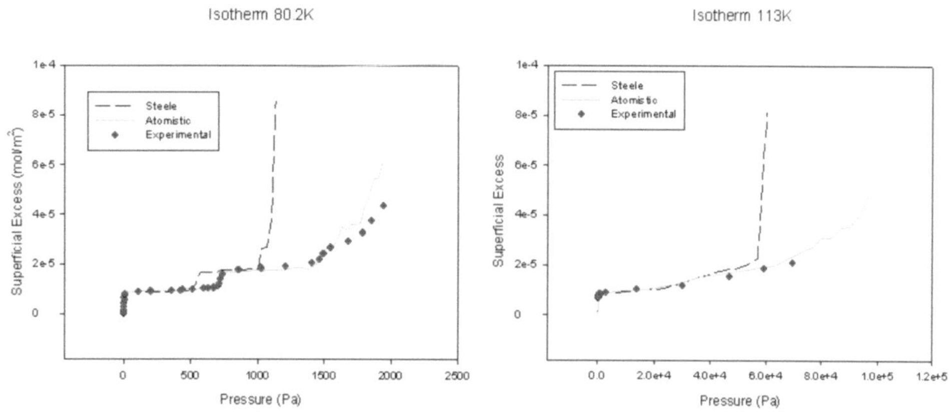

Fig. 2. Isotherms Circles are obtained from experimental results, and simulations are represented by dashed black lines for the analytic model and solid gray lines for the atomistic model. (a) 80.2K (b) 113K (Adapted from Albesa et al., 2008) Copyright 2008 American Chemical Society

In Figure 3 the density profile ρ*, as a function of the distance to the surface and the degree of coverage θ (adsorbed volume divided the monolayer volume), showed a series of peaks that allow to rationalized the mentioned layer-by-layer adsorption mechanism on one hand, and on the other hand exhibits differences between the results obtained from both models, because, the analytic model gives a denser (approximately 30%) phase that the atomistic model (see Figure 3). It is due to the fact that in the analytical approximation the surface is completely flat and smooth and methane molecules packing is more effective than in the atomistic model, which means that surface inhomogeneity becomes important. The difference becomes evident as pressure increase, because homogeneous approximation predicts condensation in advance after the third monolayer is formed (see Fig. 2a) instead of follow the real behavior, which is better fit by the atomistic model. In Fig. 3 one can also note that for the analytic model the layer equilibrium distances (mean density value) are closer to the surface than the corresponding to the atomistic model.

Among the thermodynamic aspects of gas adsorption on carbonaceous surfaces, enthalpy of adsorption study gives interesting and useful information about the process being accessible to experimental and theoretical explorations. From the experimental side both: direct methods like adsorption (Gravelle, 1978), flow (Groszek, 1998), differential scanning (Baudu et al., 1993), and immersion calorimetry (Zettlemoyer & Narayan, 1966); and indirect methods as inverse gas chromatography and adsorption isotherm have been widely employed last years. When the adsorbed phase is a rare gas, nitrogen, carbon dioxide or, methane on simple surfaces (graphene, single walled nanotubes, or nanocones), the use of adsorption isotherm measurements to determine the variation of adsorbed amount with temperature gives enough results to obtain accurate and reliable values of the heat of adsorption Q_{st} (Do, 1998).

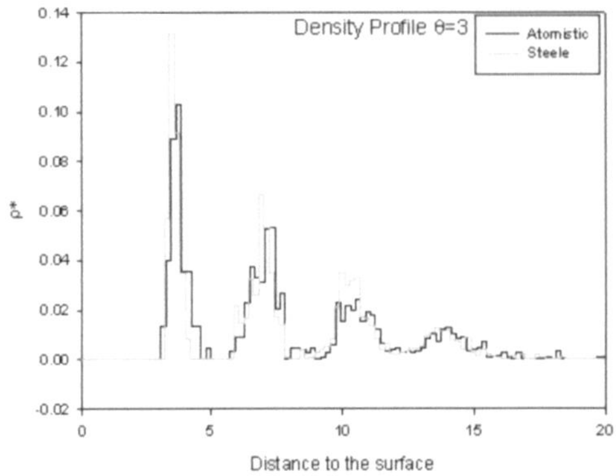

Fig. 3. Density profiles ρ* as a function of the distance to the surface in angstroms and the degree of coverage θ at 80.2 K when the isotherm is completed. Solid lines represent the results from the analytical model, and broken lines represent the atomistic model at θ = 3.0. (Adapted from Albesa et al., 2008) Copyright 2008 American Chemical Society

From two isotherms determined at similar but different temperatures, T_1 and T_2, a classical thermodynamic calculation gives:

$$Q_{st} = \frac{R\,T^2}{T_2 - T_1} ln\left(\frac{p_2}{p_1}\right) \tag{4}$$

where R is the gas constant, p_1 and p_2 are the equilibrium pressures at temperatures T_1 and T_2, respectively, when the adsorbed volume is constant and T is the corresponding mean temperature.

From the theoretical side the Computer Monte Carlo simulations (MCS) with the grand canonical ensemble not only gives the average number of molecules in the simulation box but also yields information about the isosteric heat, discriminating the two contributions, i.e. the solid-fluid interaction, and the fluid-fluid interaction. The isosteric heat of adsorption, $-\Delta H^0$, is defined as the difference between the molar enthalpy of adsorbate in the vapor phase, h_g, and the partial molar enthalpy of the adsorbed phase ($h_a = \left(\partial H_a / \partial N_a\right)_{p,T}$) (Pascual et al., 2003).

$$Q_{st} = \left\{ RT - \frac{\left[\langle E^g N\rangle - \langle E^g\rangle\langle N\rangle\right]}{\left[\langle N^2\rangle - \langle N\rangle^2\right]} \right\} - \frac{\left[\langle E^s N\rangle - \langle E^s\rangle\langle N\rangle\right]}{\left[\langle N^2\rangle - \langle N\rangle^2\right]} \tag{5}$$

The square-bracketed term in eq. 5 is the contribution of the fluid-fluid interaction to the isosteric heat of adsorption, whereas the last term is the contribution from the fluid-solid interaction. The isosteric heat of adsorption Q_{st} as a function of the coverage θ, for the system methane - graphite at lower temperatures was obtained from simulations by using equation 8 and is shown in Figure 4. We note, from both models, two clear peaks corresponding to the first and second layer completion. The Q_{st} values resulting from the atomistic and analytic models were 12.6 KJ/mol and 13.5 KJ/mol respectively, in good agreement with others values reported by Do and Do (D. Do & H. Do, 2005) (12.6 KJ/mol), and Piper and Morrison (Piper & Morrison, 1984) (13.4 KJ/mol). We can also see agreement between the analytic and atomistic models over the whole range of coverage except near the firs monolayer completion, where the isosteric heat of adsorption given by the analytic model is 3 KJ/mol greater than the atomistic model value. In this region, there is a better agreement with the experimental results for the analytic model. The difference between both models is due to the fluid-solid contribution because the fluid-fluid contribution is the same in both models. To explain this one can take into account that the effective packing, obtained from the analytic model, gives a very well defined distance of the first layer to the surface (see. Fig. 3), as a consequence at very low coverage all molecules lay almost at the same distance from the surface, with a well defined fluid-solid interaction value. On the contrary the constrains of atomistic model determines greater dispersion of the first layer distance to the surface (see Fig. 3), and taking into account that there is a minimum distance that a molecule can approach to the surface, the layer becomes pushed away from the surface and the fluid-solid interaction, as well as the isosteric heat value, becomes lower.

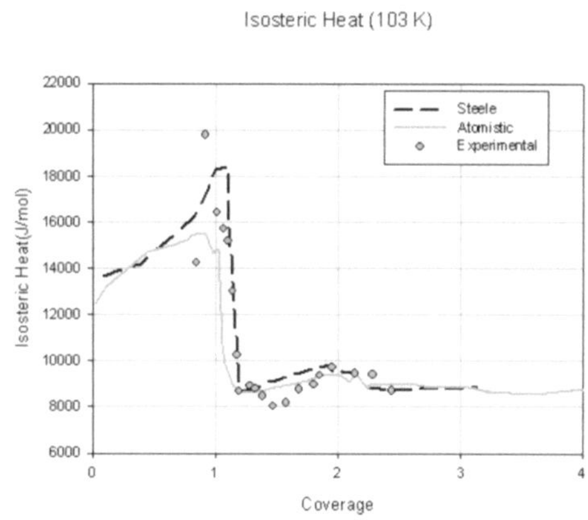

Fig. 4. Isosteric heat of adsorption as a function of the degree of coverage θ at 103 K. Circles are obtained from experimental results, and simulations are represented by dashed black lines for the analytical model and solid gray lines for the atomistic model. (Adapted from Albesa et al., 2008) Copyright 2008 American Chemical Society)

To describe the adsorbed system, aside quantum mechanical calculations, that gives answers at atomic level, where only a few molecules are analyzed; and Monte Carlo simulations (MCS) with the grand canonical ensemble, which explore at molecular level, where a more representative number of molecules is taken into account; the mean-field approximation (MFA) of the lattice model (De Oliveira & Griffiths, 1978) is other theoretical tool that that tell, at mesoscopic level, the behavior of adsorption isotherms of several layer of rare or simple gases adsorbed on graphite. According to the differences in the molecular density of the first layer obtain with the two mentioned models employed in MCS, the MFA could elucidate about this controversy. The mesoscopic level, where MFA operates, gives a global description of both the surface and the gas, but ignores what happens on the molecular level, and also assumes, a priori, a layer by layer adsorption process, being each layer constitutes by molecules adsorbed on a fixed lattice. These three mentioned techniques, in their respective ranges, give complementary information of the same phenomena.

Following the description of methane - graphite system, in the MFA one assumes, as in the lattice gas model (Burley, 1972), that the region above the substrate accessible to methane molecules can be divided into cells whose centers forma a regular lattice. No more that one molecule is permitted in a cell, and each par of gas molecules in adjacent cells contributes an amount $-\epsilon$ to the potential energy. All molecules in the *j*th layer experience an additional potential energy $-v_j$ due to the substrate, with

$$v_j = \in \left[\delta_{j1} D + \frac{(1-\delta_{j1})C}{j^3} \right] \tag{6}$$

The occupation number, n_{jk}, corresponding to the *k*th cell in the *j*th layer, take the values *0* or *1* if the cell is empty or occupied, respectively, then δ_{jk} is *1* when $k = j$ and *0* otherwise. *C* and *D* are proportional to the minimum energies for the interaction of a methane molecule with the semi - infinite continuous slab (9 - 3 potential) (Steele, 1974) and to the potential summation over different surfaces sites, respectively.
For each configuration $\{ n_{j\,k} \}$, the Gibbs - Boltzmann probability has to be proportional to $exp(-\beta H)$, where

$$H = -\in \sum_{(j,k),(j',k')} n_{jk}\, n_{j'k'} - \sum_j (\mu + v_j) \sum_k n_{jk} \tag{7}$$

$\beta = (k_B T)^{-1}$, k_B the Boltzmann constant, *T* the absolute temperature, μ the gas chemical potential, apart from a temperature–dependent constant, and *(j,k)* denotes nearest neighbor pair of cells.
Calling ρ_j the average value of n_{jk} in the *j*th layer, the grand potential can be obtained, according to the MFA (Burley, 1972), by minimizing Ω, as a function of $\rho_1, \rho_2, \ldots$, where

$$\Omega L^2 = k_B T \sum \{ \rho_j \ln \rho_j + (1-\rho_j) \ln (1-\rho_j) \} - \sum (\mu + v_j) \rho_j - \in \left[\frac{1}{2} a \sum \rho_j + b \sum \rho_j \rho_{j+1} \right] \tag{8}$$

L^2 is the number of cells in a single layer, and each cell has a number *a* of nearest neighbors in the same layer (related with the density of the layer), and *b* nearest neighbors just above it. At a minimum of Ω, the coupled set of equations

$$m_j = tanh \left\{ \frac{1}{2} \beta \left[\Delta\mu + v_j + \frac{1}{2} \in \left(a\, m_j + b\, m_{j-1} + b\, m_{j+1} \right) \right] \right\} \tag{9}$$

is satisfied for $j = 1, 2, \ldots$. Here, $m_j = 2\rho_j - 1$ ($m_0 = -1$) and $\Delta\mu = \mu + \epsilon (a + 2b)/2$ is the chemical potential minus the resulting value for an adsorbed layer of infinite thickness.
To compute adsorption isotherms, eq. (9) is truncated at $j = 20$ ($m_{21} = m_\infty$), where m_∞ is the negative solution of

$$m_\infty = tanh \left\{ \frac{1}{2} \beta \left[\Delta\mu + \frac{1}{2} \in \left(a + 2b \right) m_\infty \right] \right\} = 2\rho_\infty - 1 \tag{10}$$

The equation can be solved numerically for different values of *a*, *b* that are appropriate for each lattice configurations. For each value of β and $\Delta\mu$, the solutions that minimize eq. (7) are used.
Figures 5a and 5b show the results obtained taking $C = 12$ and $D = 20$, for the temperatures $T = 103$ *K* and $T = 171$ *K*, respectively. It is noted that at low temperatures the isotherms

show greater dependence on the density of the condensed phase (associated with *a* and *b* values), and the better approach to the experimental results correspond to *a* = 6, and *b* = 3, corresponding to a triangular arrange compatible with results obtained with mechanical quantum calculation with three methane molecules. In addition the density associated with *a* = 6, and *b* = 3, fits experimental isotherms better than other values corresponding to higher densities (see Fig. 5a). When the temperature increases and approaches the critical value, this dependence disappears as is shown in Fig. 5b with different values of *a* and *b*.

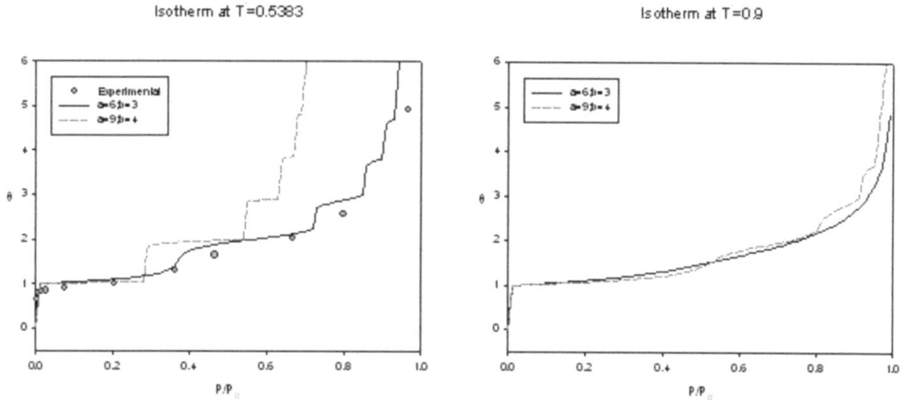

a) b)

Fig. 5. Experimental and lattice gas model isotherms at (a)103 K.(b) Experimental and lattice gas model isotherms at 171 K. (Adapted from Albesa et al., 2008) Copyright 2008 American Chemical Society

3.2 Simple gas adsorption on graphitic curved surfaces

Carbon nanotubes and in general microporous carbonaceous materials offer interesting properties from the technological point of view, as a mean to store hydrogen and methane (Bhatia & Myers, 2006), that are energy sources environmentally favorable, or as sieves to gas separations like oxygen and nitrogen (Arora & Sandler). Adsorptive properties of nanotubes and fullerenes differ from other graphitic carbons because of their carbon surface curvature (C-C-C bonding angles). Carbon atoms have to adopt quasi sp^2 hybridization due to their highly curved structure. The level of curvature of valence orbital depends on the material radius: great radius lead to hybridization near sp^2 pure, while small radius leads to sp^3 (Niyogi, 2002).

To describe phenomena that occur at molecular level one employs simulations, Monte Carlo grand canonical (Albesa et al., 2008) to study adsorption at equilibrium, or Molecular Dynamics (Cheng et al., 2005) when one is interested transport properties far from the equilibrium. However, the accuracy of these descriptions depend on the potential models adopted, choosing generally the 12-6 Lennard Jones potential, or the Crowell - Brown potential that don't take into account the surface curvature (Wang & Johnson, 1999). A correct potential has to identify the equilibrium position and to give a good approximation in its neighborhood, at least up to the curve inflexion point. To reach this it is necessary to have with a well description of the molecular bonds, because the approximations would neglect the effect of the hybridization differences of the carbon atomic orbitals, induced by

the curvature of the surface. To solve this problem force fields were derived (Kostov et al., 2002) where the curvature effect was taking into account, however, the parameters calculated for these fields overestimate the interaction of the nanotubes of smaller radius, because they are considered as a model molecule with free radicals.
In order to describe the potential energy that suffer a molecule near the surface of different nanotubes, theoretical calculations at DFT level were performed (Albesa et al., 2009) for hydrogen, nitrogen and methane, were the surface was mimicked by deforming a coronene molecule (see Fig 6). The nanotube diameters d can be described by the Hammada's indices (m,n), as $d = a/\pi\,(m^2 + mn + n^2)^{1/2}$ and, $a = 0.246\ nm$, employing "armchair" type ($n = m$), taking into account the independence of quirality for simple gases adsorption (Vilaplana, 2005).

Fig. 6. Cluster carbon atoms employed in the calculations

The exploration involves diameters according to experimental values, that means, $d = 0.407$, 0.949, and $2.034\ nm$, corresponding in the Hammada's notation to $(3,3)$, $(7,7)$, and $(15,15)$ respectively. In spite of that results obtained from DFT (see Fig. 7), as almost all Generalized Gradient Approximation calculations, give energy values lesser that those obtained experimentally, i.e., $3.3\ KJ/mol$ for hydrogen (Okamoto & Miyamoto, 2001), and $10.5\ KJ/mol$ for methane (Talapatra & Migone, 2002), it is interesting to explore the relative values and their deviation when nanotubes diameter change. The interest on DFT results is because, beyond this approach gives lesser energy values than those experimentally obtained and a bad description very far from the surface, it allow to describe very well the relative energy potential behavior near the equilibrium point. As is shown in Figure 7a, the interaction energy of hydrogen as a function of the distance to the surface, resulting from DFT calculation, determines very low values on the nanotube external face, which tell as that storage this gas on closed nanotubes is rather difficult. We can also see that the values corresponding to the two smaller nanotubes are almost equal. In case of methane, showed in Figure 7b, one can see that the nanotube of the intermediate diameter has a fewer attractive potential with regard to the other two kind of nanotubes. If the curves are fitted with the classical 12-6 Lennard-Jones potential, the values obtained for the position of the potential minimum (σ) is almost the same for different nanotubes, around $0.37\ nm$, for hydrogen; and $0.402\ nm$, for methane, being $0.32\ nm$, and $0.36\ nm$ the values corresponding to the adsorption on graphite; and the deep potential well (ε) is, as the nanotube diameter increases, 7.4, 12.6, and $9.36\ K^{-1}$, for hydrogen; and 14.7, 12.6, and $14.9\ K^{-1}$, for methane. From these results one can conclude that the potential energy obtained doesn't show an important dependence on the curvature of the nanotubes, at least in the explored range.
The experimental isotherms for the adsorption of methane, measured at 77.3 K, unlike the case of the adsorption on graphite that has only one step in the monolayer regime (0.2 kPa), have one sub step below this pressure and other one above it (Albesa et al., 2010). As in the case of graphite when the temperature increases, this stepwise adsorption behavior is less pronounced, and above 103 K the substeps disappear.

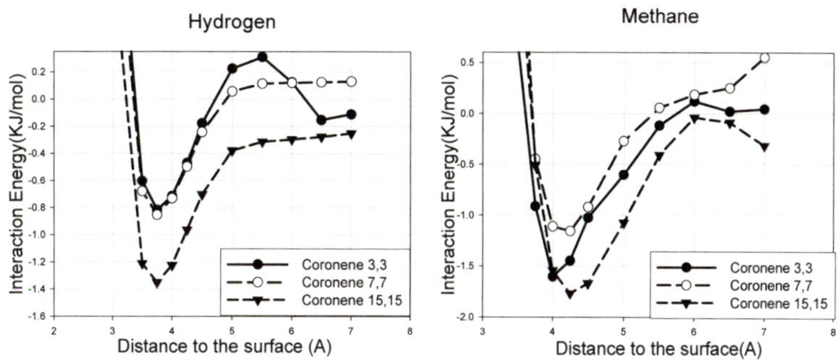

Fig. 7. Energy potential curve for (a) hydrogen molecule (b) methane molecule

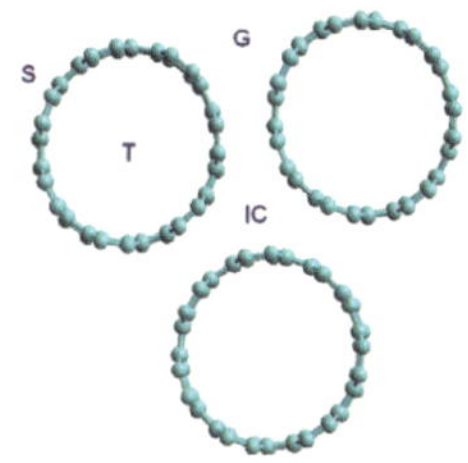

Fig. 8. Smaller simulation cluster, showing adsorption sites: the grooves (G) between two tubes; on nanotube surface (S); in interstitial channels (IC); and in the interior of a nanotube (T), for open tubes. (Adapted from Albesa et al., 2010) Copyright 2010 American Chemical Society

To explain this behavior, beyond the theoretical interest to explore the adsorption of simple molecules on the curved surface, that mimic the external environment of a nanotube, these kind of carbonaceous structures don't appear alone but forming different bundles of numerous tubes. These kind of arrange generate new potential adsorption sites, besides the mentioned sites on curved graphene, corresponding to sites on the external phase of the bundle external tubes. For single walled nanotubes (SWNT) four different adsorption sites has been identified (see Fig. 8), i.e. the previous mentioned convex external walls or outer surface sites (S), the interstitial channels (ICs), the grooves between two adjacent outer nanotubes of the bundle (G), and, in case of SWNTs open at the ends, the inner sites (T).
To describe the adsorption process, in closed SWNT, it has been proposed (Bienfait et al., 2004; Kuznetsova et al., 2000) that it starts as linear chains at the strongest binding energy sites: grooves on the outside surface of the bundles and some larger, accessible, defect - induced interstitial channels. After these sites are filled, adsorption proceeds on the external surface of the bundles. The 2D adsorbate structure on the external bundle surface initially builds up adjacent to the occupied grooves until the entire external surface is covered by a single monolayer. The binding energies for this latter stage of adsorption are comparable but somewhat smaller than for adsorption on the basal plane of graphite, whereas the binding energy on the preferential adsorption sites (G, IC) populated during the initial stage of adsorption is considerably larger that tat on the planar graphite. In order to elucidate between the more attractive sites like groove (G) and interstitial channels (ICs), which one begins to fill up first, it has been performed Ab Initio and Molecular Mechanical Calculations, for the case of methane. In order to mimic different sites (S, IC, G, and T) of

the substrate, a triangular array consisting of three identical tubes was used, this is the minimum bundle configuration that exhibits the four (S, IC, G) characteristic sites (see Fig. 8). The distance between the tubes or van der Waals gap always is taken equal to *0.34 nm*. Three different nanotube diameters were explored, i.e. *0.949*, *1.628*, and *2.034 nm*, or Hamada indices *(7,7)*, *(12,12)*, and *(15,15)* respectively. Although the last diameter (*2.034 nm*) is not realistic, it's enough large that the IC sites are able to accommodate a molecule and can mimic a general intertube channel that can appear in general tube bundles. In other words, the two first diameters were employed to analyze G, S and T (in case of open nanotubes) sites, and the last one to IC sites. As the molecular dynamics simulations, using MM+ force field, as the calculation employing ab initio density functional theory (DFT) (Albesa et al., 2010), tell that IC sites begin to fill before the G sites.
A complementary picture at molecular level, can be performed by Monte Carlo simulations (MCS) with the grand canonical ensemble, in order to take into account the behavior of a representative number of molecules in the adsorbed phase. Results showed in Figure 9, for tubes with a diameter of 2.034 nm, confirm the sequence of filling of different sites in the bundles (is spite of not being realistic, this diameter allows adsorption in IC sites, and the other sites follow the same steps, independent of the kind of nanotubes employed).
The isotherms of nanotubes of 1.628 nm of diameter at different temperatures (83, 93, and 113 K) are compared between closed and opened end in Figures 10 and 11. Unlike the closed end (Fig. 10), the opened end tube (Fig. 11) shows a knee when a monolayer starts forming on the inside surface of the nanotubes, and also when this interior layer is completed (Tasca et al., 2002), and a new high energy site appears in the axial phase (see Fig. 12). It can also mentioned the isotherms corresponding to smaller tubes (0.949 nm of diameter) exhibits only the first knee because there is no possibility of an inner axial phase.

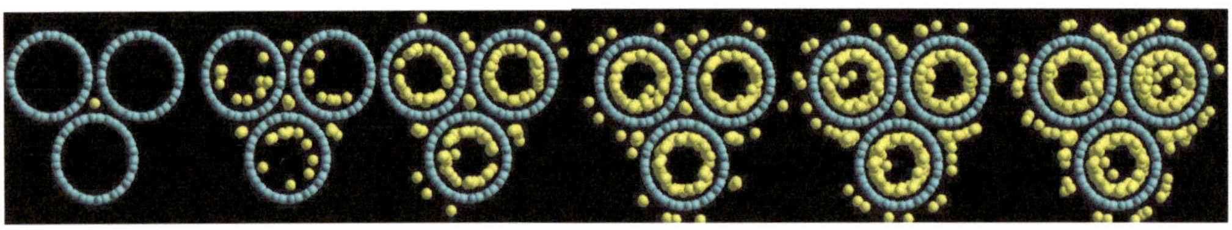

Fig. 9. Upper view of filling evolution corresponding to isotherm at 113 K and pressures of 0.0113, 5.09, 22.6, 50.9, 226, and 409 Pa for opened end tubes of 2.034 nm, (15,15) SWNT. (Adapted from Albesa et al., 2010) Copyright 2010 American Chemical Society

This dependence, of the adsorption process, on the nanotube diameter, suggests that it would be useful to classify the bundles according to the diameter of the tubes involved, analogous to the method used for porous substrates classification employing porous sizes.
Comparing simulated isotherms with the adsorption on graphite we note that the monolayer on graphite is completed at a pressure below that corresponding to the nanotubes, and that the isosteric heat of adsorption is greater for graphite. This is due to the greater gas - solid interaction for the graphite case, because the nanotubes are single - walled, while graphite is formed by several graphene layers. The gas - gas interaction on the nanotubes is weaker than on graphite, because the curvature of the nanotubes reduces the number of neighboring adsorbate molecules in this case, relative to those present in the graphite case.

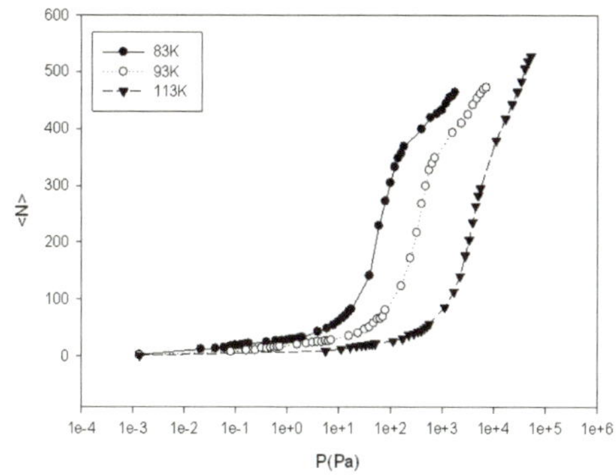

Fig. 10. Simulated isotherm at 83, 93, and 113 K for closed end tubes of diameter 9.52nm (7,7) SWNT. (Adapted from Albesa et al., 2010) Copyright 2010 American Chemical Society

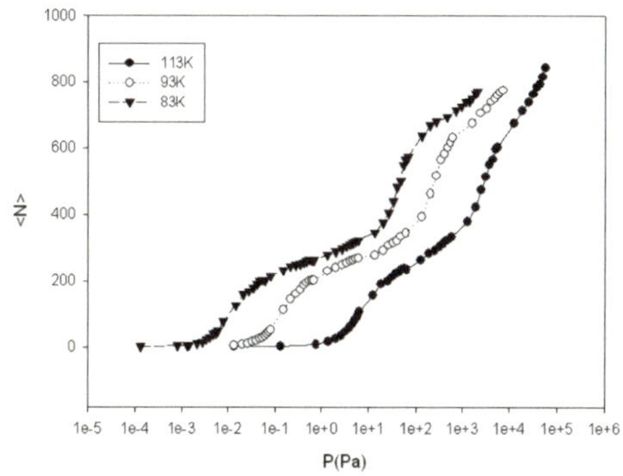

Fig. 11. Simulated isotherm at 83, 93, and 113 K for closed end tubes of diameter 1.628 nm (12,12) SWNT (Adapted from Albesa et al., 2010) Copyright 2010 American Chemical Society

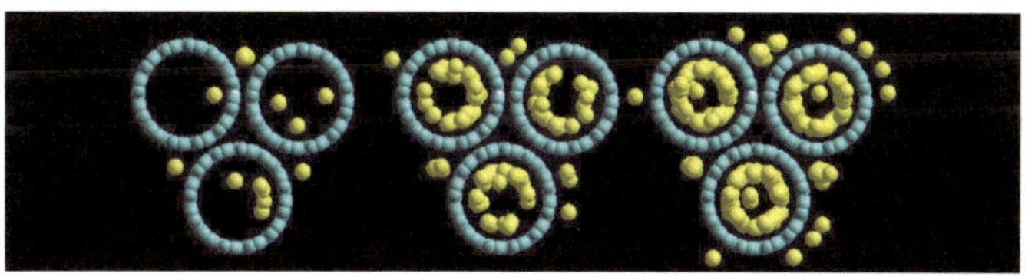

Fig. 12. Upper view of filling evolution corresponding to isotherm at 113 K and pressures of 1.38, 6.27, 56.3 Pa for opened end tubes of 1.628 nm, (12,12) SWNT. (Adapted from Albesa et al., 2010) Copyright 2010 American Chemical Society

4. Gas separation of nitrogen-oxygen and ethane-ethylene mixtures

Most commercial processes used in air separation employed synthetic zeolites. These materials are selective with respect to nitrogen, that is, they are able to adsorb more nitrogen than oxygen in a rate 4:1, due to the interaction between nitrogen molecule and oxygen cation to compensate charges. However find some kind of adsorbent that prefer oxygen is also interesting, because to reach the same separation than zeolites will only require a quarter of work (Jayaraman & Yang, 2005). Among different substrates that have greater affinity with oxygen, carbonaceous materials are promising options (Arora & Sandler, 2007), being their principal use as selective sieves. Beside the isotherms, and the isosteric heat of adsorption, selectivity is other quantity of interest, selectivity of specie 1 respect to specie 2 is defined as:

$$S_{12} = (\theta_1/\theta_2)/(p_1/p_2) \tag{14}$$

where θ_i, and p_i, (i = 1, 2) are the relative coverage and partial pressure respectively, and, when there is no interaction between adsorbed species, it only depends on temperature , T, otherwise it is function of θ_1, θ_2 and T. In previous sections it was mention that adsorption explorations at molecular levels, i.e. by MCS, imply to select a model, and in case of gas mixtures, choose one approach enough simply to reduce computing time, but preserving the essential characteristics, becomes critical. One possibility is consider the molecules as pseudo spheres, that is only one Lennard - Jones interaction site per molecule (1CLJ), and a more realistic one would be consider molecules as composed by two interaction sites (2CLJ). We make a comparison of these two models for the adsorption of oxygen and nitrogen mix similar to the air composition, i.e. 80% and 20 % respectively. A remarkable difference appears in selectivity (Fig.13) because 1CLJ approximation gives fewer adsorption of nitrogen than oxygen, although the difference between adsorbed quantities is not high, reflects an important increase of the 1CLJ model selectivity. In Figure 14 we also note that the adsorption for 2CLJ model begins a lesser pressures than in the case of 1CLJ model. The density profiles (Figs. 15 and 16) show that the increase of pressure in 1CLJ model moves some molecules from the first layer to the second layer, so for nitrogen as for oxygen. In 2CLJ approximation this moving occurs basically with oxygen, while the density of the nitrogen first layer increase, this behavior, opposite to what one would hope from the entropic point of view according to the molecules size, is due to temperature considered, because greater temperatures favor nitrogen adsorption
Given that carbon nanotubes have internal sites that, after activation, are available for adsorption, and these have a narrow pore size distribution, these materials are presented as a promising alternative as adsorbents capable of achieving a good separation between N_2 and O_2.

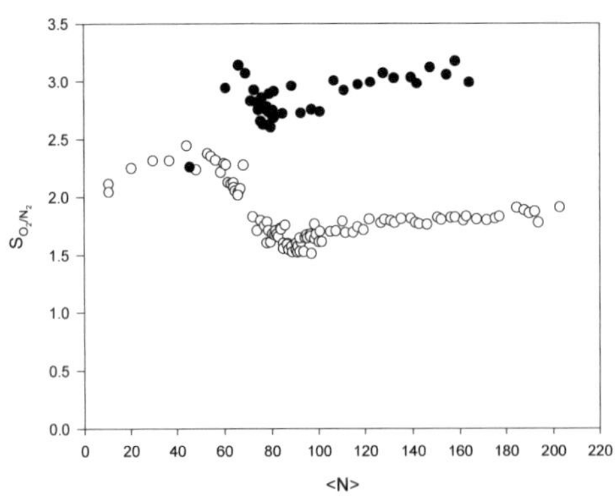

Fig. 13. Simulated selectivity at 100 K of air on graphite for 1CLJ (black) and 2CLJ (white) models. Total: squares, oxygen: triangles, and nitrogen: circles

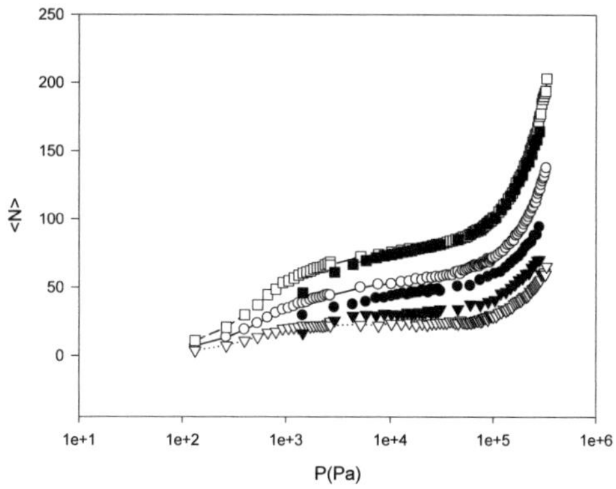

Fig. 14. Simulated adsorption isotherm at 100 K of air on graphite for 1CLJ (black) and 2CLJ (white) models. Total: squares, oxygen: triangles, and nitrogen: circles

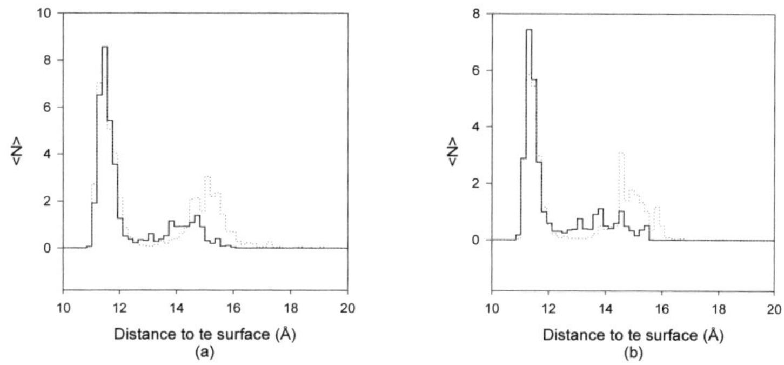

Fig. 15. Density profiles ρ as a function of the distance to the surface for 1CLJ model. Solid lines represent the results at 5866.1680 Pa and dotted lines at 45462 Pa (a)Nitrogen (b) Oxygen

By comparing the results obtained by Monte Carlo simulations in nanotubes (Fig. 17 a-c) with those obtained in graphite is observed that both materials show similar trends: there is a maximum degree of separation every time a monolayer is completed. As the nanotubes can complete an inner and outer monolayer, there are two peaks in the separation. Nanotubes with greater separation power are the *(10,10)*, diameter *1.356 nm*. As shown in the figures, the separation process takes place inside the nanotubes.

The degree of separation on the outer surface is negligible. From this arises the importance of obtaining good methods of activation, since the presence of interstitial defects (which are the size of the order of the nanotubes *(7,7)*, diameter *0.949 nm*) decreases significantly the separation of these mixtures.

Olefin - paraffin separation represents one of the most important separations in chemical and petrochemical industries. The production of plastics, rubbers, films, and other chemicals from olefins such as ethylene, requires high purity olefins (> 99%). Ethylene is generally produced by cracking or thermal decomposition of ethane, and during these processes the gas is an ethane and ethylene mix. The conventional cryogenic distillation is as efficient and reliable, and continues being the leading technology for ethane/ethylene separation; however, it is also energetic expensive due to the similar volatility of ethane and ethylene. In a typical ethylene plant, the cracking apparatus represent approximately 25% of the unit const, while the other 75% is due to compression, calefaction, dehydration, recovering and refrigeration systems (Anson et al., 2008).

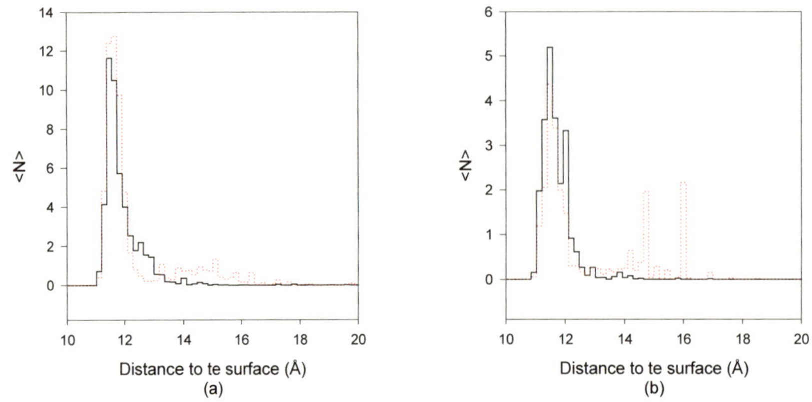

Fig. 16. Density profiles ρ as a function of the distance to the surface for 2CLJ model. Solid lines represent the results at 5866.1680 Pa and dotted lines at 45462 Pa(a)Nitrogen (b) Oxygen

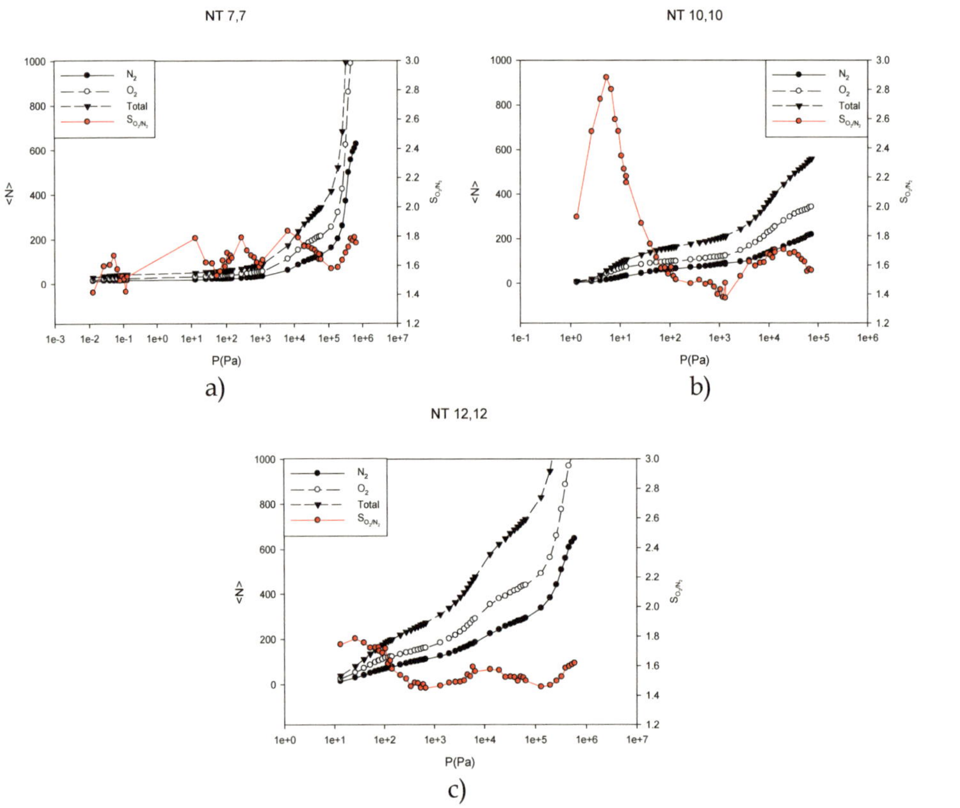

Fig. 17. Simulated isotherm 100 K and separation degree (right offset) for open nanotubes (a) (7,7) ; (b) (10,10); (c) (12,12)

An effective separation method that produces raw materials with highly ethylene enrichment without using cryodistillation would reduce the energetic and equipment costs associated with paraffin/olefin separation. Therefore exist interest to develop new separation techniques (Fuentes & Mendez, 2002). Many alternative separations have been investigated, including chemical, extractive distillation, physical adsorption, and separation of the base membrane (Eldridge, 1993). Adsorption is one alternative to separation, which is till under develop. Adsorbent materials are generally incorporated joined to transition metals (copper or silver), and as a result the preferential adsorption of olefin is obtained. This selective adsorption is due to the strong interaction between non saturated olefins bonds and the metal ion on the surface, forming a π complex. Because of the preferential way in which olefin is adsorbed (Anson et al., 2010) by these substance it is possible to obtain paraffin pure. However, when paraffin selective adsorbed, olefin pure can be obtained, which becomes adsorption as efficient alternative. Unfortunately there are few materials able to reach this separation (Herdes et al., 2007). This preferential adsorption is due fundamentally to the interaction between methyl sites of the adsorbent and the adsorbat (Kroon & Vega, 2009).

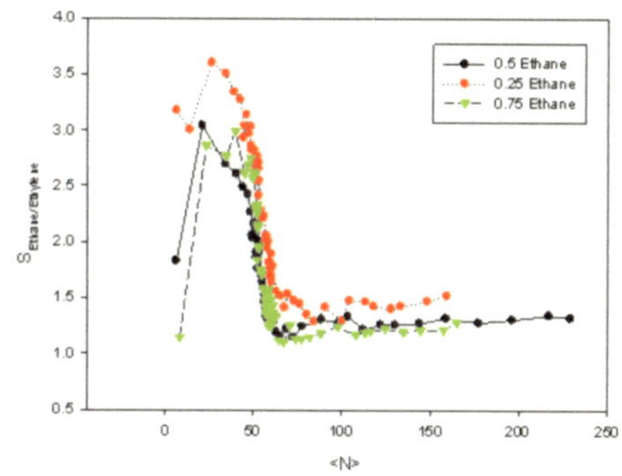

Fig. 18. Selectivity for ethane/ethylene in graphite as a function of the number of adsorbed molecules

It is well known that for only one component, the adsorption on carbonaceous materials is greater for ethane that ethylene. That is why these materials are studying when one is looking for a highly pure ethylene.
The simulation results show that the selectivity (Fig. 18), as a function of the number of adsorbed molecules, reaches a maximum when the monolayer is completed, as it happen with nitrogen - oxygen mixtures.
The profile is the same for all the compositions studied. The decrease in the degree of separation occurs because, once filled the monolayer, the adsorption of ethylene occurs more rapidly than ethane (Fig. 19 a-c). At high concentrations of ethylene, the number of molecules of ethane decreases, i.e., ethylene displaces ethane (Fig. 19 a). Here we see two phenomena, the principle of selective adsorption of ethane is due to energetic effects. As pressure increases the entropic effects are greater and that is why, because of their size, increases the adsorption of ethylene
Isosteric heat of adsorption of the mix is quite similar to the corresponding values of pure substances. It is also observed that there isn't a great variation of isosteric heat of adsorption with composition.(Fig 20 a-b)
A similar situation is observed in the adsorption of mixtures of nanotubes *(7,7)*; where the main adsorption takes place on the surface. Although in this case the coefficient of ethane-ethylene separation (Fig 21) is less than in the case of graphite.

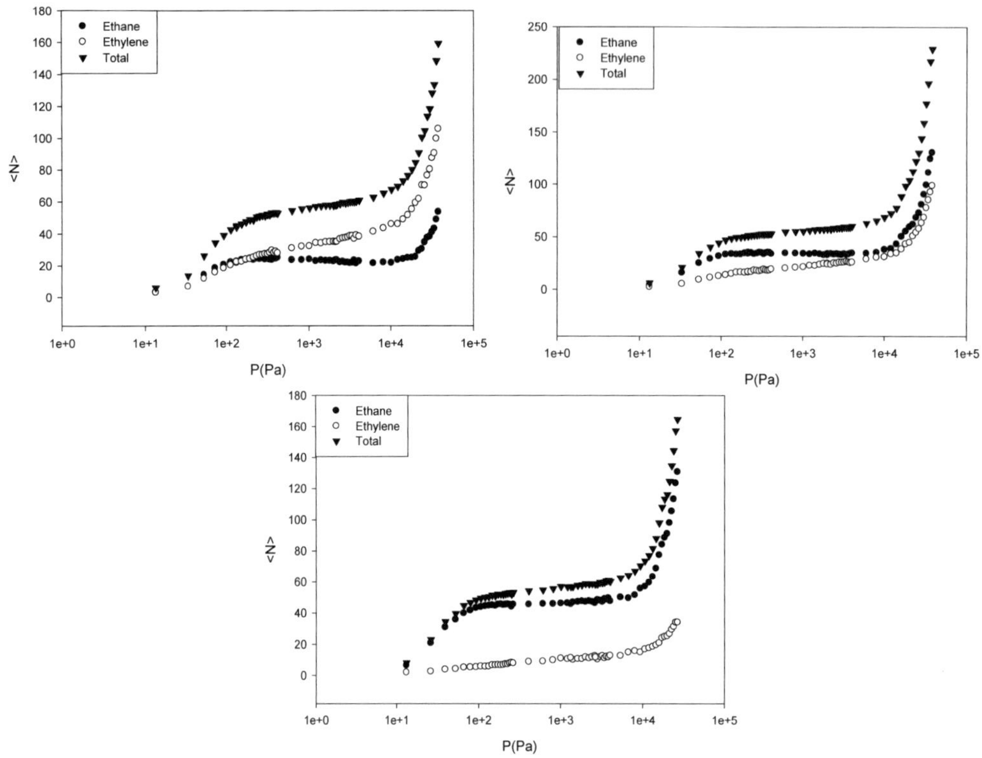

Fig. 19. Simulated adsorption isotherm at 153 K of ethane/ethylene mixtures on graphite for different compositions (a) 25%Ethane (b)50%Ethane (c)75%Ethane

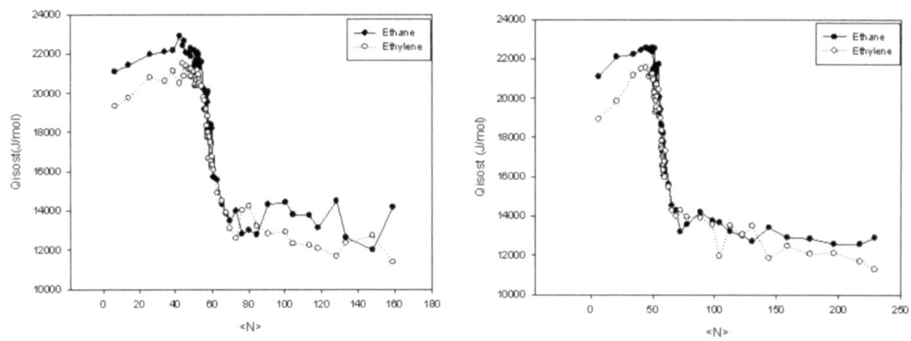

Fig. 20. Isosteric heat of adsorption at 153 K for ethane/ethylene mixtures on graphite for different compositions (a) 25%Ethane (b)75%Ethane

In this substrate the displacement of ethane molecules by ethylene which is observed in graphite is not present. The effect can be attributed to the curvature of this kind of nanotubes; this is, the small diameter provokes a less compact arrangement of the molecules, and therefore a less important effect of the entropic factors.

If we now consider nanotubes with larger diameters, so that adsorption occurs in the interior of them, we note that, as in the case of mixtures of nitrogen and oxygen, the

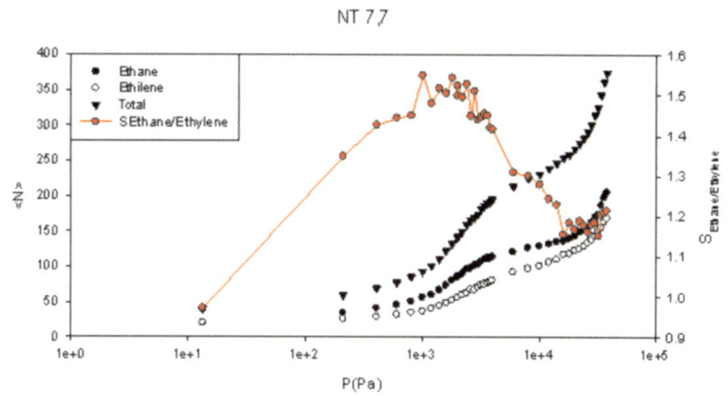

Fig. 21. Isotherms and Selectivity (right offset) for ethane/ethylene in *(7,7)* nanotubes as a function of the pressure

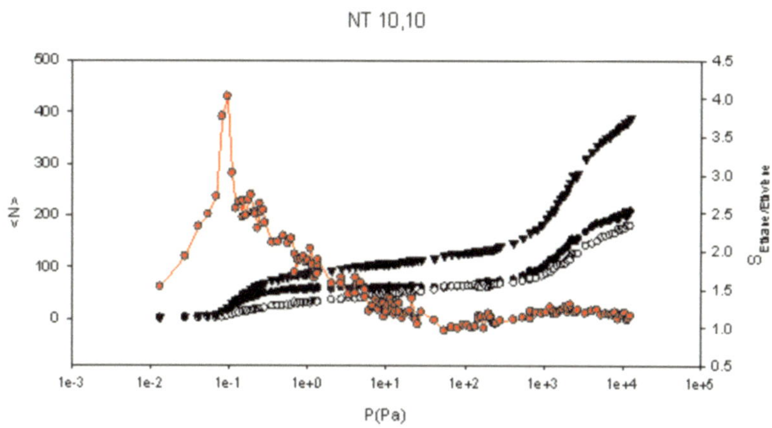

a)

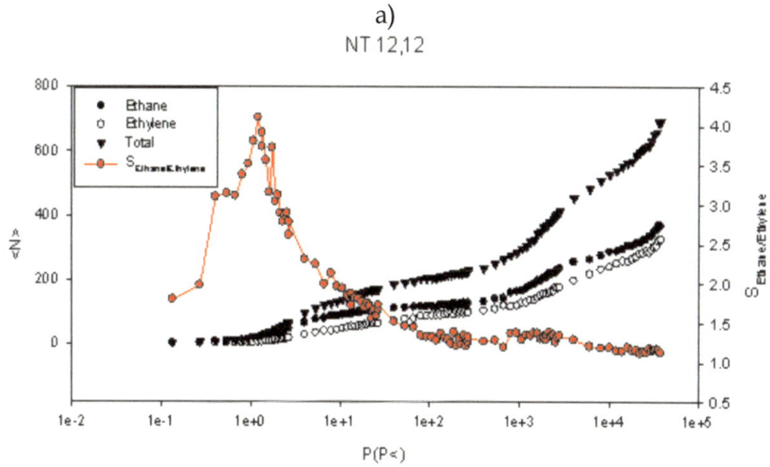

b)

Fig. 22. Isotherms and Selectivity (right offset) for ethane/ethylene as a function of the pressure in a) *(10,10)* nanotubes b) in *(12,12)* nanotubes

separation on the outside is very small, being slightly greater than one. Also note that the degree of separation in nanotubes *(10,10)* and the *(12,12)* (diameters *1.356 nm* and *1.628 nm* respectively) is practically the same (Figs. 22a and 22b). However, we can see that the positions of the maxima occur at different pressures. In this way, having a size distribution of nanotubes can enhance the range of pressures that are working.
Of particular interest is that the results obtained in systems nitrogen / oxygen and ethane / ethylene for the case of graphite, show similar trends to those obtained from more complex substrates, such as carbon nanotubes. This is why the above mentioned studies provide rich information on new systems, because, thanks to its structural simplicity, experiments can be carried out with great precision.

5. Conclusions

Experimental results of methane adsorption at low temperature, confirm a layer-by-layer filling mechanism, where the first two layers are clearly defined, but above 113 K this mechanism disappears and all layers are available to be filled. MCS confirm the necessity to employ atomist models in order to describe the adsorption a higher pressures, beyond that at very low pressures simple analytic models give good results with cheaper computational cost. The production of changes on the arrangement of adsorbed molecules as surface coverage increases is also corroborated, at microscopic level, by quantum mechanical descriptions. Although the heat of adsorption constituted an experimental and theoretical useful tool to explore thermodynamic aspects of the adsorption process, and from can be easily determined from the experimental side, theoretical descriptions have to be considered with some details. For instance if the models deal with isotherms associated to ideal monolayer, with molecules laying in a plane at a fix distance of the surface, heat of adsorptions values obtained are well defined, but by using other models, perhaps more realistic, when molecules are not arranged on a plane but around a mean distance of the surface, the number of molecules needed to have values with low dispersion is higher than the usual employed in MCS. The comparison, between experimental results and those obtained from MCS with different model approximations, confirms previous asseveration. In addition, Mean Field Approximation confirms that lower density values, similar to those derived from MCS by employing atomistic model, fit better experimental isotherms than the higher densities obtained from other models.
By considering curved graphene surfaces one can explore the adsorption on other types of graphitic structures (carbon nanotubes, nanocones, and others kinds of fullerenes). In spite DFT calculations of the interaction energy between the curved graphene substrates and simple molecules, reveals that surface potential energy is almost independent on the curvature, at least when these curvatures range values corresponding to common nanotubes; these kind of carbonaceous structures have many others adsorption sites that have to be explored. As an example, single walled nanotubes (SWNT) form bundles of various tubes, and then, besides the previous mentioned convex external wall sites (S), where the adsorption is a two dimensional process, other kinds of sites can appear. One are sites lying between two adjacent outer nanotubes of the bundle or groove sites (G), associated to one dimensional adsorption, other sites are similar to those founded in cylindrical pores, and emerge when tubes diameter are so higher that they left interstitial

channels (IC) with enough space in order to accommodate adsorbed molecules in them; and finally, and eventually they have inner sites (T) when there are SWNTs with their ends open. Beyond DFT calculations, where the energy of each sites confirm the postulated adsorption sequence: starts at the linear chains of G sites, and then in IC, if there are any one with enough space to accommodate molecules inside them, and finally on S sites. In case of open sites T sites have to be filled before G sites. MCS gives information about the shape of the isotherms, for instance, in case of open tubes, on the curve emerge a knee when a monolayer starts forming on the inside surface of the tubes, and also when this interior layer is completed.

By comparing these results with adsorption on graphite, one observes two facts: the monolayer of graphite is completed at a pressure below that corresponding to the nanotubes, and the isosteric heat of adsorption is greater for graphite, the reason is because the nanotubes considered are single walled (one graphene layer, rolled up) while graphite is formed by several graphene layers.

Results obtained from the study of gas mixtures adsorption (nitrogen / oxygen and ethane / ethylene) on different carbonaceous structures, show that graphite and more complex surfaces (such as carbon nanotubes) present similar trends. In this way, the behavior above described will provide valuable information on new systems.

6. Acknowledgment

The authors gratefully acknowledge financial support from the UNLP (Universidad Nacional de La Plata), CICPBA (Comisión de Investigaciones Científicas de la Provincia de Buenos Aires) and CONICET (Consejo de Investigaciones Científicas y Tecnológicas)

7. References

Albesa, A.G.; Llanos, J.L. & Vicente, J.L. (2008). Comparative study of methane adsorption on graphite. *Langmuir*, Vol. 24, pp. 3836-3840

Albesa, A.G. & Vicente, J.L. (2008). Theoretical study of methane adsorption on graphite. *The Journal of the Argentine Chemical Society*, Vol. 95, No. 1-2, pp. 10-22

Albesa, A.G.; Filippin, A. & Vicente, J.L. (2009). Effects of the Surface Curvature on the Adsorption of Simple Gases over Carbonaceous Materials using the Density Functional Theory (DFT) *CIT Technologic Information Review*, Vol. 29, pp. 31-37

Albesa, A.G.; Fertitta, E.A. & Vicente, J.L. (2010). Comparative Study of Methane Adsorption on Single-Walled Carbon Nanotubes. *Langmuir*, Vol. 26, pp. 786-795

Anson, A.; Wang, Y.; Lin, C.C.H.; Kuznicki, T.M. & Kuznicki, S.M. (2008). Adsorption of ethane and ethylene on modified ETS-10. *Chemical Engineering Science*, Vol. 63, pp. 4171-4175

Anson, A.; Lin, C.C.H.; Kuznicki, T.M. & Kuznicki, S.M. (2010). Separation of ethylene/ethane mixtures by adsorption on small-pored titanosilicate molecular sieves. *Chemical Engineering Science*, Vol. 65, pp. 807-811

Arora, G. & Sandler, S. (2007). Nanoporous carbon membranes for separation of nitrogen and oxygen: Insight from molecular simulations. *Fluid Phase Equilibria*, Vol. 259, pp. 3-8

Azami, T.; Kasuya, D.; Yuge, R.; Yudasaka, M.; Iijima, S.; Yoshitake, T. & Kubo, Y. (2008). Large-Scale Production of Single-Wall Carbon Nanohorns with High Purity. *J. Phys. Chem.*, Vol. 112, pp. 1330-1334

Bandosz, T.J.; Biggs, M.J.; Gubbins, K.E.; Hattori, Y.; Iiyima, T.; Kaneko, K.; Pikunic, J. & Thomson, K.T. (2003). Molecular Models of Porous Carbons, In: *Chemistry and Physics of Carbon*, Vol. 28, Marcel Dekker, Inc., L.R. Radovic, (Ed), pp. 41-228, ISBN 0-8247-0987-X, New York, USA

Baudu, M.; LeCloirec, P. & Martin, G. (1993), First approach of desorption energies of water and organic molecules onto activated carbon by differential scanning calorimetry studies. *Water Res.*, Vol. 27, pp. 69-76

Beaver, M. & Sircar, S. (2010), Methane Reforming Using a Novel Sorption Enhanced Reaction Concept, In: *Clean Energy Systems and Experiences*, InTech, K. Eguchi (Ed.), Viena, Austria

Bertoncini, C.; Odetti, H. & Bottani, E.J. (2000). Computer Simulation of Phenol Physisorption on Graphite. *Langmuir*, Vol. 16, pp. 7457-7463

Bhatia, S.K. & Myers, A.L. (2006). Optimal Conditions for Adsorptive Storage. *Langmuir*, Vol. 22, pp. 1688-1700

Bienfait, M.; Zeppenfield, P.; Dupont - Pavlovsky, N.; Muris, M.; Johnson, M.R.; DePie, M. & Vilches, O.E. (2004). Thermodynamics and structure of hydrogen, methane, argon, oxygen, and carbon dioxide adsorbed on single-wall carbon nanotube bundles. *Phys. Rev. B*, Vol. 70, pp. 035410

Bottani, E.J. & Bakaev, V.A. (1994), The Grand Canonical Ensemble Monte Carlo Simulation of Nitrogen on Graphite. *Langmuir*, Vol. 10, pp. 1550-1555

Burley, D.M. (1972) Col.2, pp. 329, In: *Phase Transitions and Critical Phenomena*; Academic Press, C. Domb & M.S. Green (Eds.), London, Great Britain

Cheng, A. & Steele, W.A. (1990). Computer Simulation Study of the Chemical Potential of Argon Adsorbed on Graphite. *Mol. Simulation*, Vol. 4, pp. 349-359

Cheng, H.; Cooper, A.C.; Pez, G.P.; Kostov, M.K.; Piotrowski, P. & Stuart, S.J. (2005), Molecular Dynamics Simulations on the Effects of Diameter and Chirality on Hydrogen Adsorption in Single Walled Carbon Nanotubes *J. Phys. B*, Vol. 109, pp. 3780-3786

De Oliveira, M.J. & Griffiths, R.S. (1978). Lattice-gas model of multiple layer adsorption. *Surf. Sci.*, Vol. 71, pp. 687-694

Do, D.D. & Do, H.D. (2005). Evaluation of 1-Site and 5-Site Models of Methane on Its Adsorption on Graphite and in Graphitic Slit Pores. *J. Phys. Chem. B*, Vol. 109, pp. 19288-19295

Dresselhaus, M.S.; Dresselhaus, G.D.; Eklund, P.C. (1996), *Science of Fullerenes and Carbon Nanotubes*, Academic Press (Ed.), New York, USA

Eldridge, A. (1993). Olefin/paraffin separation technology: a review. *Industrial & Engineering Chemistry Research*, Vol. 32, pp. 2208

Fuentes, A.B. & Mendez, I. (2002), Separation and Purification Technology, Vol. 28, pp. 29

Gravelle, P.C.J. (1978), Methods for the determination of heats of adsorption. *Therm. Anal.*, Vol. 14, pp 53-77

Groszek, A.J. (1998). Flow adsorption microcalorimetry *Thermochim. Acta*, Vol. 312, pp. 133-143

Hamilton, J.J. & Goodstein, D.L. (1983),.Thermodynamic study of methane multilayers adsorbed on graphite.*Phys. Rev. B*, Vol. 28, pp. 3838-3848

Harris, P.J.F. (2003). Impact of the Discovery of Fullerenes on Carbon Science, In: *Chemistry and Physics of Carbon*, Vol. 28, Marcel Dekker, Inc., L.R. Radovic, (Ed), pp. 1-40, ISBN 0-8247-0987-X, New York, USA

Herdes, C.; Valente, A.; Lin, Z.; Rocha, J.; Coutinho, J.A.P.; Medina, R.; Vega, L.F. (2007). Selective Adsorption of Volatile Organic Compounds in Micropore Aluminum Methylphosphonate-α: A Combined Molecular Simulation−Experimental Approach. *Langmuir*, Vol. 23, pp. 7299

Jayaraman, A. & Yang, R.T. (2005). Stable oxygen-selective sorbents for air separation . *Chemical Engineering Science*, Vol. 60, pp. 625-634

Kim, H.Y & Steele, W.A. (1992) Computer-simulation study of the phase diagram of the CH_4 monolayer on graphite: Corrugation effects. *Phys. Rev. B*, Vol. 45, pp. 6226-6233

Kostov, M.K.; Cheng, H.; Cooper, A.C. & Pez, G.P. (2002), Influence of Carbon Curvature on Molecular Adsorptions in Carbon-Based Materials: A Force Field Approach. *Phys. Rev. Lett.*, Vol. 89, pp 146105- (1-4)

Kroon, M.C. & Vega, L.F. (2009). Selective Paraffin Removal from Ethane/Ethylene Mixtures by Adsorption into Aluminum Methylphosphonate-α: A Molecular Simulation Study. *Langmuir*, Vol. 25, pp. 2148-2152

Krungleviciute, V; Calvi, M.M.; Wagner, J.A.; Migone, A.D.; Yudasaka, M. & Iijima, S. (2008). Probing the Structure of Carbon Nanohorn Aggregates by Adsorbing Gases of Different Sizes. *J. Phys. Chem.*, Vol. 112, pp. 5742-5746

Krungleviciute, V; Migone, A.D. & Pepka, M. (2009). Characterization of single-walled carbon nanohorns using neon adsorption isotherms. *Carbon*, Vol. 47, pp. 769-774

Kuznetsova, A.; Yates, J.; Liu, J. & Smalley, R.E. (2000) *J. Chem. Phys. 112*, 9590−9598

Llanos, J.L.; Fertita, A.E.; Flores, E.S. & Bottani, E.J. (2003), SO_2 Physisorption on Exfoliated Graphite. *J. Phys. B*, Vol. 107, pp. 8448-8453

Myers, A.L. & Prausnitz, J.M. (1965), Thermodynamics of mixed-gas adsorption.AIChE, J., Vol. 11, pp. 121-127

Nicholson, D. & Parosonage, N.G. (1982), *Computer Simulation and the Statistical Mechanics of Adsorption*, Academic Press (Ed.), London, Great Britain

Niyogi, S.; Hamon, M.A. ; Hu, H. ; Zhao, B. & Bownik, S. (2002), Chemistry of Single-Walled Carbon Nanotubes. *Acc. Chem. Res.*, Vol. 35, pp. 1105-1113

Okamoto, Y. & Miyamoto, Y. (2001), Ab Initio Investigation of Physisorption of Molecular Hydrogen on Planar and Curved Graphenes.*J. Phys. Chem. B*, Vol. 105 (17), pp. 3470-3474

Parr, R.G. & Perdew, J.P. (1989), *Density Functional Theory of Atoms and Molecules*, Oxford University Press (Ed.), Oxford, Great Britain

Pascual, P.; Ungerer, P.; Taviatan, B.; Pernot, P. & Boutin, A. (2003). Development of a transferable guest-host force field for adsorption of hydrocarbons in zeolites I. Reinvestigation of alkane adsorption in silicalite by grand canonical Monte Carlo simulation. *Phys. Chem. Chem. Phys.*, Vol. 5, pp. 3684-3693

Piper, J. & Morrison, J.A. (1984). Heats of adsorption of methane multilayers on graphite *Phys. Rev. B*, Vol. 30, pp. 3486-3489

Sabzyan, H. & Babajani, M. (2005), A semiempirical quantum mechanical study of methane–graphite interaction. *J. Molec. Struct. : THEOCHEM*, Vol. 726, pp. 155-160

Scheele, C.W. (1777). *Chemische Abhandlung fon der Luft un dem Feuer*, Akad.d.Wiessenschaft Mitgliedes, Leipzig, Germany

Steele, W.A. (1973). The physical interaction of gases with crystalline solids: I. Gas-solid energies and properties of isolated adsorbed atoms. *Surf. Sci.*, Vol. 36, pp. 317-352

Steele, W.A. (1974), *The interaction of Gases with Solid Surfaces*, Pergamon Press (Ed.), ISBN 10-0080-177247, Oxford, Great Britain

Talapatra, S. & Migone, A.D. (2002). Adsorption of methane on bundles of closed-ended single-wall carbon nanotubes. *Phys. Rev. B*, Vol. 65, pp. 045416(1-6)

Tasca, R.A.; Calbi, M.M. & Cole, M.W. (2002), Lattice model of gas condensation within nanopores. *Phys. Rev. E*, Vol. 65, pp. 061607(1-9)

Vilaplana, A.F. (2005) Ab initio computational investigation of physisorption of molecular hydrogen on achiral single - walled carbon nanotubes, *J. Chem. Phys.*: 122, 214724-1-7

Wagner, K. (1996). Methane, In: *International thermodynamics tables of the fluids state*, Vol. 13, IUPAC chemical data series, Blacwell Sciece (Ed.), Boston, USA

Wang, Y. & Perdew, J.P. (1991), Correlation hole of the spin-polarized electron gas, with exact small-wave-vector y high-density scaling, *Phys. Rev B*: 44(24), 13298-13307

Wang, Y. & Johnson, J. (1999). Molecular simulation of hydrogen adsorption in single-walled carbon nanotubes y idealized carbon slit pores, *J. Chem. Phys.*: 110 577-586

Wuebbles, D.J. & Hyhoe K. (2000), Atmospheric Methane: Trends and Impacts, In: *Non CO_2 greenhouse gases*, Springer Verlag, J. van Ham (Ed.), Amsterdam, Holland

Yudasaka, M.: Iijima, S. & Crespi, V.H. (2008), Single Wall Carbon Nanohorns and Nanocones, In: *Carbon Nanotubes, Topics Appl. Physics*, 111, 605, Springer Verlag, A. Jorio; G.D. Dresselhaus & M.S. Dresselhaus (Eds.), Berlin, Germany

Zettlemoyer, A.C. & Narayan, K.S. (1966), *Chem. Phys. Carbon*, Vol. 2, pp. 197

6

Electronic States of Graphene-Based Ferromagnets

Masashi Hatanaka
Department of Green and Sustainable Chemistry,
School of Engineering, Tokyo Denki University,
2-2 Kanda Nishiki-cho, Chiyoda-ku, Tokyo
Japan

1. Introduction

Graphenes have attracted many physicists and/or chemists because of the beautiful structure and tunable electronic states. One of the remarkable properties is the high carrier mobility due to the famous Dirac point, in which the effective mass is theoretically zero (Novoselov, et al., 2004, 2005). The motion of electrons is described by pseudo relativistic effect, and new aspects on physics and chemistry of two-dimensional systems have been developed as nano-scale technology. Nowadays, there has been increasing interest of modified graphenes toward thin films, nano particles, adsorbance agents, and so on, as well as possible micro-electronic devises.

On the other hand, magnetism of graphenes is another interesting theme, related to the early studies on so-called graphene ribbons. That is, some graphene derivatives are promising candidates for organic ferromagnets. The magnetic properties depend on topological conditions such as edges, pores, and defects. Toward a new type of ferromagnets, chemical modification of graphene is a highly challenging theme. Paramagnetism of graphitic polymers itself has been theoretically predicted, relating to the edge states, pores, and defects. However, it was not until band structures of modified graphenes revealed the existence of flat bands at the frontier levels that robust ferromagnetism has been highly expected from graphene-based skeletons. The edges, pores, and defects in these systems should be ordered so as to cause completely flat bands in the Hückel level. Hückel analysis on modified graphenes gives a good perspective toward the flat-band ferromagnetism. In this chapter, graphene-based ferromagnetism is analyzed by crystal orbital method. Recent advances in magnetic graphenes are reviewed in view of their electronic states.

2. Methylene-edged graphenes

It is well known that graphene ribbons with peculiar type of edges have polyradical character, of which flat bands cause ferromagnetic interactions. Fig. 1 shows three types of graphene ribbons. Fig. 1a is the famous graphene ribbons with two-sided acene (zigzag) edges. The magnetic ordering has been predicted based on the band structures. The HOCO (highest occupied crystal orbital) and LUCO (lowest unoccupied crystal orbital) contact and become flat at the wavenumber region $|k|>2\pi/3$ (Fujita et al., 1996). Thus, at least within

the same edge, ferromagnetic interactions are expected. However, total magnetism in the acene-edged graphene ribbons is probably small, because spin moments at both sides tend to cancel each other. This is intuitively deduced from the conventional Heisenberg model. To avoid the spin-cancellation, hydrogen, oxygen and fluorine passivations have been suggested (Kusakabe & Maruyama, 2003; Maruyama & Kusakabe, 2004).
Instead, Klein suggested methylene-edged graphene ribbons shown in Fig. 1b (Klein, 1994; Klein & Bytautas, 1999). Nowadays, these are called Klein edges. At least within the same edge, the Klein-edged graphenes are also expected to show ferromagnetic interactions due to the flatness of frontier bands at $|k|<2\pi/3$ (not $|k|>2\pi/3$). In this case, however, spin cancellation is also expected due to spin alternation in the bipartite lattice. While two-sided Klein edges cause spin cancellation, one-sided Klein edges shown in Fig. 1c are expected to show robust ferromagnetic interactions, because ferrimagnetic spin-polarized structures can be drawn by classical valence-bond theory. Although some studies predicted the antiferromagnetic states of the one-sided Klein edges with twisted or partially substituted methylene groups (Maruyama & Kusakabe, 2004), conservation of the planarity probably makes them ferromagnetic non-Kekulé polymers, and the electronic states are described by Wannier functions rather than conventional Bloch functions (Hatanaka, 2010a). Then, there appears one non-bonding band at the frontier level. The non-bonding band is completely flat within the Hückel approximation. Each Wannner function spans common atoms between the adjacent cells. This is necessary and sufficient condition for ferromagnetic interactions in conjugated π systems, similar to triplet biradicals. Non-cancelled spin alignment and itinerant ferromagnetism are expected due to the flat band.

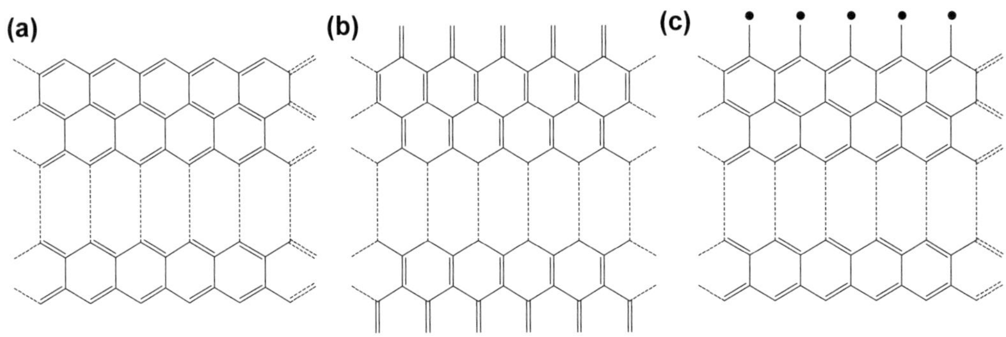

Fig. 1. Graphenes with (a) acene (zigzag)-, (b) two-sided Klein-, and (c) one-sided Klein edges.

For simplicity, we first consider a small non-Kekulé polymer shown in Fig. 2. There are non-bonding crystal orbitals (NBCOs) at the frontier level. They are completely degenerate under Hückel approximation. Each NBCO can be transformed into Wannier functions localized around each cell, as formulated below.
We consider Bloch functions corresponding to the NBCO band:

$$\varphi_k = \frac{1}{\sqrt{N}} \sum_{\mu}^{N} \sum_{r}^{cell} \exp(\ ik\mu) C_r(k) \chi_{\mu,r}, \tag{1}$$

where the wavenumber k runs from $-\pi$ to π, μ is the cell index, N is the number of cells, and r is the index of atomic orbitals. Here we adopt the real part of $C_r(k)$:

$$C'_r(k) = \frac{1}{2}\{C_r(k) + C_r(k)^*\} = \frac{1}{2}\{C_r(k) + C_r(-k)\}. \tag{2}$$

This procedure minimizes the exchange integral of the system. The Wannier function localized at the ν-th call is expressed by Equations (3) and (4):

$$\psi_\nu = \sum_{\mu}^{N} \sum_{r}^{cell} a_r(\mu - \nu)\chi_{\mu,r}, \tag{3}$$

$$a_r(\mu - \nu) = \frac{1}{2\pi}\int_{-\pi}^{\pi} \exp\{i(\mu - \nu)k\}C'_r(k)dk, \tag{4}$$

where

$$\tau = \mu - \nu. \tag{5}$$

Each Wannier coefficient is a function of the integer τ, which represents the difference from the ν-th cell. The Wannier functions should be normalized using a proper normalization factor, because the linear combinations of Bloch coefficients in Equation (2) are not always normalized to unity. Under Hückel approximation, the normalization factor is:

$$C' = \frac{1}{\sqrt{\sum_{\tau}\sum_{r} |a_r(\tau)|^2}}. \tag{6}$$

Wannier functions in the original paper are not normalized (Hatanaka, M. & Shiba, R., 2007; Hatanaka, 2010a). In the present review, however, we adopt the renormalized Wannier functions, and the exchange integrals are recalculated. Since each Wannier function coefficient $a_r(\tau)$ localizes at one or a few unit cells around the ν-th cell, $a_r(\tau)$ with $|\tau|\geq 2$ can be ignored. Then, the exchange integral K_{ij} between the i-th and j-th Wannier functions is nontrivial only when $|i-j| = 1$:

$$\begin{aligned} K_{ij} &= \iint \psi_i(1)\psi_j(1)\frac{e^2}{r_{12}}\psi_i(2)\psi_j(2)d\tau_1 d\tau_2 \\ &= \sum_r \sum_s \sum_t \sum_u a_r(i)a_s(j)a_t(i)a_u(j)(rs|tu) \\ &\cong \left\{ \sum_r^{i-th\ cell}\{a_r(0)^2 a_r(-1)^2\} + \sum_r^{j-th\ cell}\{a_r(0)^2 a_r(1)^2\}\right\}(rr|rr) \\ &\propto 2\sum_r^{i-th\ cell}\{a_r(0)^2 a_r(1)^2\}, \end{aligned} \tag{7}$$

where $(rs\,|\,tu)$ denotes the electron-repulsion integrals, and only one-centered integrals were taken into account in the approximation. The last expression results from Equation (2). From Equation (7), the exchange integrals of non-bonding extended systems are deduced from the amplitude pattern of their Wannier functions. The Bloch and Wannier functions can be calculated by usual secular equations and numerical integrals (Hatanaka, M. & Shiba, R., 2007).

The ν-th Wannier function ψ_ν is schematically shown in Fig. 2. In general, Wannier functions decay rapidly with increase in distance from the central cell. However, in this case, we see that the ν-th and (ν+1)-th Wannier functions span common atoms between the adjacent region. Then, the two-electron wavefunction $\psi_\nu\psi_{\nu+1}$ contains so-called ionic terms, in which two electrons are accommodated to the same atomic orbitals. While simultaneous occupation of two electrons with antiparallel spins is allowed in the same atomic orbital, occupation of electrons with parallel spins is forbidden due to the Pauli principle. Thus, in this system, high-spin state is preferred rather than low-spin state due to reduction of Coulomb repulsion.

Klein-edged graphenes with n ladders are similarly analyzed as non-Kekulé polymers, as shown in Fig. 3. In these systems, amplitudes of the Wannier functions are mainly spread at upper and lower edges. When the number of ladders n increases, the amplitudes inside the graphene plane become small, and highly localize at the edges. Nevertheless, even if the number of ladders becomes infinite, each Wannier function spans common atoms between the adjacent cells, and thus, the ferromagnetic interactions are always positive. This means that the high-spin stabilities are attributed to itinerant character of edge states, which is described by the Wannier functions. Nowadays, it has been theoretically shown that the Wannier functions in such degenerate systems should be symmetric with respective to the central cell, and minimize the exchange integral of the systems (Hatanaka, 2011). Then, the exchange integral is calculated from square of the amplitude pattern of the two-electron wavefunction $\psi_\nu\psi_{\nu+1}$, as schematically shown in Fig. 4. The exchange integral K versus the number of ladders n is shown in Fig. 5. The exchange integral is calculated from the Wannier coefficients, taking count of one-centered integrals. The exchange integral converges to a positive value, and thus, the system is expected to be ferromagnetic in the balk order. For chemical stability, carbonyl edges isoelectonic with methylene groups are promising.

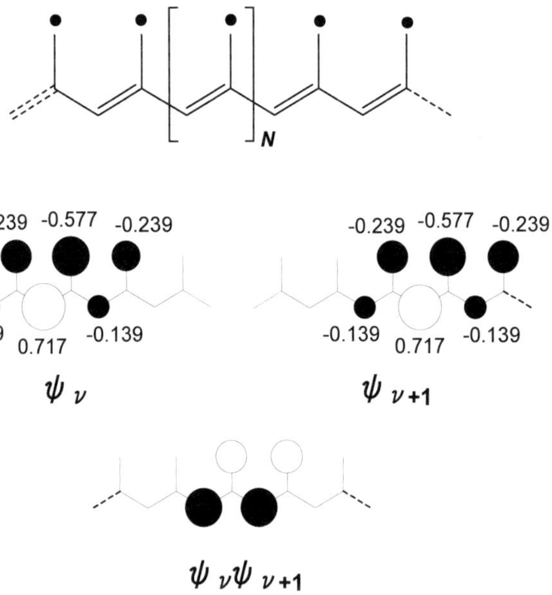

Fig. 2. Wannier functions of the simplest non-Kekulé polymer. This is a simple model for Klein-edged graphenes.

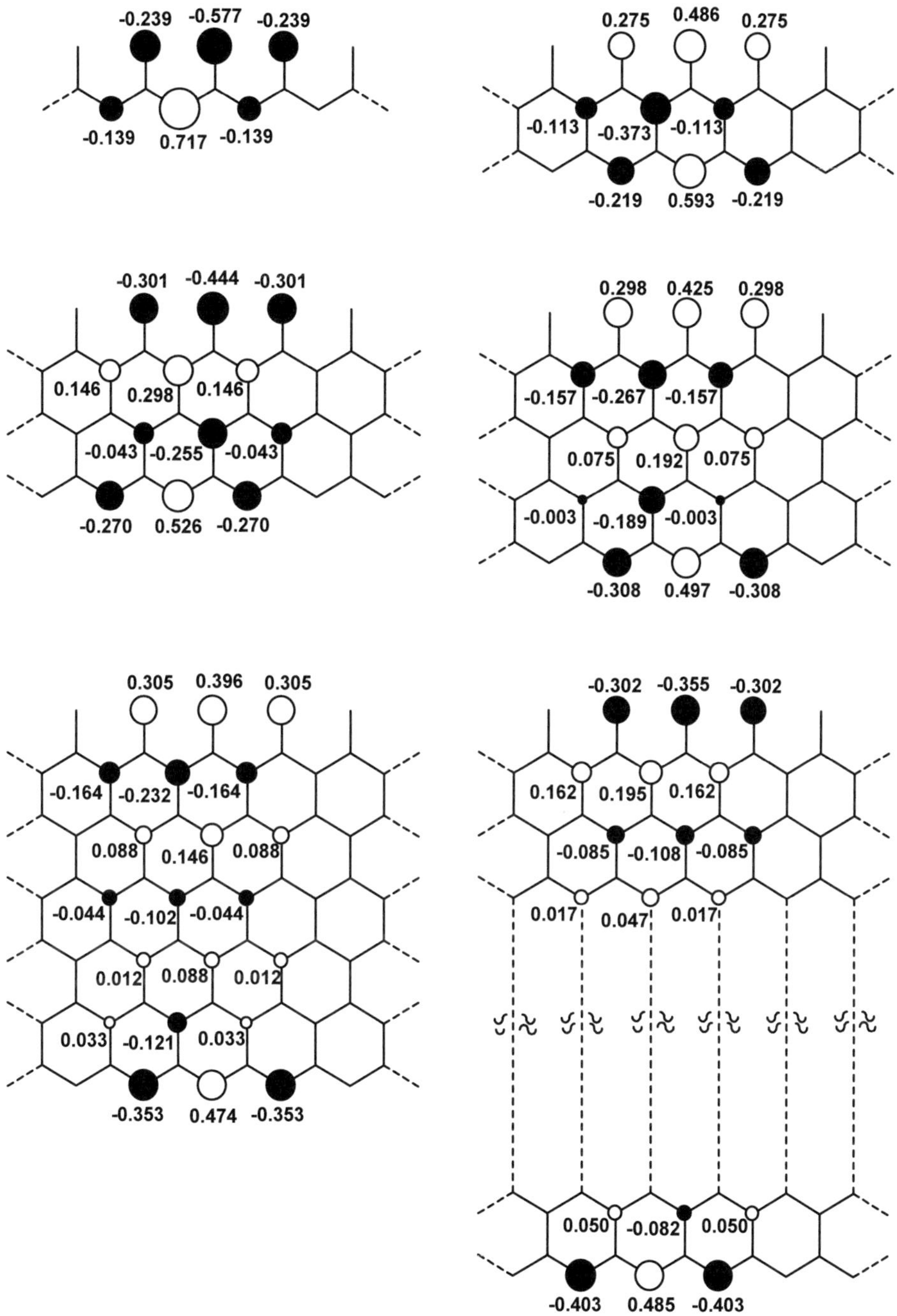

Fig. 3. Wannier functions of one-sided Klein edges with the number of ladders n=0, 1, 2, 3, 5, and 10.

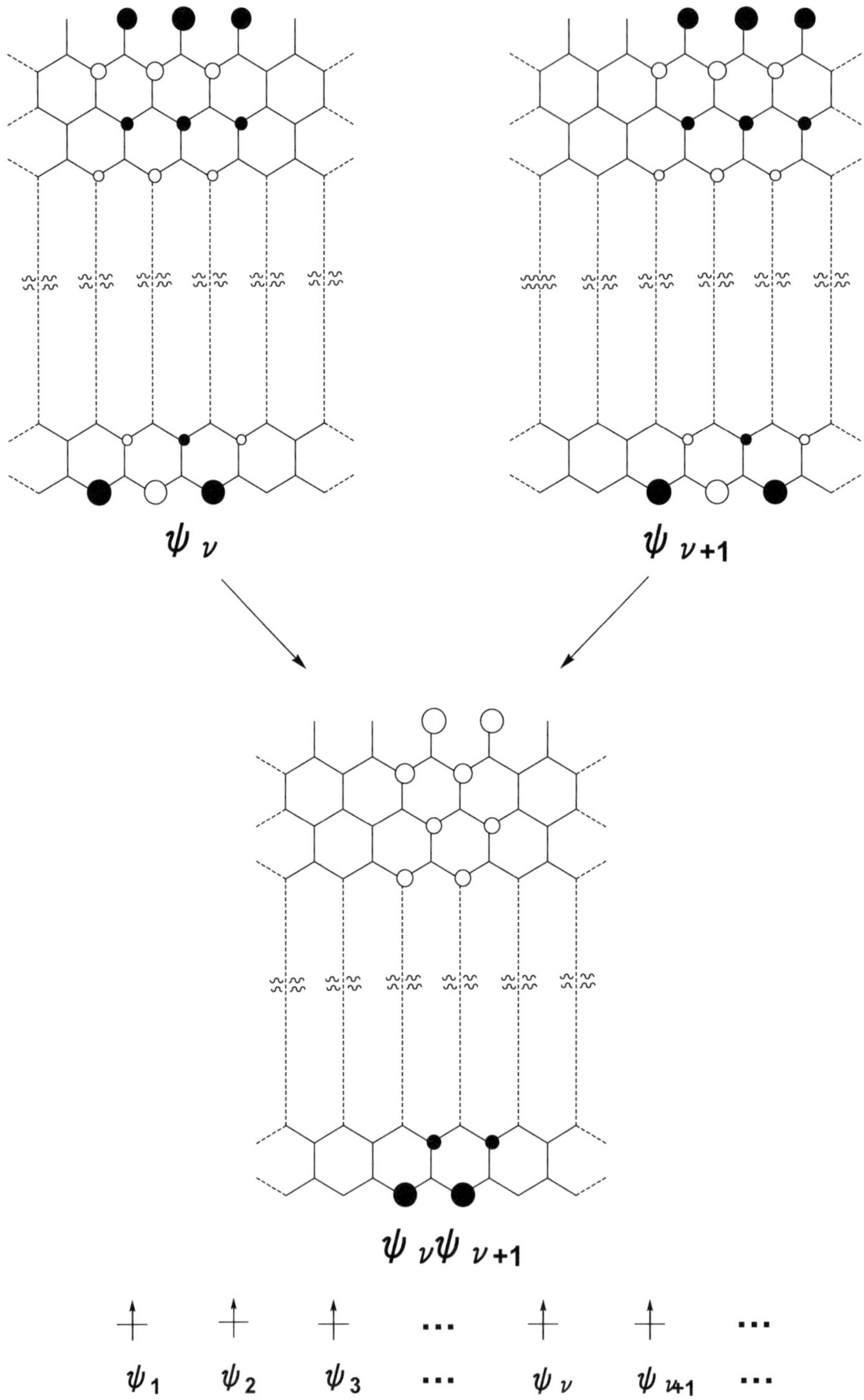

Fig. 4. Construction of two-electron wavefunstions from adjacent Wannier functions of one-sided Klein edges.

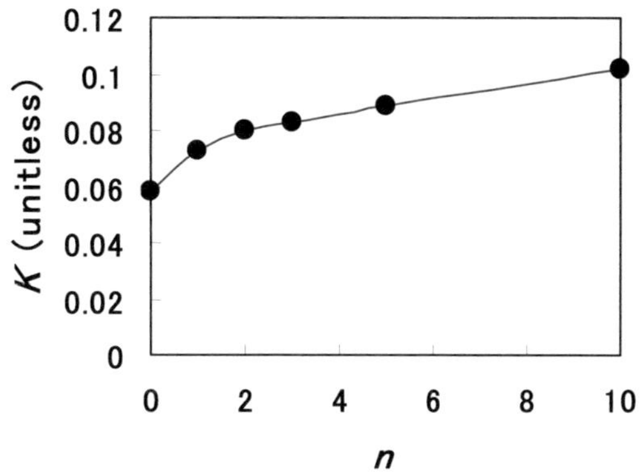

Fig. 5. Exchange integral K of one-sided Klein-edges per unit versus the number of ladders n.

3. Porous graphenes

Fig. 6 shows porous graphene. This compound was synthesized as a two-dimensional nanostructure (Bieri et al., 2009). They synthesized porous graphene by aryl-aryl coupling reactions on Ag (111) surface, and observed STM (scanning tunneling microscope) image of the honeycomb structures. In 2010, band structures of porous graphenes were investigated by several workers (Du et al., 2010; Hatanaka, 2010b; Li et al., 2010). The Hückel-level dispersion and DOS (density of states) are shown in Figs. 6 and 7 (Hatanaka, 2010b). The dispersion suggests semi-conductive band gaps, and the frontier bands are so flat as to be available for ferromagnetism. It is interesting that both HOCO and LUCO become flat at all the wavenumber region. Thus, this material is expected to show ferromagnetism when it is oxidized or reduced by proper dopants.

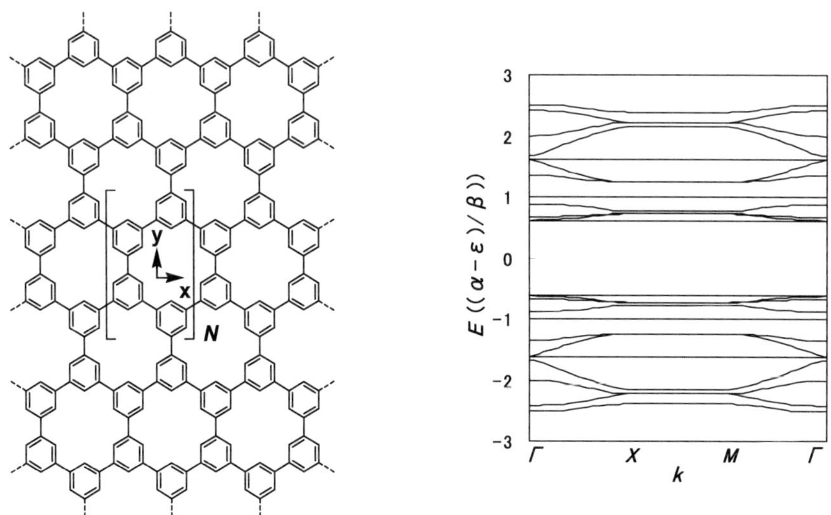

Fig. 6. Porous graphene and the dispersion based on Hückel level of theory.

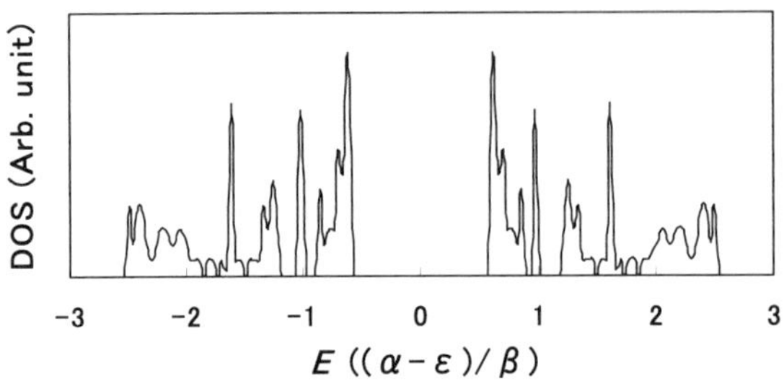

Fig. 7. DOS (density of states) of porous graphene based on Hückel level of theory.

The flat bands result from nodal character of phenylene units. Fig. 8 shows amplitude patterns of selected Hückel molecular orbitals of cyclohexa-*m*-phenylene, which is considered as a unit group in the porous graphene. We see that the HOMO and LUMO have nodes at each phenylene unit. Orbital interactions between node-node linkages are zero within the Hückel approximation. Thus, both HOCO and LUCO in porous graphene become flat at the frontier levels. It is interesting that eigenvalues ε of HOMO (α+0.618β) and LUMO (α-0.618β) in cyclohexa-*m*-phenylene are identical to those of butadiene. Indeed,

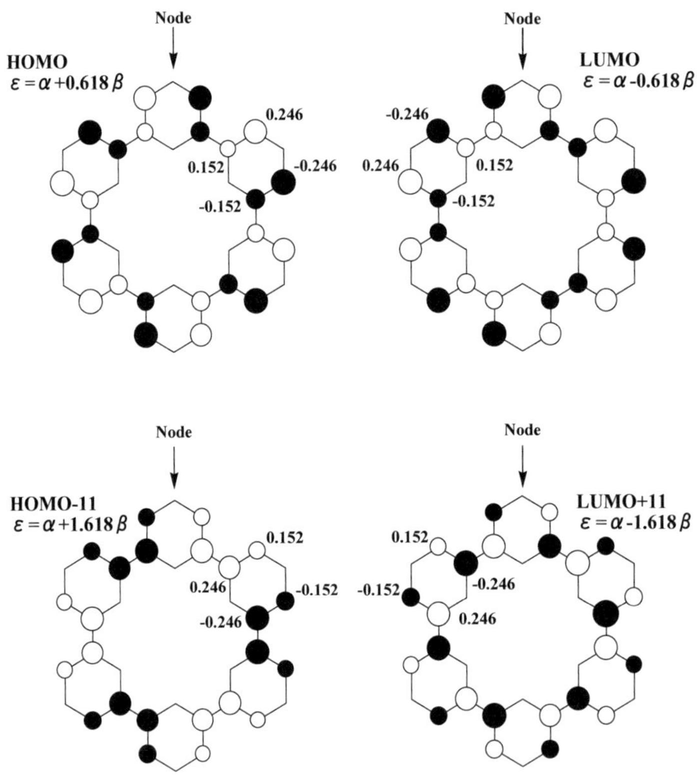

Fig. 8. Selected molecular orbitals of cyclohexa-*m*-phenylene. There are nodes at the peripheral sites. α and β are Coulomb and resonance integral, respectively.

apart from the normalization factors, amplitude patterns of butadiene's HOMO and LUMO spread between each phenylene unit. Moreover, eigenvalues of HOMO-11 (α+1.618β) and LUMO+11(α-1.618β) of cyclohexa-*m*-phenylene are identical to those of the first and fourth orbitals in butadiene. These orbitals also cause flat bands in porous graphene at eigenvelues (α±1.618β). In porous graphene, additional flat bands coincidently appear at eigenvalues (α±1.000β). These bands come from nodal character of e_{1g} orbitals in each benzene fragment.

Porous graphene ribbons with various edges are also of interest, despite the coupling directions are not unique and some defects of the coupling reactions may cause diversity of molecular-weight distribution. Fortunately, it has been proved that the frontier non-bonding level of any porous oligomer is invariant with respect to molecular weight and/or coupling direction due to the zero-overlap interactions of the phenylene units. Thus, porous graphenes including porous ribbons are promising precursors toward organic ferromagnets.

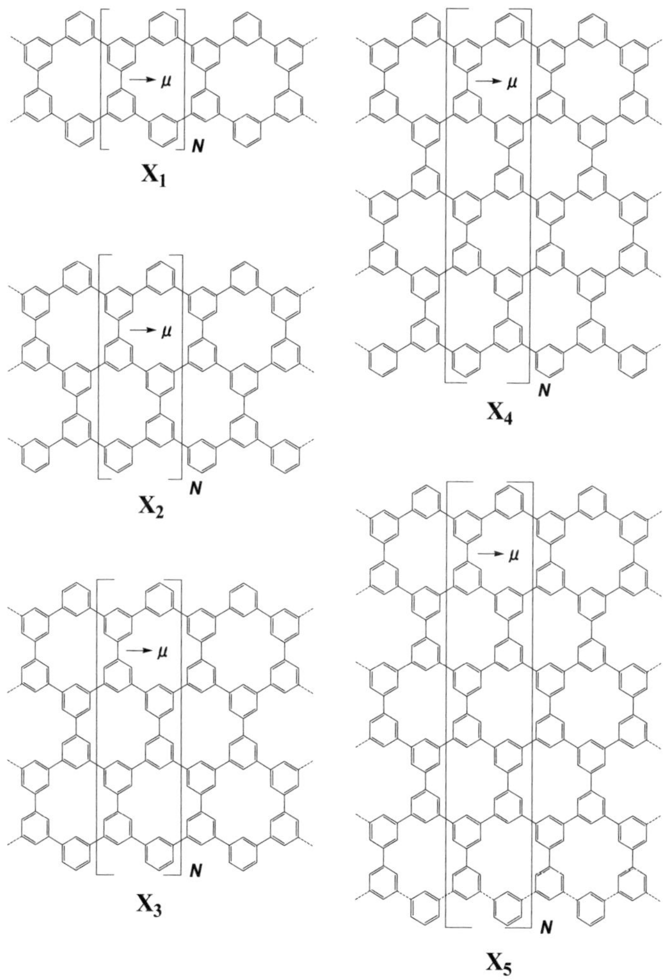

Fig. 9. Porous graphene ribbons $\mathbf{X}_n$ (n=1-5) with n ladders cut along the x axis. μ is the lattice vector.

Porous structures including boron and/or nitrogen are also interesting in that the hetero atoms serve as dopants, which increase or decrease the number of electrons in the frontier levels.

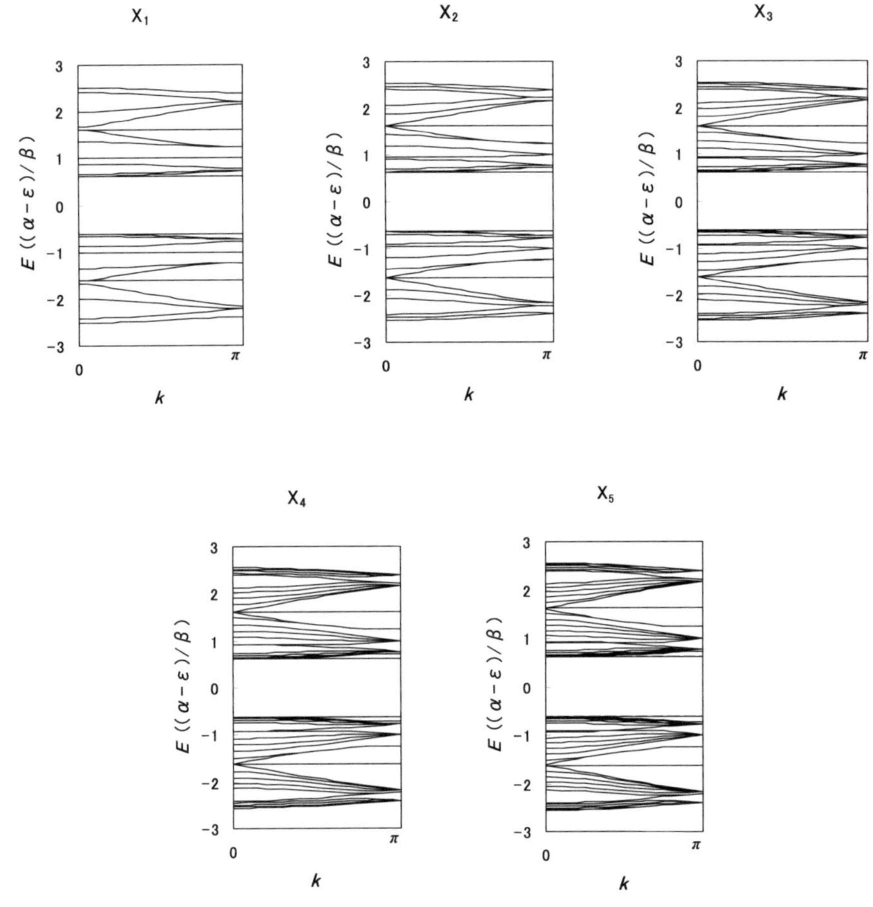

Fig. 10. Dispersions of $\mathbf{X}_n$ (n=1-5) under Hückel level of theory.

Fig. 9 shows porous graphene ribbons cut along the x axis. The systems are classified by the number of porous ladders n, as denoted as $\mathbf{X}_1$-$\mathbf{X}_5$. The edges have bay areas consisting of seven carbon atoms, and resemble acene edges of conventional graphene. Fig. 10 shows dispersion of $\mathbf{X}_1$-$\mathbf{X}_5$. We see that the HOCOs and LUCOs are completely flat, similar to the two-dimensional porous graphene. Flat bands at ($\alpha\pm1.618\beta$) are also conserved due to the nodal character of phenylene units. Interestingly, HOCOs and LUCOs in each system are n-fold degenerate, respectively. That is, HOCO and LUCO in $\mathbf{X}_1$ are single flat bands, and HOCO and LUCO in $\mathbf{X}_2$ are doubly degenerate. Similarly, three-, four-, and five-fold degenerate HOCOs and LUCOs appear in $\mathbf{X}_3$, $\mathbf{X}_4$, $\mathbf{X}_5$, respectively. These degeneracies come from the systematic increase of the porous ladders. The resultant flat bands are also available for possible ferromagnetic interactions.

Fig. 11 shows porous graphene ribbons cut along the y axis. The systems are also classified by the number of porous ladders n, as denoted as $\mathbf{Y_1}$-$\mathbf{Y_5}$. The edges have bay areas consisting of twelve carbon atoms, and resemble so-called phenanthrene edges of conventional graphene. Fig. 12 shows dispersion of $\mathbf{Y_1}$-$\mathbf{Y_5}$. The HOCOs and LUCOs are completely flat, similar to $\mathbf{X_1}$-$\mathbf{X_5}$. Flat bands at ($\alpha \pm 1.618\beta$) also appear. HOCOs and LUCOs in each system are also n-fold degenerate, respectively.

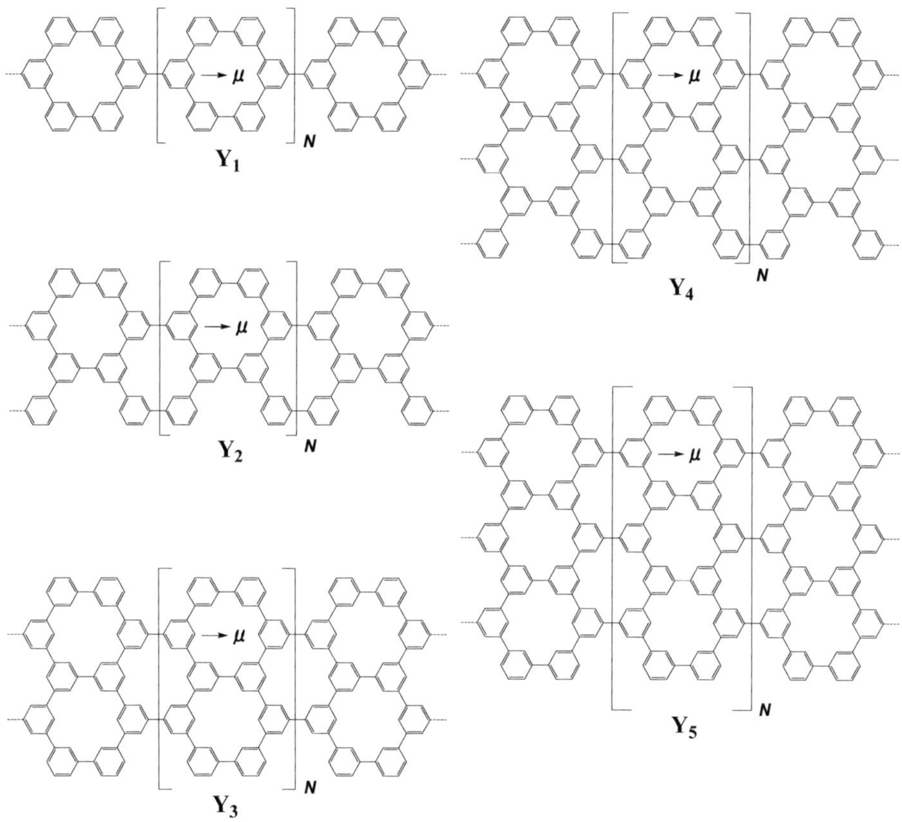

Fig. 11. Porous graphene ribbons $\mathbf{Y}_n$ (n=1-5) with n ladders cut along the y axis. $\boldsymbol{\mu}$ is the lattice vector.

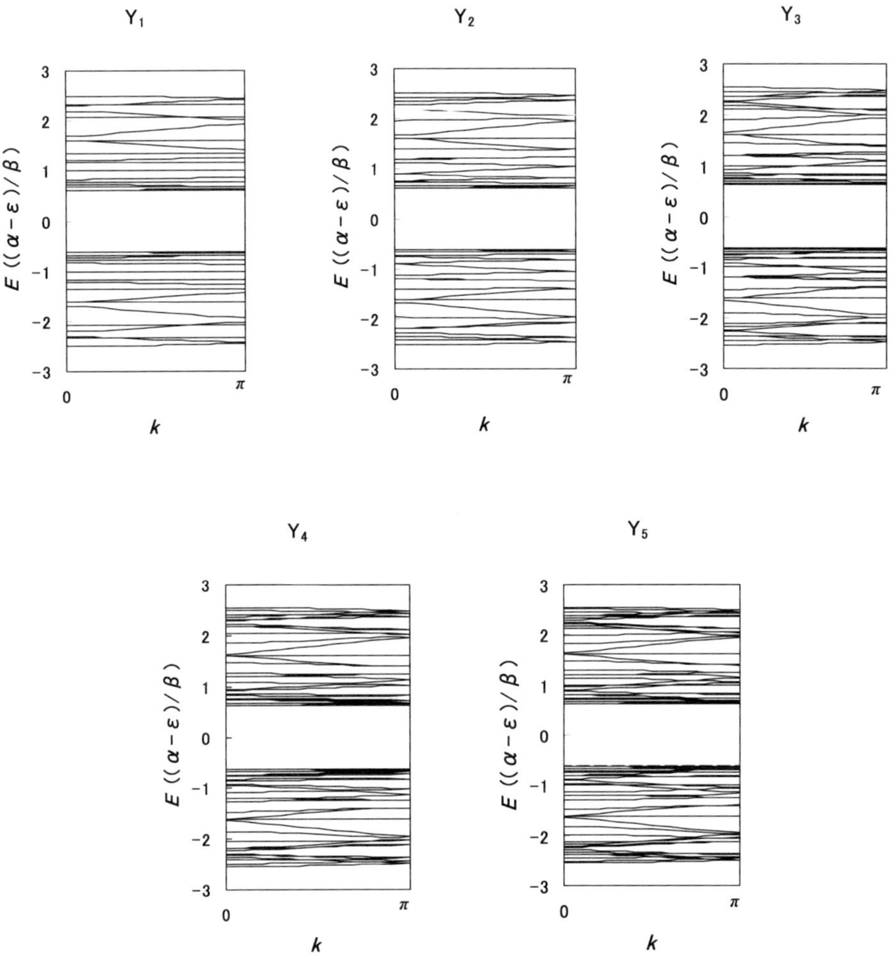

Fig. 12. Dispersions of $\mathbf{Y}_n$ (n=1-5) under Hückel level of theory.

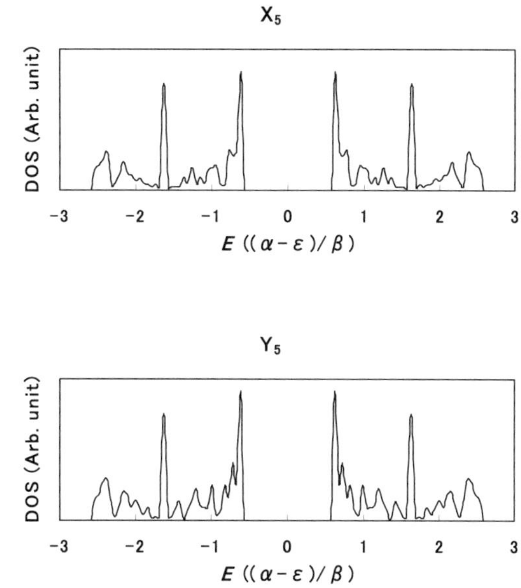

Fig. 13. DOS of $\mathbf{X_5}$ and $\mathbf{Y_5}$ under Hückel level of theory.

Fig. 13 shows DOS of $\mathbf{X_5}$ and $\mathbf{Y_5}$. There are main peaks corresponding to the flat bands. The largest peaks are due to the systematic degeneracies of HOCOs and LUCOs, and thus, photoelectron spectroscopy experiments will give a major peak at ca. 9.1 eV. Theoretical predictions on the magnetism of porous graphenes await experimental confirmations.

4. Defective graphenes

Recently, ferromagnetism of HOPG (highly oriented pyrolytic graphite) was found at room temperature (Červenka, et al., 2009; Esquinazi & Höhne, 2005; Mombrú et al., 2005). HOPGs often have defects, which form grain boundaries between the polycrystals. Laterally and longitudinally slipped defects in acene-edged HOPGs are particularly interesting in that the resultant defects are analyzable by graphene-based model (Hatanaka, 2010c). Fig. 14 shows definition of defects, in which the lateral and longitudinal displacements are represented by Δx and Δz. Fig. 15 shows extended-Hückel dispersion of acene-edged graphenes with laterally and longitudinally slipped defects. The number of honeycomb ladders is fixed to be 19. The bold lines and dotted lines represent π and σ bands, respectively. In laterally slipped defects ($\Delta x > 0$), flatness of the frontier bands is lost with increase in the displacements, and the resultant double occupation of electrons leads to antiferromagnetic ground state. On the other hand, in longitudinally slipped defects ($\Delta z > 0$), flatness of the frontier bands is conserved even if the displacement becomes infinity, and the resultant electronic state resembles that of polycarbene, in which π and σ spins coexist with ferromagnetic interactions.

Here we consider simple models of defects by using graphene ribbons with 3 ladders. Fig. 16 shows change of orbital pattern for the laterally slipped graphenes. We note that the effective displacement of lateral slip is smaller than the lattice period a. Frontier Bloch functions (HOCO and LUCO at $k = \pi/a$) of non-distorted graphene is localized at the edges (Fig. 16a), and contact at $k = \pi/a$. π-Frontier Bloch functions in laterally slipped graphene are doubly degenerate and also contact at $k = \pi/a$ under all the displacements (Fig. 16b). The amplitudes are localized inside the fault as well as at the acene edges. With increase in the lateral displacement, there appears a peculiar σ band, of which amplitude is shown in Fig. 16c. We see that this band consists of weak interacted dangling bonds in the slipped fault. The weak interaction causes crossing of frontier bands, as shown in the middle of Fig. 15, and thus, the magnetism due to acene-edged graphenes is considered to disappear.

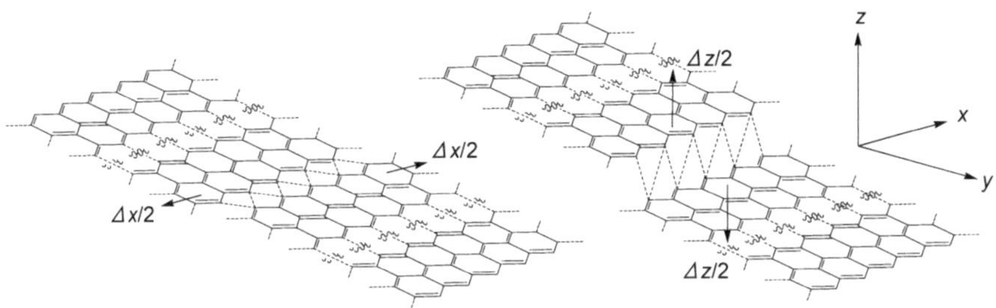

Fig. 14. Laterally and longitudinally slipped defective graphenes.

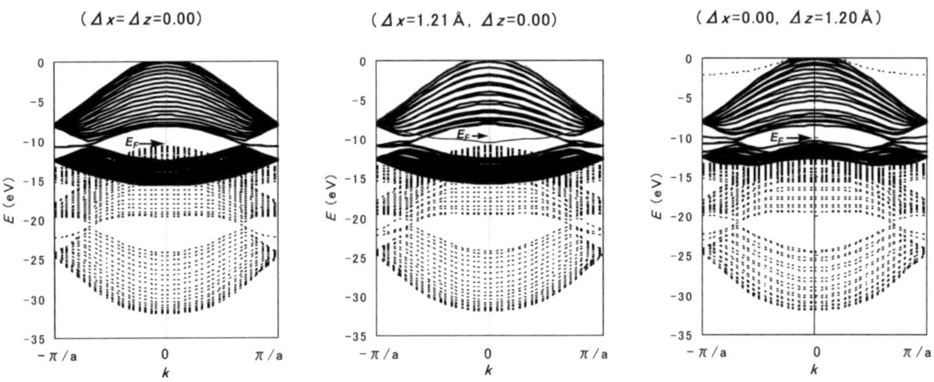

Fig. 15. Dispersion of defective graphenes based on extended Hückel level of theory. E_F is the Fermi levels.

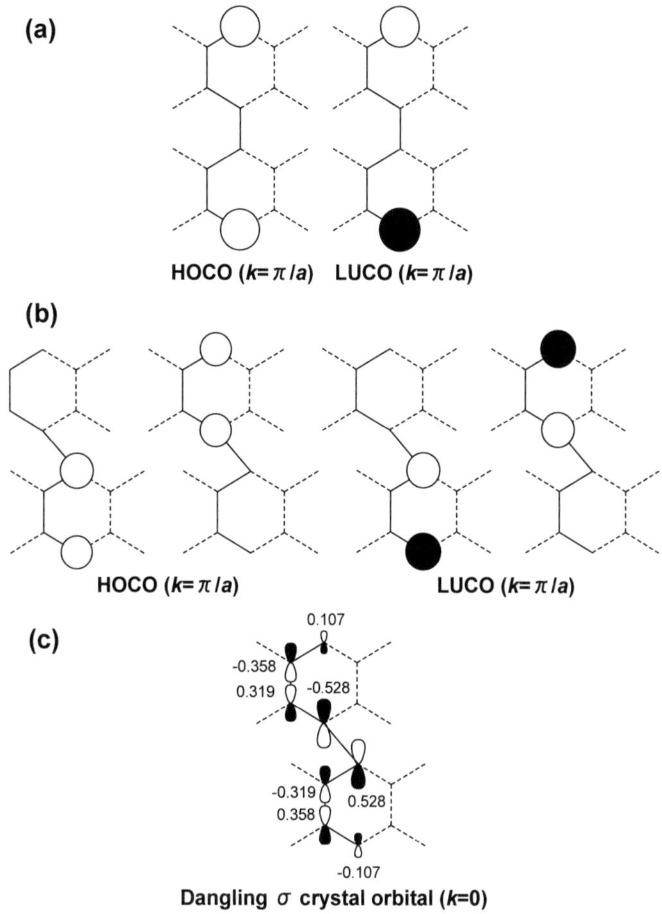

Fig. 16. Selected Bloch functions of acene-edged graphene with laterally slipped faults. (a) Δx=0, HOCO and LUCO at k=π/a, (b) Δx=1.21 Å, HOCOs and LUCOs at k=π/a, and (c) Δx=1.21 Å, dangling σ crystal orbital at k=0.

Fig. 17 shows π-frontier Bloch functions of longitudinally slipped graphene. Contrary to the laterally slipped graphene, amplitudes of the HOCO and LUCO always spread at the edges, and thus, their dispersions contact at k= π/a (Fig. 17a). When the longitudinal displacement becomes large, carbene-type σ bands also appear at the frontier level (Fig. 17b). These orbitals accommodate magnetic electrons of the carbene sites, and thus, longitudinally displacement causes another type of magnetism within the fault. This is considered to be the reason why HOPGs often exhibit weak ferromagnetism. When the displacement is small, it has been shown that energy cost for longitudinal slip is smaller than that of lateral slip (Hatanaka, 2010c), and the energy cost is perhaps compensated by the high-spin stability.
Thus, as rough estimation, we guess that ferromagnetism observed in HOPGs is attributed to longitudinally slipped defects, which form grain boundaries including multilayers of graphene planes. Indeed, topographies corresponding to the longitudinally slipped defects were recorded in HOPGs by AFM (atomic force microscope) and MFM (magnetic force microscope) techniques (Červenka et al., 2009).

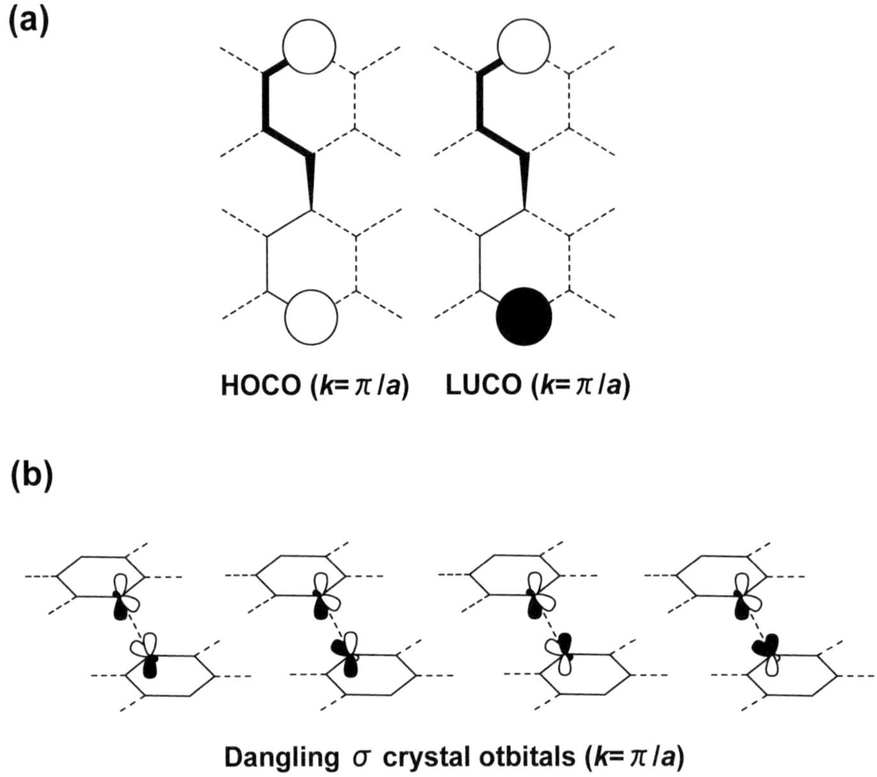

Fig. 17. Selected Bloch functions of acene-edged graphene with longitudinally slipped faults. (a) Δz=1.20 Å, HOCO and LUCO at $k=\pi/a$, and (b) $\Delta z\rightarrow\infty$, dangling σ crystal orbitals at $k=\pi/a$.

5. Conclusion

Some graphene-based ferromagnets were analyzed in view of their electronic states. There appear flat bands at the frontier levels, in which the Wannier functions span common atoms between the adjacent cells. If geometry conditions such as edges, pores, and defects are well controlled by chemical modification, graphene-based ferromagnets will be realized through the flat bands. Amplitudes of the Wannier functions have non-bonding character, and the frontier electrons are itinerant around each central cell. The degeneracy of frontier flat bands and the positive exchange integrals play key roles for ferromagnetic interactions of graphene-based ferromagnets.

6. Acknowledgements

Thanks are due to Elsevier B. V. for permission to use figures in the original papers (Hatanaka 2010a, 2010b, 2010c).

7. References

Bieri, M; Treier, M; Cai, J; Aït-Mansour, K; Ruffieux, P; Gröning, O; Gröning, P; Kastler, M; Rieger, R; Feng, X; Müllen, K & Fasel, R. (2009). Porous graphenes: two-dimensional polymer synthesis with atomic precision. *Chemical Communications*, pp. 6919-6921

Červenka, J; Katsnelson, M. I & Flipse, C. F. J. (2009). Room-temperature ferromagnetism in graphite driven by two-dimensional networks of point defects. *Nature Physics*, Vol. 5, pp. 840-844

Du, A.; Zhu, Z & Smith, S. C. (2010). Multifunctional porous graphene for nanoelectronics and hydrogen storage: New properties revealed by first principle calculations. *Journal of the American Chemical Society*, Vol. 132, pp. 2876-2877

Esquinazi, P & Höhne, R. (2005). Magnetism in carbon structures. *Journal of Magnetism and Magnetic Materials*, Vol. 290-291, pp. 20-27.

Fujita, M; Wakabayashi, K; Nakada, K & Kusakabe, K. (1996). Peculiar localized state at zigzag graphite edge. *Journal of the Physical Society of Japan*, Vol. 65, pp.1920-1923

Hatanaka, M. (2010a). Wannier analysis of magnetic graphenes. *Chemical Physics Letters*, Vol. 484, pp. 276-282

Hatanaka, M. (2010b). Band structures of porous graphenes. *Chemical Physics Letters*, Vol.488, pp. 187-192

Hatanaka, M. (2010c). Band structures of defective graphenes. *Journal of Magnetism and Magnetic Materials*, Vol. 323, pp. 539-546

Hatanaka, M. (2011). Stability criterion for organic ferromagnetism. *Theoretical Chemistry Accounts*, Vol. 129, pp. 151-160

Hatanaka, M. & Shiba, R. (2007). Ferromagnetic interactions in non-Kekulé polymers. *Bulletin of the Chemical Society of Japan*, Vol. 80, pp. 2342--2349

Klein D. J. (1994). Graphitic polymer stripes with edge states. *Chemical Physics Lettes*, Vol. 217, pp. 261-265

Klein D. J. & Bytautas L. (1999). Graphitic edges and unpaired π-electron spins. *The Journal of Physical Chemistry A*, Vol. 103, pp. 5196-5210

Kusakabe, K & Maruyama, M. (2003). Magnetic nanographite. *Physical Review B*, Vol. 67, pp. 092406-1-4

Li, Y.; Zhou, Z.; Shen, P. & Chen, Z. (2010). Two-dimensional polyphenylene: experimentally available porous graphene as a hydrogen purification membrane. *Chemical Communications*, pp. 3672-3674

Maruyama, M & Kusakabe, K. (2004). Theoretical predication of synthesis methods to create magnetic nanographite. *Journal of the Physical Society of Japan*, Vol. 75, pp.656-663

Mombrú, A. W; Pardo, H; Faccio, R; de Lima, O. F; Leite, E. R; Zanelatto, G; Lanfredi, A. J. C; Cardoso, C. A. & Araújo-Moreira, F. M. (2005). Multilevel ferromagnetic behavior of room-temperature bulk magnetic graphite. *Physical Review B*, Vol. 71, pp.100404(R)-1-4

Novoselov, K. S; Geim, A. K; Morozov, S. V; Jiang, D; Zhang, Y; Dubonos, S. V; Grigorieva, I. V. & Firsov, A. A. (2004). Electronic field effect in atomically thin carbon films. *Science*, Vol. 306, pp. 666-669

Novoselov, K. S; Geim, A. K; Morozov, S. V; Jiang, D; Katsnelson, M. I; Grigorieva, I. V.; Dubonos, S. V. & Firsov, A. A. (2005). Two-dimensional gas of massless Dirac fermions in graphene. *Nature*, Vol. 438, pp. 197-200

7

Nonlinear Transport Through Ultra Narrow Zigzag Graphene Naoribbons

Hosein Cheraghchi
Damghan University
Iran

1. Introduction

Graphene, a monolayer of a honeycomb lattice of carbon atoms has been attracted a great amount of attention from both experimental and theoretical points of view Novoselov et al. (2006). Flat structure of graphene makes its fabrication more straightforward than carbon nanotubes. Moreover, dreams of carbon nanoelectronic approach to the reality based on planar graphene structures. This structure overcomes some difficulties of nanoelectronics based on carbon nanotubes, by using lithography, one-dimensional ribbon patterns on graphene sheets Liu et al. (2009). Experiments in graphene-based devices Ozyilmaz et al. (2007) have shown the possibility of controlling their electrical properties by the application of an external gate voltage. For achieving realistic nanoelectronic applications based on graphene nanoribbons (GNR), width of ribbon have to be narrow enough that a transport gap is opened Han el al. (2007); Li et al. (2008); Wang et al. (2008). Using a chemical process, sub-10 nm GNR field-effect-transistors with very smooth edges have been obtained in Ref.[Li et al. (2008); Wang et al. (2008)] and demonstrated to be semiconductors with band-gap inversely proportional to the width and on/off ratio of current up to 10^6 at room temperature. By connecting GNRs with different types of edges and widths, it is applicable to fabricate electronic devices based on graphene nanoribbons.

The origin of transport gap which is opened in a gate voltage region of suppressed nonlinear conductance is still not well understood Molitor et al. (2009); Son et al. (2006); Sols et al. (2007). Two factors are responsible for transport gap: the edge disorder leading to localization Mucciolo et al. (2009) and the confinement Nakada et al. (1996); Brey & Fertig-a (2006); Bery&Fertig-b (2006); Zheng et al. (2007); Malysheva& Onipko (2008). However, in nonlinear regime, transport gap is also opened by transition selection rules which originates from the reflection symmetry Duan et al. (2008).

Based on the tight-binding approach, GNRs with armchair shaped edges are either metal or Semiconductor Son et al. (2006); Nakada et al. (1996); Brey & Fertig-a (2006); Bery&Fertig-b (2006); Zheng et al. (2007). Moreover, in this approach, zigzag edge ribbons are metal regardless of their widths Malysheva& Onipko (2008). While *ab initio* calculations Son et al. (2006) predict that regardless of the shape of the edges, GNRs are semiconductor. In zigzag GNRs, the bands are partially flat around the Fermi energy, which means that the group velocity of conduction electrons is close to zero. Their transport properties are dominated by edge states.

Similar to carbon nanotubes, electronic transition through a ZGNRs follows from some selection rules. The rotational symmetry of the incoming electron wave function with respect

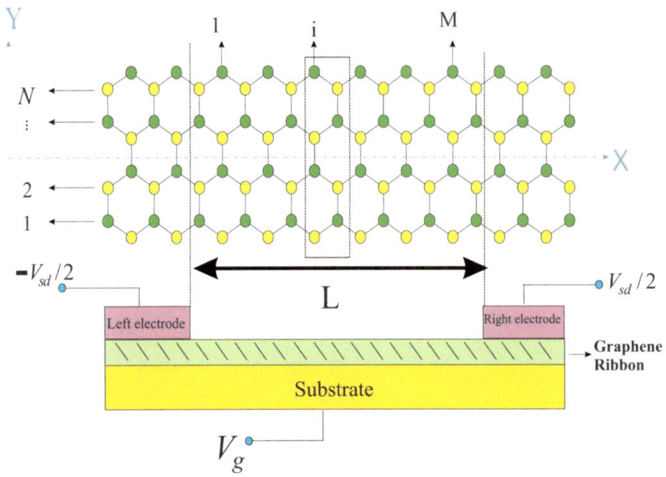

Fig. 1. Gated Zigzag graphene nanoribbon which is divided into three regions: left, right and central region. Dotted rectangular is the unit cell which is used for finding the band structure of graphene ribbons. Lower panel shows a field-effect transistor structure based on graphene ribbons where the gate voltage is applied on the whole system.

to the tube axis is conserved while passing through nanotubes Farajian& Esfarjani& Kawazoe (1999). Correspondingly, the transverse reflection symmetry of the incoming and outgoing wave functions results in the parity conservation in ZGNRs with even number of zigzag chains Duan et al. (2008); Cresti et al.-a (2008); Cresti et al.-b (2008); Akhmerov et al. (2008); Nakabayashi et al. (2009); Wakabayashi& Aoki (2002); Wang et al. (2008). As a consequence of the even-odd effect, a negative differential resistance (NDR) region appears in the I-V characteristic curve of P-N even ZGNR junctions Wang et al. (2008). However, in experiment, edges of graphene ribbons can simply absorb some chemical compounds Kobayashi et al. (2006). Even any small asymmetry is enough to destroy the blocked transitions induced by the parity conservation rule. Although there is no parity selection rules in armchair GNR's, NDR also can be found in their I-V characteristic curves. Such NDR originates from the interaction between the narrow density of states of the doped leads and the discrete states in the scattering region Ren et al. (2010). NDR has also been observed in GNR nanojunctions which its NDR's origin is traced back to the electrostatic profile. Enhanced localization of the HOMO and LUMO states induced by a charge depletion reduces overlap of states and current Cheraghchi& Esfarjani (2008). A NDR behavior has been also reported in P-N nanotube junctions Farajian& Esfarjani& Kawazoe (1999). Historically, NDR was first observed in the degenerated N-P diode junctions Esaki (1958). Nowadays, NDR has been reported in many other molecular devices Dragoman et al. (2007); Cheraghchi& Esfarjani (2008); Cheraghchi& Esmailzadeh (2010). To shed light on the experimental work in Ref.[Li et al. (2008); Wang et al. (2008)] and also introducing new origin of transport gap, in this chapter, we investigate nonlinear transport through ZGNRs by using non-equilibrium Green's function (NEGF) approach. The on/off ratio of the current in the NDR region for gated even ZGNR's reaches up to 10^6. It is also shown that a stable NDR against electrostatic interaction up to 10^5 is appeared in ultra narrow odd ZGNRs around $\pm 1V$ in both positive and negative polarity. These NDRs

are induced by transport gaps which are opened by two selection rules governing electron transition through ZGNRs: (i) the parity conservation, and (ii) that allowed transition are between connected bands Cresti et al.-a (2008); Cresti et al.-b (2008). Based on band structure analyzing, we show that transport gap opened by the second selection rule is filled for ribbons wider than $10nm$. So, sub-$10nm$ ribbons with long enough length provide experimental manifestation of the NDR phenomenon in I-V curve of GNRs. On the other hand, the gate voltage regulates the current flow by shifting the blocked energy regions with respect to the Fermi level. Moreover, for gated even ZGNRs, on/off ratio of the current displays a power law behavior as a function of ribbon length as $M^{7.5}$. However, on-off current ratio for odd ZGNRs increases exponentially with the ribbon length.
This NDR is still robust against edge impurities when the edges are doped by slightly amount of impurity. These edge impurities can significantly affect the electronic structure and transition selection rules of even ZGNRs. Edge states with energies about -0.1 to $0.2eV$ have been observed Kobayashi et al. (2006).
Our calculations show that the details of the electrostatic potential profile along the ribbon can not affect the emergence of NDR. The same conclusion has been reported by Ref.[Wang et al. (2008)], but they have not elaborated on the physical reason behind this robustness. By following the self-consistent charge and potential profiles at different voltages, we demonstrate that at low voltages, strong screening of the external potential at contacts results in a flat electrostatic potential along the ribbon. Subsequently, the e-e interaction at a mean field level, does not change the magnitude of $I_{\rm on}$. However, for voltages higher than the NDR threshold $V_{\rm on}$, the transfer of charge along the edges, leads to more reduction in $I_{\rm off}$ which improves the switch performance.
This chapter is organized as follows: in section 2, we briefly explain transition selection rules in symmetric and asymmetric ZGNRs which govern transport properties of GNRs. In section 3, we present Hamiltonian and a short review of NEGF. The origin of NDR seen in the I-V curve of even and odd ZGNRs is explained in section 4. We demonstrate in section 5 that the e-e interaction does not have a significant effect on the phenomena of NDR in the I-V curve. The last section concludes our results.

2. Transition rules

2.1 Transition selection rules in symmetric ZGNR

Presence of the reflection symmetry in the incoming and outgoing wave functions leads to the transition selection rule which regulates the current flow through ZGNRs Duan et al. (2008); Wang et al. (2008). However, there is no reflection symmetry in ZGNRs with odd zigzag chains along the width while even ZGNRs have a mirror symmetric plate which bisects the ribbon plane (Fig.(1)). By application of the symmetry operator on the wave function of even ZGNRs, odd or even parity can be realized for each subband.

$$P\psi_{n,k_x}(x,y) = \psi_{n,k_x}(x,-y) = \eta\psi_{n,k_x}(x,y) \tag{1}$$

where P is the parity operator and ψ_{n,k_x} is the eigen function of nth subband with the wave vector k_x along the longitudinal direction. The eigenvector contains two parts: a plane wave ($e^{ik_x x}$) along the longitudinal axis, and a constraint solution along the transverse direction with free boundary condition. So the eigenvector can be shown as $e^{ik_x x}\phi_n(y)$ Rainis et al. (2009). Eigenvalues of the parity operator are as $\eta = (-1)^{n+1}$, where $n = 1, 2, ..., 2N$ counts bands from the bottom to the top of the band structure, respectively. Eigenvalues of the parity

operator are independent of the value of k_x. As an example, in even ZGNR, lower subband of two central bands has negative parity and upper one has positive parity (Fig.2).
Using the tight-binding approximation, eigenvectors and eigenvalues of the ribbon is derived by diagonalization of the following Hamiltonian Ezawa (2006):

$$H\psi_{n,k_x}(x,y) = E(n,k_x)\psi_{n,k_x}(x,y)$$
$$H = H_0 + V_{0,+1}e^{ik_xx} + V_{0,-1}e^{-ik_xx} \tag{2}$$

where H_0 is the tight-binding Hamiltonian of the unit cell shown in Fig.(1.a). In the nearest neighbor approximation, $V_{0,+1}, V_{0,-1}$ are the overlap matrices between the marked unit cell ('0') and its left ('-1') and right ('+1') unit cells. Having eigenvalues of this Hamiltonian $E(n,k_x)$ results in the energy spectrum of graphene ribbons.

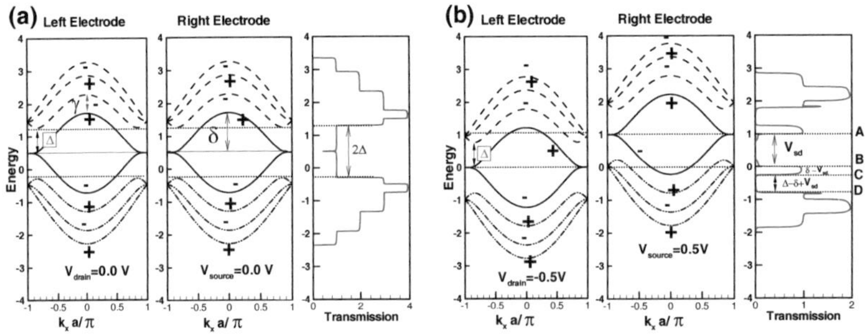

(a) (b)

Fig. 2. transmission (right panel) and band structure of right (center panel) and left (left panel) electrodes for ZGNR with 6 unit cells in length and 4 zigzag chains. Applied bias is considered to be **a)** $V_{\rm sd} = 0$ and **b)** $V_{\rm sd} = 1.0V$. Gate voltage is $V_g = 0.5V$. Here, $\triangle$ and γ are the energy separation of upper/lower group of bands from the central bands at the Dirac point ($k_x = \pm 2\pi/3a$) and $k_x = 0$, respectively. Moreover, δ is the half-width of the central bands at $k_x = 0$.

As a consequence of Eqs.(1,3), in even ZGNRs, we have $[P, H] = 0$. Therefore, parity is conserved during coherent transport. Parity conservation in even ZGNRs results in the blocked transition of the incoming wave function with positive parity into the outgoing wave function with negative parity of the central subbands.
Electronic transition is controlled by two selection rules. First, the parity is conserved in tunneling of electron through even-ZGNRs. Therefore, at zero source-drain voltage, one can expect full transmission which is shown in Fig.(2.a). In this case, all bands with the same parity are energetically aligned and there is no gap in the transmission curve. In Fig.(2.a), the energy of transmission curve is shifted by the gate voltage ($0.5V$). Parity of each band is indicated by plus/minus signs. In the range of 2Δ, there is one conducting channel which results in unit transmission coefficient.
Fig.(2.b) represents band structures of electrodes which are shifted with respect to each other due to the source-drain voltage $V_{\rm sd} = 1.0V$. The gap in the transmission curve of Fig.(2.b), AB region, indicates that transport between bands of opposite parity is blocked. In the energy regions in which back scattering of electrons increases due to the selection rules, it is established that the transmission decays exponentially with the length as $e^{-\gamma L}$.

The second selection rule which governs electron transport through ZGNRs, is that, electron transition is allowed between connected bands. Figs.(2.a,b) show the band structure classified in three different groups; namely, central, upper and lower bands which are indicated by solid, dashed and dashed-dot-dot lines, respectively. The common feature of bands in each group is that, they are connected at the zone boundary, while distinct groups are disconnected. When one considers the electron transport, the longitudinal momentum k_x, of electrons changes as a result of applied $V_{\rm sd}$. The precise form of this variation in k_x, crucially depends on profile of the superimposed longitudinal potential. These groups are disconnected from each other from the point of longitudinal momentum. Variation of momentum of electron k_x depends on the shape of superimposed longitudinal potential. The transport properties for smoothly varying $V_{\rm sd}$, are significantly different from $V_{\rm sd}$ profiles with sharp spatial variations. The electronic transition between an eigenstate (m_1, k) in the right electrode and an eigenstate (m_2, q) in the left electrode is proportional to Fourier transform of longitudinal voltage and structure factor Cresti et al.-a (2008); Cresti et al.-b (2008),

$$\langle \psi_{m_1}(k) \mid V_{sd}(x) \mid \psi_{m_2}(q) \rangle = S\tilde{V}_{sd}(k-q), \tag{3}$$

where structure factor of S is equal to $[1 + (-1)^{P_{m_1}+P_{m_2}}]$ for even ZGNRs and parity of band m is equal to $P_m = (-1)^{(m+1)}$. Parity selection rule in even ZGNRs originates from this structure factor. This parity selection rule is mesoscopic analogue of chirality factors governing transport of Dirac electrons in planar graphene Ulloa& Kirczenow (1987).

If a constant gate voltage is applied on the whole system, since there is no longitudinal potential variation, momentum of electron remains unaffected against the gate voltage. However, linear variation of the applied source-drain bias (with the slope $V_{\rm sd}/L$) changes the electron momentum. So, smooth variation of the potential in longer ribbons results in a small momentum variation of electron. Consequently, transition of electron between disconnected bands is forbidden when the length of ribbon is so large that one can assume $\tilde{V}_{sd}(k-q) \to \delta(k-q)$. Therefore, a smooth potential in the longitudinal direction can just scatter the electron among the class of states belonging to the same group of connected energy bands.

Now let us focus on the two transport gap regions: AB and CD in Fig.(2.b). The AB gap is a consequence of parity selection rules, while the CD gap is due to blockage of transition between disconnected groups. As can be seen in Fig.(2.b), the AB gap is proportional to the source-drain voltage, V_{sd}. Moreover, this gap is independent of the ribbon width. Of course, in the wide ribbons, the upper and the lower band groups approach to the central group, especially at the point $k_x = 0$, where γ in Fig.(2.a) tends to zero as log-normal.

When the ribbon width is increased, the separation γ between the upper/lower and central groups of bands, is reduced, which tends to loosen the second selection rule based on band groups; hence filling in the gaps. However when we increase the ribbon length, our classification of bands into connected groups is recovered. Therefore the AB gap is essentially governed by the aspect ratio of ZGNR.

The CD gap is equal to $\Delta - \delta + V_{\rm sd}$, where Δ and δ are the energy separation of upper/lower group from the central bands at the Dirac point, and the half width of the central bands at $k_x = 0$, respectively. The dependence of Δ on width N is: $\Delta \propto (2.13 \pm 0.02)N^{-(0.864 \pm 0.003)}$, while δ has a Log-Normal behavior which asymptotically approaches to the constant value of 0.9738 ± 0.0002 as N goes up to 10. The conducting region BC in Fig.(2.b) can exist only, when $\delta - \Delta < V_{sd} < \delta$. From the dependence of Δ and δ on N, the CD gap exists if N is less than 30. Hence, NDR is estimated to be observable for ribbon width $\leq 7nm$. The lowest

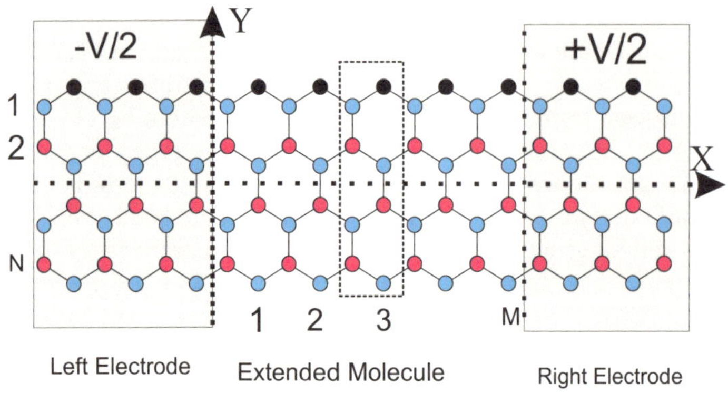

Fig. 3. Even asymmetric zigzag graphene nanoribbon considered as a central region attached to two electrodes. Transverse zigzag carbon chains and unit cells inside the central region are labeled by N and M. Darker edge atoms are absorbed atoms which make ribbon asymmetric in respect to the X-axis.

achieved ribbon width is sub-$10nm$-wide ($\sim 2 \pm 0.5nm$) Li et al. (2008); Wang et al. (2008) with the length $\sim 236nm$. Such long ribbons with small width provide fascinating experimental manifestation of the selection rules in transport properties.

One of the experimental requirements of the nanoribbon fabrication is the presence of some absorbed edge impurities. So, reflection symmetry could be simply failed during designing of electronic devices based on ZGNRs. Operation of an electronic device should not be affected by small asymmetry. Recently Wang et al. (2008), based on even ZGNR substrate and also the parity selection rule, an N-P diode junction was designed. This diode shows an NDR in positive polarity of its I-V curve. According to the previous discussions, this device is strongly sensitive to any asymmetry. Electronic devices designed by the transition rule arising from disconnecting band groups is much reliable than those which are based on the symmetry. In the next section, we introduce a nanoswitch which its operation is not so sensitive to the asymmetry. Fig.(3) shows asymmetric zigzag graphene nanoribbon with even number of zigzag chains in width (N).

2.2 Transition selection rules in asymmetric ZGNR

Those selection rules controlling the electron transition through even ZGNRs will be modified when impurity absorption changes onsite energies of atoms located in one edge of the ribbon. Application of the edge asymmetry influences the selection rules as the following:

1) Transition between those subbands which have been already forbidden based on the parity conservation, would be now allowed.

2) As shown in Fig.(4.b), in the spectrum of central bands of even ZGNRs, an energy gap is induced by the edge impurity. The opening of the gap depends directly on the onsite energy attributed to the edge impurity. As a result, in this case, the first conduction and the last valance bands detach from each other so that they do not belong to the same band group. Therefore, transition between these subbands is not allowed now.

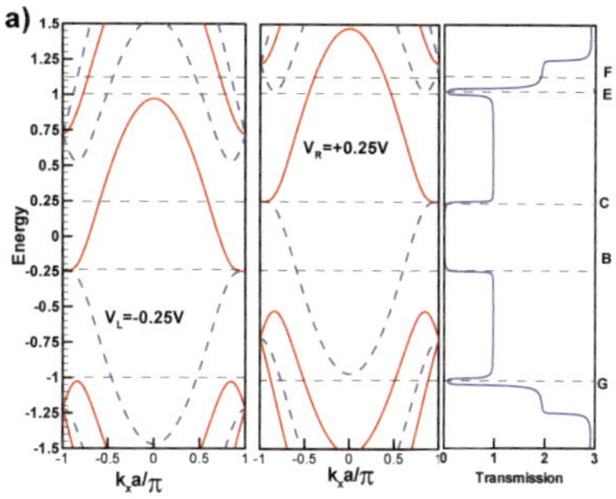

(a) a

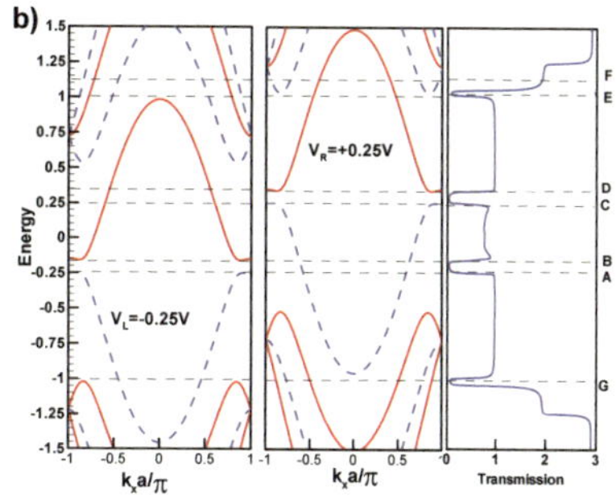

(b) a

Fig. 4. Energy spectrum of the left and right electrodes of graphene ribbons with 4 zigzag chains $(M, N) = (12, 4)$ under $0.5V$ applied bias a) without b) with the presence of the edge impurity. In figure (b), it is supposed that the doping impurities adsorbed by one of the ribbon edges can induce additional onsite energy on the edge atoms as $\varepsilon_\alpha = 0.1$. To investigate the selection rules, transmission also is plotted. Bands with solid and dashed lines have positive and negative parity, respectively.

These two selection rules have two opposite operations so that depending on the system length and edge impurity, they compete with each other. One of them increases transmission and the other one decreases it. Fig.(4.b) shows the energy spectrum and transmission through a ZGNR (with $N = 4$) at the applied bias of $0.5V$. It is supposed that the edge impurity changes onsite energy of the edge atoms as the value of $\varepsilon_\alpha = 0.1$. The energy gap induced by the edge impurities also keeps its trace in the transmission curve. Due to the asymmetry-induced gap, transmission in the energy ranges of AB and also CD shown in Fig.(4.b) is zero. However, one conducting channel is allowed in the energy range of DE. The only allowed transition in this energy range is the transition from the first conduction band of the right electrode to the same band of the left electrode. The same as symmetric case, at the energy points corresponding to E and G, the flow of the current is blocked due to the transition between two different band groups. By application of the edge impurity, it is interesting to note that in the energy range of BC, one conducting channel is opened and contributes into the transport. According to the parity selection rule, in this energy range, the electronic transition even ZGNRs must be blocked. However, parity is not conserved in asymmetric even ZGNRs and the transition is permissive. On the other hand, because of the asymmetry-induced energy gap, from the viewpoint of longitudinal momentum, the upper band of central group belonging to the left electrode is detached from the lower band of the central group belonging to the right electrode. So, the electronic transition between detached bands is forbidden. As it was explained in the previous section, if the ribbon length is long enough, the electron transition between these two detached subbands is effectively blocked. However, if the system length is considered to be very short (e.g. M=1), the transition between disconnecting bands is permissive. In the ribbons shorter than a critical length, application of the asymmetric edge impurity can not effectively destroy the parity conservation. As shown in Fig.(5.a), the parity conservation however gradually fails when the ribbon length approaches the critical length. Therefore, in the energy range of $B - C$, transmission increases with the ribbon length. On the other hand, in a ribbon longer than the critical length (here $M_{cr.} = 12$), transition of electrons between the separated central bands belonging to the different electrodes is gradually blocked

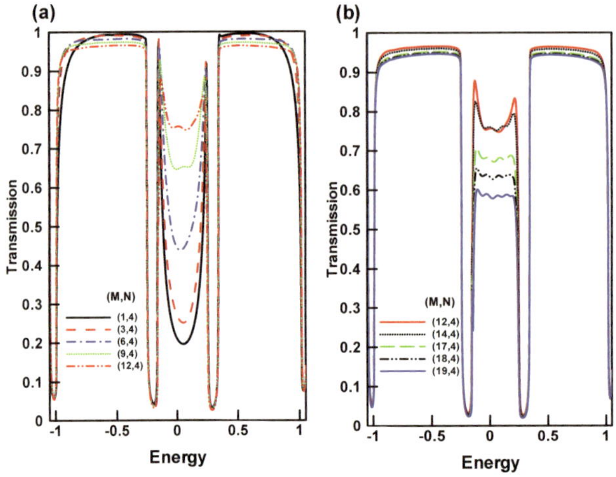

Fig. 5. Transmission through graphene ribbon with 4 zigzag chains in width for different ribbon lengths: a) $M \leq 12$ b) $M \geq 12$. Applied bias is considered to be $0.5V$. Edge impurity changes onsite energy of the edge atoms as $\varepsilon_\alpha = 0.1t$.

and so transmission in Fig.(5.b) decreases with the length. Based on this competition between two selection rules, one can explain non-linear transport through asymmetric even ZGNRs.

3. Hamiltonian and formalism

Fig.(1) shows schematic side view of graphene nanoribbon. In presence of source-drain applied potential, ribbon is divided into three regions; left, right electrodes and also central interacting region. Gate voltage is applied by means of substrate on the graphene plate. The interacting Hamiltonian is written in the tight-binding approximation. This Hamiltonian is a functional of charge density:

$$H\{n\} = \sum_i(\varepsilon_i + [(x_i - x_0)/L - 0.5]V_{sd} + \sum_j U_{ij}\delta n_j])c_i^\dagger c_i + \sum_{<ij>} t(c_i^\dagger c_j + c_i c_j^\dagger), \tag{4}$$

ε_i shows onsite energy of ith carbon atom and t represents the hopping integral between nearest neighbor atoms. One π orbital is considered per each site for graphene system. Without losing any generality, we set onsite energies (ε_i) of all sites equal to zero. All energies are in units of $t_{C-C} = 2.5eV$. Application of a gate voltage is achieved by shifting atomic onsite energies in all three regions. The applied source-drain potential, V_{sd}, and the gate voltage, V_g, preserve transverse symmetry with respect to the ribbon axis (X direction in Fig.(1)). Linear variation of the source-drain voltage along the ribbon is the solution of the Laplace equation with Dirichlete boundary condition on the contacts. U_{ij} is the electrostatic Green's function and $\delta n_i = n_i - n_i^0$ is the change in the self-consistent charge n_i from its initial equilibrium zero-bias value. This third term is the direct Coulomb interaction created by the bias-induced charges at a mean field level which is the solution of Poisson equation. The electrostatic Green's function for a distribution of charges between two parallel conducting planes located at $x = 0, L$ which are held at zero potential Jackson (1975), has the following form:

$$U(\vec{r},\vec{r}') = 2\int_0^\infty dk J_0(\alpha k)\frac{\sinh(kz_<)\sinh(k(L-z_>))}{\sinh(kL)},$$
$$\alpha = \sqrt{(x-x')^2+(y-y')^2+U_H^{-2}}, \tag{5}$$

where U_H is the Hubbard parameter whose semi-empirical value for carbon Esfarjani& Kawazoe (1998) is about $4t_{C-C}$. This parameter determines the strength of electron-electron interaction. This electrostatic Green's function is appropriate for the kernel of Ohno-Klopmann model Ohno& Klopman (1964).
For self-containing, we present a very brief review of the NEGF formalism. Charge density in non-equilibrium situation is calculated by $[-iG^<]$ as the occupation number in the presence of the two electrodes with an applied source-drain bias Taylor et al. (2001).

$$n_i = \frac{-1}{\pi}\int_{-\infty}^{E_F-\frac{V}{2}} Im[G^r(E)]_{ii}dE + n_i^{non-eq} \tag{6}$$

where non-equilibrium part of charge can be calculated by the following integral,

$$n_i^{non-eq} = \frac{1}{2\pi}\int_{E_F-\frac{V}{2}}^{E_F+\frac{V}{2}} [-iG^<(E)]_{ii}dE \tag{7}$$

where within a one-particle theory ,

$$-iG^< = G^r(\Gamma_L f_L + \Gamma_R f_R)G^a \tag{8}$$

Here $f_{L/R}$ is the Fermi-Dirac distribution function of electrodes and G^r/a is the retarded/advanced Green's function defining as the following:

$$G^{r/a} = [(E\pm\eta)I - H\{n\} - \Sigma_L^{r/a} - \Sigma_R^{ra}]^{-1} \tag{9}$$

and Γ is the escaping rate of electrons to the electrodes which is related to the self-energies as $\Gamma_p = i[\Sigma_p^r - \Sigma_p^a]$ with $p = L/R$ Munoz (1998); Datta (1995); Taylor et al. (2001). Here $\eta \to 0^+$. Solving equations 6 and 9 self-consistently results in a self consistent charge and Green's functions. Finally, the current passing through the molecule is calculated by the Landauer formula for zero temperature Datta (1995) which is valid for coherent transport.

$$I(V) = \frac{2e}{h}\int_{E_F-V/2}^{E_F+V/2} dE\, T(E,V) \tag{10}$$

where $T(E,V)$ are the bias dependent transmission coefficient.

$$T = \mathrm{Tr}[G^r\Gamma_R G^a \Gamma_L] \tag{11}$$

4. Negative differential resistance

In this section, we will present our results about nonlinear transport properties which emerges in current-voltage characteristic curves.

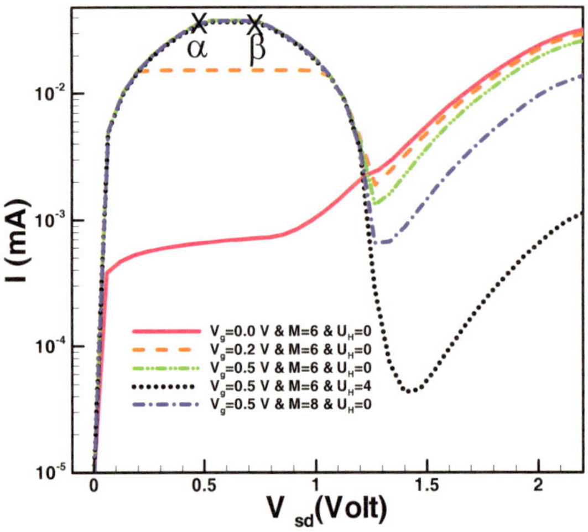

Fig. 6. Current-voltage characteristic curve for different gate voltages when the ribbon size is as $(N, M) = (4, 6)$. The effect of different parameters such as size effect, electrostatic potential and also gate voltage is investigated on I-V curve. Hubbard parameter (U_H) is on site Coulomb repulsive.

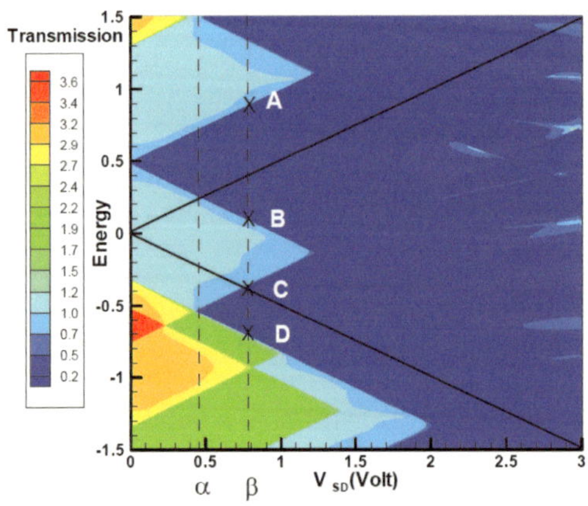

Fig. 7. Contour plot of transmission with respect to the energy and V_{sd} for the system size (N,M)=(4,6) and the applied gate voltage $V_g = 0.5V$. Darked Oblique lines shows the current integration window. Points marked by A, B, C, D correspond to the horizontal lines with the same name in Fig.(2). Lines of α, β are the trace of points with similar names in Fig.(6). Fermi energy was fixed at zero $E_F = 0$.

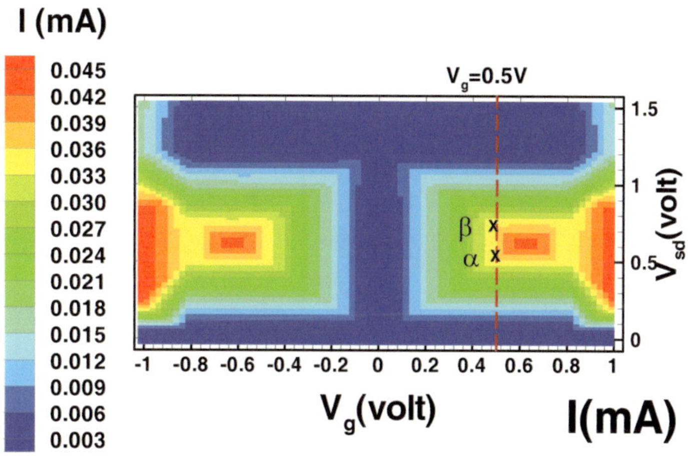

Fig. 8. Contour plot of the current in terms of V_{sd} and V_g for a zigzag graphene nanoribbon with (N,M)=(4,6). Vertical dotted line corresponds to $V_g = 0.5V$. α and β are those points which was shown in Figs(6,7).

4.1 NDR in gated-EVEN zigzag graphene nanoribbon

Fig.(6) shows current-voltage characteristic curve of a ZGNR with 4 zigzag chains and 6 unit cells in length. In the case of zero gate voltage, flow of current is blocked due to the parity selection rule, while at a given V_{sd}, gate bias turns the current on. After a range of V_{sd} in which the current remains unchanged, current begins to reduce with increasing V_{sd}. NDR threshold voltage V_{on} decreases with gate voltage for $V_g < 0.6V$. Dependence of NDR threshold voltage on the gate voltage can also be seen in Fig.(8). This NDR also symmetrically appears in the negative polarity of V_{sd}. The NDR threshold voltage and I_{on} remain unchanged in the presence of the electron-electron interaction (with a given Hubbard term $U = 4t_{C-C}$). However, reduction of the current in off state, I_{off}, is intensified when one takes electrostatic potential into account.
To understand the origin of NDR, it is helpful to look at the 3D contour-plot of transmission in plane of energy and V_{sd} which is presented in Fig.(7). Blocked energy intervals AB and CD, which are indicated in Fig.(7), correspond to those intervals shown in Fig.(2). For voltages lower than the vertical line α ($V < V_\alpha$), transmission is a nonzero constant for the whole region of the conduction window represented in Eq.(10). As a result, current increases proportionally to V_{sd}. In the voltage interval $[V_\alpha, V_\beta]$, the blocked region AB, originating from the parity selection rule, contributes to the current integration window of Eq.(10). However, nonzero range of transmission remains unchanged along with V_{sd} resulting in the fixed current in the voltage range $[V_\alpha, V_\beta]$. So, current remains unchanged in this range. For voltages $V \geq V_\beta$, the CD gap contributes in the current integration window, and consequently the NDR phenomenon emerges.
Regarding the importance of gate voltage in the current flow, let us investigate the effect of the gate voltage on $I - V_{\mathrm{sd}}$ curve by contour plotting of the current with respect to V_{sd} and V_g in Fig.(7). For gate voltages $|V_g| < 0.1V$, shift of transmission is not remarkable enough to contribute to conducting channels in the current integration. So, current is blocked by the

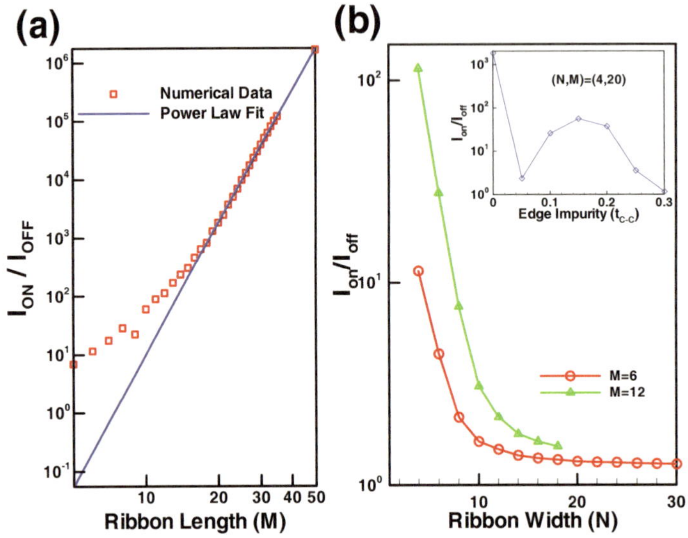

Fig. 9. a) On-off ratio of the current increases as a power law with the ribbon length for $N = 4$ and $V_g = 0.5V$. b) On-off ratio of the current decreases with the ribbon width (circle and triangular points) while it is disappeared for edge impurity (diamond points) stronger than $0.3t_{C-C}$ (inset figure).

parity selection rule. In the range $0.1V < |V_g| < 0.6V$, contribution of conducting region BC in transport is accompanied with the blockage arising from AB and CD gaps in voltages $V > V_{\rm on}$. As a consequence, current reduces after a threshold voltage. In this range, on/off ratio of the current increases and $V_{\rm on}$ reduces with increasing the gate voltage.

As can be seen in I-V curves of Fig.(6), off current reduces for longer ribbons which enables us to achieve high performance switches by increasing aspect ratio of the ribbon length/width. The reason is connected to smooth variation of the applied potential along the ribbon such that during transport, electrons are scattered among those states which belong to continuous bands. As a consequence, blockage originating from electronic transition between disconnected bands is intensified by increasing the ribbon length. In fact, when the length of ribbon increases, transmission in the AB and CD gaps decreases exponentially.

Since off current is induced by contribution of the gaps in the current integration, $I_{\rm off}$ efficiently decreases with increasing the ribbon length (M). Fig.(9.a) shows that $I_{\rm on}/I_{\rm off}$ displays a power law behavior as a function of the ribbon length for large M: $I_{\rm on}/I_{\rm off} \propto M^{\eta}$ where $\eta = 7.5061 \pm 0.03505$. As an example, for $M = 50$, on/off ratio goes up to 10^6 which suggests experimental fabrication of high performance switches based on the GNR nanoelectronics.

Experimentally, it was observed in Ref.[Li et al. (2008); Wang et al. (2008)] that the room-temperature on/off ratio induced by the gate voltage increases exponentially as the GNR width decreases. They observed that $I_{\rm on}/I_{\rm off}$ is equal to 1, 5, 100 and $> 10^5$ for $W = 50nm, 20nm, 10nm$ and sub-$10nm$, respectively. Similarly, as shown in Fig.(9.b), on/off ratio calculated for the set up considered in this paper, also decreases with the ribbon width, while reduction of on/off ratio can be compensated by considering longer ribbons. However, NDR phenomenon is disappeared for the ribbons wider than $7nm$.

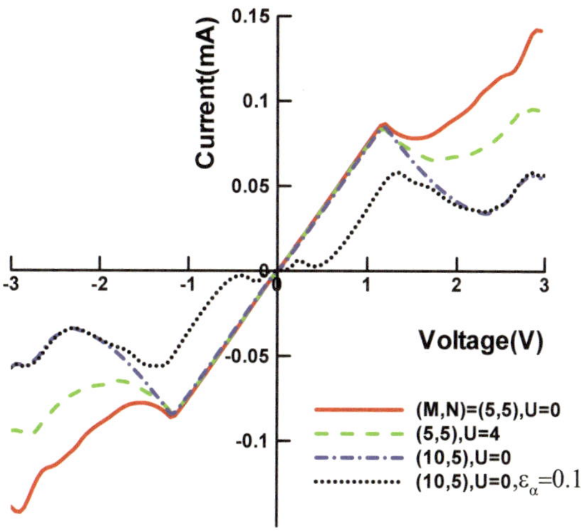

Fig. 10. Current-voltage characteristic curve of an odd zigzag graphene nanoribbon with $N = 5$ (zigzag chains). I-V curves are compared for two ribbon lengths; $M = 5$ and 10. For the case of $(M, N) = (5, 5)$ and for comparison purpose, I-V curve is also plotted in the presence of electron-electron interaction (U). NDR phenomena is preserved when one of the ribbon edges is doped by slight impurity (as ε_α).

In *ab initio* calculations Son et al. (2006), by using hydrogen-termination of zigzag edges, mirror-symmetry of ZGNRs and consequently parity conservation could be retained. Correspondingly, by several repetition of the heat treatments and hydrogenation, it is also possible to create well-ordered H-terminated edges in experiment Kobayashi et al. (2006). However, the edge states with energies about -0.1 to 0.2 eV have been experimentally observed Kobayashi et al. (2006) that emerge at hydrogen-terminated zigzag edges. To simulate the edge states and the effect of symmetry breaking on NDR phenomenon, it is assumed to dope one of the ZGNR edges by slight impurity. Edge impurity is considered to apply as a change in the on-site energy of the edge atoms (ε_α) with respect to on-site energy of the other atoms. In case of edge disorder, ε_α plays the role of the averaged on-site energy of the edge atoms. Inset figure indicated in Fig.(9.b) shows that on/off switching reduces with the edge impurity strength, however, NDR still emerges for $\varepsilon_\alpha < 0.3t_{C-C}$.

4.2 NDR in ODD zigzag graphene nanoribbon

Current-voltage characteristic curve of an odd ZGNR with 5 zigzag chains ($N = 5$) is shown in Fig.(10). Lower than the external bias $1.2V$, current increases linearly with the applied bias as an *Ohmic* device. After a threshold voltage ($1.2V$), NDR occurs at both positive and negative polarity.

The origin of NDR seen in odd ZGNRs is interpreted by analyzing their energy spectrum accompanied to transmission curve. Fig.(11) shows energy spectrum $E(n, k_x)$ and transmission through 5 ZGNR at $V = 1.4V$.

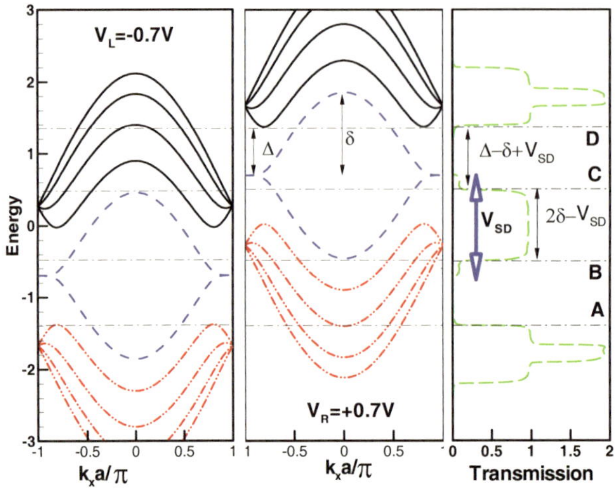

Fig. 11. Energy spectrum of the left and right electrodes and transmission through zigzag graphene nanoribbon with $(M, N) = (10, 5)$ at voltage $(V_{SD} = 1.4V > V_T)$. The band structure is divided into three groups which are called by upper , centeral and lower band groups. These groups are classified based on the bands which are connected in terms of k_x. The bold hollow arrows show the current integration window which based on Eq. 10, is proportional to V_{SD}. The Fermi level is set to be as $E_f = 0$. The half-width of the central bands at $k_x = 0$ is called as δ which is equal to the threshold NDR voltage. Δ is energy separation of the upper bands from the central bands at the Dirac points. Transport gaps AB and CD are equal to $\Delta - \delta + V_{SD}$.

In odd ZGNRs, parity has noncommutative relation with the Hamiltonian. Therefore, parity has no conservation and consequently transmission is not blocked by the parity selection rule, while parity conservation in even zigzag nanoribbons opens transmission gap around Fermi level Cheraghchi& Esmailzadeh (2010). In the range of BC of Fig.(11), there exists one conducting channel which results in the unity transmission around the Fermi level. So at low biases, current increases linearly with bias. This single-channel transport around the Fermi level remains unchanged even for high voltages. However, for voltages greater than the NDR threshold voltage $(V > V_T)$, blocked regions marked by the ranges of AB and CD comes into the current integration window. The current integration window is proportional to V_{SD} and is shown with the bold hollow arrows in Fig.(11). Therefore, when source-drain applied bias increases, current begins to decrease.

Blocked regions (AB and CD) arise from a selection rule which increases back scattering in the lengthy ribbons. According to this rule, the electronic transition between those bands which are disconnected from the view point of longitudinal momentum, decreases exponentially with the length.

Band structure analyzing demonstrates that the threshold voltage is equal to the half-width of the central bands at $k_x = 0$ as $V_T = \delta = [E(N+1, k_x = 0) - E(N, k_x = 0)]/2$. As shown in Fig.(11), Δ is energy separation of the upper bands from the central bands at the Dirac points. There is a Log-Normal behavior of δ versus number of zigzag chains (N) such that as $N \to \infty$, the NDR threshold voltage asymptotically approaches to the value of 0.9738 ± 0.0002. So the NDR threshold voltage slightly decreases with the ribbon width. Analyzing transport gaps appeared in the band structure shows that they are equal to $\Delta - \delta + V_{SD}$ where $\Delta \propto$

N^{-1}. Since δ approaches to a constant values when $N \to \infty$, in a given voltage, transport gap is disappeared for $N > 30$ which is nearly equivalent to $10nm$. The other effect which enhances performance of this electronic switch, is the ribbon length. Fig.(10) shows an increase of on/off ratio with the ribbon length. Moreover, NDR region ($V_{off} - V_{on}$) occurs in a more extended range of the I-V curve. Exponential decay of transmission with the length in the gap regions develops quality of switching. In odd ZGNRs, I_{on}/I_{off} increases exponentially with the ribbon length.
If one of the ribbon edges is doped by small impurity such as $\varepsilon_\alpha = 0.1t$, because of a band gap which is induced by the edge impurity in low biases, current decreases. Fig.(10) represents that even with the presence of edge impurity, still the region containing NDR still exists. However, asymmetry decreases on/off ratio of the current. Furthermore, asymmetric ZGNR behaves as a semiconductor while symmetric ZGNRs behave as an Ohmic devices Ren et al. (2010). The effect of asymmetry on NDR competes with the ribbon length. Since asymmetry can not be ignored in experiment, longer ribbons are in favor of keeping NDR in the I-V curve. On the other hand, additional to edge impurity, this asymmetry can be assigned to a sublattice symmetry breaking induced by spontaneous ferromagnetic spin ordering of the electrons localized at the zigzag edges Son et al. (2006). In fact, the border atoms at the two opposite zigzag edges belong to different sublattices. So spin orientation along the edges induces different magnetic potentials at the edges. As a result, a small band gap is opened around Fermi level which depending on the ribbon width, is about 0.15 eV. The asymmetry which we have considered is about 0.3 eV which is stronger than the gap opened by spin-polarization of the edges. We can conclude that spin-polarization along the zigzag edges can not affect emerging of this NDR phenomenon Ren et al. (2010).

5. Electrostatic potential and charging effect on NDR

Emerging phenomenon of negative differential resistance in I-V curve is not destroyed by the e-e interaction and is independent of the details of electrostatic potential profile. However, interaction reduces off-current as shown in the I-V curves of Fig.(6). In this section, we present the reason for robustness of NDR against e-e interaction for the gated even ZGNR. The same results can be extracted for odd ZGNRs.
To substantiate the above claim, comparison of transmission curves in the presence and absence of the e-e interaction is useful. It is apparent from Fig.(12.a) that for voltages less than V_{on}, transmission in conducting channels is robust against the e-e interaction while transmission increases in the gaps with respect to the non-interacting case. But this enhancement is slight enough which can not affect the emergence of NDR. However, for voltages $V > V_{\text{on}}$, interaction lowers transmission coefficient in the conducting channels (such as BC region) in which higher subbands participate in transport. Such behavior is corroborated in Fig.(12.b) which indicates transmission at that voltage corresponding to the off-current, V_{off}. Reduction in the transmission coefficient of the conducting channels results in further reduction of the off-current. To explain the reason for such phenomenon, it is necessary to study the potential and charge profiles. Electrostatic potential averaged on each unit cell is represented in Fig.(13.a) in terms of the ribbon length. For voltages less than V_{on}, potential sharply drops only at the contact regions which connects the system to electrodes. In such a case, external potential is strongly screened by redistributed electrons and, electrostatic potential of the central atoms remains close to zero. Screening is performed by discharging of electrons from the area connected to the source and their accumulation around the drain electrode. These facts are obvious from transferred charge and electrostatic

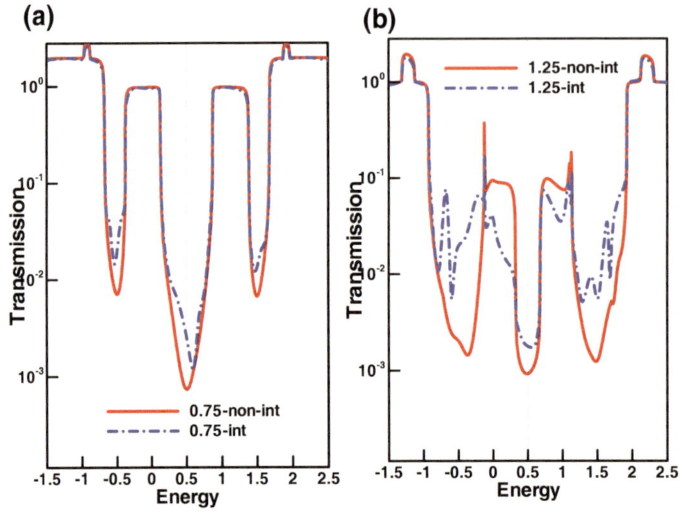

Fig. 12. Transmission curve of zigzag graphene nanoribbon with 4 zigzag chains in width for voltages a) 0.75 volt b) 1.25 volt. Interacting and non-interacting curves are compared with each other.

potential profiles represented in Fig.14. Since $U(n-n_0)$ determines electrostatic behavior of the potential, discharging of electrons weakens the external potential penetrated from the source electrode. Moreover, charge accumulation around the drain electrode prevents potential drop in the central part of the system. So in the case of strong screening, potential drops only at the contacts. However, when the applied bias goes beyond 1 volt, screening is being weakened and external potential can penetrate inside the central region. The reason as to why screening is weak, can be sought in charge distribution. Figure (13.b) illustrates that for voltages less than 0.5 volt, in and out flow of charge are balanced with each other such that the total transferred charge remains close to zero. However, around the voltage V_{on} and voltages above, the charge is mainly transferred from the edges of the ribbon, so that the source electrode does not inject further charge to middle atoms of the ribbon. As a consequence, by increasing the applied bias and so gradient of the potential along the central region, charge depletion is mainly enhanced in the middle bar area of ZGNR. On the other hand, since the only way for transporting electrons is the edge atoms, significant accumulation of charge appears along the two edge lines of ZGNR.

In summary, at voltages less than V_{on}, electrostatic potential is only dropped at the contacts and therefore momentum of electrons is only varying in the area where the potential drops, while longitudinal momentum of electron remains unchanged across the central portion. In other words, potential steeply drops in the low-area district around the contacts which results in violation of the blockage rule which governs on transition between disconnected energy bands. Subsequently, transmission coefficient slightly increases in the blocked energy ranges. In other words, in this case, an increase in gradient of the potential facilitates electronic transport in the blocked energies. Note that interaction preserves transverse symmetry, so the parity selection rule still governs electronic transport. Therefore, the AB gap induced by the parity conservation still survives for voltages larger than V_{on}. For voltages $V > V_{on}$, electrostatic potential gradually penetrates into the whole system so that the potential of the

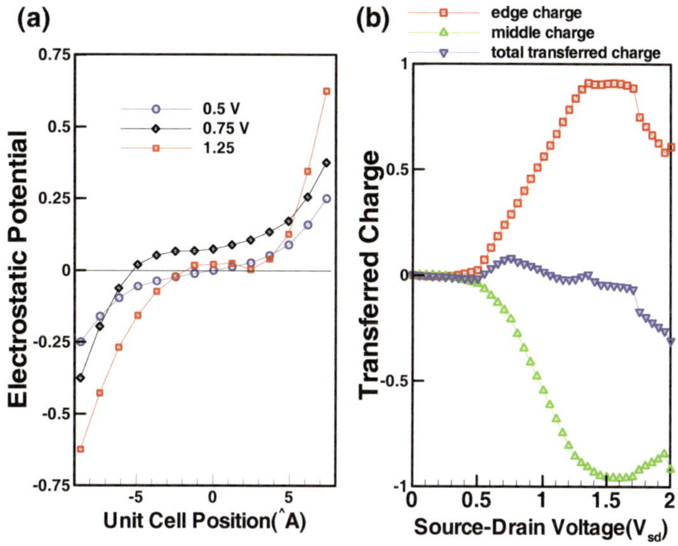

Fig. 13. a) Edge, middle and total transferred charge in terms of source-drain applied bias. b) Electrostatic potential per unit cell in terms of unit cell position for source-drain voltages 0.5, 0.75, 1.25 volts. The gate voltage is for all curves equal to $V_g = 0.5$ V.

central region is not flat. In addition, because the edge transport of electrons dominates, the transverse potential is deeper in the middle of ZGNR than its edges. Therefore, the band structure of the interacting central region differs from the band structure of electrodes. As a consequence, for voltages $V > V_{\text{on}}$, transmission of conducting channels and also off current reduces.

Comparison of odd and even ZGNRs: There are some interesting differences between results arising from *odd* ZGNRs with those results belonging to *even* ZGNRs Cheraghchi& Esmailzadeh (2010) which is kind of odd-even effect. In odd ZGNRs, NDR appears in voltages upper than $1V$ while in even ZGNR, NDR occurs for voltages lower than $1V$. On/off ratio of the current in gated even ZGNRs increases as a power law with the function of the ribbon ribbon length while in odd ZGNRs, on/off increases exponentially. Screening of the external bias by electrons of system in even ZGNRs is so stronger than screening effects in odd ZGNRs. As a consequence, the effect of electrostatic interaction on increase of the on/off ratio in even ZGNRs is much effective than in odd ZGNRs. Furthermore, transferred charge from/into the central portion of graphene nanoribbon depends on odd or even zigzag chains in width.

6. Conclusion

As a conclusion, based on a model calculation and non-equilibrium Green's function formalism, we found that there exists a region of negative differential resistance in I-V curve of ultra narrow (lower than $10nm$) zigzag graphene nanoribbons with odd or even number of zigzag chains in width. In this range of widths, two selection rules govern on the electronic transition are: (i) the parity conservation, and (ii) the allowed transition between connected bands. On/off ratio of the current increases up to 10^5 as a function of the ribbon length which proposes possibility of manipulation of ZGNRs as high quality switch in nanoelectronic based on graphene nanoribbons. Emergence of the NDR phenomenon is not sensitive to

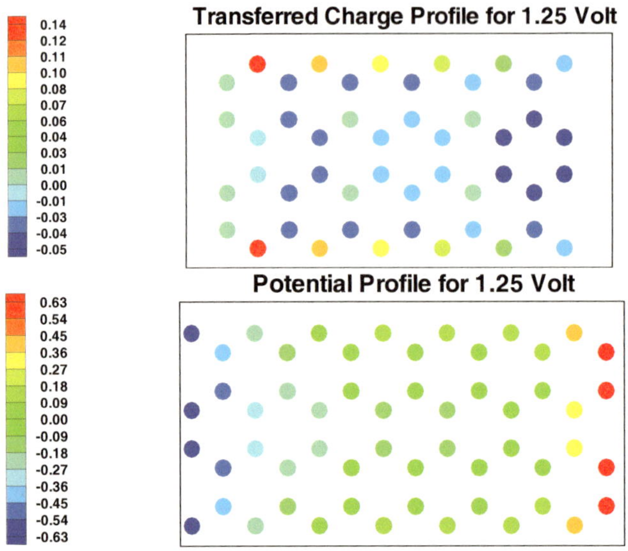

Fig. 14. Transferred charge and Electrostatic potential profiles for $V_{sd} = 1.25$ volt and $V_g = 0.5$ volt in the weak screening case. Due to charge transfer through the ribbon edges, screening is so weak that the external potential can penetrate inside the central portion.

details of the electrostatic potential profile. Because of strong screening in low voltages, the major potential drop takes place at the contacts. However, in voltages larger than the NDR threshold, due to charge transfer through the ribbon edges, screening is so weak that the external potential can penetrate inside the central portion. As a consequence, off current reduces in comparison to non-interacting ribbons.

In addition, e-e interaction enhances on-off ratio of the current which originates from a flat electrostatic potential deep inside the ribbon due to screening of the external bias by electrons close to the junctions. Furthermore, this NDR is not much sensitive to the edge asymmetry. So emerging of this NDR is robust against spin orientation along the edges.

7. References

Novoselov, K. S.& Geim, A. K.& Morozov, S. V.& Jiang, D.& Zhang,Y.& Dubonos, S. V.& Grigorieva, I. V.& Firsov, A. A. (2004). Electric Field Effect in Atomically Thin Carbon Films, In:*Science*, Vol. 306, No. 5696, pp. (666-669).

Liu, Z. & Suenaga, K. & Harris, P. J. F. & Iijima, S. (2009). Open and Closed Edges of Graphene Layers, In:*Phys. Rev. Lett.*, Vol. 102, page. (015501).

Ozyilmaz. B. &Jarillo-Herrero. P. &Efetov. D. & Abanin. D. A.& Levitov. L. S.&Kim. P. (2007). Electronic Transport and Quantum Hall Effect in Bipolar Graphene p-n-p Junctions, In:*Phys. Rev. Lett.*, Vol. 99, page. (166804).

Han, M. Y.& ozyilmaz, B.&Zhang, Y.& and Kim, P. (2007). Energy Band-Gap Engineering of Graphene Nanoribbons, In:*Phys. Rev. Lett.*, Vol. 98, pp (206805).

Molitor, F. &Jacobsen, A.& Stampfer,C.& Güttinger, J.& Ihn, T. and Ensslin, K. (2009). Transport gap in side-gated graphene constrictions, In:*Phys. Rev. B.*, Vol. 79, pp. (075426).

Li, X.& Wang, X.& Zhang, L.& Lee, S.& Dai, H. (2008). Chemically Derived, Ultrasmooth Graphene Nanoribbon Semiconductors, In:*Science*, Vol 319, pp. (1229-1232).

Wang, X.& Ouyang, Y.& Li, X.& Wang, H.& Guo, J.& Dai, H. (2008). Room-Temperature All-Semiconducting Sub-10-nm Graphene Nanoribbon Field-Effect Transistors, In:*Phys. Rev. Lett.*, Vol. 100, pp. (206803).

Son, Y. W.& Cohen, M. L.& Louie, S. G. (2006). Energy Gaps in Graphene Nanoribbons, In:*Phys. Rev. Lett.*, Vol. 97, pp. (216803).

Sols, F. &Guinea, F.& Castro Neto, A. H. (2007). Coulomb Blockade in Graphene Nanoribbons, In: *Phys. Rev. Lett.*, Vol. 99, pp. (166803).

K. Nakada, K.& Fujita, M.& Dresselhaus, G.& Dresselhaus, M. S. (1996). Edge state in graphene ribbons: Nanometer size effect and edge shape dependence, In: *Phys. Rev. B.*, Vol. 54, pp. (17954Ű17961).

Brey, L.& Fertig, H. (2006). Edge states and the quantized Hall effect in graphene, In. *Phys. Rev. B.*, Vol. 73, pp. (195408-195413)

Brey, L.& Fertig, H. (2006). Electronic states of graphene nanoribbons studied with the Dirac equation, *Phys. Rev. B.*, Vol. 73, pp.(235411-235416).

Zheng, H.& Wang, Z. F.& Luo, T.& Shi, Q. W.& Chen, J. (2007). Analytical study of electronic structure in armchair graphene nanoribbons, In: *Phys. Rev. B.*, Vol. 75, pp. (165414-165420).

Malysheva, L.& Onipko, A. (2008). Spectrum of π Electrons in Graphene as a Macromolecule, In: *Phys. Rev. Lett.*, Vol. 100, pp. 186806.

Mucciolo, E. R.& Castro Neto, A.& Lewenkopf, C. H. (2009). Conductance quantization and transport gaps in disordered graphene nanoribbons, In: *Phys. Rev. B.*, Vol. 79, pp. (075407-075412).

Li, Z.& Qian, H.& Wu, J.& Gu, B-L.& Duan, W. (2008). Role of Symmetry in the Transport Properties of Graphene Nanoribbons under Bias, In: *Phys. Rev. Lett.*, Vol. 100, pp. 206802.

Farajian, A. A.& Esfarjani, K.& Kawazoe, Y. (1999). Nonlinear Coherent Transport Through Doped Nanotube Junctions, In: *Phys. Rev. Lett.*, Vol. 82, pp. (5084).

Cresti, A.& Grosso, G.& Parravicini, G. P. (2008). Field-effect resistance of gated graphitic polymeric ribbons: Numerical simulations, In: *Phys. Rev. B.*, Vol. 78, pp. (115433-115440).

Cresti, A.& Grosso, G.& Parravicini, G. P. (2008). Valley-valve effect and even-odd chain parity in p-n graphene junctions, In: *Phys. Rev. B.*, Vol. 77, pp. (233402).

Akhmerov, A. R.& Bardarson, J. H.& Rycerz, A.& Beenakker, C. W. J. (2008). Theory of the valley-valve effect in graphene nanoribbons, In: *Phys. Rev. B.*, Vol. 77, pp. (205416-205421).

Nakabayashi, J.& Yamamoto, D.& Kurihara, S. (2009). Band-Selective Filter in a Zigzag Graphene Nanoribbon, In: *Phys. Rev. Lett.*, Vol. 102, pp. (066803).

Wakabayashi, K.& Aoki, T. (2002). Electrical conductance of zigzag nanographite ribbons with locally applied gate voltage, In: *Int. J. Mod. Phys. B.*, Vol. 16, pp. (4897-4909).

Wang, Z. F.& Li, q.& Shi, Q. W.& Wang, X.& Yang, J.& Hou, J. G.& Chen, J. (2008). Chiral selective tunneling induced negative differential resistance in zigzag graphene nanoribbon: A theoretical study, In: *Appl. Phys. Lett.*, Vol. 92, pp. (133114).

kobayashi, Y.& Fukui, K. I.& Enoki, T.& Kusakabe, k. (2006). Edge state on hydrogen-terminated graphite edges investigated by scanning tunneling microscopy, In: *Phys. Rev. B.*, Vol. 73, pp. (125415-125423).

Niimi, Y.& Matsui, T.& Kambara, H.& Tagami, K.& Fukuyama, H. (2006). Scanning tunneling microscopy and spectroscopy of the electronic local density of states of graphite surfaces near monoatomic step edges, In: *Phys. Rev. B.*, Vol. 73, pp. (085421).

Ren, Y.& Chen, K-Q. (2010). Effects of symmetry and StoneŰWales defect on spin-dependent electronic transport in zigzag graphene nanoribbons, In: *J. Appl. Phys.*, Vol. 107, pp. (044514).

Esaki, L. (1958). New Phenomenon in Narrow Germanium p-n Junctions, In: *Phys. Rev.*, Vol. 109, pp. (603-604).

Dragoman, D.& Dragoman, M. (2007). Negative differential resistance of electrons in graphene barrier, In: *Appl. Phys. Lett.*, Vol. 90, pp. (143111).

Cheraghchi, H.& Esfarjani, K. (2008). Negative differential resistance in molecular junctions: Application to graphene ribbon junctions, *Phys. Rev. B.*, Vol. 78, pp. (085123).

Ulloa, S. E.& Kirczenow, G. (1987). Electronic structure of staging dislocations, electron scattering states, and the residual resistance of graphite intercalation compounds, iN: *Phys. Rev. B.*, Vol. 35, pp. (795).

Jackson, J. D. (1975). *Classical Electrodynamics* (New York: John Wiley and Sons), 2 nd Ed.

Esfarjani, K.& Kawazoe, Y. (1998). Self-consistent tight-binding formalism for charged systems, In: *J. Phys.: Condens. Matter*, Vol. 10, p. (8257).

Ohno, K. (1964). Some remarks on the Pariser-Parr-Pople method , Theor. Chim. Acta. Vol. 2, p. (219); Klopman, G. (1964). A Semiempirical Treatment of molecular Structures. II. Molecular Terms and Application to diatomic Molecules, J. Am. Chem. Soc. Vol. 86, pp. (4550Ű4557).

Nardelli, M. B. (1999). Electronic transport in extended systems: Application to carbon nanotubes, *Phys. Rev. B.*, Vol. 60, pp. (7828Ű7833).

Broyden, C. G. (1965). A class of methods for solving nonlinear simultaneous equations , Math. Comput. Vol. 19, pp.(577-593), also See, e.g., Ohno, K. & Esfarjani, K. & Kawazoe, Y. *Computational Materials Science from Ab Initio to Monte Carlo Methods* (Springer, Berlin, 1999).

Munoz, M. C. *et al.*, (1988). In: *Prog. Surf. Sci.*, Vol. 26, p. (117); Garcia-Moliner F. & Velasco, V. R. *Theory of Single and Multiple interfaces* (World Scientific, Singapore, 1992).

Datta, S. *Electronic Transport in Mesoscopic Systems* (Cambridge U.P, Cambridge, 1995).see chapter 2 and references.

Liang, G. C. & Ghosh, A. W. & Paulsson, M. & Datta, S. (2004). Electrostatic potential profiles of molecular conductors, In: *Phys. Rev. B.*, Vol. 69, pp. (115302).

Ezawa, M. (2006). Peculiar width dependence of the electronic properties of carbon nanoribbons, In: *Phys. Rev. B.*, Vol. 73, pp. (045432).

Rainis, D. & Taddei, F. & Dolcini, F. & Polini, M. & Fazio, R. (2009). Andreev reflection in graphene nanoribbons, In: *Phys. Rev. B.*, Vol. 79, pp. (115131).

Taylor, J. & Guo, H. & Wang, J. (2001). Ab initio modeling of quantum transport properties of molecular electronic devices, In: *Phys. Rev. B.*, Vol. 63, pp. (24540).

Cheraghchi, H.& Esmailzadeh, H. (2010). A gate-induced switch in zigzag graphene nanoribbons and charging effects, In: *Nanotechnology*, Vol. 21, pp. (205306).

8

Field Emission from Graphene Nanosheets

Takahiro Matsumoto[1,4], Tomonori Nakamura[3,5],
Yoichiro Neo[3], Hidenori Mimura[3] and Makoto Tomita[6]
[1]*Research and Development Center, Stanley Electric Corporation*
[2]*Center for Quantum Science and Technology, Osaka University*
[3]*Research Institute of Electronics, Shizuoka University*
[4]*National Institute of Advanced Industrial Science and Technology*
[5]*Research Laboratory, Onizuka Glass Corporation*
[6]*Department of Physics, Shizuoka University*
Japan

1. Introduction

The 2D graphene sheet structure is promising for field emission of electrons, because the carrier mobility and electron mass were reported to have exceptionally large and small values, respectively, due to the quantum relativistic effect (Novoselov et al., 2005; Zhang et al., 2005; Geim & Novoselov, 2007).

In this chapter, we overview the superior aspects of field emitter by using a multi layered graphene nanosheet structure (GNS) and give quantitative description of the physics of electron field emission from this GNS. Firstly (section 2&3), we briefly overview the fabrication method, structure characterization, and field emission current characteristics of GNS. Secondly (section 4), we describe the correlation between the emission current fluctuation and molecular adsorption, because the emission current is affected both by the adsorption and desorption of molecules and by the temperature of the emission sites onto the graphene sheet. However, there is no quantitative discussion to describe the adsorption process of field emission. Therefore, we construct a stochastic model based on the physisorption of atoms and/or molecules onto the emission sites, which can be applied to other types of field emitters. Here macroscale statistical discussion is shown to give insights for microscale emission mechanism. Thirdly (section 5), we will proceed atomic level studies by using a field emission microscopy, which gives fruitful information on the microscopic origin of the emission from graphene. Symmetrical emission patterns such as two-fold, four-fold, and ring patterns were observed. These symmetrical patterns are experimentally correlated with the excitation mode of π conjugated bonds and are theoretically explained from the excited modes of six membered ring in the graphene sheet. These results and theoretical analyses give the solution to the mystery of FEM patterns obtained by other types of carbon materials such as phthalocyanine argued for more than 60 years. After the clarification of the fundamental aspects of GNS, we overview the application aspects of the GNS field emitter (section 6). The high performance of the field emitter is presented by the demonstration of a time resolved x-ray radiography system and a field emission scanning electron microscope (FE-SEM) system. We will also present our recent progress in miniature sized x-ray tubes with GNS field emitter. The tube has both the lifetime more than 10,000 hours and the stability better than 1 %; therefore, we believe GNS is best suited for field emitters. We will put the x-ray tube into practical use in coming years. Finally (section 7), we will give our future prospects for the GNS field emitter.

2. Fabrication and nanostructure characterization

The GNS field emitters are fabricated by hydrogen plasma etching of a carbon rod in a microwave (a frequency of 2.45 GHz) plasma chemical vapor deposition equipment as shown in Fig. 1. Typical etching condition is as follows; a microwave power of 800 W, a gas pressure of 1.3 kPa, a H_2 gas flow rate of 80 sccm, a substrate temperature of 600 ˚C, a substrate bias of -200 V, and an etching time of 30 min (Matsumoto & Mimura, 2004; Matsumoto & Mimura, 2005).

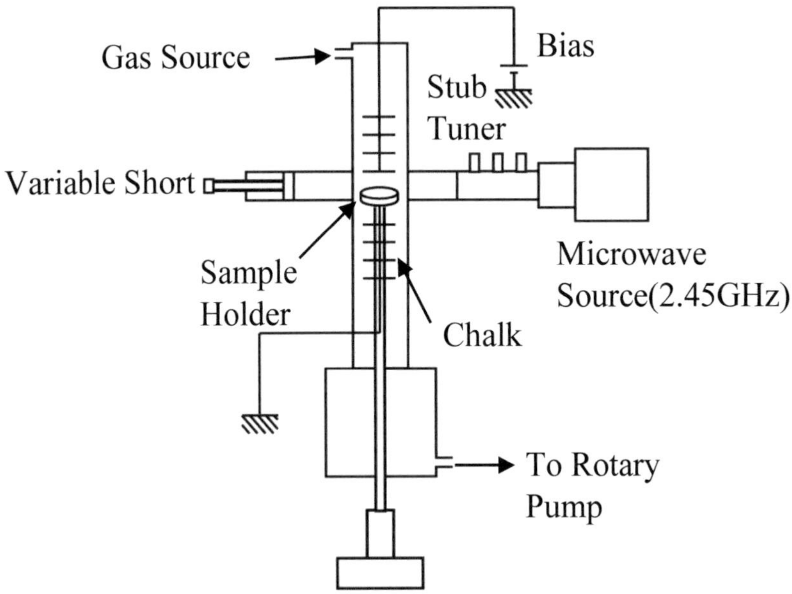

Fig. 1. Schematic of a microwave plasma CVD equipment for hydrogen plasma etching. (Reprinted with permission, Matsumoto & Mimura 2005, American Institute of Physics.)

A carbon rod with a diameter of 0.5 mm ϕ is mechanically sharpened at one end to less than a 10 μm ϕ diameter as shown in Fig. 2 (a) before the plasma etching and then nanoneedles are fabricated on the tip by the hydrogen plasma etching. An SEM image of a carbon rod after hydrogen plasma etching is shown in Fig. 2 (b). A lot of nanoneedles are fabricated on the spearhead region of the graphite rod. A transmission electron microscope (TEM) image of one nanoneedle is shown in Fig. 2 (c). Typical aspect ratio of the nanoneedle is on the order of 1000. The radius of curvature in the top region of the needle is less than 5 nm. This small radius and the high aspect ratio make it suitable for a field electron emission cathode. A high resolution TEM image of the nanoneedle is shown in Fig. 2 (d). A lattice fringe pattern is clearly observed from the bottom to the top of the needle as shown in Fig. 2 (d). Based on the lattice fringe and diffraction patterns (c axis) shown in Fig. 2 (d), the top region of the nanoneedle consists of a two-dimensional graphene sheet with an interplanar spacing of 0.36 nm (Matsumoto et al., 2007; Matsumoto et al., 2008). This value is larger than that of the hexagonal graphite structure (0.34 nm), which indicates that the c-axis lattice is relaxed. Another diffraction pattern (a axis) whose direction is orthogonal to the interplanar direction is observed. Based on the distance of a-axis diffraction patterns, we can determine the atomic level of spacing, and this is 0.21 nm. This value corresponds to (010) plane spacing of the six membered ring in the graphene sheet. Therefore, the interaction between the graphene sheets is weakened due to the expansion of the interplanar lattice spacing, thus the GNS possesses the two dimensional (2D) aspects compared to the three dimensional graphite structure. As well as the weakened interplaner interaction, the 2D graphene sheet structure with the lattice fringes from the bottom to the top direction is promising for field emission, because the carrier mobility and electron mass are shown to have exceptionally large (μ= 15000 cm^2 V^{-1} s^{-1}) and small (0.007m_e, m_e: free electron mass) values in this 2D system (Novoselov et al., 2005; Geim & Novoselov, 2007).

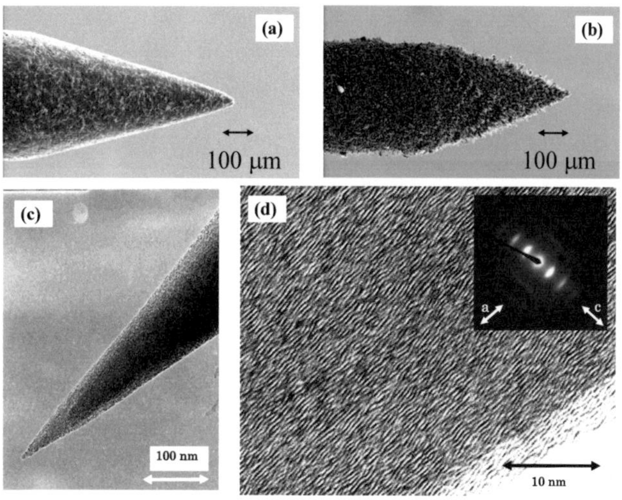

Fig. 2. SEM Image of a carbon rod: (a) before hydrogen plasma etching and (b) after hydrogen plasma etching. (c) TEM image of a single GNS emitter. (d) High-resolution TEM image. Inset shows the selected area electron diffraction pattern of a single GNS emitter. (Reprinted with permission, Matsumoto et al. 2007, American Institute of Physics.)

3. Field emission characteristics

Fowler and Nordheim (FN) first derived a semiclassical theory of field emission currents from cold metals in 1928 (Fowler & Nordheim, 1928). In this theory, the system is simplified as a one-dimensional structure along the direction of the external field. The emission tip is modeled as a semi-infinite quantum well. By employing the Wentzel-Kramers-Brillouin approximation, following FN plot is given.

$$ln\ (I/V^2) \propto -1/V \quad (1)$$

where I is in amperes per square centimeter of emitting surface and V is the applied voltage.

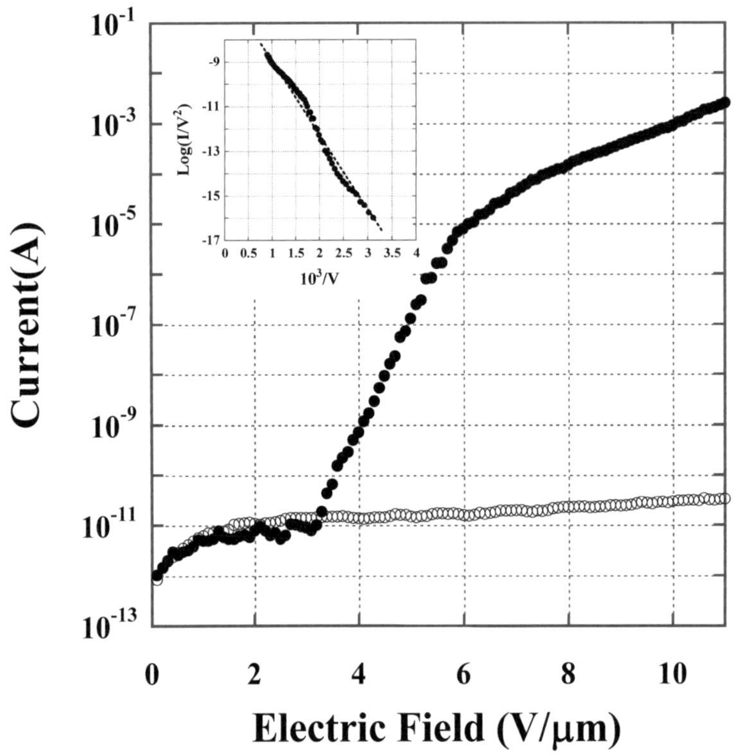

Fig. 3. Current-voltage characteristics for a mechanically sharpened carbon rod (open circles) and a GNS cathode (solid circles). Inset shows the Fowler-Nordheim plots ($\ln(I/V^2)$ vs 1/V) for the I-V characteristics of the GNS cathode. (Reprinted with permission, Matsumoto et al. 2007, American Institute of Physics.)

The above FN theory should not be applied to nanometer sized emitters such as GNS and CNT because the geometrical size of the tip is comparable to the electron wavelength. Much sophisticated emission theory is developed by several authors (He et al., 1991; Liang & Chen, 2008; Forbes, 2001), and these theories will explain the difference between the straight line obtained by the simple FN plot and the experimentally obtained slightly curved feature (dotted line) in the inset of Fig. 3. For the precise fitting by using these sophisticated theories, it is necessary to determine physical values such as a shape and a size as well as the electronic properties; e.g. the defect density and work function of the tip. However, as shown in Fig. 2, it is difficult to determine these values for our tip. Therefore, here we estimate the field emission characteristics by using the FN plots.

Figure 3 shows typical logarithmic current-voltage (I-V) characteristic of the GNS field emitter. The figure also shows I-V characteristic of a sharpened graphite rod without nanostructure. The currents were collected on a 3 mm ϕ diameter aluminum anode, which was located at 100 μm in front of the cathode. The measurements were carried out in a vacuum chamber with a residual pressure of 1 x 10^{-6} Pa. The mechanically sharpened graphite rod without nanostructure showed little field emission current (open circles), while the GNS emitter starts to emit electrons at an average electric field of about 3 V/μm and the emission current exceeds 2 mA at an applied electric field of 11 V/μm (solid circles). The FN plot of the emission current from the GNS emitter is shown in the inset of Fig. 3. Linear dependence of the Fowler-Nordheim plot suggests that the electron emission is dominated by the Fowler-Nordheim tunneling process as described in Eq. (1).

4. Emission current fluctuation and stochastic birth and death model

The electronic properties of nanomaterials are very sensitive to the adsorption of molecules (Kong et al., 2000; Zhao et al., 2002; Grujicic et al., 2003; Andzelm et al., 2006). This characteristic is clearly observed when measuring field emission-current fluctuations of carbon nanostructures, where the emission current stability is easily lost under high residual pressure and the stability is obtained by thermally desorbing adsorbed molecules onto the emission sites. This phenomenon can be explained as the adsorption and desorption of molecules onto emission sites of carbon nanostructures with the molecules adsorbing onto the nanostructures affecting the electronic properties, such as charge transfer and tunneling probability (Dyke & Dolan, 1956; Gadzuk & Plummer, 1973).

In this section, we show a model in which the field emission current fluctuation originates from the adsorption and desorption of molecules onto the emission sites, which is well described using a stochastic birth and death model (Feller, 1957). The emission current fluctuation was analyzed as a cathode temperature using differential equations obtained from the model. We derive the Poisson distribution for the deviation of the fluctuating emission current and finally, we show the method of the determination of physisorption energy of various molecules using the field emission current fluctuation.

Figure 4 shows the fluctuating emission current distribution measured in a 10^{-4} Pa H_2 atmosphere at different temperatures. The current intensity distribution at 300 K shown in Fig. 4 (solid circles) gives the current fluctuation deviation, $\Delta I/I_p = 0.31$, where ΔI is the full width at half maximum (FWHM) of the fluctuation and I_p is the peak value of the current. The deviation can be reduced by heating the cathode: $\Delta I/I_p = 0.21$ at 700 K (solid triangles), and $\Delta I/I_p = 0.16$ at 925 K (solid squares). This reduction in the deviation suggests that heating is an effective way to stabilize the field emission current. Along with the reduction of the deviation, the peak of the current distribution shifts to the lower current side with increasing cathode temperature. Both the reduction of the deviation and the lower peak shift of the current distribution can be interpreted qualitatively as the adsorption and desorption of atoms or ions on or off surface of the emission sites (Hosoki et al., 1979; Yamamoto et al., 1979).

Figure 5 illustrates the model in which the emission current fluctuation originates from the adsorption and desorption of atoms and/or ions. The current fluctuation occurs due to the occupation of the emission sites by the adsorbed atoms. Here, we postulate that the magnitude of the current (I) is proportional to the number of occupied states (n); *e.g.*, $I(n)=I_0+\eta n$, where I_0 is the emission current of un-occupied state, and η is the magnitude of the current-hop due to the adsorption of single molecule, to explain both the reduction of the deviation and the lower peak shift of the current distribution. For this model, we define the transition probability of the number of adsorbed atoms from state E_i to state E_j as

$$P_{ij}(t) = P(X(s+t) = j \mid X(s) = i) \tag{2}$$

where $X(s)$ is a random variable at time s. We postulate that the system changes only through transitions from states to their nearest neighbors. If at time t, the system is in state E_i, the probability $[P_{ii+1}(h)]$ that between t and $t+h$ the transition $E_i \rightarrow E_{i+1}$ occurs equals

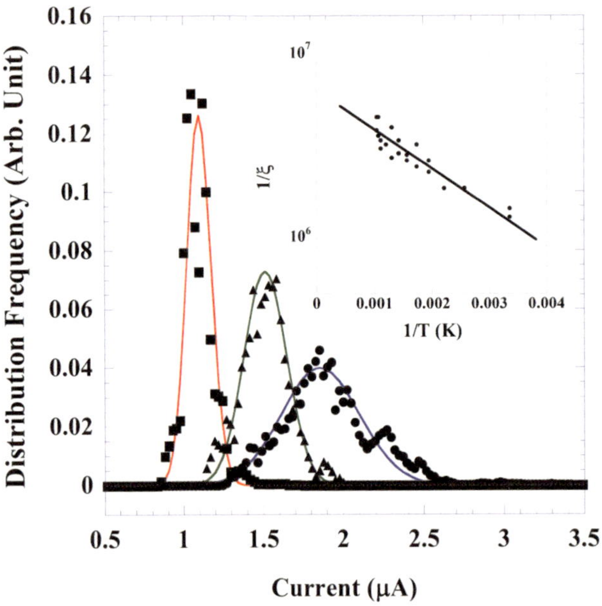

Fig. 4. Histogram of the emission current intensity at 300 K (solid circles), 700 K (solid triangles), and 925 K (solid squares) in a H_2 atmosphere; the solid lines are the theoretically fitted curves. The inverse of the variance for each temperature obtained by the theoretical fitting is shown as solid circles in the inset. The solid line in the inset is the fitted curve with a physical adsorption energy of 45 meV. (Reprinted with permission, Matsumoto et al. 2008, American Physical Society.)

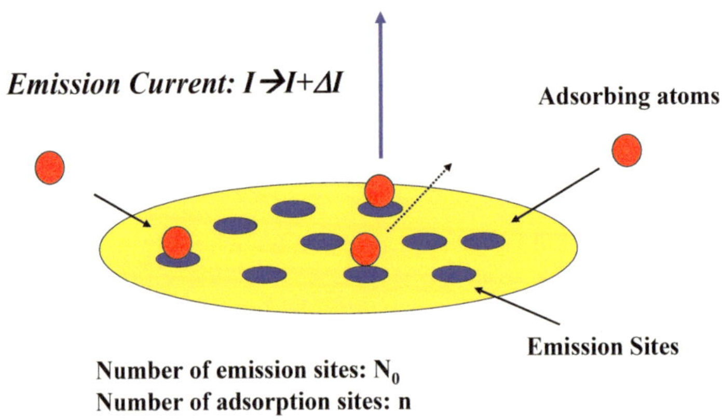

Fig. 5. Physical desorption model in which fluctuation of the emission current originates from the adsorption and desorption of atoms and/or molecules onto the emission sites. (Reprinted with permission, Matsumoto et al. 2008, American Physical Society.)

$\lambda_i h+o(h)$, and the probability $[P_{ii-1}(h)]$ of $E_i \rightarrow E_{i-1}$ equals $\mu_i+o(h)$. The probability that during $(t, t+h)$ more than one change occurs is $o(h)$, where λ_i corresponds to the adsorption rate, μ_i corresponds to the desorption rate, $o(h)$ denotes a small quantity of order of magnitude h, and h is the subinterval of time duration $h=1/N$ (N: total interval, and generally $h\rightarrow 0$). That is; the adsorption and desorption rates in state E_i are equal to the sum of three distinct ways as shown in Fig. 6 and they can be written as

$$P_{ii+1}(h) = \lambda_i h + o(h) \tag{3}$$

$$P_{ii-1}(h) = \mu_i h + o(h) \tag{4}$$

$$P_{ii}(h) = 1 - (\lambda_i + \mu_i)h + o(h) \tag{5}$$

$$P_{ik}(h) = o(h) \qquad (|i-k| \geq 2) \tag{6}$$

$$P_{ij}(h) = \delta_{ij} \tag{7}$$

We consider that the adsorption and desorption processes shown in Fig. 5 can be described quantitatively as time-homogeneous Markov processes satisfying the following Chapman-Kolmogorov equation (Feller, 1957);

$$P_{ij}(t+h) = \sum_k P_{ik}(t) P_{kj}(h) \tag{8}$$

By taking $k=j-1, j, j+1$ for the summation in Eq. (8) and using Eqs. (2) to (7), we obtain the following differential equations:

$$\frac{dP_{ij}(t)}{dt} = \mu_{j+1} P_{ij+1}(t) + \lambda_{j-1} P_{ij-1}(t) - (\lambda_j + \mu_j) P_{ij}(t) \tag{9}$$

$$\frac{dP_{i0}(t)}{dt} = \mu_1 P_{i1}(t) - \lambda_0 P_{i0}(t) \tag{10}$$

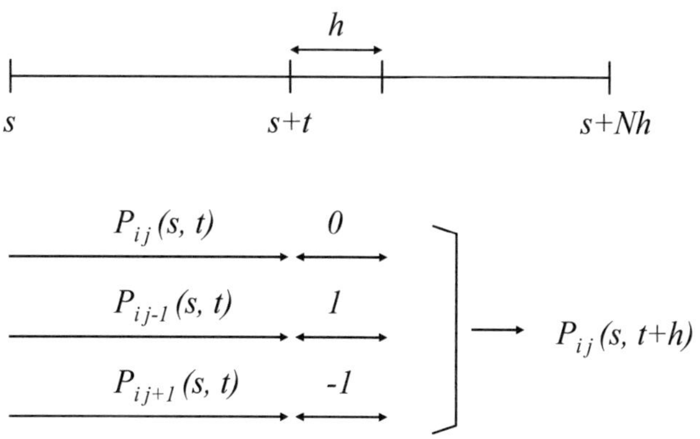

Fig. 6. Time intervals used in the calculation of $P_{ij}(t)$ showing the three ways in which j states are adsorbed in the period between s and $s+t+h$.

To describe the adsorption and desorption processes shown in Fig. 5, we assume that the rate λ_j is proportional to the number of unoccupied emission sites, while the rate μ_j is proportional to the number of occupied sites. Defining N_0 as the total number of emission sites and j as the number of occupied sites, we set

$$\lambda_j = \alpha(N_0 - j) \tag{11}$$

$$\mu_j = \beta j \tag{12}$$

where α and β are constants, such that α depends linearly on both the current density and the residual pressure and β depends on the temperature of the cathode. For example, $\beta \propto exp[-E_{ad}/k_B T]$, where E_{ad} is the physical adsorption energy, T is the temperature, and k_B is the Boltzmann constant. Considering the stationary distribution for Eqs. (9) to (12), the limits $\lim P_{ij}(t) = p_j$, $t \to \infty$, exist and are independent of the initial conditions (Matsumoto et al., 2008). Therefore, Eqs. (9) to (12) can be combined to express the stationary distribution p_n as the following Poisson distribution:

$$p_n = \frac{\xi^n}{n!} \exp(-\xi) \tag{13}$$

where $\xi = N_0 \alpha / (\alpha + \beta)$ and we assume the desorption ratio, β, is much larger than the adsorption ratio, α.
The solid lines in Fig. 4 are the theoretically fitted curves given by Eq. (13), where the best fits were obtained using the fitted values, $\xi = 2.76\times10^{-7}$ for 920 K (red line), $\xi = 3.45\times10^{-7}$ for 700 K (green line), and $\xi = 7.72\times10^{-7}$ for 300 K (blue line). By fitting the histogram of the current fluctuation at various cathode temperatures with Eq. (13), and then plotting these theoretically determined values logarithmically [$ln(1/\xi)$] as a function of the inverse of the temperature, $1/T$,

$$ln(1/\xi) \propto - Ead \cdot (1/k_B T) \tag{14}$$

we can determine the physical adsorption energy, E_{ad}, from the slope, as shown in the inset of Fig. 4. The solid line shown in the inset of Fig. 4 is the theoretical line computed with a physisorption energy of E_{ad} = 45 meV. The adsorption energies of H_2 molecules on graphite-like surfaces show a wide range of binding energies from 20 to 80 meV (Jacobson et al., 2002; Tran et al., 2002) because the adsorption energy of molecules on the carbon nanostructures varies as the size (*e.g.*, the tube diameter), coordinate sites, and the surface structure of the carbon nanomaterials change. We cannot distinguish the adsorption sites of the H_2 molecules onto the emission sites of the carbon nanostructures; however, the physisorption energy of E_{ad} = 45 meV is similar to the energy onto a graphite surface (Vidali et al., 1991).
The error obtained by this method is within ±5 meV, which allows the determination of the physisorption energy of various molecules to a high degree of accuracy. Figure 7 shows the values of $1/\xi$ obtained by measuring the emission current fluctuations as a function of various temperatures for CO (green circles), Ar (blue squares), and He molecules (red triangles). The slope of the solid line shows the theoretically determined physical adsorption energies, E_{ad}: 110 meV for CO molecules, 100 meV for Ar, and 15 meV for He. These graphene nanoneedle physisorption energies are similar to the physisorption energies for the graphite surface (Vidali et al., 1991).

We cannot rule out the possibility that the current fluctuation occurs due to variations in the adsorbate coverage on step edges and defects, where the enhancement of physisorption energy occurs (Ulbricht et al., 2002; Ulbricht et al., 2006). However, polar adsorbates such as CO molecules on the graphite nanostructure showed similar adsorption energy to those on the graphite basal surface, and we did not observe enhancement of the binding energy due to a dipole interaction between polar molecules and step edges or defects. Therefore, we consider that the electron emission occurs not from the edge of the nanoneedle, but from the basal plane of the graphene sheet. Next section, we will study the emission mechanism of graphene nanoneedles from the point of view of the nanometer sized region by using field emission and field ion microscopes.

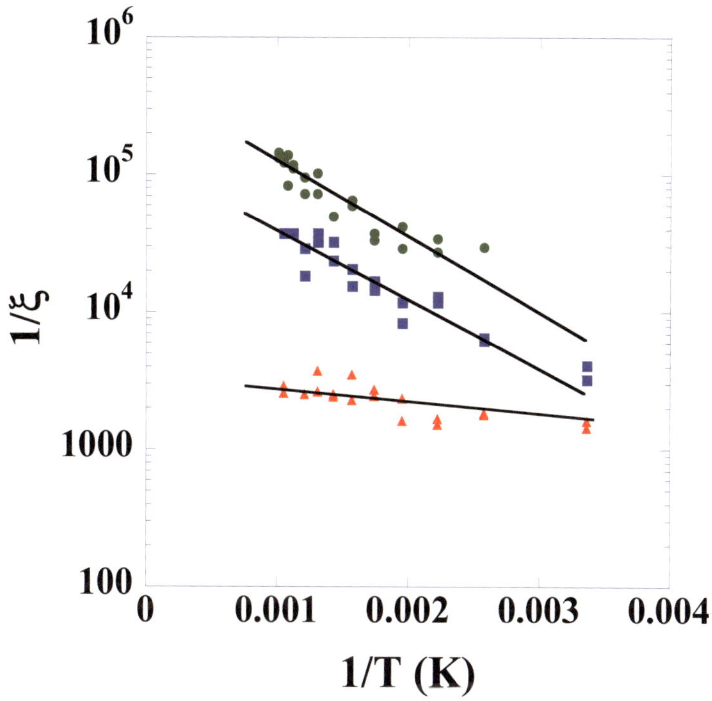

Fig. 7. Inverse of the variance, $1/\xi$, for CO molecules (green circles), Ar molecules (blue squares), and He molecules (red triangles) as a function of GNS cathode temperature. The solid lines are the fitted curves with physical adsorption energies of 110 meV for CO molecules, 100 meV for Ar, and 15 meV for He. (Reprinted with permission, Matsumoto et al. 2008, American Physical Society.)

5. Microscope images of emission patterns and their quantum states

5.1 Field emission and field ion microscope images

In section 4, we simply measured the total current (the ensemble of the emission sites) and discussed the statistical dynamics by using a stochastic differential equation. In this section, the ensemble of the emission sites will be resolved into a number of single emission site by using a field emission microscope (FEM) and a field ion microscope (FIM). We show that each emission site of GNS creates many beautiful symmetrical patterns. These symmetrical patterns are known as cloverleaf patterns of organic molecules and the origin of the cloverleaf patterns has been discussed more than 60 years; however, this is still open question (Müller, 1953; Ashworth, 1951; Dyke & Dolan, 1956; Becker, 1955; Wolf, 1955; Dean, 2002). In order to solve this difficult problem, we try to reproduce the same cloverleaf patterns by using historically used organic molecules such as phthalocyanine and pentacene. By considering phase information for the obtained symmetrical patterns, which is a first attempt to explain these symmetrical patterns, and by applying the diffraction analysis to the tunneling electron observed by FIM (near field wave function), we succeed in explaining how the near field pattern obtained by FIM changes into the far field pattern obtained by FEM. Our final conclusion for the cloverleaf pattern is that the electron emission originates from anti-bonding states (lowest unoccupied molecular orbital) of π-conjugated bonds (six-membered ring) in the graphene sheet (Neo et al., 2010).

Our ultra-high-vacuum experimental apparatus includes both FEM and FIM chambers. In order to obtain the same information for FEM and FIM, FEM and FIM measurements were successively performed through a load-lock and a preparation chamber in order to avoid exposure to air. The base pressures in the FEM and FIM chambers are below 1×10^{-7} Pa and 2×10^{-8} Pa, respectively. The FEM apparatus consists of a field emission cathode mounted on an x-y-z-manipulator, a phosphor screen, and an evaporation boat to adsorb organic molecules onto a tungsten tip. A GNS structure and an electrically-polished tungsten (011) tip were attached to 0.15 mm diameter tungsten filaments. It is possible to heat the cathode by resistive-heating of a tungsten filament. A phosphor screen on an ITO glass substrate was placed in front of the tip in the FEM chamber, with a 50 mm gap. The adsorption processes of the organic molecules were monitored in-situ by observations of FEM images. The adsorption conditions were easily controlled by adjusting the boat temperature and hence its vapor pressure. The FIM chamber consists of a helium cryo-cold head and a micro-channel plate (MCP). The MCP with a phosphor screen was placed 50 mm in front of the cathode. The cathode was cooled by a cold head up to 9 K. Two variable-leak valves were included in order to introduce helium and hydrogen gas molecules as imaging particles under FIM operation. The FEM and FIM images were recorded using a charge-coupled device (CCD) camera.

After flash-heating for a few minutes at around 1000 K, field emission could typically be initiated in our FEM apparatus, even when low voltages of less than -2 kV were applied to the cathode. Figure 8 (a) shows a phosphor screen image of the electrons emitted from a GNS cathode, and it can be observed that each of the bright-spots shows symmetrical patterns such as two-lobed, four-lobed, circle and ring patterns as shown in Fig. 8 (b). These patterns exhibit some of the following tendencies; (i) each of the patterns could rotate discontinuously and (ii) could change reversibly into another pattern. (iii) Interestingly, no patterns with an odd number of lobes (e.g. three-lobed patterns) were observed. (iv) The two parts of each two-lobed pattern and the four parts of each four-lobed pattern were always symmetrical, had the same brightness, and vanished at the same time. These behaviors of the cloverleaf patterns were similar to those observed for other types of organic molecules such as phthalocyanine and pentacene, and we will show similar symmetrical patterns obtained by these molecules in Fig. 10.

After the FEM investigations, the GNS emitter was transferred to the FIM chamber. The emitter was cooled to 25 K, which was above the hydrogen triple point. Hydrogen up to a pressure of 1×10^{-2} Pa was introduced into the FIM chamber as an imaging gas, and then high positive voltages were applied to the GNS cathode. Figures 9 (a), (b) and (c) show the successive FIM images in a sequence in which the tip voltage was increased up to +10 kV. In the high-voltage region (over +10 kV) bright lines due to reflected multi-layers structures (Williams, 1968; Murr & Inal, 1971) were observed, as shown in Fig. 9 (c). However, in the weak-field region, the cloverleaf patterns observed in the FEM images, were also appeared in the FIM measurements. Figure 9 (d) shows a collection of the patterns such as those observed in the weak region (around +5 kV). These patterns were almost the same as the cloverleaf patterns obtained in the field emission regime. The same symmetrical patterns obtained by both FEM and FIM are interesting because the FIM gives the near field images such as the wave function of tunneling electron, while the FEM gives the far field images.

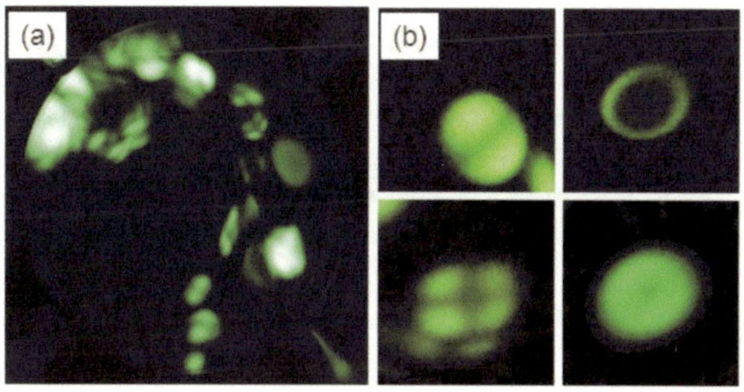

Fig. 8. (a) FEM images of a GNS cathode on a phosphor screen at 2.2 kV applied voltage, (b) the four typical kinds of cloverleaf patterns observed with GNS; two-lobed, four-lobed, ring and circle. (Reprinted with permission, Neo et al. 2010, American Institute of Physics.)

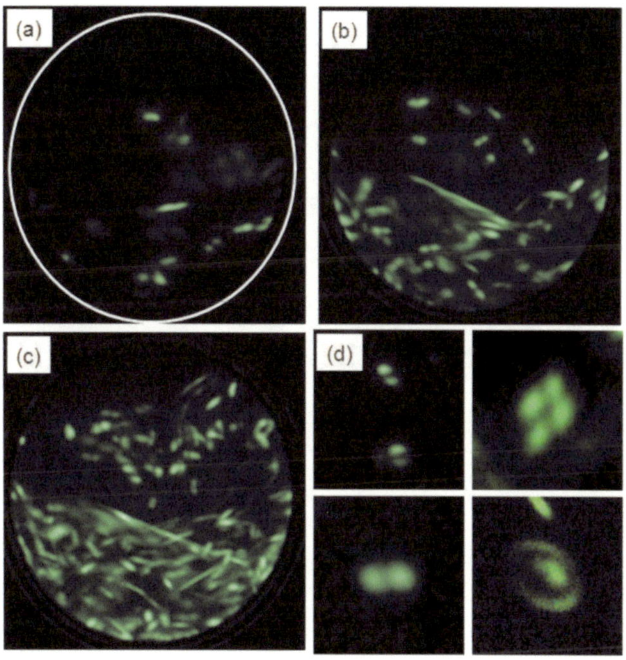

Fig. 9. A series of FIM images obtained with GNS in a sequence of increasing applied voltages; (a) 4 kV, (b) 5 kV, (c) 10 kV. (d) Collection of typical FIM patterns. (Reprinted with permission, Neo et al. 2010, American Institute of Physics.)

5.2 Field emission and field ion microscope images of Cu-phthalocyanine (Cu-Pc) and pentacene

We observed various symmetrical FEM and FIM patterns in GNS emitters such as two-lobed, four-lobed, circle and ring patterns. The patterns observed in GNS are very similar to those previously-observed in organic small molecules such as pentacene and Cu-Pc. Here we reproduce the cloverleaf patterns by the evaporation of pentacene and Cu-Pc onto a tungsten tip.

An electrically-polished tungsten (011) tip was mounted and an evaporation boat containing small quantities of these molecules was equipped in the FEM chamber. After cleaning the tungsten surface by heating to around 2200 K for 30 seconds, the evaporation boat was heated electrically, and then the field emission occurred from the clean tungsten surface. When several cloverleaf patterns appeared on the phosphor screen, the electrical heating of the evaporation boat was stopped. Figures 10 (a) and (b) show the cloverleaf patterns of FEM obtained in pentacene and Cu-Pc, respectively. The insets in the figure indicate the typical patterns that were observed for each molecule. As was observed in GNS, the same two-lobed, four-lobed, circle and ring patterns were observed in both organic molecules, especially clearly observed in Cu-Pc (Müller, 1953). The FEM images of these molecules, representing far field pattern of electron wave, show the same symmetrical patterns as those observed in GNS structures.

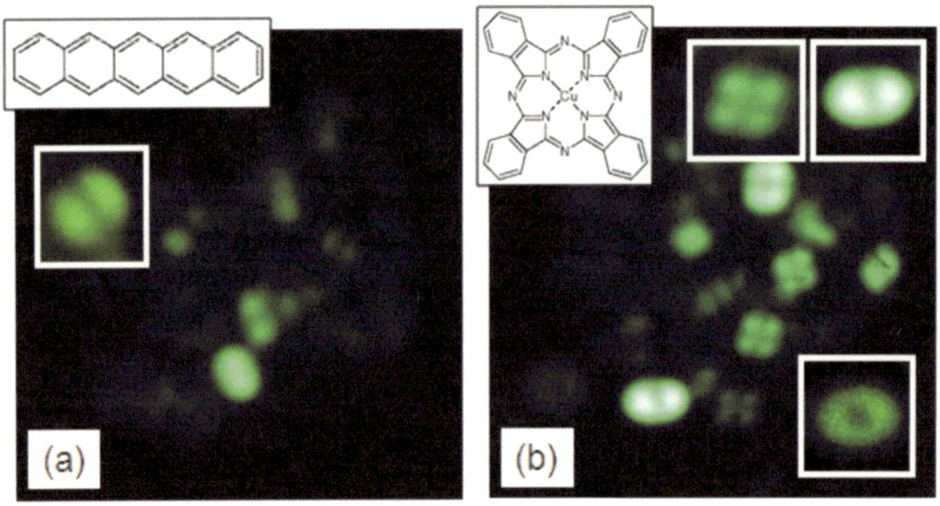

Fig. 10. FEM images with (a) pentacene and (b) Cu-Pc, respectively. The insets indicate the molecular structures and the cloverleaf patterns typically observed in FEM. (Reprinted with permission, Neo et al. 2010, American Institute of Physics.)

The electronic states of the emission sites (near field images) for pentacene and Cu-Pc were also investigated using the FIM method. Figures 11 (a) and (b) show the collected FIM patterns that were observed with pentacene and Cu-Pc, respectively. The observation of the cloverleaf patterns were performed at low electric field region around +4 kV. This value was much lower than that observed for the FIM images of tungsten surface (over +10 kV) because the adsorbed molecules are easily desorbed by the high electrical field. For pentacene, two-lobed patterns were mainly observed; on the other hand, two- and four-lobed patterns were clearly observed for Cu-Pc. These results agree well with the tendencies that were observed in a series of FEM measurements. However, the agreement between the near field FIM patterns and the far field FEM patterns is not obvious and contains interesting information because the electron wave function tunneling through these molecules will be distorted by diffraction. That is, the FIM probes the electronic states of the near field emission region, while the FEM shows a far field pattern of the diffracted electron wave function emitted from the nanometer sized region. It looks like a contradiction; however, if we consider phase information for the symmetrical patterns, we show that this agreement is physically natural when we consider the emission originating from the anti-bonding states. Next, we will theoretically analyze the symmetrical patterns (wave function) by adding the phase information. We will analyze the near field and the far field patterns based both on the principle of least action and on the diffraction of the electron wave optics.

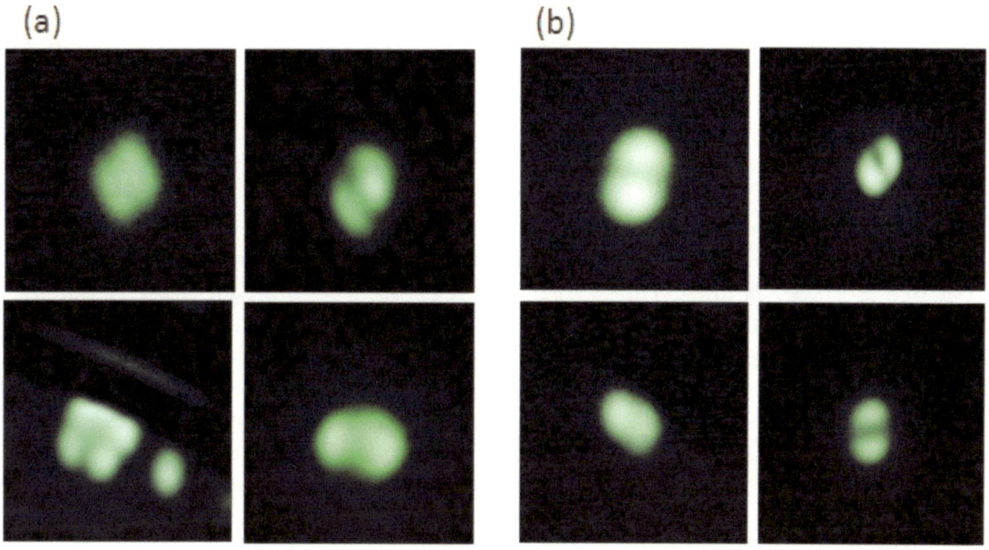

Fig. 11. Collection of cloverleaf patterns observed at around +4 kV with (a) Cu-Pc and (b) pentacene. (Reprinted with permission, Neo et al. 2010, American Institute of Physics.)

5.3 Diffraction of tunneling electron

Here electron wave optics is applied to analyze both the FEM and FIM patterns. The theoretical description of electron wave optics is almost the same as that of electromagnetic wave optics (Born & Wolf, 1999). The difference between the electron optics and optics is the treatment of wavevector, and in the electron optics, wavenumber depends on the potential. However, qualitatively, both the electron optics and optics give almost the same formalism in spite of the difference of the phase (wavenumber) treatment. The key to understand diffraction images is a popularly known Fourier transformation. The detail of the electron wave optics was given in the following papers (Couley, 1995; Ximen, 1994) and here we show final expression. The dffracted electron beam can be expressed by the convolution of the near field wave function and the spherical wave as

$$\Psi(x_2, y_2) = A(\Phi, z)\iint dx_1 dy_1\, \Psi(x_1, y_1)\, \frac{z_{12}}{r_{12}} \frac{e^{i\vartheta}}{r_{12}} \tag{15}$$

where $\Psi(x_1, y_1)$ is the electron wave function obtained by FIM pattern (near field image at x_1 and y_1 with phase information), $\Psi(x_2, y_2)$ is the electron wave function obtained by FEM pattern (far field image at x_2 and y_2 with phase information), $A(\Phi,z)$ is the amplitude determined by Schrödinger equation with electron potential Φ, z_{12}/r_{12} is the obliquity factor where r_{12} is the distance between (x_2, y_2, z_2) and (x_1, y_1, z_1), and $z_{12}=|z_2-z_1|$, and $e^{i\vartheta}/r_{12}$ is the spherical wave of which phase ϑ is determined by the motion of electron in a potential. By applying the principle of least action, this phase can be expressed as a simple form like $\vartheta=k_0r_{12}+\varphi(z)$ where $k_0 = [2em\Phi(z_1)]^{1/2}/\hbar$ is a wavenumber of electron wave at $z=z_1$ and $\varphi(z)$ can be treated as a constant value in this diffraction process. At Fraunhofer region as shown in Fig. 12, Eq. (15) becomes a Fourier transform relation between the near field pattern and the far field pattern as

$$\Psi(x_2, y_2) \propto \iint dx_1dy_1\, \Psi(x_1, y_1)\, exp\,[\frac{ik_0}{z_{12}}(x_1x_2 + y_1y_2)] \tag{16}$$

Therefore, we can calculate the near field pattern from the far field pattern or vice versa easily by using the Fourier transform relation obtained in Eq. (16).

Next we consider the phase of the wave function because the intensity measurements of FEM and FIM do not contain phase information. If we consider the phase for the obtained two-lobe patterns of FIM, we can consider two cases; each lobe has the same in-phase or different anti-phase. Figure 13 shows theoretically computed results for two-lobe pattern of (a) in-phase and (b) anti-phase. When the two-lobes have the same phase, calculated far field image becomes an intense circular spot with side bands ringing. On the other hand, when the two lobes have the anti-phase, interestingly, calculated far field image becomes topologically the same as that of near field image. We also show the calculated results for four-lobe patterns in Fig. 13 (c) in-phase and (d) anti-phase. The agreement between the near field FIM patterns and the far field FEM patterns indicates that the electron wave function tunneling through these molecules has the anti-symmetrical phase. Based on our series of FIM and FEM measurements for various molecules including Cu-Pc and pentacene, π conjugated bonds are essential to form the symmetrical patterns. The anti-symmetry of π conjugated bonds naturally suggests the conclusion that the symmetrical leaf pattern originates from the lowest unoccupied molecular orbitals.

Many symmetrical patterns were observed in various organic molecules with π conjugated bonds, such as six-lobed and ring patterns as well as the two- and four-lobed patterns. It is interesting to consider these patterns by correlating with the π conjugated bonds of six membered ring. By considering the confinement states in a cylindrical coordinate, that is, by solving the Schrödinger equation in this boundary condition, we can obtain following wave function described by Bessel function as

$$\Psi_{nm}(r,\theta) \propto \frac{1}{|J_{n+1}(\xi_{n,m})|} J_n(\xi_{n,m}\cdot r / r_0)\, exp\,(in\theta) \tag{17}$$

where coordinate is transformed from Cartesian coordinate to the cylindrical coordinate, n is the mode number, m is the order number, $J_n(\xi_{nm})$ is the nth order Bessel function and ξ_{nm} is the mth value of the solution of $J_n(\xi_{nm})=0$. By solving Eq. (17), we can obtain wave functions at the near field region. Figure 14 shows the experimentally observed symmetrical patterns, the solution of the wave function for lower quantum numbers, its intensity (square of the wave function), and the calculated far field intensity obtained by Fourier transformation. Almost all the patterns obtained by both FEM and FIM measurements were reproduced by this computing analysis of Eq. (17). The difference of the observed patterns originates from the difference of the energy level of π-electron at excited states (LUMO states).

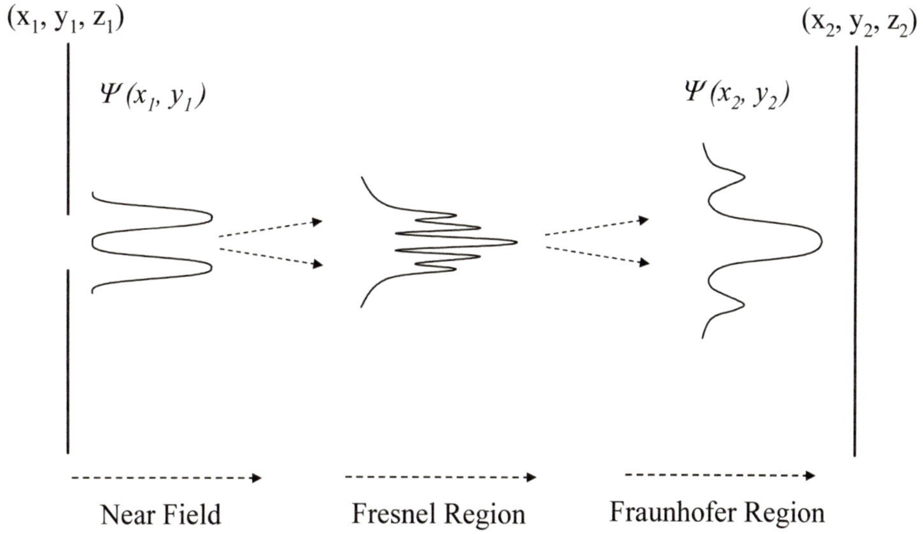

Fig. 12. Schematic image of the diffraction of electron wave and the change of the wave function from near field to far field (Fraunhofer region).

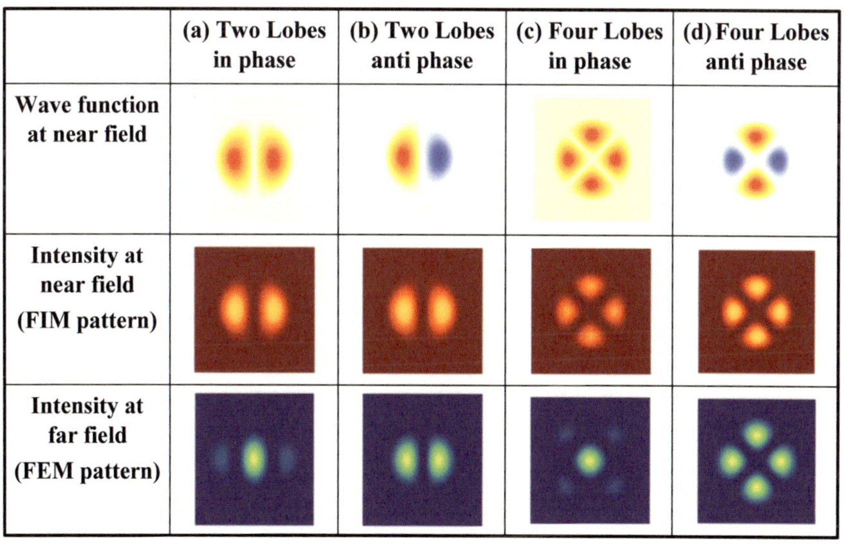

Fig. 13. Theoretically calculated near field and far field intensities for two-lobes and four-lobes patterns. The phase of the wave function is introduced as in-phase [(a) and (c)] or anti-phase [(b) and (d)].

Quantitative discussion related with the magnitude of the energy difference for the different quantum number is necessary to be clarified; however, qualitatively, adsorbed or desorbed atoms on and off π conjugated ring cavity (emission sites) would change the energy level of the cavity modes, thus the patterns rotate discontinuously and change reversibly.

We note here, the quantum number n is degenerated; however, when the degeneracy is solved by the existence of a magnetic moment and/or a localized spin near the π conjugated cylindrical cavity, we can expect topologically interesting patterns such as an entangled pattern of electron wave function for FEM measurements (Müller, 1953; Ashworth, 1951; Dyke & Dolan, 1956; Becker, 1955).

The mystery of the same topological patterns between FIM and FEM were clearly solved by considering the phase of the electron wave function for both FIM and FEM patterns, and quantitatively analyzed by using the electron wave optics. The various emission patterns observed in GNS have the same origin as those observed in organic molecules, where the symmetrical patterns can be understood from the point of view of the excited states of the π electron of six membered ring in a two dimensional graphene sheet. Therefore, the patterns have the same brightness, and vanish at the same time. Furthermore, odd number of the lobe-pattern is prohibited to be appeared due to the confinement of cylindrical cavity. The change of the patterns can be understood from the point of view of the change of the energy level by the adsorption of atoms or molecules on and off the six membered ring. Coherence of the tunnelling electron from a two dimensional graphene sheet is regarded as high because there is no inelastic scattering process. Thus, we consider that the graphene is a promising candidate for a coherent electron source. Many new physics unfold before us such as the superposition of coherent electron emission from a graphene sheet, opening a way for making a plane wave and an amplification of electron beam.

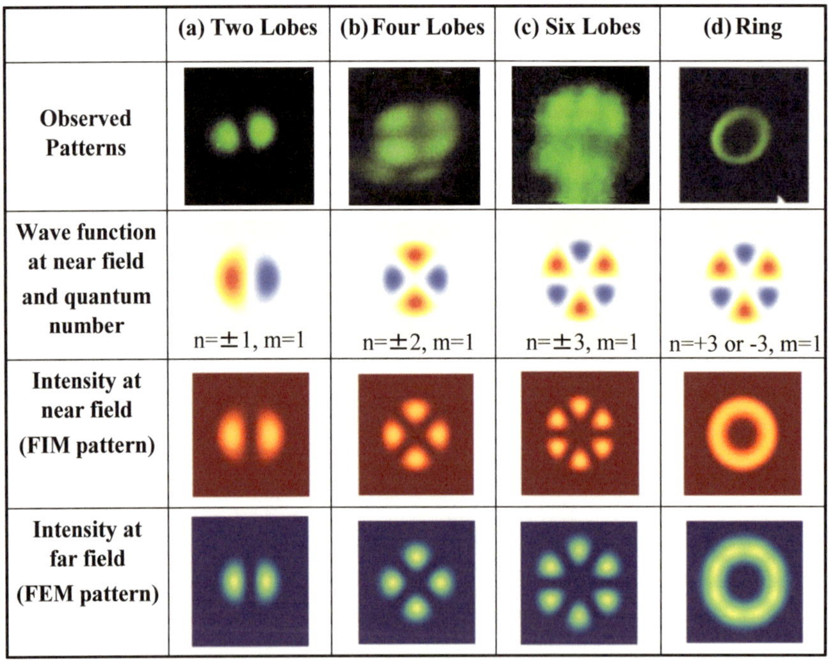

Fig. 14. Experimentally observed cloverleaf patterns and theoretically calculated wave functions, near field intensity, and far field intensity.

6. Applications

6.1 X-ray applications

Recently achieved electron emission from carbon nanotubes (CNT) and diamond films show promising aspects due to their possible use in practical cold cathode for the applications such as flat panel displays, x-ray tubes, and microwave amplifier. However, these emission sources still have problems related to adherence of the interface between carbon layers and its substrate. Furthermore, the advantage of a high gas storage capacity of CNT acts as disadvantage for the fabrication of highly evacuated vacuum devices.

We overviewed superior aspects of field emission properties in GNS cold cathode. Especially, unsaturated behavior of electron emission and stable emission in a high residual pressure region of the order of 10^{-4} Pa (We will introduce the results of the emission stability later in the FE-SEM section.) are distinguishing characteristics compared to other types of carbon emitters. This is due to the fabrication of carbon nanostructure onto a carbon substrate. Both the high current density and the stable emission characteristic are desirable for high power operation of devices such as x-rays, electron microscopes, and microwave generators. In this section, we show the performance of the GNS as a cold cathode was demonstrated by various applications. Firstly, we show the results of high intensity and short pulse x-ray generation and then introduce the demonstration of stop motion of x-ray transmission images. Fast motions of the order of 10 μs are obtained by using GNS cathode in a convenient manner. Secondly, we will introduce the fabrication of miniature sized x-ray tubes with the GNS emitter. Long lifetime of the order of 10,000 hours and the stability better than 1 % are emphasized. We believe that this x-ray tube is the first commercial base application using a graphene technology. Lastly, we show the results of the stable emission from GNS in a high residual pressure region, which is suitable for the construction of field emission type scanning electron microscope with a convenient manner.

Figure 15 (a) shows a schematic of a triode-type field emission x-ray tube composed of a GNS cathode, a metal grid (100-mesh placed 0.5 mm from the cathode), and a Cu metal target. The x-ray tube was evacuated by a turbo-molecular pump to a base pressure of 10^{-4} Pa. For pulse x-ray generation, negative pulse voltage with a peak height of 1-10 kV and pulse duration between 1 ms and 10 μs (repetition 1-100 Hz) was applied to the cathode. The metal grid was grounded and a constant positive bias of about 20 kV was applied to the anode. Figure 15 (b) shows the pulse voltage applied to the GNS cathode (pulse duration of 1 ms; dotted line) and generated x-ray pulses (solid line) detected by Gd_2O_2S:Eu phosphor with a photomultiplier. High intensity x-ray pulse was obtained by applying a negative pulse voltage to the cathode. A 1-ms-pulse duration for the applied voltage was used, because the response of our detection system was limited by the phosphor decay, which was about 500 μs. A much shorter x-ray pulse, on the order of 10 μs, was generated and by using this pulse we demonstrated in the following two applications that the GNS cold cathode can be used for high speed x-ray radiography.

In the first application, Fig. 16 (a) shows single-shot x-ray transmission image of a rotating chopper (7500 rpm) obtained by placing the chopper between the pulse x-ray emission source and a cooling type charge coupled device camera (CCD) with a Gd_2O_2S:Eu phosphor. The image was obtained at an applied anode bias of 25 kV (DC) and a cathode bias of -10 kV. Based on the angular velocity of the rotating chopper and the sharpness of the obtained image, the generated x-ray pulse duration was estimated to be about 10 μs.

In the second application [Fig. 16 (b)], in situ image of a rotating drill and the process of making a hole inside a wood plate were obtained using the same condition used in the first application. For this demonstration, we used a 2 mm-diameter drill (2600 rpm) and a 5-mm-thick wood plate. Images were obtained of the rotating drill moving inside of the wood plate, where this frame was detected by a single shot x-ray flush (10 μs duration). The advantage of the single shot x-ray detection is that clear dynamical transmission images of a fast motion can be obtained without the use of a sophisticated high-speed camera.

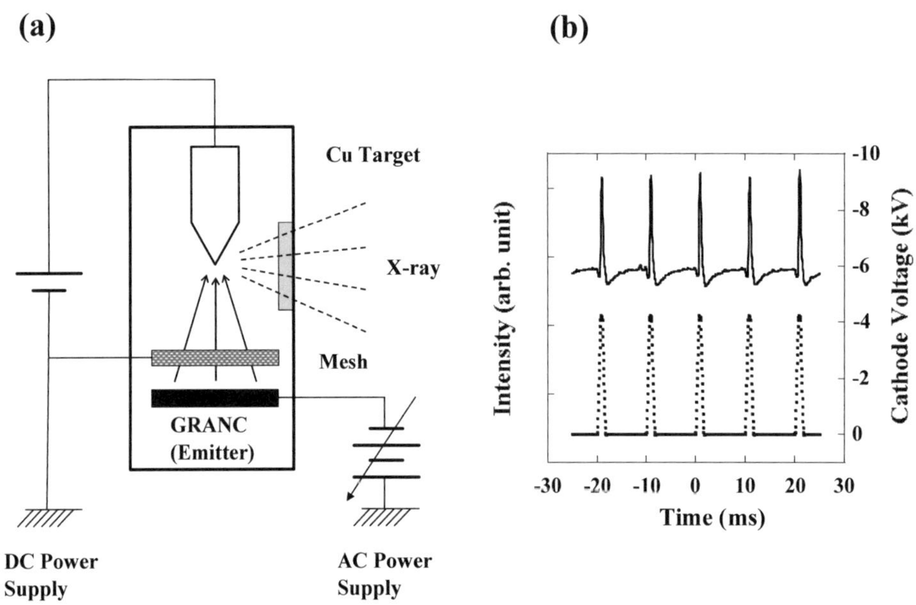

Fig. 15. (a) Schematic of a triode-type field emission x-ray tube, and (b) generated x-ray pulses. (Reprinted with permission, Matsumoto et al. 2004, American Institute of Physics.)

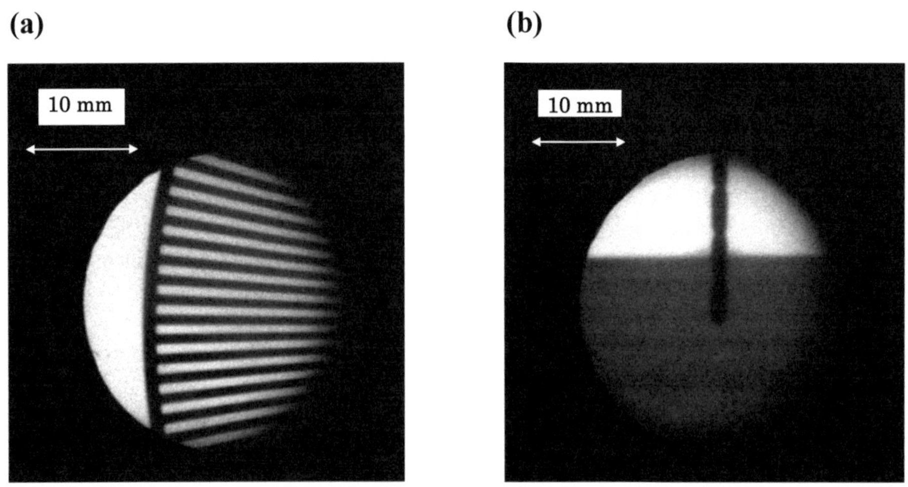

Fig. 16. Single shot x-ray transmission images of (a) rotating chopper (7500 rpm) and (b) rotating drill (2600 rpm) moving inside of the wood plate. (Reprinted with permission, Matsumoto et al. 2004, American Institute of Physics.)

6.2 X-ray tube fabrications

Bright and stable field electron emission from GNS seems to be promising to construct an x-ray vacuum tube from the point of view of a long longevity and a low power consumption as well as a high spatial resolution and the high speed transmission images compared to a conventionally used x-ray tube with thermal cathode (filament). In this section, we introduce our recent progress in the fabrication of a miniature sized x-ray tube with the GNS (Jyouzuka et al., 2010).

Figure 17 (a) shows a photograph of a fabricated miniature sized x-ray tube with GNS cathode. The size of the x-ray tube was 2.4 cm ϕ x 7 cm long. X-ray radiation was obtained through a Be window. The evacuation of the x-ray tube is important and this is performed by a turbo-molecular pump to a base pressure of less than 10^{-7} Pa. The bakeout temperature was above 300 °C. After the aging process, the x-ray tube was tipped off from the evacuation system. Figure 17 (b) shows the current-voltage characteristics of a typical x-ray tube with a GNS cathode. The GNS cathode was grounded and positive bias was applied to the Be window (anode). The distance between the cathode and the anode is about 1 mm. The emission current increased drastically at the applied voltage of 5 kV and the emission current exceeds 1 mA at the applied voltage of 8.5 kV. A high current from GNS offers a high power x-ray tube in a compact manner.

(a)

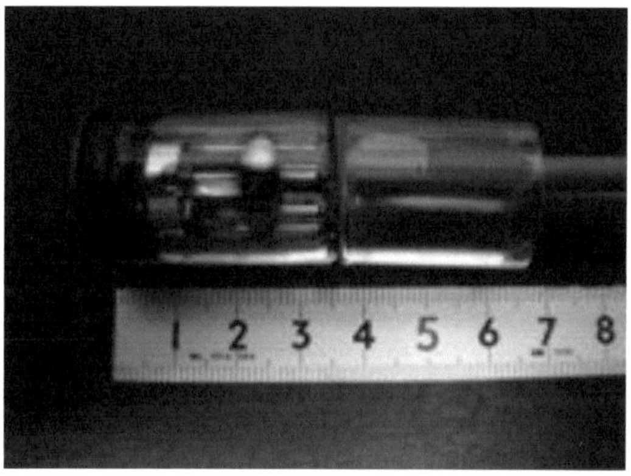

(b)

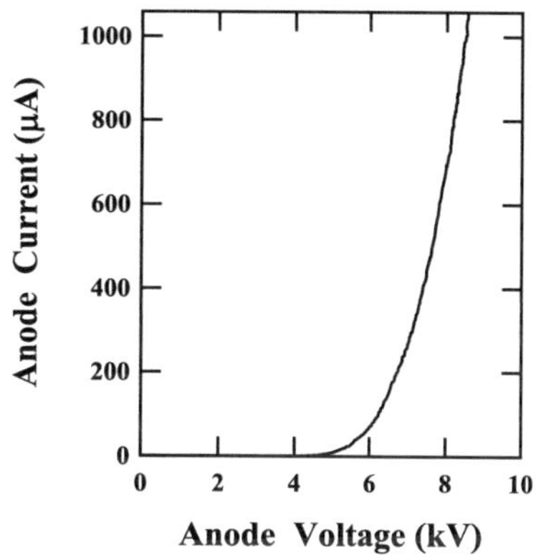

Fig. 17. (a) Photograph of a miniature sized x-ray tube with GNS cathode. (b) Current-voltage characteristics of an x-ray tube with GNS cathode. (Reprinted with permission, Jyouzuka et al. 2010, American Institute of Physics.)

As described in section 4, the current stability depends on the degree of vacuum level. To stabilize the emission current technically, we here employ a feedback system. The electric feedback diagram of the power supply is shown in Fig. 18 (a). The electric circuit supplies a negative high voltage to the GNS and the anode is grounded. The emission current from GNS is monitored by the current monitor connected to the GNS cathode. In order to stabilize the emission current, the feedback circuit receives the signal from the current monitor and then controls the control gate voltage. We note here that the control gate supplies a negative voltage which is possible to control the field enhancement factor of the GNS cathode. Figure 18 (b) shows the time dependence of the emission current (-20 kV) without (upper) and with (lower) the feedback control described above. Here we set an emission current less than 100 μA in order to make a noisy state of the emission current. In the measurement without the feedback control, the initial emission current was set at 50 μA.

A large noise current was observed and the deviation is estimated to be 7.7 μA. By operating the feedback control, we succeeded in reducing the noise current drastically, as shown in the lower figure of 18 (b) where the same voltage of -20 kV was applied to GNS. The emission current from GNS was set to be the same value of 50 μA. The large noise was significantly reduced and the deviation becomes 0.54 μA. Furthermore, no current flowing into the control gate was observed, implying that the control of the electric field (field enhancement factor) offers a very promising (efficient and fast speed) feedback method to stabilize the emission current in a simple manner.

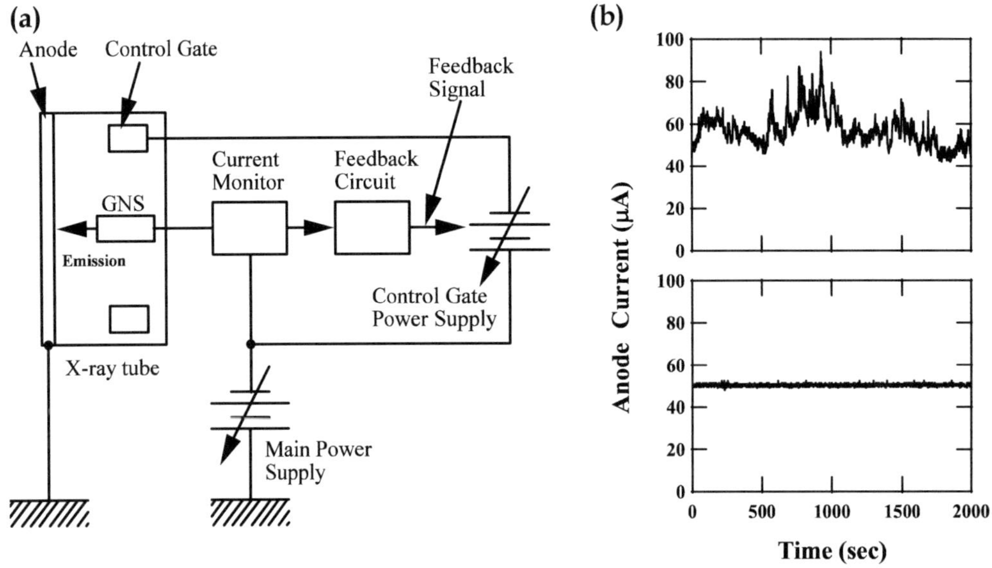

Fig. 18. (a) Feedback circuit to stabilize the emission current. (b) Time dependence of the emission current fluctuation without (upper) and with (lower) the feedback control. (Reprinted with permission, Jyouzuka et al. 2010, American Institute of Physics)

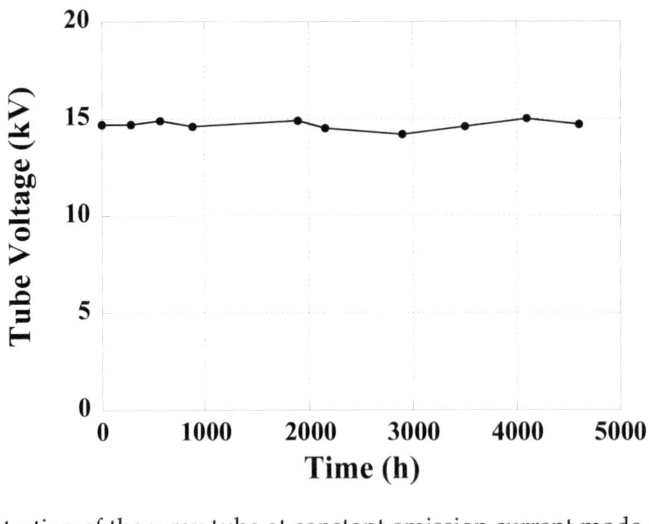

Fig. 19. Lifetime testing of the x-ray tube at constant emission current mode of 500 μA.

Lastly, we show the lifetime of the x-ray tube. Figure 19 shows the time dependence of the applied voltage at the constant current condition of 500 μA. We observed no degradation in the anode bias voltage to maintain the constant current and the lifetime of our x-ray tubes exceeds 5,000 h. The lifetime testing of the x-ray tube is now successively undergoing and it would exceed more than 10,000 h. Both the high emission current more than 1 mA, lifetime more than 10,000 hours, and the stability better than 1 % show very promising aspects of graphene emitters.

6.3 FE SEM applications

Intense electron emission at a high residual pressure of 10^{-4} Pa seems to be promising to construct a high-resolution electron microscope system without the need for a massive ultrahigh vacuum system. In this section, we show the performance of the GNS as a cold cathode was demonstrated by the construction of a compact FE scanning electron microscope (FE-SEM) system, where the brightness of the GNS cathode was determined to be on the order of 10^{12} $Asr^{-1}m^{-2}$ (Neo et al., 2006; Matsumoto et al, 2007).

A schematic of an SEM optical system equipped with a GNS cathode is shown in Fig. 20. The SEM experiments were performed under a residual pressure of about 10^{-4} Pa. This pressure was much higher than that of typical FE-SEM with a tungsten tip, which generally requires a low residual pressure, below 10^{-7} Pa. Either GNS or a tungsten filament thermal emission (TE) cathode was mounted on the SEM system. Electrode 1, shown in Fig. 20, was used as a wehnelt for a TE cathode and as an extracting gate electrode (0.1-1 kV) for a FE cathode. Other electrodes, such as a suppressor to focus the electron beam, were not included in this SEM system. A single final objective lens was used to focus the crossover on the target (sample holder), where the electron beam diameter and convergence angle were measured. The objective lens was composed of a permanent magnet and assistant coil and was designed to focus at 3.0 kV acceleration voltage. The aperture diameter was 0.3 mm-ϕ. The spatial resolution was evaluated by obtaining the images of a 4 μm-wide copper mesh located on the sample holder. In this lens configuration, the source size was reduced to 0.182 times at the sample target. In addition, a Faraday cup on the sample holder was used to collect the probe current.

Figure 21 (a) shows the SEM image of the copper mesh obtained using the TE cathode to compare both the resolution and brightness with those of the GNS cathode. The resolution of the image obtained using the TE cathode was smeared because of its large source size. The brightness B was calculated as

$$B = \frac{I}{dSd\Omega} = \frac{I}{\pi r^2 \pi \alpha^2} \tag{18}$$

where I is the probe current, dS is the source size, $d\Omega$ is the solid angle, r is the radius of the beam, and α is the open angle of the electron optics. For the TE cathode, the beam diameter $2r$ was estimated at about 4 μm and the measured I was about 0.6 μA. Therefore, the brightness B of the TE cathode was estimated to be about 2×10^8 $Am^{-2}sr^{-1}$, which coincides with the brightness reported for a tungsten filament TE cathode (Joy et al., 1986). This B value of the TE cathode shows that this method is appropriate for estimating B.

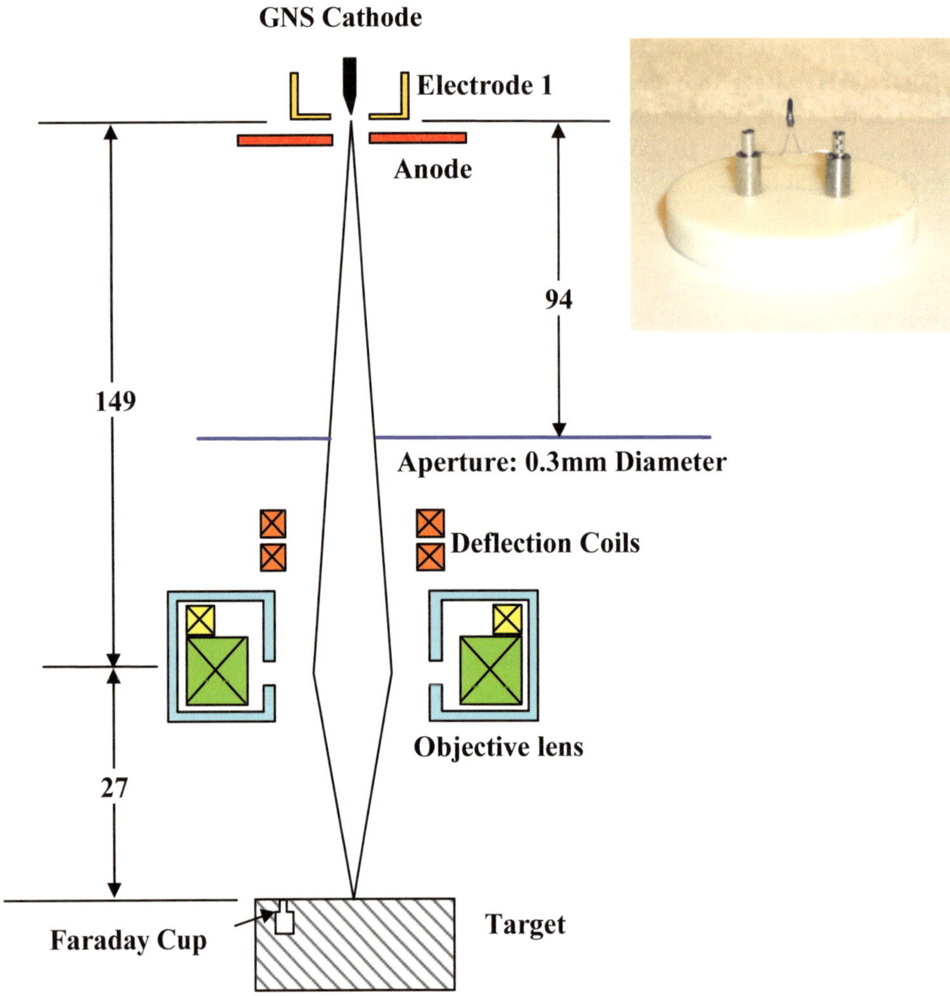

Fig. 20. FE-SEM system equipped with a GNS cathode. (Reprinted with permission, Matsumoto et al. 2007, American Institute of Physics.)

Figure 21 (b) shows the SEM image of the copper mesh obtained using the GNS cathode. The spatial resolution was clearly improved compared to that obtained using the tungsten filament TE cathode. This result shows that an SEM image was obtained using the GNS cathode despite a high residual pressure on the order of 10^{-4} Pa. However, as shown in Fig. 21 (b), many horizontal noise lines were observed. This is because the electron beam fluctuation occurred during the scanning of the electron beam to obtain SEM images, which is due to ion bombardment and/or atom adsorption as was described in section 4.

The fluctuation of the emission current originates from the adsorption and desorption of atoms and/or molecules onto emission sites of GNS cathode. To stabilize the emission current, generally, an evacuation of the residual gas to the degree of ultrahigh vacuum (10^{-9} Pa) is performed, and this can be achieved with a massive and costly vacuum system. On the other hand, in section 4, we showed that the number of adsorbed atoms per unit time depends on the temperature of the cathode. Next, we show the reduction of the current fluctuation by heating the GNS cathode, and show clearer FE-SEM images compared to those obtained with the FE-SEM without heating of GNS cathode at a residual pressure above 10^{-3} Pa.

FE-SEM optical system equipped with a GNS-gun designed is the same as described in Fig. 20, where the GNS cathode was attached on the W-filament by using a graphite dispersion in order to heat the cathode, where the temperature of the GNS cathode was measured by a piro- or a radiation-thermometer.

Figure 22 (a) shows FE-SEM images obtained using the GNS cathode at 1200 K. The stability of the emission current was improved compared to that of the GNS cathode without heating, thus leading to acquire clearer images compared to those obtained using the GNS cathode at room temperature. Especially, copper grains are clearly evident, which indicates that the emission source size was reduced compared to TE cathode and that the heated GNS cathode was stable during the acquisition of SEM image (60 s). Figure 22 (b) shows the highest resolution image obtained by using the GNS cathode. Based on this result, the maximum spatial resolution of this SEM optical system was analyzed by the distance between Cu grains, and it was estimated to be 30 nm, which is indicated by the solid white lines shown in Fig. 22 (b). Both the stability and the spatial resolution obtained by the GNS cathode are promising to construct a compact and high resolution FE-SEM system, because it is possible to obtain higher resolution images less than 1 nm by using a commonly used 200-power magnification lens at this high residual pressure region.

The source size was estimated to be 160 nm. The maximum emission current measured by the Faraday cup was about 70 nA. Based on these experimentally determined parameters, the brightness of the GNS cathode was estimated as 5×10^{11} $Asr^{-1}m^{-2}$, which is similar to that of CNTs (de Jonge et al., 2002). A higher brightness of the order of 10^{13} $Asr^{-1}m^{-2}$ should be attainable if we apply higher extraction voltage to the cathode.

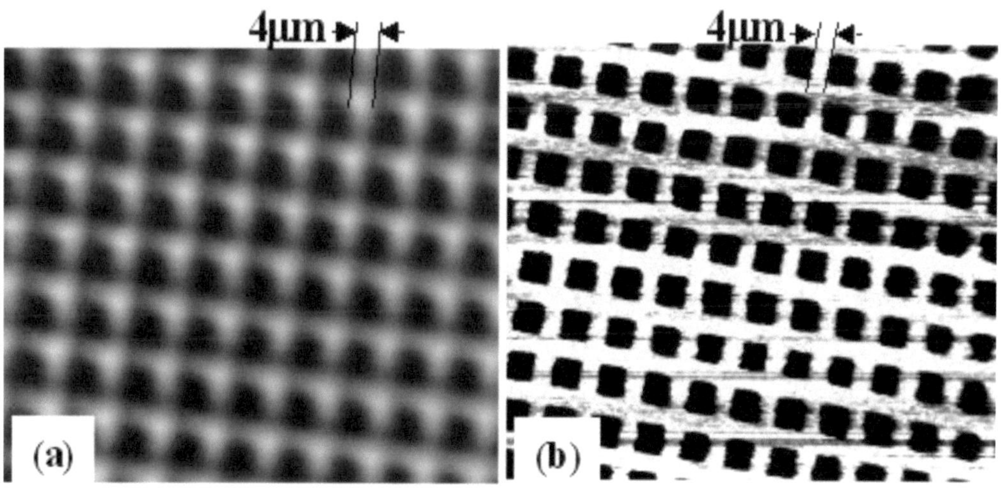

Fig. 21. SEM images of a copper mesh obtained using a (a) TE cathode and (b) GNS cathode. (Reprinted with permission, Neo et al. 2006, American Institute of Physics.)

(a) (b)

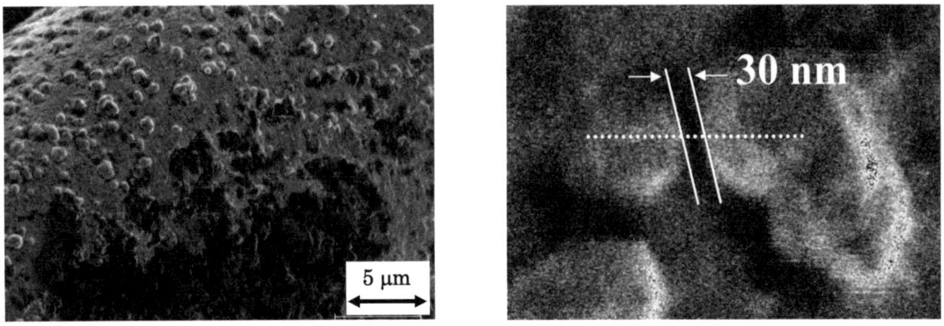

Fig. 22. (a) Copper mesh images obtained using the GNS cathode. (b) Highest resolution image obtained by using the GNS cathode. (Reprinted with permission, Matsumoto et al. 2007, American Institute of Physics.)

7. Conclusion

We showed novel properties of field emission from graphene nanosheets and found that the 2D graphene sheet structure is promising for field emission of electrons owing to the exceptionally large values of carrier mobility and small values of electron mass. From the point of view of emission physics, the macroscale statistical analysis by using the birth and death model gave the insight for the microscale emission mechanism. By proceeding into the atomic level analysis using FEM and FIM, we clarified the emission mechanism both of graphene sheets and of organic materials, where π electron plays an important role for the emission. This new finding not only gives us the solution of the mystery of symmetrical emission patterns discussed for more than 60 years but also paves a way toward a new coherent source of electrons.

After the clarification of the fundamental aspects of GNS, we overviewed the application aspects of the GNS field emitter. The recent progress in the miniature sized x-ray tube clearly shows that GNS is best suited for field emitters. We believe electron emission from graphene will open a new world for classical electron vacuum devices.

8. References

Andzelm, J; Govind, N. & Maiti, A. (2006). Nanotube-based gas sensors-role of structure defects. *Chemical Physics Letters*, Vol. 421, pp. 58-62.

Ashworth, F. (1951). Field emission microscopy. *Advances in Electronics*, Vol. 3, pp. 1-42.

Becker, J. (1955). Adsorption on metal surfaces and its bearing on catalysis. *Advances in catalysis*, Vol. 7, pp. 135-211.

Born, M. & Wolf, E. (1999). *Principles of optics; electromagnetic theory of propergation, interference and diffraction of light*. Cambridge University Press. Seventh Edition.

Cowley, J. (1995). *Diffraction Physics*. North-Holland Personal Library, Third Revised Edition.

Dean, K.; Chalamala, B. & Coll, B. et al. (2002). Carbon nanotube field emission electron sources. *New Diamond and Frontier Carbon Technology*, Vol. 12, pp. 165-180.

Dyke, W. & Dolan, W. (1956). Field emission. *Advances in electronics and electron physics*, Vol. 8, pp. 89-185.

Feller, W. (1957). *An introduction to probability theory and its applications*. John Wiley & Sons, New York, Vol. 1, pp. 444-482.

Forbes, R. (2001). Low macroscopic-field electron emission from carbon films and other electrically nanostructured heterogeneous materials: hypotheses about emission mechanism. *Solid State Electronics*, Vol. 45, 779-808.

Fowler, R. & Nordheim, L. (1928). Electron emission in intense electric fields. *Proceedings of the Royal Society (London) A*, Vol. 119, pp. 173-181.

Gadzuk, J. & Plummer, E. (1973). Field emission energy distribution. *Review of modern physics*, Vol. 45, pp. 487-548.

Geim, A. & Novoselov, S. (2007). The rise of graphene. *Nature Material*, Vol. 6, pp. 183-191.

Grujicis, M.; Cao, G. & Singh, R. (2003). The effect of topological defects and oxygen adsorption on the electronic transport properties of single-walled carbon-nanotubes. *Applied Surface Science*, Vol. 211, pp. 166-183.

He, J.; Cutler, P. & Minkovsky, N. (1991). Generalization of Fowler-Nordheim field emission theory for nonplanar metal emitters. *Applied Physics Letters*, Vol. 59, pp. 1644-1646.

Hosoki, S.; Yamamoto, S. & Futamoto, M. et al. (1979). Field emission characteristics of carbon tips. *Surface Science*, Vol. 86, pp. 723-733.

Jacobson, N.; Tegner, B. & Schröder, E. et al. (2002). Hydrogen dynamics in magnesium and graphite. *Computational Materials Science*, Vol. 24, pp. 273-277.

Joy, D.; Roming, Jr., A. & Goldstein, J. (1986). *Principles of analytical electron microscopy*. First Edition, Plenum Press, New York.

Jyouzuka, A.; Nakamura, T. & Onizuka Y. et al. (2010). Emission characteristics and application of graphite nanospine cathode. *The Journal Vacuum Science & Technology B*, Vol. 28, pp. C2C31-C2C34.

Kong, J.; Franklin, N. & Zhou, C. et al. (2000). Nanotube molecular wires as chemical sensors. *Science*, Vol. 287, pp. 622-625.

Liang, S. & Chen, L. (2008). Generalized Fowler-Nordheim theory of field emission of carbon nanotubes. *Physical Review Letters*, Vol. 101, 027602-1-027602-3.

Matsumoto, T. & Mimura, H. (2004). Intense electron emission from graphite nanocraters and their application to time-resolved x-ray radiography. *Applied Physics Letters*, Vol. 84, pp. 1804-1806.

Matsumoto, T. & Mimura, H. (2005). High intensity pulse x-ray generation by using graphite-nanocrater cold cathode. *The Journal Vacuum Science & Technology B*, Vol. 23, pp. 831-835.

Matsumoto, T.; Neo, Y. & Mimura, H. et al. (2007). Stabilization of electron emission from nanoneedles with two dimensional graphene sheet structure in a high residual pressure region. *Applied Physics Letters*, Vol. 90, pp. 103516-1-103516-3.

Matsumoto, T.; Neo, Y. & Mimura, H. et al. (2008). Determining the physisorption energies of molecules on graphene nanostructures by measuring the stochastic emission-current fluctuation. *Physical Review E*, Vol. 77, pp. 031611-1-031611-4.

Murr, L. & Inal, O. (1971). Field ion microscopy of graphite fibers. *Journal of Applied Physics*, Vol. 42, pp. 3487-3493.

Müller, E. (1953). Feldemission. *Ergebnisse d. Exakt Naturwiss*, Vol. XXVII, pp. 290-360.

Neo, Y; Matsumoto, T. & Tomita, M. et al. (2010). Revealing real images of cloverleaf pattern emission sites by using field ion microscopy. *The Journal Vacuum Science & Technology B*, Vol. 28, pp. C2A1-C2A4.

Neo,Y.; Mimura, H. & Matsumoto, T. (2006). Field emision characteristics of a graphite nanoneedle cathode and its application to scanning electron microscopy. *Applied Physics Letters*. Vol. 88, pp. 073511-1-073511-3.

Novoselov, S.; Geim, A., & Morozov, S., et al. (2005). Two-dimensional gas of massless Dirac fermions in graphene. *Nature*, Vol. 438, pp. 197-200.

Tran, F.; Weber, J. & Wesolowski, T. et al. (2002). Physisorption of molecular hydrogen on polycyclic aromatic hydrocarbons: A theoretical study. *The Journal of Physical Chemistry B*. Vol. 106, pp. 8689-8696.

Ulbricht, H.; Gunnar, M. & Hertel, T. (2002). Physisorption of molecular oxygen on single-wall carbon nanotube bundles and graphite. *Physical Review B*, Vol. 66, pp. 075404-1-075404-7.

Ulbricht, H.; Zacharia, R. & Cindir, N. et al. (2006). Thermal desorption of gases and solvents from graphite and carbon nanotube surfaces. *Carbon*, Vol. 44, pp. 2931-2942.

Vidali, G.; Ihm, G. & Kim, H. et al. (1991). Potentials of physical adsorption. *Surface Science Reports*, Vol. 12, pp. 133-181.

Williams, W. (1968). Field ion-microscopy of graphite. *Journal of Applied Physics*, Vol. 39, pp. 2131-2132.

Wolf, P. (1954). Die abbildung molekularer objekte mit dem Feldelektronenmikroskop. *Zeitshrift für angewandte physik*, Vol. 12, pp529-535.

Ximen, J. (1994). Fundamentals of electron optics. *Advances in Optical and electron microscopy*. Vol. 13, pp. 243-301.

Yamamoto, S.; Hosoki, S. & Fukuhara, S. et al. (1979). Stability of carbon field emission current. *Surface Science*, Vol. 86, pp. 734-742.

Zhang, Y.; Tan, Y. & Stormer, H. et al. (2005). Experimental observation of the quantum hall effect and Berry's phase in graphene. *Nature*, Vol. 438, pp. 201-204.

Zhao, J.; Buldum, A. & Han, J. et al. (2002). Gas molecule adsorption in carbon nanotubes and nanotube bundles. *Nanotechnology*, Vol. 13, pp. 195-200.

de Jonge, N.; Lamy, Y. & Schoots, K. et al. (2002). High brightness electron beam from a muti-walled carbon nanotube. *Nature*, Vol. 420, pp. 393-395.

Part 2

Molecular Dynamics Simulation

9

Theory of Defect Dynamics in Graphene

L.L. Bonilla[1] and A. Carpio[2]
[1]*Universidad Carlos III de Madrid*
[2]*Universidad Complutense de Madrid*
Spain

1. Introduction

The experimental discovery of graphene (single monolayers of graphite) (Novoselov et al, 2004; 2005) and its extraordinary properties due to the Dirac-like spectrum of its charge carriers have fostered an enormous literature reviewed by a number of authors (Castro Neto et al., 2009; Geim & Novoselov, 2007; Geim, 2009; Vozmediano et al., 2010). The electronic, chemical, thermal and mechanical properties of graphene are exceptionally sensitive to lattice imperfections (Castro Neto et al., 2009; Geim & Novoselov, 2007). These defects and even the ripples that always cover suspended graphene sheets (Fasolino et al., 2007; Meyer et al., 2007) induce pseudo-magnetic gauge fields (Vozmediano et al., 2010) and have given rise to the notion of strain engineering (Guinea et al., 2009). Thus the study of defects and ripples in graphene is crucial and it has generated important experimental work (Coleman et al., 2008; Gómez-Navarro et al., 2010; Meyer et al., 2008; Wang et al., 2008). Real time observation of defect dynamics is possible using Transmission Electron Microscopes (TEM) corrected for aberration that have single atom resolution (Meyer et al., 2008). In these studies, defects are induced by irradiation and their evolution is observed in a time scale of seconds (Meyer et al., 2008), much longer than sub-picosecond time scales typical of sound propagation in a primitive cell.

The graphene lattice is hexagonal and defects appear in different forms: pentagon-heptagon (5-7) pairs, Stone-Wales (SW) defects (two adjacent pentagon-heptagon pairs in which the heptagons share one side, denoted in short as 5-7-7-5 defects) (Meyer et al., 2008), pentagon-octagon-pentagon (5-8-5) divacancies (Coleman et al., 2008), asymmetric vacancies (nonagon-pentagon or 9-5 pairs) and more complicated groupings such as 5-7-7-5 and 7-5-5-7 adjacent pairs or defects comprising three pentagons, three heptagons and one hexagon (Meyer et al., 2008). In other two dimensional (two-dimensional) crystals such as Boron Nitride (hBN) symmetric vacancies in which one atom is missing have been observed (Meyer et al., 2009). On the long time scale of seconds and for unstressed graphene, Stone-Wales defects are unstable: their two pentagon-heptagon pairs glide towards each other and annihilate, and the same occurs to defects comprising three pentagons, three heptagons and one hexagon, whereas 5-7-7-5 and 7-5-5-7 adjacent pairs remain stable (Meyer et al., 2008). In stressed graphene oxide samples, SW defects split into their component 5-7 pairs which then move apart (Gómez-Navarro et al., 2010). While most theoretical studies on the influence of defects in electronic properties assume a given defect configuration and then proceed to analyze its effects (Castro Neto et al., 2009; Vozmediano et al., 2010), it is important to predict defect stability and evolution.

In recent work, we have explained the observed long time defect dynamics in graphene by considering defects as the core of edge dislocations or dislocation dipoles in a planar two-dimensional hexagonal lattice (Carpio & Bonilla, 2008; Carpio et al., 2008). Our theory is a computationally efficient alternative to ab initio approaches such as molecular dynamics or density functional theory (Abedpour et al., 2007; Fasolino et al., 2007; Segall, 2002; Thompson-Flagg et al., 2009). Our top-down approach starts from linear elasticity. We discretize continuum linear elasticity on a hexagonal lattice and replace differences of vector displacements along primitive directions by periodic functions thereof which are linear for small differences. Our *periodized discrete elasticity* allows dislocation gliding along primitive directions and it reduces to continuum linear elasticity very far from dislocation cores (Carpio & Bonilla, 2005). Introducing a large damping in the resulting equations of motion and solving them numerically, we are able to predict the stable cores corresponding to a given dislocation configuration. Using this theory, we have predicted the stability of pentagon-heptagon defects (that are the cores of dislocations) (Carpio & Bonilla, 2008; Carpio et al., 2008). Similarly, a study of dislocation dipoles in unstressed samples (Carpio & Bonilla, 2008; Carpio et al., 2008) predicts that SW are unstable whereas symmetric vacancies, divacancies and 7-5-5-7 defects are stable. In stressed samples, our theory predicts that Stone-Wales defects split into two 5-7 pairs that move apart (Carpio & Bonilla, 2008), as confirmed later by experiments (Gómez-Navarro et al., 2010).

In this chapter, we first describe periodized discrete elasticity for a planar graphene sheet and explain the evolution of several defects considered as cores of dislocations or dislocation dipoles. We then explain how an extension of our theory may describe a suspended graphene sheet in three dimensions, which is able to bend away from the planar configuration. We also discuss how to incorporate a mechanism for the formation and evolution of ripples. The rest of the chapter is as follows. Periodized discrete elasticity and its equations of motion are explained in Section 2 for a planar graphene sheet. The stable cores corresponding to the far field of a single edge dislocation and a single dislocation dipole are used in Section 3 to illustrate the way defects are constructed numerically. Our results are also compared to available experiments in graphene and other two-dimensional crystals, in particular to those by Meyer et al. (2008). Section 4 contains the extension to three space dimensions. We assume that the suspended sheet has a trend to locally bend upwards or downwards represented by an Ising spin. These spins are coupled to the carbon atoms in the sheet, are in contact with a thermal bath and evolve stochastically according to Glauber dynamics. Damping is caused by coupling to the bath and by Glauber dynamics. The formation of ripples in suspended graphene sheet is explained as a phase transition from the planar sheet that occurs below a certain critical temperature. Numerical solutions of the equations of motion illustrate the theoretical results and in particular show ripples and curvature of the sheet near a pentagon-heptagon defect. The last section is devoted to our conclusions.

2. Periodized discrete elasticity of planar graphene

In the continuum limit, small elastic deformations of graphene sheets have a free energy

$$F_g = \frac{\kappa}{2}\int (\nabla^2 w)^2 dxdy + \frac{1}{2}\int (\lambda u_{ii}^2 + 2\mu u_{ik}^2)\, dxdy, \tag{1}$$

corresponding to a membrane (Nelson, 2002), in which u_{ik}, $\mathbf{u} = (u_1, u_2) = (u(x,y), v(x,y))$, $w(x,y)$, κ, $\lambda = C_{12}$ and $\mu = C_{66}$ are the in-plane linearized deformation tensor, the in-plane displacement vector, the vertical deflection of the membrane, the bending stiffness (measured in units of energy) and the Lamé coefficients (Landau & Lifshitz, 1986) of graphene (graphite is isotropic in its basal plane so that $C_{11} = \lambda + 2\mu$), respectively. We have used the convention of sum over repeated indices. In (1), $\nabla^2 w = \partial_x^2 w + \partial_y^2 w$ is the two-dimensional laplacian and the two-dimensional Lamé coefficients have units of energy per unit area. Dividing by the distance between graphene planes in graphite, 0.335 nm, they can be converted to the usual units for 3D graphite. The in-plane linearized deformation tensor is

$$u_{ik} = \frac{1}{2}\left(\partial_{x_k} u_i + \partial_{x_i} u_k + \partial_{x_i} w \partial_{x_k} w\right), \quad i,k = 1,2, \tag{2}$$

in which we have ignored the small in-plane nonlinear terms $\partial_{x_i} u \partial_{x_k} u + \partial_{x_i} v \partial_{x_k} v$.
In this section, we shall consider that only in-plane deformations are possible so that $w = 0$. Then the equations of motion derived from (1)-(2) are the Navier equations of linear elasticity for (u, v) (Landau & Lifshitz, 1986). If we add to these equations a phenomenological damping with coefficient γ (to be fitted to experiments), we have

$$\rho_2 \partial_t^2 u + \gamma \partial_t u = (\lambda + 2\mu)\, \partial_x^2 u + \mu\, \partial_y^2 u + (\lambda + \mu)\, \partial_x \partial_y v, \tag{3}$$

$$\rho_2 \partial_t^2 v + \gamma \partial_t v = \mu\, \partial_x^2 v + (\lambda + 2\mu)\, \partial_y^2 v + (\lambda + \mu)\, \partial_x \partial_y u, \tag{4}$$

where ρ_2 is the two-dimensional mass density. The governing equations of our theory are obtained from (3)-(4) in a three step process (Carpio & Bonilla, 2008): (i) discretize the equations on the hexagonal graphene lattice, (ii) rewrite the discretized equations in primitive coordinates, and (iii) replace finite differences appearing in the equations by periodic functions thereof in such a way that the equations remain invariant if we displace the atoms one step along any of the primitive directions. The last step allows dislocation gliding.

2.1 Discrete elasticity

Let us assign the coordinates (x, y) to the atom A in sublattice 1 (see Figure 1). The origin of coordinates, $(0,0)$, is also an atom of sublattice 1 at the center of the graphene sheet. The three nearest neighbors of A belong to sublattice 2 and their cartesian coordinates are n_1, n_2 and n_3 below. Its six next-nearest neighbors belong to sublattice 1 and their cartesian coordinates are $n_i, i = 4, \ldots, 9$:

$$n_1 = \left(x - \frac{a}{2}, y - \frac{a}{2\sqrt{3}}\right), \; n_2 = \left(x + \frac{a}{2}, y - \frac{a}{2\sqrt{3}}\right), \; n_3 = \left(x, y + \frac{a}{\sqrt{3}}\right),$$

$$n_4 = \left(x - \frac{a}{2}, y - \frac{a\sqrt{3}}{2}\right), \; n_5 = \left(x + \frac{a}{2}, y - \frac{a\sqrt{3}}{2}\right), \; n_6 = (x - a, y),$$

$$n_7 = (x + a, y), \; n_8 = \left(x - \frac{a}{2}, y + \frac{a\sqrt{3}}{2}\right), \; n_9 = \left(x + \frac{a}{2}, y + \frac{a\sqrt{3}}{2}\right). \tag{5}$$

In Fig. 1, atoms n_6 and n_7 are separated from A by the primitive vector $\pm\mathbf{a}$ and atoms n_4 and n_9 are separated from A by the primitive vector $\pm\mathbf{b}$. Instead of choosing the primitive vector

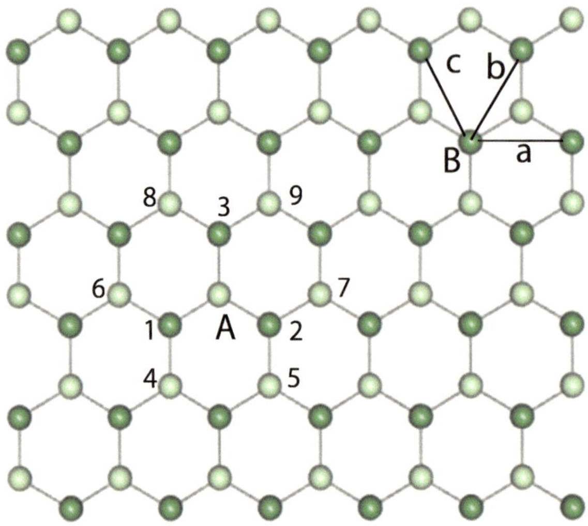

Fig. 1. Neighbors of a given atom A in sublattice 1 (light colored atoms).

$\pm\mathbf{b}$, we could have selected the primitive direction $\pm\mathbf{c}$ along which atoms n_8, A and n_5 lie. Let us define the following operators acting on functions of the coordinates (x, y) of node A:

$$Tu = [u(n_1) - u(A)] + [u(n_2) - u(A)] + [u(n_3) - u(A)] \sim \left(\partial_x^2 u + \partial_y^2 u\right) \frac{a^2}{4}, \tag{6}$$

$$Hu = [u(n_6) - u(A)] + [u(n_7) - u(A)] \sim (\partial_x^2 u)\, a^2, \tag{7}$$

$$D_1 u = [u(n_4) - u(A)] + [u(n_9) - u(A)] \sim \left(\frac{1}{4}\,\partial_x^2 u + \frac{\sqrt{3}}{2}\,\partial_x\partial_y u + \frac{3}{4}\,\partial_y^2 u\right) a^2, \tag{8}$$

$$D_2 u = [u(n_5) - u(A)] + [u(n_8) - u(A)] \sim \left(\frac{1}{4}\,\partial_x^2 u - \frac{\sqrt{3}}{2}\,\partial_x\partial_y u + \frac{3}{4}\,\partial_y^2 u\right) a^2, \tag{9}$$

as the lattice constant a tends to zero. Similar operators can be defined if we replace the point A in sublattice 1 by a point belonging to the sublattice 2. Now we replace in (3) and (4), Hu/a^2, $(4T - H)u/a^2$ and $(D_1 - D_2)u/(\sqrt{3}a^2)$ instead of $\partial_x^2 u$, $\partial_y^2 u$ and $\partial_x\partial_y u$, respectively, with similar substitutions for the derivatives of v, thereby obtaining the following equations at each point of the lattice:

$$\rho_2 a^2 \partial_t^2 u + \gamma\, \partial_t u = 4\mu\, Tu + (\lambda + \mu)\, Hu + \frac{\lambda + \mu}{\sqrt{3}}\, (D_1 - D_2) v, \tag{10}$$

$$\rho_2 a^2 \partial_t^2 v + \gamma\, \partial_t v = 4(\lambda + 2\mu)\, Tv - (\lambda + \mu) Hv + \frac{\lambda + \mu}{\sqrt{3}}\, (D_1 - D_2) u. \tag{11}$$

These equations have two characteristics time scales, the time $t_s = \sqrt{\rho_2 a^2/(\lambda + 2\mu)}$ it takes a longitudinal sound wave to traverse a distance a, and the characteristic damping time, $t_d = \gamma a^2/(\lambda + 2\mu)$. Using the known values of the Lamé coefficients (Lee et al., 2008; Zakharchenko et al., 2009), $t_s \approx 10^{-14}$ s. Our simulations show that it takes $0.4 t_d$ a SW to disappear after it is created by irradiation which, compared with the measured time of 4 s (Meyer et al., 2008), gives $t_d \approx 10$ s. On a t_d time scale, we can ignore inertia in (10)-(11).

2.2 Nondimensional equations in primitive coordinates

We now transform (10)-(11) to the nondimensional primitive coordinates u', v' taking $u = a(u' + v'/2)$, $v = \sqrt{3}av'/2$, use the nondimensional time scale $t' = t/t_d$ and ignore inertia, thereby obtaining

$$\partial_{t'} u' = \frac{4\mu T u'}{\lambda + 2\mu} + \frac{\lambda + \mu}{\lambda + 2\mu}\left[\left(H - \frac{D_1 - D_2}{3}\right) u' + \left(H + \frac{D_1 - D_2}{3} - 2T\right) v'\right], \quad (12)$$

$$\partial_{t'} v' = \frac{2}{3}\frac{\lambda + \mu}{\lambda + 2\mu}(D_1 - D_2)u' + 4Tv' + \frac{\lambda + \mu}{\lambda + 2\mu}\left(\frac{D_1 - D_2}{3} - H\right) v'. \quad (13)$$

2.3 Periodized discrete elasticity

Equations (12) - (13) do not allow for the changes of neighbors involved in defect motion. One way to achieve these changes is to update neighbors as a defect moves. Then (12) and (13) would have the same appearance, but the neighbors n_i would be given by (5) only at the start. At each time step, we keep track of the position of the different atoms and update the coordinates of the n_i. This is commonly done in Molecular Dynamics, as computations are actually carried out with only a certain number of neighbors. Convenient as updating is, its computational cost is high and analytical studies thereof are not easy.

In simple geometries, we can avoid updating by introducing a periodic function of differences in the primitive directions that automatically describes link breakup and union associated with defect motion. Besides greatly reducing computational cost, the resulting periodized discrete elasticity models allow analytical studies of defect depinning and motion (Carpio & Bonilla, 2003; 2005). Another advantage of periodized discrete elasticity is that boundary conditions can be controlled efficiently to avoid spurious numerical reflections at boundaries.

To restore crystal periodicity, we replace the linear operators T, H, D_1 and D_2 in (12) and (13) by their periodic versions:

$$\begin{aligned}
T_p u' &= g(u'(n_1) - u'(A)) + g(u'(n_2) - u'(A)) + g(u'(n_3) - u'(A)),\\
H_p u' &= g(u'(n_6) - u'(A)) + g(u'(n_7) - u'(A)),\\
D_{1p} u' &= g(u'(n_4) - u'(A)) + g(u'(n_9) - u'(A)),\\
D_{2p} u' &= g(u'(n_5) - u'(A)) + g(u'(n_8) - u'(A)), \quad (14)
\end{aligned}$$

where g is a periodic function, with period one, and such that $g(x) \sim x$ as $x \to 0$. In this work, g is a periodic piecewise linear continuous function:

$$g(x) = \begin{cases} x, & -\alpha \le x \le \alpha, \\ -\frac{2\alpha}{1-2\alpha}x + \frac{\alpha}{1-2\alpha}, & \alpha \le x \le 1 - \alpha. \end{cases} \quad (15)$$

The parameter α controls defect stability and mobility under applied stress. It should be sufficiently large for elementary defects (dislocations, vacancies) to be stable at zero applied stress, and sufficiently small for dislocations to move under reasonable applied stress (Carpio & Bonilla, 2005). We use $\alpha = 0.4$ to account for experimentally observed stability properties of the defects. For lower values, the stable defect described in section 4 loses the Stone-Wales component. The periodic function g can be replaced by a different type of periodic function to achieve a better fit to available experimental or numerical data.

3. Stable cores of dislocations and dislocation dipoles in planar graphene

3.1 Boundary and initial conditions for a single dislocation

We solve (12)-(13) with the periodic operators T_p, H_p, D_{1p} and D_{2p}, using as initial and boundary conditions the far field of appropriate dislocations which are the stationary solutions of the linear elasticity equations (Landau & Lifshitz, 1986). Since the latter are a good approximation four spacings away from the core of SW defects in graphene, and our model equations seamlessly reduce to linear elasticity in the far field, we use a relatively small lattice with 19×19 spacings ($2 \times 18 \times 18$ carbon atoms) in our numerical simulations (Carpio & Bonilla, 2008). Consider first the case of a single edge dislocation with Burgers vector $(a, 0)$ and displacement vector $\mathbf{u} = (u(x,y), v(x,y))$

$$u = \frac{a}{2\pi}\left[\tan^{-1}\left(\frac{y}{x}\right) + \frac{xy}{2(1-\nu)(x^2+y^2)}\right],$$
$$v = \frac{a}{2\pi}\left[-\frac{1-2\nu}{4(1-\nu)}\ln\left(\frac{x^2+y^2}{a^2}\right) + \frac{y^2}{2(1-\nu)(x^2+y^2)}\right], \quad (16)$$

where $\nu = \lambda/[2(\lambda+\mu)]$ is dimensionless; cf. page 114 of (Landau & Lifshitz, 1986). (16) has a singularity $\propto (x^2+y^2)^{-1/2}$ at the origin of coordinates and it satisfies $\int_C (d\mathbf{x}\cdot\nabla)\mathbf{u} = -(a, 0)$, for any closed curve C encircling the origin. Using (16), we write $\mathbf{u} = (u, v)$ in primitive coordinates, $U'(l,m) = [u(x-x_0, y-y_0) - v(x-x_0, y-y_0)/\sqrt{3}]/a$, $V'(l,m) = 2v(x-x_0, y-y_0)/(a\sqrt{3})$, where $x = (x' + y'/2)a$, $y = \sqrt{3}\,ay'/2$, $x' = l$, $y' = m$ (integers) and $(x_0, y_0) \neq (0,0)$ to avoid that the singularity in (16) be placed at a lattice point. To find defects, we solve the periodized versions of the discrete elasticity equations (12)-(13) with the initial and boundary conditions:

$$\mathbf{u}'(l,m;0) = \mathbf{U}'(l,m), \quad \text{and} \quad \mathbf{u}'(l,m;t) = \mathbf{U}'(l,m) + F(m,0) \quad \text{at lattice boundaries.} \quad (17)$$

Here F is a dimensionless applied shear stress. For $|F| < F_c$ (Peierls stress), the solution of the periodized version of (12)-(13) relaxes to a stable dislocation $(u'(l,m), v'(l,m))$ with appropriate far field, which is (16) if $F = 0$.

Numerical simulations give us the location of carbon atoms at each time t. We represent atoms by spheres of arbitrary size. As a guide to the eye and to visualize defects more easily, we have attached fictitious bonds to these spheres (Carpio & Bonilla, 2008; Carpio et al., 2008). Depending on the location of the singularity (x_0, y_0), there are two possible configurations corresponding to the same edge dislocation in the continuum limit. If (x_0, y_0) is placed between two atoms that form any non-vertical side of a given hexagon, the core of the deformed lattice $(l + u'(l,m), m + v'(l,m))$ is a 5-7 (pentagon-heptagon) defect. If (x_0, y_0) is placed in any other location different from a lattice point, the core of the singularity forms an octagon having one atom with a dangling bond (Carpio & Bonilla, 2008; Carpio et al., 2008). Stable 5-7 defects are commonly observed in experiments (Gómez-Navarro et al., 2010; Meyer et al., 2008; 2009), whereas adsorbed atoms (not considered in our model) may attach to a dangling bond thereby destroying the octagon configuration.

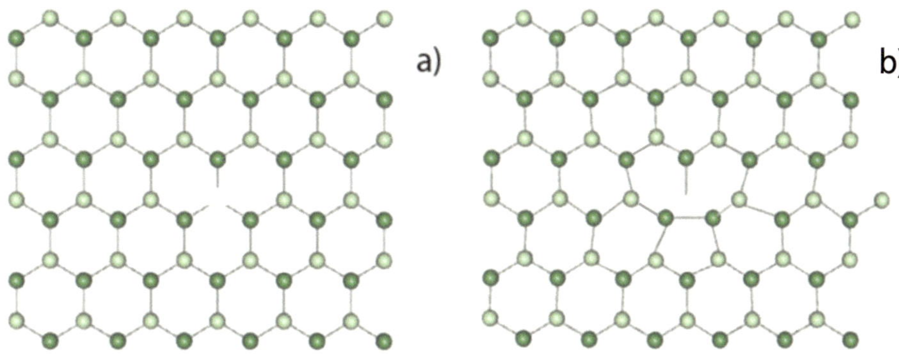

Fig. 2. (a) Symmetric vacancy. (b) Asymmetric vacancy (nonagon-pentagon defect).

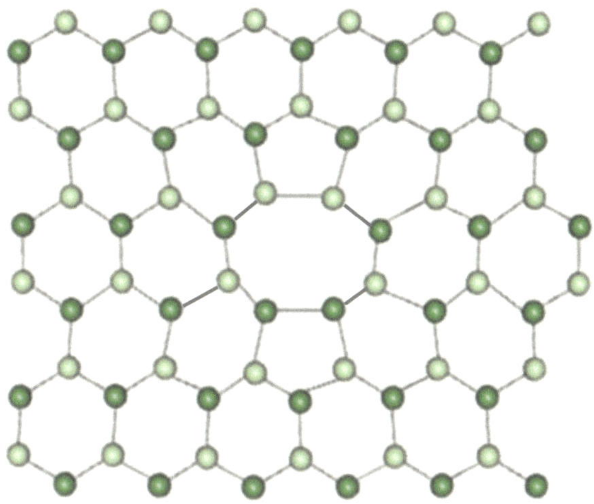

Fig. 3. Pentagon-octagon-pentagon divacancy.

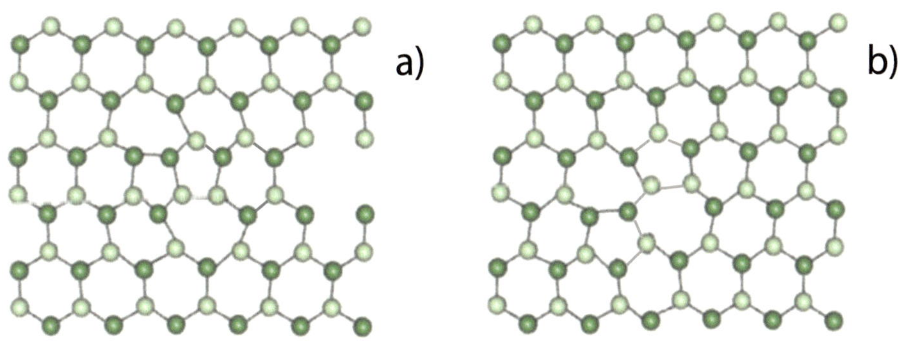

Fig. 4. (a) Stable 7-5-5-7 defect. (b) Unstable 5-7-7-5 Stone Wales defect.

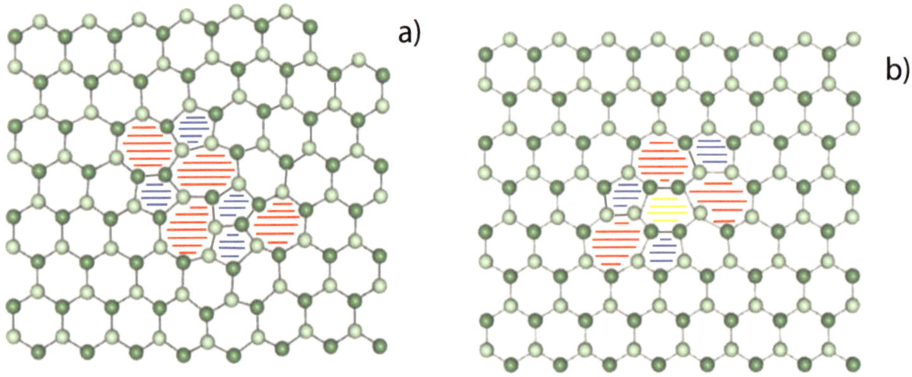

Fig. 5. Defects corresponding to two dislocation dipoles: (a) Stable configuration comprising a pair of 5-7-7-5 and 7-5-5-7 defects. (b) Metastable defect comprising three heptagons, three pentagons and one hexagon.

3.2 Dislocation dipoles

A dislocation dipole is formed by two dislocations with opposite Burgers vectors, $\pm\mathbf{a}$. Depending on how we place the origin of coordinates, different dipole configurations result. Let $\mathbf{U}(x,y)$ be the displacement vector (16) of a single dislocation. We find the dipole cores by selecting as zero stress initial and boundary conditions $\mathbf{U}(x-x_0^+, y-y_0^+) - \mathbf{U}(x-x_0^-, y-y_0^-)$, with different $(x_0^\pm, y_0^\pm)$. Let a, $L = a/\sqrt{3}$, $H = 3L/2$ and $h = L/2$ be the lattice constant, the hexagon side, the vertical distance between two nearest neighbor atoms belonging to the same sublattice separated by one primitive vector **b** or **c**, and the vertical distance between nearest neighbor atoms having different ordinate, respectively. We get:

- Vacancies: $x_0^+ = -0.25a$, $y_0^+ = -0.8h + H/2$ and $x_0^- = -0.25a$, $y_0^- = -0.8h$. This initial configuration is the asymmetric vacancy (9-5 defect) of Fig. 2(b), which evolves to the symmetric vacancy of Fig. 2(a) under overdamped dynamics.
- Stable divacancy: $x_0^+ = -0.25a$, $y_0^+ = -0.8h + H$ and $x_0^- = -0.25a$, $y_0^- = -0.8h$. Fig. 3.
- Stable 7-5-5-7 defect: $x_0^+ = -0.25a + a$, $y_0^+ = -0.8h$ and $x_0^- = -0.25a$, $y_0^- = -0.8h + H$. Fig. 4(a).
- Unstable Stone-Wales 5-7-7-5 defect: $x_0^+ = -0.25a + a$, $y_0^+ = -0.8h$ and $x_0^- = -0.25a$, $y_0^- = -0.8h$. Fig. 4(b). For $F = 0$, this initial configuration corresponds to two dislocations with opposite Burgers vectors that share the same glide line, and it evolves to the undisturbed lattice when the dislocations move towards each other and annihilate.

3.3 Dislocation dipole pairs

We now study the evolution of configurations comprising two dislocation dipoles each. Fig. 5(a) depicts a stable defect consisting of a 5-7-7-5 SW defect adjacent to a rotated 7-5-5-7 defect:

$$\mathbf{U}(x-x_0^+, y-y_0^+) - \mathbf{U}(x-x_0^-, y-y_0^-) + \tilde{\mathbf{U}}(\tilde{x}-\tilde{x}_0^+, \tilde{y}-\tilde{y}_0^+) - \tilde{\mathbf{U}}(\tilde{x}-\tilde{x}_0^-, \tilde{y}-\tilde{y}_0^-), \quad (18)$$

$$x_0^+ = -0.3a,\ y_0^+ = -0.7h + 2H, \quad x_0^- = -0.3a - a,\ y_0^- = -0.7h + 2H,$$
$$\tilde{x}_0^+ = 0.3a + a,\ \tilde{y}_0^+ = 0.3h - H, \quad \tilde{x}_0^- = 0.15a,\ \tilde{y}_0^+ = 0.5h. \quad (19)$$

Here $\mathbf{U}(x,y)$ is the edge dislocation (16) with origin of coordinates at a central atom of type A in Figure 1 and Burgers vector $\mathbf{a}$ (in units of the lattice constant a). $\tilde{\mathbf{U}}(\tilde{x},\tilde{y})$ is an edge dislocation with Burgers vector $\mathbf{b}$. To obtain $\tilde{\mathbf{U}}(\tilde{x},\tilde{y})$ in (18), we consider the axes $(\tilde{x},\tilde{y})$ rotated a $\pi/3$ angle from the axes (x,y). Next we form a 7-5-5-7 defect by combining a *positive* dislocation with Burgers vector $(a,0)$ centered at $(\tilde{x}_0^+,\tilde{y}_0^+)$ and a *negative* dislocation with Burgers vector $(-a,0)$ centered at $(\tilde{x}_0^-,\tilde{y}_0^-)$. Then the result is rewritten in the original coordinates (x,y). This is the same defect as reported in Figures 3(h) and (i) of Meyer et al's experiments (Meyer et al., 2008) and it remains stable under overdamped dynamics. The 7-5-5-7 defect is stable (Carpio & Bonilla, 2008; Carpio et al., 2008), and this apparently stabilizes our pair of dislocation dipoles for the selected initial configuration. Other nearby configurations evolve to two octagons corresponding to a dipole comprising two edge dislocations with opposite Burgers vectors. As explained before, adsorbed atoms may be attached to the dangling bonds thereby eliminating these configurations and restoring the undisturbed hexagonal lattice.

Let us consider now two dislocation dipoles all whose component dislocations have Burgers vectors directed along the x axis according to the initial and boundary condition depicted in Figure 5(b) (a defect comprising three pentagons and three heptagons):

$$\mathbf{U}(x-x_0^+,y-y_0^+)-\mathbf{U}(x-x_0^-,y-y_0^-)+\mathbf{U}(x-\hat{x}_0^+,y-\hat{y}_0^+)-\mathbf{U}(x-\hat{x}_0^-,y-\hat{y}_0^-), \quad (20)$$

$$x_0^+=-0.3a+a,\ y_0^+=-0.7h,\quad x_0^-=-0.3a,\ y_0^-=-0.7h,$$
$$\hat{x}_0^+=-0.3a-a,\ \hat{y}_0^+=-0.7h-H,\quad \hat{x}_0^-=-0.3a,\ \hat{y}_0^+=-0.7h-H. \quad (21)$$

Starting from a negative dislocation centered at $(x_0^-,y_0^-)=(-0.3a,-0.7h)$, the first dipole adds a positive dislocation shifted one lattice constant to the right. The second dipole consists of a negative dislocation shifted vertically downwards a distance $H=3L/2=\sqrt{3}a/2$ (1.5 times the hexagon side, or $\sqrt{3}/2$ times the lattice constant) from (x_0^-,y_0^-) and a positive dislocation which shifts horizontally to the left the previous one a distance equal to one lattice constant. Under overdamped dynamics, this defect disappears as the positive and negative dislocations comprising each dipole glide towards each other.

3.4 Comparison with results of experiments

Carbon atoms and defects in graphene sheets are visualized by operating at low voltage (≤ 80 kV, to avoid irradiation damage to the sample) a transmission electron aberration-corrected microscope (TEAM) with appropriate optics (Meyer et al., 2008). This microscope is capable of sub-Ångstrom resolution even at 80 kV and can produce real time images of carbon atoms on a scale of seconds: each frame averages 1s of exposure and the frames themselves are 4 s apart (Girit et al., 2009; Meyer et al., 2008). The images obtained in experiments can be used to determine the time evolution of defects in graphene created by irradiation or sample treatment (Girit et al., 2009; Gómez-Navarro et al., 2010; Meyer et al., 2008; 2009).

As explained before, stable pentagon-heptagon defects but not isolated octagons with a dangling bond are observed in experiments (Gómez-Navarro et al., 2010; Meyer et al., 2008; 2009). The octagon configuration is quite reactive and may be destroyed by attaching an adsorbed carbon atom, which is not contemplated in our model. Our theory indicates that symmetric vacancies are stable and asymmetric ones are unstable. In experiments, both symmetric and asymmetric vacancies are observed in unstressed graphene (Meyer et al., 2008), whereas in single layers of hexagonal Boron Nitride (hBN) only symmetric vacancies are observed (Meyer et al., 2009), as predicted by our theory. See Carpio & Bonilla (2008) for a possible mechanism to produce stable asymmetric vacancies in graphene. Stable 5-8-5 divacancies are also observed (Coleman et al., 2008). The annihilation of the 5-7-7-5 SW defect in 4(b) (the heptagons share one side) 4 s after its creation is seen in Figures 3(c) and (d) of the paper by Meyer et al. (2008). Our model predicts that SW under sufficient strain split in their two 5-7 pairs that move apart (cf Fig 6 in Carpio & Bonilla (2008)), which has been observed very recently; cf Fig. 4(a) and (b) in Gómez-Navarro et al. (2010). Of the two pairs of dislocation dipoles, the defect comprising four pentagons and four heptagons was observed to be stable in Figures 3(h) and (i) in Meyer et al. (2008), whereas the defect comprising three pentagons, three heptagons and one hexagon disappeared after a short time (Meyer et al., 2008), as predicted by the numerical simulations of our model equations.

4. Ripples and defects in a three dimensional suspended graphene sheet

The first observations of suspended graphene sheets showed evidence of ripples: ondulations of the sheet with characteristic amplitudes and wave lengths (Meyer et al., 2007). Monte Carlo calculations also showed ripple formation (Fasolino et al., 2007) and explored the possibility that the ripples are due to thermal fluctuations; see also Abedpour et al. (2007). According to Thompson-Flagg et al. (2009), thermal fluctuations would produce much smaller ripples than observed in experiments and therefore this explanation does not seem likely. Thompson-Flagg et al. (2009) explain ripples as a consequence of adsorbed OH molecules on random sites. More recent experiments on graphene sheets suspended on substrate trenches have shown that ripples can be thermally induced and controlled by thermal cycling in a clean atmosphere (Bao et al., 2009) which seems to preclude the explanation based on adsorbed OH molecules (Thompson-Flagg et al., 2009). Bao et al. (2009) use the difference in the thermal expansion coefficients of the graphene sheet (negative coefficient) and its pinning substrate (positive coefficient) to vary the tension in the sheet which, in turn, governs the ripples (Bao et al., 2009; Bunch et al., 2007).

In this section, we present a theory of ripples and defects in suspended graphene sheets based on periodized discrete elasticity (Bonilla & Carpio, 2011). A suspended graphene sheet may bend upwards or downwards and therefore it may be described by the free energy of the membrane (1) with $w \neq 0$ in the continuum limit. Carbon atoms in the graphene sheet have σ bond orbitals constructed from sp^2 hybrid states oriented in the direction of the bond that accommodate three electrons per atom. The remaining electrons go to p states oriented perpendicularly to the sheet. These orbitals bind covalently with neighboring atoms and form a narrow π band that is half-filled. The presence of bending and ripples in graphene modifies its electronic structure (Castro Neto et al., 2009). Out-of-plane convex or concave deformations of the sheet have in principle equal probability and transitions between these deformations are associated with the bending energy of the sheet. A simple way to model this situation to consider that out-of-plane deformations are described by the values of an Ising spin associated to each carbon atom. Then the vertical coordinate of each atom in the graphene sheet is coupled to an Ising spin $\sigma = \pm 1$ in contact with a bath at temperature T and that the spin system has an energy

$$F_s = -f \int \sigma(x,y)\, w(x,y)\, dxdy, \tag{22}$$

in the continuum limit. f has units of force per unit area. The spins flip stochastically according to Glauber dynamics at temperature θ (measured in units of energy). Thus at any time t, the system may experience a transition from $(\boldsymbol{u}, \boldsymbol{v}, \boldsymbol{w}, \boldsymbol{\sigma})$ to $(\boldsymbol{u}, \boldsymbol{v}, \boldsymbol{w}, R_{(x,y)}\boldsymbol{\sigma})$ at a rate given by

$$W_{(x,y)}(\boldsymbol{\sigma}|\boldsymbol{u}, \boldsymbol{v}, \boldsymbol{w}) = \frac{\zeta}{2}\left[1 - \beta(x,y)\,\sigma(x,y)\right], \qquad \beta(x,y) = \tanh\left(\frac{fa^2 w(x,y)}{\theta}\right), \tag{23}$$

(Glauber, 1963), where $R_{(x,y)}\boldsymbol{\sigma}$ is the configuration obtained from $\boldsymbol{\sigma} = \{\sigma(x,y)\}$ (for all points (x,y) on the hexagonal lattice) by flipping the spin at the lattice point (x,y). The parameter ζ gives the characteristic attempt rate for the transitions in the Ising system. Since the bending energy of the graphene sheet is κ, the attempt rate should be proportional to an Arrhenius factor:

$$\zeta = \zeta_0 \exp\left(-\frac{\kappa}{\theta}\right), \tag{24}$$

where ζ_0 is a constant.
The total free energy,

$$F = F_g + F_s = \int \left[\frac{\kappa}{2}(\nabla^2 w)^2 + \frac{\lambda}{2}u_{ii}^2 + \mu u_{ik}^2 - f\sigma w\right] dxdy, \tag{25}$$

is obtained from (1) and (22) and it provides the equations of motion (Bonilla & Carpio, 2011):

$$\begin{aligned}\rho_2 \partial_t^2 u =& \lambda\, \partial_x \left(\partial_x u + \partial_y v + \frac{(\partial_x w)^2 + (\partial_y w)^2}{2}\right) + \mu\, \partial_x[2\partial_x u + (\partial_x w)^2] \\ &+ \mu\, \partial_y \left(\partial_y u + \partial_x v + \partial_x w \partial_y w\right),\end{aligned} \tag{26}$$

$$\begin{aligned}\rho_2 \partial_t^2 v =& \lambda\, \partial_y \left(\partial_x u + \partial_y v + \frac{(\partial_x w)^2 + (\partial_y w)^2}{2}\right) + \mu\, \partial_y[2\partial_y v + (\partial_y w)^2] \\ &+ \mu\, \partial_x \left(\partial_y u + \partial_x v + \partial_x w \partial_y w\right),\end{aligned} \tag{27}$$

instead of (3)-(4) and

$$\begin{aligned}\rho_2 \partial_t^2 w =& -\kappa(\nabla^2)^2 w + \lambda \partial_x \left[\left(\partial_x u + \partial_y v + \frac{(\partial_x w)^2 + (\partial_y w)^2}{2}\right)\partial_x w\right] \\ &+ \lambda \partial_y \left[\left(\partial_x u + \partial_y v + \frac{(\partial_x w)^2 + (\partial_y w)^2}{2}\right)\partial_y w\right] \\ &+ \mu \partial_x \left\{2\partial_x u \partial_x w + (\partial_y u + \partial_x v)\partial_y w + [(\partial_x w)^2 + (\partial_y w)^2]\partial_x w\right\} \\ &+ \mu \partial_y \left\{(\partial_y u + \partial_x v)\partial_x w + 2\partial_y v \partial_y w + [(\partial_x w)^2 + (\partial_y w)^2]\partial_y w\right\} + f\sigma,\end{aligned} \tag{28}$$

for w. In this section, we omit the phenomenological friction (with coefficient γ) in the equations of motion because the stochastic Glauber dynamics already provides an effective friction for the spins which try to reach equilibrium at the bath temperature θ. The periodized discrete elasticity corresponding to these equations is derived in the same fashion as for planar graphene in Section 3, except that we need to add the new difference operators to those in

(6)-(9):

$$\Delta_h u = u(n_7) - u(A) \sim (\partial_x u)\, a, \tag{29}$$

$$\Delta_v u = u(n_3) - u(A) \sim (\partial_y u)\, \frac{a}{\sqrt{3}}, \tag{30}$$

$$Bw = [Tw(n_1) - Tw(A)] + [Tw(n_2) - Tw(A)] + [Tw(n_3) - Tw(A)] \sim \frac{a^4}{16}(\nabla^2)^2 w, \tag{31}$$

as $a \to 0$. The discrete elasticity equations are now (Bonilla & Carpio, 2011):

$$\rho_2 a^2 \partial_t^2 u = 4\mu\, Tu + (\lambda+\mu)\, Hu + \frac{\lambda+\mu}{\sqrt{3}}\,(D_1 - D_2)v + \frac{4\mu}{a}\,\Delta_h w\, Tw + \frac{\lambda+\mu}{a}\,[\Delta_h w\, Hw + \Delta_v w\,(D_1 - D_2)w], \tag{32}$$

$$\rho_2 a^2 \partial_t^2 v = 4(\lambda+2\mu)\, Tv - (\lambda+\mu)Hv + \frac{\lambda+\mu}{\sqrt{3}}\,(D_1 - D_2)u + \frac{4\sqrt{3}}{a}(\lambda+2\mu)\Delta_v w\, Tw + \frac{\lambda+\mu}{a\sqrt{3}}\,[\Delta_h w\,(D_1 - D_2)w - 3\Delta_v w\, Hw], \tag{33}$$

$$\begin{aligned} \rho_2 a^2 \partial_t^2 w = {} & \frac{\lambda+2\mu}{a}\left\{\left[Hu + \frac{\Delta_h w}{a} Hw + \frac{2\Delta_v w}{a}(D_1 - D_2)w\right]\Delta_h w + \left[\sqrt{3}(4T - H)v \right.\right. \\ & \left.\left. + \frac{3\Delta_v w}{a}(4T - H)w\right]\Delta_v w\right\} + \frac{\lambda+\mu}{a}(D_1 - D_2)u\Delta_v w + \frac{\lambda\Delta_h w}{\sqrt{3}a}(D_1 - D_2)v \\ & + \frac{\mu\Delta_v w}{a}[\sqrt{3}(4T - H)u + (D_1 - D_2)v + \sqrt{3}Hv] + \frac{2\Delta_h u}{a}(2\lambda T + \mu H)w \\ & + \frac{2\sqrt{3}\Delta_v v}{a}[2\lambda Tw + \mu(4T - H)w] + \frac{2\mu}{a}\left(\Delta_v u + \frac{\Delta_h v}{\sqrt{3}}\right)(D_1 - D_2)w \\ & + \frac{(\Delta_h w)^2 + (\Delta_v w)^2}{a^2}(2\lambda T + \mu H)w + \frac{4\mu}{a^2}Tw(4T - H)w - \frac{16\kappa}{a^2}Bw + fa^2\sigma. \end{aligned} \tag{34}$$

We have periodized these equations after writing them in primitive coordinates, adopted the nondimensional time scale $t' = t/[t]$ as in (12)-(13) and solved them numerically. In our simulations, we have used the numerical values of the isothermal elastic moduli at 300K provided by Zakharchenko et al. (2009), $\lambda = 2.57$ eV/Å^2, $\mu = 9.95$ eV/Å^2, which agree well with experiments (Lee et al., 2008) and with the values for graphite at the basal plane (Blakslee et al., 1970). The bending coefficient at 300 K is $\kappa = 1.08$ eV (Zakharchenko et al., 2009). The time constant is $[t] = a\sqrt{\rho_2/(\lambda+2\mu)} = 1.1033 \times 10^{-14}$ s because $a = 2.461$ Å and we get $\rho_2 = 2160 \times 3.35 \times 10^{-10} = 7.236 \times 10^{-7}$ kg/m^2 from the three-dimensional mass density of graphite (the distance between graphene planes in graphite is 3.35 Å). We use $f = 0.378$ meV and $1/\zeta = 10$ s (Bonilla & Carpio, 2011). $1/\zeta$ is the characteristic time it takes a spin to flip, i.e., the time it takes a portion of the graphene sheet to change its concavity. We equal this time to $t_d = 10$ s, the characteristic time for defect motion mentioned in Section 2; see also Carpio & Bonilla (2008). The ratio between $[t]$ and the flipping time is very small $\delta = \zeta[t] = 1.1033 \times 10^{-15}$. In our simulations, we have used a larger value of δ which shortens

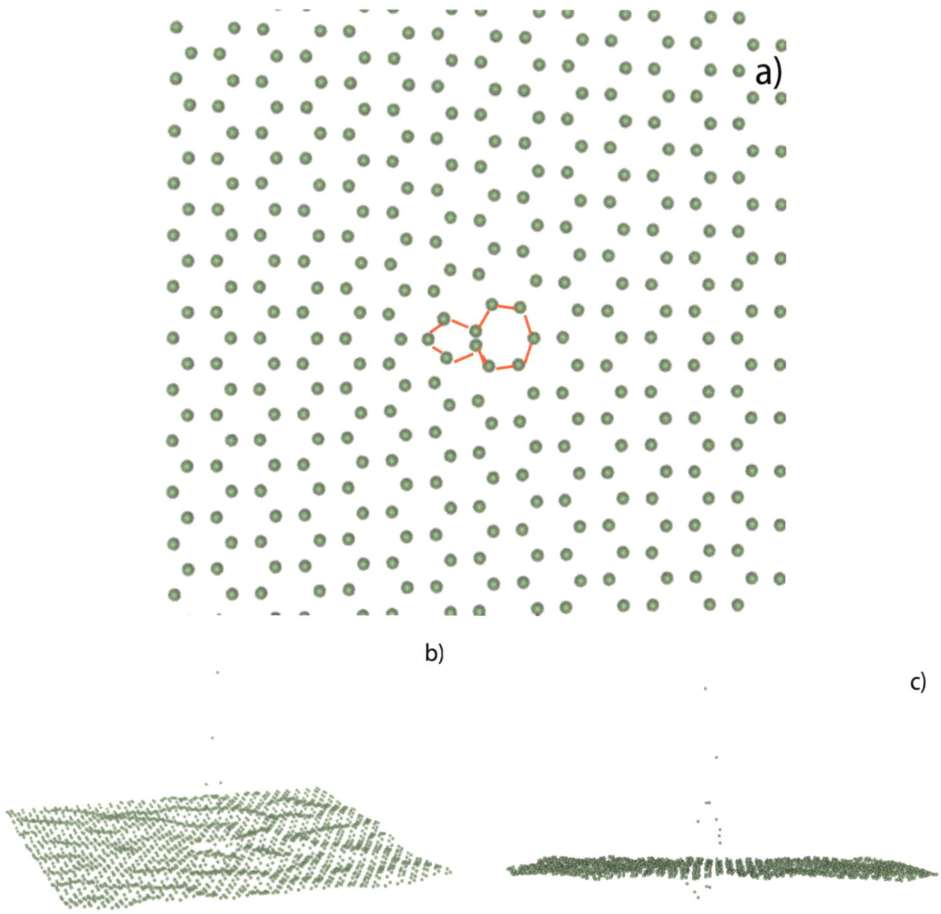

Fig. 6. Suspended graphene sheet containing one pentagon-heptagon defect. (a) View from above the sheet; (b) side view showing that the atoms forming the pentagon move above the plane of the sheet; (c) farther side view showing that there are displaced atoms above the sheet (pentagon) and below it (heptagon). (b) and (c) show that the height of the ripples is smaller than the vertical displacement of the atoms near the defect.

the relaxation time it takes the graphene sheet to reach a stable configuration without changing the latter.

We solve the equations of motion with the same boundary and initial conditions for the in-plane displacement vector (u, v) corresponding to a single pentagon-heptagon defect as explained in Section 3. We also impose that the vertical displacement w vanishes at the border atoms and for fictitious border atoms whose displacements are required when solving the equations of motion. Figure 6 shows a suspended graphene sheet with a pentagon-heptagon defect. At 300 K, the graphene sheet is below its critical temperature and therefore a rippled state is stable (Bonilla & Carpio, 2011). Thus ripples similar to those observed in experiments (Meyer et al., 2007) coexist with a local curvature near the defect as shown in Figure 6.

5. Conclusions

We have presented a theory of defects in suspended graphene sheets based on periodized discrete elasticity. The equations of linear elasticity are discretized on a hexagonal lattice, written in primitive coordinates and finite differences along primitive directions are replaced for periodic functions thereof. The latter allow for gliding of dislocations along primitive directions because spatial periodicity along them is restored in the equations of motion. Ignoring vertical displacement, we solve these equations numerically assuming that the initial and boundary conditions are given by the known linear elasticity expressions corresponding to one or several dislocations. If the equations of motion include an appropriate damping coefficient, we obtain the stable cores of the dislocations which are defects in the hexagonal lattice. The damping coefficient is fitted to experiments. The stable cores predicted by our theory have been observed in experiments.

We have also proposed a mechanism to explain ripples and curvature in suspended graphene sheets. We assume that the local trend of the sheet to bend upward or downward is represented by an Ising spin coupled to the carbon atoms. These spins are in contact with a thermal bath at the lattice temperature and flip stochastically according to Glauber dynamics. Our simulations show the appearance of ripples in the graphene sheet and the local curvature that appears near a defect, for instance near a pentagon-heptagon defect which is the core of an edge dislocation.

This work has been financed by the Spanish Ministry of Science and Innovation (MICINN) under grants FIS2008-04921-C02-01 (LLB), FIS2008-04921-C02-02 and UCM/BSCH CM 910143 (AC).

6. References

Abedpour, N.; Neek-Amal, M.; Asgari, R.; Shahbazi, F.; Nafari, N. & Rahimi Tabar, M. R. (2007). Roughness of undoped graphene and its short-range induced gauge field, *Physical Review B* Vol. 76: 195407.

Bao, W.; Miao, F.; Chen, Z.; Zhang, H.; Jang, W.; Dames, C. & Lau, C.-N. (2009). Controlled Ripple Texturing of Suspended Graphene and Ultrathin Graphite Membranes, *Nature Nanotechnology* Vol. 4: 562-566.

Blakslee, O. L.; Proctor, D. G.; Seldin, E. J.; Spence, G. B. & Weng, T. (1970). Elastic Constants of Compression-Annealed Pyrolytic Graphite, *Journal of Applied Physics* Vol. 41: 3373-3382.

Bonilla, L.L. & Carpio, A. (2011). Theory of ripples in graphene, *Preprint*.

Bunch, J.S.; van der Zande, A. M.; Verbridge, S.S.; Tanenbaum, D.M.; Parpia, J.M.; Craighead, H.G. & McEuen, P.L. (2007). Electromechanical Resonators from Graphene Sheets, *Science* Vol. 315: 490-493.

Carpio, A. & Bonilla, L.L. (2003). Edge dislocations in crystal structures considered as traveling waves of discrete models, *Physical Review Letters* Vol. 90: 135502.

Carpio, A. & Bonilla, L.L. (2005). Discrete models of dislocations and their motion in cubic crystals, *Physical Review B* Vol. 71: 134105.

Carpio, A. & Bonilla, L.L. (2008). Periodized discrete elasticity models for defects in graphene, *Physical Review B* Vol. 78: 085406.

Carpio, A.; Bonilla, L.L.; Juan, F. de & Vozmediano, M.A.H. (2008). Dislocations in graphene, *New Journal of Physics* Vol. 10: 053021.

Castro Neto, A. H.; Guinea, F.; Peres, N. M. R.; Novoselov, K. S. & Geim, A. K. (2009). The electronic properties of graphene, *Reviews of Modern Physics* Vol. 81: 109-162.

Coleman, V. A.; Knut, R.; Karis, O.; Grennberg, H.; Jansson, U.; Quinlan, R.; Holloway, B. C.; Sanyal, B. & Eriksson, O. (2008). Defect formation in graphene nanosheets by acid treatment: an x-ray absorption spectroscopy and density functional theory study, *Journal of Physics D: Applied Physics* Vol. 41: 062001.

Fasolino, A.; Los, J.H. & Katsnelson, M.I. (2007). Intrinsic ripples in graphene,*Nature Materials* Vol. 6: 858-861.

Geim, A.K. & Novoselov, K.S. (2007). The rise of graphene, *Nature Materials* Vol. 6: 183-191.

Geim, A.K. (2009). Graphene: Status and Prospects, *Science* Vol. 324: 1530-1534.

Girit, C.O., Meyer, J.C., Erni, K., Rossell, M. D., Kisielowski, C., Yang, L., Park, C.-H., Crommie, M. F., Cohen, M. L., Louie, S. G., Zettl, A. (2009). Graphene at the Edge: Stability and Dynamics *Science* Vol. 323: 1705-1708.

Glauber, R. J. (1963). Time-dependent statistics of the Ising model, *Journal of Mathematical Physics* Vol. 4: 294-307.

Gómez-Navarro, C., Meyer, J.C., Sundaram, R. S., Chuvilin, A., Kurasch, S., Burghard, M., Kern, K. & Kaiser, U. (2010). Atomic Structure of Reduced Graphene Oxide, *Nano Letters* Vol. 10: 1144-1148.

Guinea, F.; Katsnelson, M.I. & Geim, A.K. (2009). Energy gaps and a zero-field quantum Hall effect in graphene by strain engineering, *Nature Physics* Vol. 6: 33-33.

Landau, L.D. & Lifshitz, E.M. (1986). *Theory of elasticity*, 3rd ed., Pergamon Press, Oxford.

Lee, C.; Wei, X.; Kysar, J.W. & J. Hone (2008). Measurement of the Elastic Properties and Intrinsic Strength of Monolayer Graphene, *Science* Vol. 321: 385-388.

Meyer, J.C.; Geim, A.K.; Katsnelson, M.I.; Novoselov, K.S.; Booth, T.J. & Roth, S. (2007). The structure of suspended graphene sheets, *Nature* Vol. 446: 60-63.

Meyer, J. C.; Kisielowski, C.; Erni, R.; Rossell, M.D.; Crommie M.F. & Zettl, A. (2008). Direct imaging of lattice atoms and topological defects in graphene membranes, *Nano Letters* Vol. 8(No. 11): 3582-3586.

Meyer, J.C., Chuvilin, A., Algara-Siller, G., Biskupek, J. & Kaiser, U. (2009). Selective Sputtering and Atomic Resolution Imaging of Atomically Thin Boron Nitride Membranes, *Nano Letters* Vol. 9: 2683-2689.

Nelson, D. R. (2002). *Defects and Geometry in Condensed Matter Physics*, Cambridge U.P., Cambridge.

Novoselov, K.S.; Geim, A. K.; Morozov, S. V.; Jiang, D.; Zhang, Y.; Dubonos, S. V.; Grigorieva, I. V. & Firsov, A. A.(2004). Electric field effect in atomically thin carbon films, *Science* Vol. 306: 666-669.

Novoselov, K.S.; Jiang,D.; Schedin, F.; Booth, T.J.; Khotkevich, V. V.; Morozov, S. V. & Geim, A. K. (2005). Two-dimensional atomic crystals, *Proceedings of the National Academy of Sciences USA* Vol.102: 10451-10453.

Segall, M.D.; Lindan, P. J. D.; Probert, M.J.; Pickard, C.J.; Hasnip, P. J.; Clark, S.J. & Payne, M. C. (2002). First-principles simulation: ideas, illustrations and the CASTEP code, *Journal of Physics: Condensed Matter* Vol. 14: 2717-2744.

Thompson-Flagg, R. C.; Moura, M. J. B. & Marder, M. (2009). Rippling of graphene, *Europhys. Letters* Vol. 85: 46002.

Vozmediano, M.A.H.; Katsnelson, M.I. & Guinea, F. (2010). Gauge fields in graphene, *Physics Reports* Vol. 496: 109-148.

Wang, X.; Tabakman, S. M. & Dai, H. (2008). Atomic Layer Deposition of Metal Oxides on Pristine and Functionalized Graphene, *Journal of the American Chemical Society* Vol. 130: 8152-8153.

Zakharchenko, K. V.; Katsnelson, M. I. & Fasolino, A. (2009). Finite Temperature Lattice Properties of Graphene beyond the Quasiharmonic Approximation, *Physical Review Letters* Vol. 102: 046808.

10

Symmetry and Lattice Dynamics

Hui Tang[1], Bing-Shen Wang[2] and Zhao-Bin Su[3]
[1]*Institute of Physics, Chinese Academy of Sciences*
[2]*Institute of Semiconductors, Chinese Academy of Sciences*
[3]*Institute of Theoretical Physics, Chinese Academy of Sciences*
China

1. Introduction

Hexagon is one of the most beautiful substance structures in nature. From snowflakes to honeycomb lattices, we can find the presence and stability of this delicate structure. The two-dimensional hexagon had been investigated in graphite materials which has an evident layered structure (Painter & Ellis, 1970) long before the discovery of graphene. Since the successful fabrication of graphene (Berger et al., 2004; Novoselov et al., 2004), a monolayer of carbon atoms tightly packed into a two-dimensional hexagonal lattice, its various properties have been widely explored and have received increasing attention (Castro Neto et al., 2009; Das et al., 2008; Ferrari et al., 2006; Gupta et al., 2006; Partoens & Peeters, 2006; 2007). Its exotic electronic property has the potential for practical applications and provides a prospect of theoretical invention. On the other hand, graphene is a basic structural element of many carbon allotropes including graphite, charcoal, carbon nanotubes and fullerenes. It is also of theoretical significance and practical guidance to study the physical properties such as the lattice dynamics and electronic structures starting from graphene and extending to other related systems.

It is well known that symmetry concept plays an important role in physics. Specifically, in condensed matter physics, the microscopic symmetry of the structures of materials determines the symmetry properties of macroscopic physical quantities. Group theory is not only a powerful tool to classify the electronic and phonon spectra, but also essential to understand the inherent physical meaning as well as their consequence. The symmetry group D_{6h} of graphene is the basic building block of all carbon allotropes in a sense that either their symmetry groups evolve from D_{6h} with a few symmetry operations lifted or added, or some symmetry operations in D_{6h} are inherited in a certain forms. The study of symmetry provides an effective way to find the common essence and differences among carbon allotropes. In this chapter we will analyze the symmetry characteristics as well as symmetry induced constrains to the lattice dynamics of graphene, multilayered graphene, and nanotubes.

Lattice dynamics studies the vibration of lattice, i.e. the phonons. Many exotic properties of graphene, such as its very high strength and thermal conductivity are directly connected to the phonon properties. Its extremely high electrical conductivity is also related to the phonon dispersion and electron-phonon interactions. We will focus on the lattice dynamics of the multilayered graphene systems, where graphene is the limit of one layer number, and the nanotube systems with different radius and charity. In particular, the role of symmetry selection in optical activeness of phonon modes in multilayer graphene systems and symmetry restriction in the phonon dispersion calculation of carbon nanotubes is emphasized.

Multilayered graphene systems can be viewed as stacking a few number of graphene layers along the direction perpendicular to planar sheets. Some symmetry operations existed in graphene such as the space reflection $\hat{\sigma}_i$ or reflection with the plane $\hat{\sigma}_h$ may lift depending on the arrangement of layers and the layer number. The breaking of the symmetry operations reduces the order of symmetry groups and causes the change of the optical activities of the phonon branches. Quantitatively, the phonon modes will be softer or harder (red or blue shift) as the number of layers changed. All these qualitative and quantitative properties will be discussed in the next two sections and calculated results are compared with the experimental measurements.

Carbon nanotubes had been discovered (Iijima, 1991) before the fabrication of graphene. At first sight, nanotubes and graphene belong to totally different symmetry groups and follow the different symmetry restrictions on their physical properties. In virtue of the picture that a single wall carbon nanotube can be viewed as a warped graphene stripe (Saito et al., 1998), we still can ask what is the effect of the added periodic edge condition on the two sides of the stripe, and what is the consequence of the three-fold rotational $\hat{C}_3$ symmetry on graphene. In section 4, we will reveal a three-theta role, an universal triple chiral angle (3θ) dependence, of nanotubes with different chiral angles, which is actually the heritage of the three-fold rotational symmetry of graphene.

As usual, the enlargement of the primitive cell in real lattice space causes Brillouin zone folding in reciprocal space. If the edge effects of stripes and the curvature effects of nanotubes are neglected, the electronic dispersions in new small Brillouin zone can be obtained directly by folding those of graphene. For phonons, the situation would be more complicated because of their vector properties. Particularly, flexure modes (a kind of vibrational modes with parabolic dispersions when surface exists (Mahan & Jeon, 2004)) can be obtained only when a correct form of the potential energy satisfying the symmetry constrains is adopted. The emergence of the twist mode is intimately connected with the geometrical structure of the rod-like cylindrical systems.

The one-dimensional (1D) single wall carbon nanotubes belong to 1D line groups (Damnjanović et al., 1999; 2000). For chiral nanotubes, the definition of the primary chiral operation is not unique. This will result in different choice of primitive cells and the different coordinates in description of the lattice structure of nanotubes. There are two commonly used coordinates for nanotubes, thus two sets of good quantum numbers. One of them will be called it Dresselhaus coordinate (Saito et al., 1998), which has the superiority in denoting the optical transition properties at long wave length limit. The other is White coordinate (Gunlycke et al., 2008; White et al., 1993), which has a clear relation with the symmetry generators. Since the quantum numbers in Dresselhaus coordinate are related to an isogonal group of the symmetry group of the nanotube, the relation between these two sets of quantum numbers is not trivial and we will discuss it in detail in section 5. With this relation, it is easy to transform electronic or phonon dispersions with one set of good quantum numbers to the others. Finally, to illustrate the similarities and differences among carbon allotropes, the lattice dynamics of single-wall carbon nanotubes is discussed in section 6.

2. The symmetry consideration in graphene and multilayer graphene systems

Graphene is a monolayer carbon atoms with the hexagonal lattice configuration which is characterized by the D_{6h} symmetry. There are two identical carbon atoms in one unit cell as shown in Fig. 1, and the bond length between two nearest-neighbor atoms in the plane is $b = 1.42$ Å(Saito et al., 1998). The primitive lattice vectors can be set as $\vec{a}_1$ and $\vec{a}_2$ with $|\vec{a}_1| = |\vec{a}_2| = 2.46$ Å. The lattice configuration of multilayer graphene systems (MLGS) is constructed by stacking N graphene sheets along perpendicular direction, z axis. The distance between two adjacent layers is about $\frac{c}{2} = 3.35$ Å which is much larger than the in-plane bond length (Manes et al., 2007). AA-stacked or AB-stacked MLGS can be obtained respectively if all layers have the same configuration or shift alternately along one of the first-nearest carbon-carbon bonds in horizontal plane as shown in Fig. 2. The three-dimensional (3D) graphite is the limitation of AB-stacked MLGS with layer number $N \to \infty$.

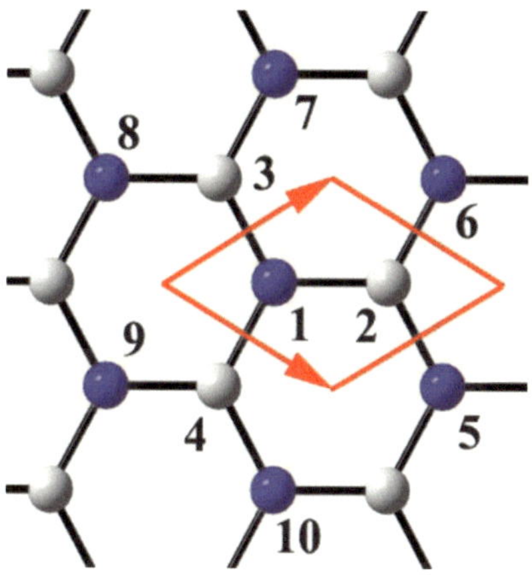

Fig. 1. The sketch of the lattice configuration of graphene. The vectors shown in the figure are the primitive lattice vectors $\vec{a}_1$ and $\vec{a}_2$.

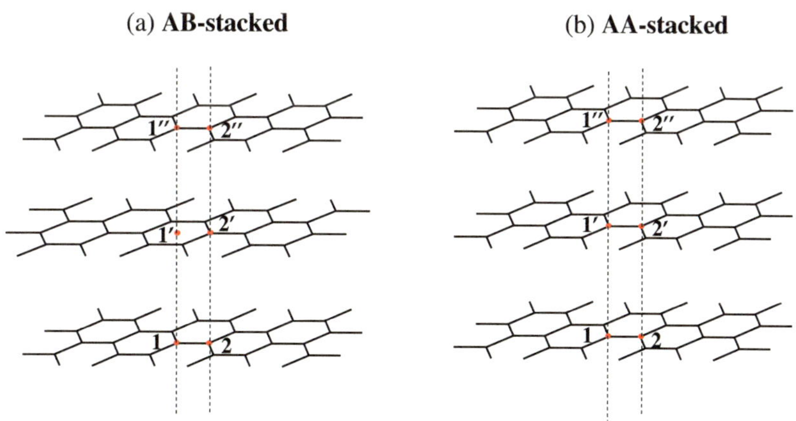

Fig. 2. The sketch of the configurations of AB-stacked in (a) and AA-stacked in (b) for multilayer graphene.

There are twelve conjugacy classes in D_{6h} group and thus twelve irreducible representations. For simplifying the discussion, D_{6h} can be expressed as the direct products of three subgroups,

$$D_{6h} = D_3 \otimes \{\hat{E}, \hat{\sigma}_h\} \otimes \{\hat{E}, \hat{\sigma}_i\} ,$$
$$D_3 = \{\hat{E}, 2\hat{C}_3, 3\hat{C}_2\} ,$$

where $\hat{\sigma}_h$ is the reflection with the plane, $\hat{\sigma}_i$ the space reflection, $\hat{C}_3$ three-fold rotation around the axis perpendicular to the graphene sheet, and $\hat{C}_2$ two-fold rotation around axis parallel to the sheet. There are three irreducible representations for D_3 group, two one-dimensional representations A and B and one two-dimensional representation E. The twelve irreducible representations of D_{6h} can be denoted by A, B, and E with the subscripts $1, 2$ and u, g, where letters are for representations of D_3, and the subscripts correspond to the parity under $\hat{\sigma}_h$ and $\hat{\sigma}_i$ respectively. The symbol for the irreducible representations we used here is the notation used in Ref. (Eyring et al., 1949) which is most commonly used in the treatment of molecules. For AB-stacked MLGS, the symmetry operations $\hat{\sigma}_h$ and $\hat{\sigma}_i$ existed in graphene cannot coexist anymore, and the point groups will depend on the layer number (Manes et al., 2007). For AB-stacked even number MLGS (EMLGS), the space reflection $\hat{\sigma}_i$ exists with respect to the middle point of pair atoms 2′ in the $\frac{N}{2}$-th layer and 1″ in the $\frac{N}{2}+1$-th layer as shown in Fig. 2(a). But there is no planar reflection $\hat{\sigma}_h$. Same as that in graphene, the two $\hat{C}_3$ and three $\hat{C}_2$ operations exist. The three two-fold axes are perpendicular to z axis with angle $\pi/3$ between each other. All the symmetry operations constitute the point group $D_{3d} = D_3 \otimes \{\hat{E}, \hat{\sigma}_i\}$. In AB-stacked odd number MLGS (OMLGS), the space reflection $\hat{\sigma}_i$ is lift instead of $\hat{\sigma}_h$ in EMLGS. The reference plane of $\hat{\sigma}_h$ is set at the middle layer of OMLGS as shown in Fig. 2(b). All operations of D_3 also exist, but the three 2-fold axes are one to one perpendicular to those of $\hat{C}_2$ in EMLGS. Consequently, the symmetry group of OMLGS is $D_{3h} = D_3 \otimes \{\hat{E}, \hat{\sigma}_h\}$.
The point group of the 3D graphite is the non-symmorphic group D^4_{6h} with non-primitive translation $\vec{\tau} = \frac{1}{2}\vec{c}$ (Brillson et al., 1971). In addition to the in-plane translation symmetry, there exists translation symmetry along z for the 3D graphite. The primitive vector along z is set as $\vec{c}$ which crosses the two adjacent layers. Thus there are four carbon atoms 1, 2, 1′, and 2′ in one unit cell as represented in Fig. 2(a).
As described in Ref. (Jiang et al., 2008, a), the environment of an atom in graphite or AB-stacked MLGS is more complicated than that in 2D graphene. The in-layer situation is the same as that of graphene, i.e. three nearest-neighbor carbon atoms and six next-nearest-neighbors for any one of carbon atom. For the inter-layer atomic positions, picking atom 2′ in the middle layer of Fig. 2(a) as an example, there are two inter-layer nearest-neighbor atoms 2 and 2″ in each of the two adjacent layers with the distance $c/2$ respectively. Furthermore, there are three inter-layer next-nearest-neighbor atoms around atom 2′ with distance $\sqrt{b^2 + (c/2)^2}$ in one of the adjacent layer (such as atom 1 is one of them). The adjacent environment of atom 1 is quite different from that of atom 2′. Because the position 1′ in Fig. 2(a) has no atom, atom 1 has no inter-layer nearest-neighbors. Instead it has six inter-layer next-nearest-neighbors in one of the adjacent layer with the same distance $\sqrt{b^2 + (c/2)^2}$.
In the AA-stacked MLGS as well as the AA-stacked 3D graphite, all layers have the same configuration. So that all of them have the same point group D_{6h} as that in graphene whatever the layer number is even or odd. Different from the graphite (AB-stacked 3D graphite), there are only two atoms in the unit cell and the primitive translation along z axis is $\vec{c}/2$ for AA-stacked 3D graphite. As shown in Fig. 2(b), the environment of a carbon atom in

the AA-stacked MLGS is simpler than that in AB-stacked MLGS. For each atom, there are one inter-layer nearest-neighbors in one of the adjacent layer with the distance $c/2$ and three inter-layer second-nearest-neighbors with distance $\sqrt{b^2 + (c/2)^2}$.

Taking the lattice displacements $\vec{u}_i$ (i runs over all atoms in one unit cell) as bases, the dynamical representation Γ^{dyn} can be expressed as the direct products of the vector representation Γ^{v} and atomic permutation representation Γ^{atom} for a given group, $\Gamma^{dyn} = \Gamma^{v} \otimes \Gamma^{atom}$. By applying the projection operator technique, the dynamical representation Γ^{dyn} is decomposed into the irreducible representations of the corresponding groups of graphene, MLGS with N even and odd, and 3D graphite respectively. Referring to Ref. (Elliott & Dawber, 1979), the infra-red (Ir) active phonon modes should be compatible with the same decomposed irreducible representations as the vector representation Γ^{v}, while the Raman active phonon modes correspond to the same irreducible representations shown up in the decomposition of a six-dimension representation with d-wave like bases: $x^2 + y^2$, z^2, $x^2 - y^2$, xy, yz, and zx. The three acoustic modes with zero frequency at the Γ point, which correspond to the vector representation Γ^{v}, are excluded in the consideration of Ir and Raman active modes.

	Group	Γ^{dyn}	Γ^{Ir}	Γ^{R}
graphene^a	D_{6h}	$A_{2u} \oplus B_{2g} \oplus E_{1u} \oplus E_{2g}$	/	E_{2g}
AB-stacked	$D_{3d}{}^b$	$N(A_{1g} \oplus A_{2u} \oplus E_g \oplus E_u)$	$(N-1)A_{2u} \oplus (N-1)E_u$	$NA_{1g} \oplus NE_g$
	$D_{3h}{}^b$	$(N-1)A_{1g} \oplus (N+1)A_{2u}$ $\oplus (N+1)E_g \oplus (N-1)E_u$	$NA_{2u} \oplus NE_g$	$(N-1)A_{1g}$ $\oplus NE_g \oplus (N-1)E_u$
	$D_{6h}^{4}{}^c$	$2(A_{2u} \oplus B_{2g} \oplus E_{1u} \oplus E_{2g})$	$A_{2u} \oplus E_{1u}$	$2E_{2g}$
AA-stacked	D_{6h} (even)	$\frac{N}{2}(A_{1g} \oplus A_{2u} \oplus B_{1u} \oplus B_{2g} \oplus E_{1u}$ $\oplus E_{1g} \oplus E_{2g} \oplus E_{2u})$	$(\frac{N}{2}-1)(A_{2u} \oplus E_{1u})$	$\frac{N}{2}(A_{1g} \oplus E_{1g} \oplus E_{2g})$
	D_{6h} (odd)	$\frac{N-1}{2}(A_{1g} \oplus B_{1u} \oplus E_{1g} \oplus E_{2u})$ $\oplus \frac{N+1}{2}(A_{2u} \oplus B_{2g} \oplus E_{1u} \oplus E_{2g})$	$\frac{N-1}{2}(A_{2u} \oplus E_{1u})$	$\frac{N-1}{2}(A_{1g} \oplus E_{1g})$ $\oplus \frac{N+1}{2}E_{2g}$
	D_{6h}	$A_{2u} \oplus B_{2g} \oplus E_{1u} \oplus E_{2g}$	/	E_{2g}

Table 1. The symmetry analysis of the phonon modes at the center point of Brillouin zone for AA- or AB-stacked MLGS. Phonon modes are classified by the decomposition of Γ^{dyn} . The irreducible representations of the Ir and Raman active modes are also listed.
[a]Reference (Saito et al., 1998); [b]Reference (Manes et al., 2007); [c]Reference (Mani et al., 1974).

We show in Table 1 the symmetry analysis (Jiang et al., 2008, a) of phonon modes and the Raman active and Ir active modes for graphene, AB- and AA-stacked MLGS, and 3D graphite. Two straightforward consequences can be find out for the Ir and Raman active modes in AB-stacked MLGS. Firstly, in the EMLGS with D_{3d} point group, as well as in graphene and 3D graphite, phonon modes cannot be Ir and Raman active simultaneously. However, in the OMLGS with D_{3h} point group, the N E_g modes are both Ir and Raman active. This is because there is no inversion center in the OMLGS. Secondly, an optical mode (denoted as the inter-layer optical mode) belongs to the A_{1g} representation with Raman active in the EMLGS (D_{3d}) and A_{2u} Ir active in the OMLGS (D_{3h}). This mode can be further identified as vibrational displacements perpendicular to the constituent layers, and oscillating with each layer as a whole but alternatively from layer to layer. These properties can be used as criterion to identify the layer number being even or odd for AB-stacked MLGS.

In the AA-stacked MLGS with N either even or odd, the symmetry group is D_{6h} which includes both σ_i and σ_h. Consequently, no phonon modes would be Ir and Raman active simultaneously in AA-stacked MLGS. Interestingly, the inter-layer optical mode mentioned above belongs to the different representations so that different active for even and odd layer numbers. In detail, it belongs to A_{1g} Raman active in the EMLGS and A_{2u} Ir active in the

OMLGS, which is the same as that in AB-stacked MLGS. Due to its vibration direction, the inter-layer optical mode favors to take the maximum advantage of the inter-layer interactions. From the next section, it is known that this optical mode would be the most sensitive mode to the layer numbers in a few layer systems and becomes a useful candidate in experiments to identify the even-oddness of the MLGS and the layer numbers.

3. Lattice dynamics of multilayer graphene systems

3.1 Vibrational potential energy

For providing a simple and straightforward method to relate the phonon modes with atomic movement modality, we choose an semi-phenomenological model of vibrational potential energy (Aizawa et al., 1990; Aizawa, 1991; Jiang et al., 2008, a) to study the lattice dynamics of graphene and multilayer graphene systems (MLGS). The lattice vibration energy in this model is expressed in quadratic terms of lattice displacements. So that it is easy to check term by term the symmetric invariance, such as the translational invariance and rigid rotational invariance, which are the essential restrictions to derive the correct phonon modes. Besides, because the potential energy is one of the key aspects in determine the vibration properties, it is possible to use the same potential terms of graphene into in-layer case of MLGS and even to carbon nanotubes, which will be discussed in more detailed in section 6.

We begin from the one graphene layer, where the vibrational potential energy can be described by five quadratic terms (Aizawa et al., 1990; Aizawa, 1991). They are the first and second nearest-neighbor radical spring stretching, the in-plane bond angle variations, the out-of-surface bond bending, and the bond twisting energies. The detailed expressions of five terms, taking atom 1 in Fig. 1 as an example, are listed in the following.

(1) V_l is the potential of the radical spring force between the nearest-neighbor atom pair,

$$V_l = \frac{k_l}{2}\sum_{i=2}^{4}[(\vec{u}_i - \vec{u}_1)\cdot\vec{e}^{\,l}_{1i}]^2\,, \tag{1}$$

where k_l is the first-order force constant and $\vec{e}^{\,l}_{1i} = \frac{\vec{r}_{1i}}{|\vec{r}_{1i}|}$.

(2) V_{sl} is similar to the first term but between a pair of next nearest-neighbored atoms illustrated as $(1, 5 \ldots 10)$ in Fig. 1,

$$V_{sl} = \frac{k_{sl}}{2}\sum_{i=5}^{10}[(\vec{u}_i - \vec{u}_1)\cdot\vec{e}^{\,l}_{1i}]^2 \tag{2}$$

with k_{sl} the second-order force constant.

(3) The potential energy for the in-plane bond bending V_{BB} is actually a term associated with bond angle variations. Three atoms are involved:

$$\begin{aligned} V_{BB} &= \frac{k_{BB}}{4}\sum_{j_i}\sum_{\substack{j'_i\\(j'_i\neq j_i)}}\left[\frac{\vec{u}_{j_i}-\vec{u}_i}{r_{ij_i}}\cdot(\vec{e}^{\,l}_{ij'_i}-\cos\theta_0\vec{e}^{\,l}_{ij_i}) + \frac{\vec{u}_{j'_i}-\vec{u}_i}{r_{ij'}}\cdot(\vec{e}^{\,l}_{ij_i}-\cos\theta_0\vec{e}^{\,l}_{ij'_i})\right]^2 \\ &= \frac{k_{BB}}{4}\sum_{j_i}\sum_{\substack{j'_i\\(j'_i\neq j_i)}}(\cos\theta'_{j_iij'_i}-\cos\theta_0)^2\,. \end{aligned} \tag{3}$$

As i is 1 (or 2), (j_i, j_i') take the sites (2, 3), (3, 4), and (4, 2) (or (1, 5), (5, 6), and (6, 1)) respectively as shown in Fig. 1. In Eq. (3), $\theta_0 = 120°$ stands for the equilibrium angle, while $\theta'_{j_i i j_i'}$ for angle in vibration between the bonds $\vec{r}_{ij_i}$ and $\vec{r}_{ij_i'}$. The expression in terms of $\vec{u}_{j_i} - \vec{u}_i$ and $\vec{u}_{j_i'} - \vec{u}_i$ is exactly same as that of $\cos\theta'$.

(4) The potential of the out-of-plane bond bending V_{rc} describes a kind of withdrawn energy of atom i by its three nearest-neighbor atoms j_i. It contains four atoms simultaneously,

$$V_{rc} = \frac{k_{rc}}{2}[(3\vec{u}_i - \sum_{j_i}\vec{u}_{j_i}) \cdot \vec{e}_z]^2, \tag{4}$$

Where i takes 1 or 2 with j_i running over the three nearest neighbors of atom i. This potential has the physical intuition as that responsible for the perpendicular optical mode and is trying to keep the four atoms on the plane.

(5) The twist potential energy for bond $\vec{r}_{1k}$ is

$$V_{tw} = \frac{k_{tw}}{2}\sum_{\langle i,j\rangle}[(\vec{u}_i - \vec{u}_j - (\vec{u}_{i'} - \vec{u}_{j'})) \cdot \vec{e}_z]^2\,, \tag{5}$$

where $\langle i, j\rangle$ represents a pair of atoms nearest-neighbored with atom 1 while k the third of its nearest neighbors. Pair $\langle i', j'\rangle$ is for a pair of atoms nearest-neighbored with atom k. It has the intuition that responsible for modes with twisted vibrations.

From the modality of atomic movements, we can classify the above in-plane vibrational potential terms into three types and extend them to the inter-layer case (Jiang et al., 2008, a). The first type is for radical stretching movements between the two atoms located in the adjacent layers. The second describes the relative movement between the two pairs of atoms with a common one as an apex. It is evolved from the above bond bending term and three atoms form one bond in a layer and another connecting the two nearest layers. The third type is new for simulating the twist force affected on the inter-layer "bond" as shown in the sketch 3, which involves more than three atoms according to the specific bond configurations. The whole of these terms is actually a generalized valence force model to including far away atoms in response to the bond charge effect (Jeon and Mahan, 2005) in certain extent. Because the inter-layer "bond" is much longer than the in-plane bond, it is resulted that all inter-layer interactions are one or two orders less than their counterparts in layer, but they themselves have comparable contributions. We give out the inter-layer terms for the AB-stacked MLGS in the following and they can be similarly generalized to the AA-stacked MLGS.

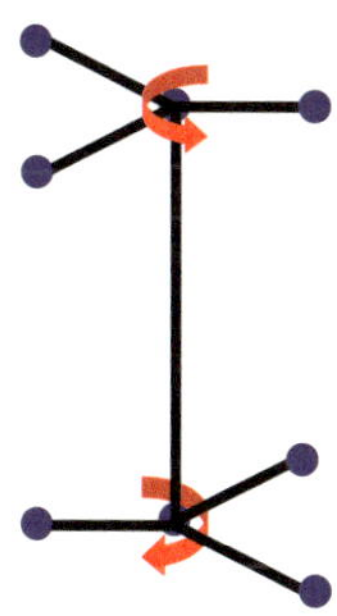

Fig. 3. The sketch of the twist force for a "bond" connecting the two adjacent layers.

(1) The inter-layer "bond" stretching energies $V_l^{(int)}$ (or $V_{sl}^{(int)}$):

$$\sum_{i,j} \frac{k'_l}{2}[(\vec{u}_i - \vec{u}_j) \cdot \vec{e}^{\,l}_{ij}]^2, \tag{6}$$

where $\vec{u}_i(\vec{u}_j)$ is the displacement vector of the atom $i(j)$ and $\vec{e}^{\,l}_{ij}$ is the unit vector from atom i to atom j. If the summation is taken over the nearest-neighbored inter-layer pair of atoms, the corresponding force constant is denoted as k'_l, while k'_{sl} for the next nearest-neighbor inter-layer pairs.
(2) For the three atoms 2, 2′ and i, where i is the in-plane nearest neighbor of atom 2 (atom 1 is one of them as shown in Fig. 2), we found by practice that under a specific configuration with atom i rather than atom 2 as an apex, a correlation term k'_{rr} has the most sensitive contribution to the layer dependence of the optical modes,

$$\frac{k'_{rr}}{2}\sum_i[(\vec{u}_2 - \vec{u}_i) \cdot \vec{e}^{\,l}_{i2} - (\vec{u}_{2'} - \vec{u}_i) \cdot \vec{e}^{\,l}_{i2'}]^2.$$

Actually the two square terms in above modality have already been accounted in the in-plane and inter-plane stretching terms respectively. Only the across term

$$V_{rr} = -k'_{rr}\sum_i[(\vec{u}_2 - \vec{u}_i) \cdot \vec{e}^{\,l}_{i2}][(\vec{u}_{2'} - \vec{u}_i) \cdot \vec{e}^{\,l}_{i2'}] \tag{7}$$

plays a real role, which weakens the interaction between the two adjacent layers. The positive definite condition for getting real frequencies is $k'_{sl} \geq k'_{rr}$.
(3) The twisting potential for an inter-layer bond between atoms 2 and 2′ is coming from the two sets of three nearest-neighbors of atoms 2 and 2′ respectively. It can be described as

$$V_{tw} = \frac{k'_{tw}}{2}[\sum_i(\vec{u}_i - \vec{u}_2) \cdot \vec{e}^{\,\theta}_i - \sum_j(\vec{u}_j - \vec{u}_{2'}) \cdot \vec{e}^{\,\theta}_j]^2, \tag{8}$$

where $\sum_i$ and $\sum_j$ represent the summation over the three in-plane first-nearest-neighbors for atoms 2 and 2′ respectively. $\vec{e}^{\,\theta}_i = \vec{e}_z \times \vec{e}^{\,l}_{2i}$ is the tangential unit vector in the plane formed by three nearest-neighbor atoms. The quadratic form as a whole ensures a proper definition for the torsion angle. For pure rotations around the bond, this expression gives zero torsion consistently. In contrast, the bond is most severely twisted when the three neighbors around atom 2 and those of atom 2′ rotate reversely.
By comparing the k'_{rr} term with V_{b-b} term (Jeon and Mahan, 2005) originated from the bond-charge model, a common point can be found that both terms have the same negative cross term. In addition, from the calculation results shown in the next subsection, two quite different frequency values in long wave length limit can be obtained to distinguish clearly the in-plane vibration and vibration along z axis. These provide an evidence that the polarization effect of graphitic materials (Saito et al., 1998) can be properly described by our vibration potential model. It can also be checked one by one that all above nine vibrational potential energy terms satisfy the rigid rotational symmetry constrain(Jiang et al., 2006; Mahan & Jeon, 2004; Popov et al., 2000) which guarantees the existence of the flexure modes with correct parabolic dispersion in low-frequency limit in low-dimensional systems. Details will be discussed in the next subsection.

3.2 Results and discussions

Because the microscopic environment of a carbon atom is almost the same in graphene, the in-layer of MLGS, and carbon nanotubes, the five intra-layer force constants we used in the following are taken from Ref. (Jiang et al., 2008, b) as $k_l = 305$ N m^{-1}, $k_{sl} = 68.25$ N m^{-1}, $k_{BB} = 1.38 \times 10^{-11}$ erg, $k_{rc} = 14.8$ N m^{-1}, $k_{tw} = 6.24$ N m^{-1}. They are originally taken references from a force-constant model for monolayer graphite formed on surfaces (Aizawa et al., 1990; Aizawa, 1991). The four inter-layer force constants are adjusted to fit the experimental values of four modes in 3D graphite as shown in Table 2. The fitting error for phonon modes is kept less than 7%. The inter-layer force constants are $k'_l = 0.77$ Nm^{-1}, $k'_{sl} = 0.95$ Nm^{-1}, $k'_{tw} = 0.64$ Nm^{-1}, $k'_{rr} = 0.9$ Nm^{-1}.

Reps	A'_1	E_{2g}	A_{2u}	E_{2g}
Experiments	30[a]	40[a]	868[b]	1586[b]
Theory	30.2	42.7	869.9	1586.6

Table 2. Comparison of several mode frequencies (in the unit of cm^{-1}) for the AB-stacked 3D graphite between our calculating results and the experimental values: [a]Ref. (Nicklow et al., 1972), [b]Ref. (Maultzsch et al., 2004).

The phonon dispersions for the AB-stacked graphite are calculated (Jiang et al., 2008, a) according to the vibration potential energy discussed in the above subsection and shown in Fig. 4. The theoretical calculations meet the experimental results not only in the low frequency region (Nicklow et al., 1972), but also in the high frequency region (Maultzsch et al., 2004; Mohr et al., 2007). The good consistency with the experimental data shows that the vibrational potential energies and parameters are reasonable and applicable. In the figure, Γ, K, M and A points are the standard notations for the specific symmetric points in the Brillouin zone hexagonal reciprocal lattice.

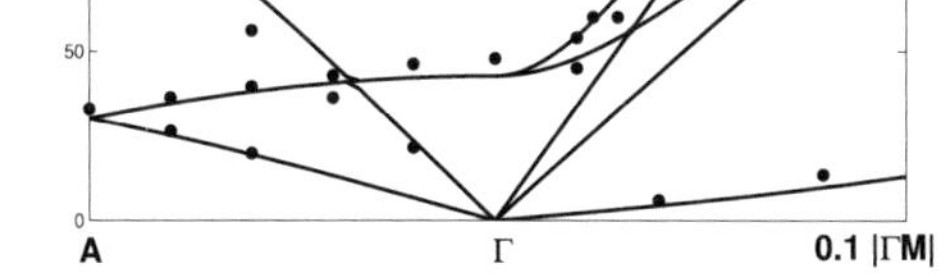

(a) The low-frequency region

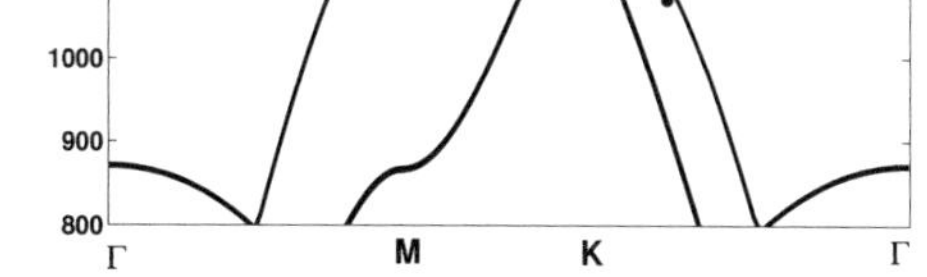

(b) The high-frequency region

Fig. 4. Phonon dispersion for the 3D graphite for the low-frequency region (a), and the high-frequency region (b). Solid dots are the experimental results of Ref. (Nicklow et al., 1972) in (a), and Ref. (Maultzsch et al., 2004; Mohr et al., 2007) in (b). Lines are the theoretical calculations. In Ref. (Maultzsch et al., 2004; Mohr et al., 2007), those phonon wave vectors, which were not exactly along the Γ-M or Γ-K-M direction, were projected onto the closest high-symmetry direction.

The phonon dispersions of graphene and AA-stacked 3D graphite are further calculated as shown in Fig. 5. Since the unit cell in both cases contains only two atoms in contrast to that of the AB-stacked graphite, there are six branches of phonon dispersion. Focus on the Γ point, a long wave length limit, we would mainly discuss three kinds of optical modes whose atomic movement modalities are sketched in Fig. 6. (1) The optical C-C in-plane stretching modes shown in Fig. 6(a) exist at about 1580 cm^{-1} for all three cases of graphene, AB-, and AA-stacked 3D graphite. These modes are doubly degenerated and Raman active with E_{2g} irreducible representation. Additional two degenerate optical in-plane modes very near them, which are Infra-red active belonging to E_{1u}, exist only in AB-stacked 3D graphite because it has four atoms in the unit cell. (2) An out-of-plane optical mode shown in Fig. 6(b) exists at about 860 cm^{-1} for graphene and AA-stacked 3D graphite, as well as for AB-stacked 3D graphite but double degenerated. The obvious difference between 1580 cm^{-1} mode and 860 cm^{-1} mode comes from the polarization effect of graphitic materials (Saito et al., 1998). (3) Except graphene, there exists an inter-layer optical mode shown in Fig. 6(c) for layer number equal to or greater than 2. Noticed that this mode exists at A point instead of Γ point for AA-stacked 3D graphite due to a phase factor difference of π needed between two adjacent layers. Asymptotic frequency of inter-layer optical mode in large number N is 150 cm^{-1} or 134 cm^{-1} for AB- or AA-stacked MLGS, respectively.

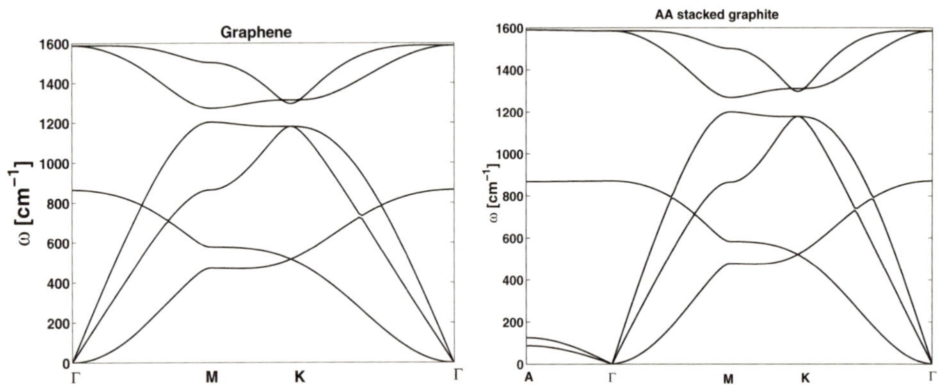

(a) Graphene (b) AA-stacked 3D graphite

Fig. 5. The calculated phonon dispersions of graphene (a), and AA-stacked 3D graphite (b).

The phonon dispersions for AA- and AB-stacked MLGS with an arbitrary layer number N can be calculated. In Fig. 7, the frequency of the intra-layer optical C-C stretching mode (Fig. 6(a))

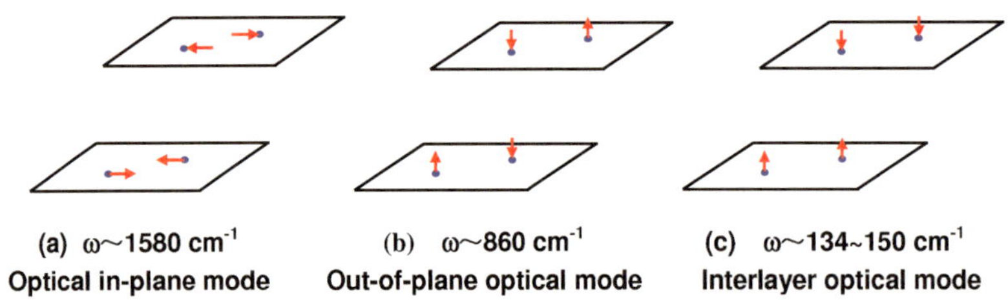

Fig. 6. Modality of atomic movements for three kinds of optical phonon modes in multilayered graphene systems.

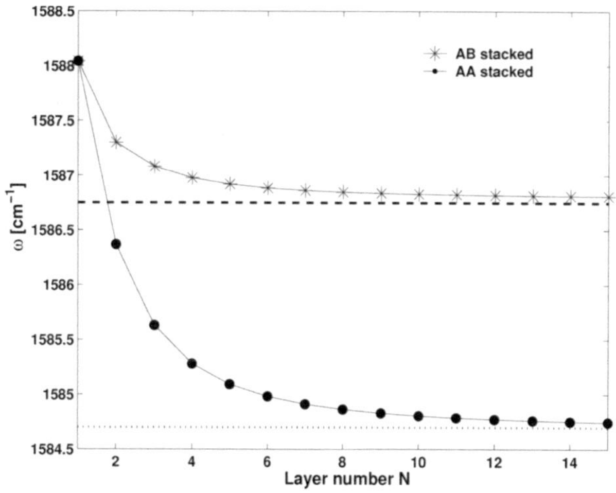

Fig. 7. The calculated frequency value for the optical C-C stretching mode vs the layer number N. Lines are draw to guide eyes.

in long wave length limit is represented as layer number N for AA- and AB-stacked MLGS. The layer number dependence of the frequency behaves a red shift as number increase which is in agreement with the experimental measurements. The frequency for this mode is about 1588 cm^{-1} in the single graphene layer, decreases with N increase, and almost saturates at about $N = 10$. The 3D limit is 1586.7 cm^{-1} (1584.7 cm^{-1}) for the AB- (AA-) stacked system. The amount of red shift value in the theoretical calculation corresponds excellently with that measured by experiments within the ranges $3 \sim 5$, $5 \sim 6$, and 8 cm^{-1} in Refs (Ferrari et al., 2006), (Gupta et al., 2006), and (Das et al., 2008), respectively.

At Γ point, the out-of-plane optical mode (Fig. 6(b)) is Ir active in the AB-stacked MLGS for both even and odd layer number (belong to A_{2u} irreducible representation). It is neither Raman active nor Ir active in the AA-stacked MLGS irrespective of the even-oddness of the layer number N. The optical activeness of the mode is useful in determining whether the MLGS is of AB- or AA-stacking. The frequency of the mode as the function of layer number N is shown in Fig. 8. In contrast to the C-C stretching optical mode, the frequency of the out-of-plane mode raises from 864.8 cm^{-1} to 872.6 cm^{-1} and exhibits a blue shift as layer number increase which could be identified with the development of the experimental technique.

For the inter-layer optical mode, the layer number dependence of long wave length frequency is shown in Fig. 9. This mode takes the greatest advantage of the inter-layer interaction and is considerably dependent on the layer number N as well as the stack style. When $N = 2$, this mode has the frequencies 106 cm^{-1} and 94.5 cm^{-1} for the AB- and AA-stacked MLGS respectively. The frequencies increase with increasing N and almost saturate at about $N = 10$. The 3D limit values are 149.8 cm^{-1} and 133.6 cm^{-1} for the AB- and AA-stacked MLGS respectively. The frequency differences as well as the alternation of Raman and Ir active of the mode might inspire considerably experimental interesting. For comparison, we plot the low-frequency phonon dispersions of the AB-stacked 2-layer system and 3D AB-stacked graphite in the same Fig. 10. The frequencies of the low-frequency optical modes in the 2-layer graphene are much smaller than their counterparts in the 3D graphite. For three acoustic modes shown in Fig. 10, because there is no $\hat{C}_4$ symmetry in the systems, two linear acoustic

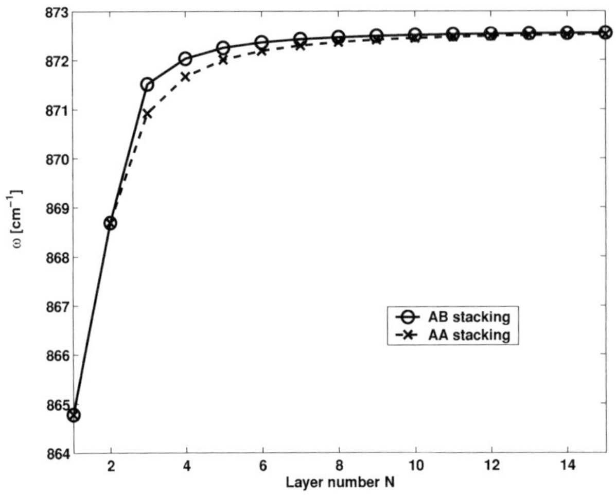

Fig. 8. The calculated frequency value for the out-of-plane optical mode vs the layer number N. This mode is Ir active in the AB stacking while it is neither Ir nor Raman active in the AA stacking. Lines are draw to guide eyes.

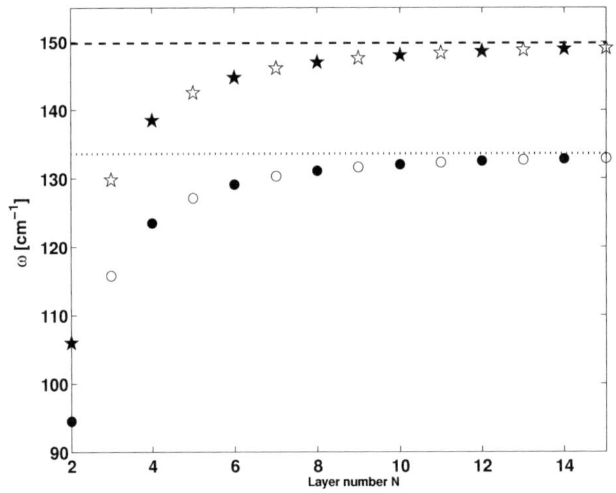

Fig. 9. The frequencies of the inter-layer optical mode vs the layer number N. Datas for the AB- and AA-stacked MLGS are designated by pentagrams and circles, respectively. The Raman and Infra-red activities for this mode are displayed by the full and empty symbols, respectively. The broken and dashed lines correspond to the frequencies of corresponding 3D graphite, respectively.

modes have different dispersions. And the third one marked in Fig. 10 is the flexure modes with quadratic dispersions near Γ point.

4. The effect of three-fold rotation symmetry of graphene on nanotubes

As constrain imposed by three-fold rotation symmetry $\hat{C}_3$, all the physical quantities of graphene have the same symmetry property. Consequently, the physical quantities of a single wall carbon nanotube (SWCNT), which can be viewed as wrapped graphene stripes,

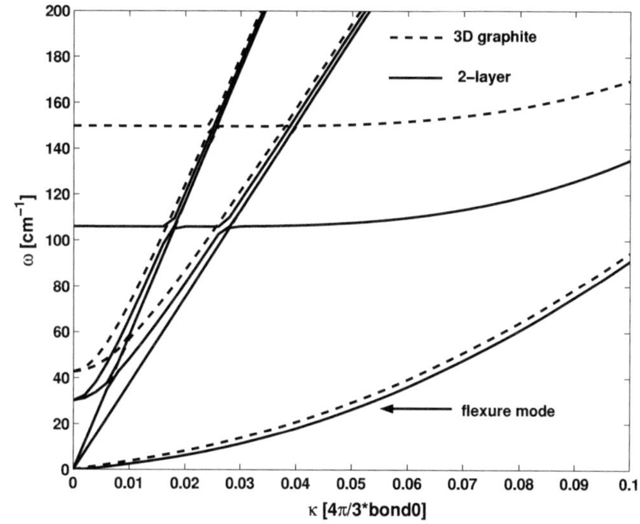

Fig. 10. In the low frequency region, there is significant difference between 3D AB-stacked graphite and the 2-layer AB-stacked graphene.

should also reserve some kind of similar symmetry property. It will manifest itself as the three-theta dependence for all physical quantities. In this section we will analysis this universal three-theta role in detail.

We start with the definition of a chiral vector for SWCNT as $\vec{R} = n_1\vec{a}_1 + n_2\vec{a}_2$ on the graphene lattice planer sheet, where $\vec{a}_1$ and $\vec{a}_2$ are the primitive lattice vectors (see Fig. 1) and (n_1, n_2) a pair of integers (Saito et al., 1998). Equivalently, a nanotube can also be denoted by the radius r and chiral angle θ (z axis is set parallel to the tube axis). They have the relations as

$$r = \frac{|\vec{a}_1|}{2\pi}\sqrt{n_1^2 + n_1 n_2 + n_2^2}, \qquad \theta = \arctan\frac{\sqrt{3}n_2}{2n_1 + n_2}. \tag{9}$$

We may establish a mapping from the space of chiral vectors on the planar sheet to that of the nanotube structure in a fixed way of wrapping

$$f : \{\text{chiral vector set}\} \mapsto \{\text{nanotube set}\}.$$

However this is not a one-to-one mapping. As mentioned in Ref. (Ye et al., 2006), a given SWCNT can be equivalently composed by three chiral vectors $\vec{R}_0$, $\vec{R}_2$ and $\vec{R}_4$ with chiral angles as θ, $\theta + 2\pi/3$ and $\theta + 4\pi/3$ respectively within the corresponding graphene sheet. Due to the $\hat{C}_6 = \hat{C}_3\hat{C}_2$ symmetry of the hexagonal graphene lattice sheet, there is another set of vectors $\vec{R}_1$, $\vec{R}_3$ and $\vec{R}_5$ with chiral angles as $\theta + \pi/3$, $\theta + \pi$ and $\theta + 5\pi/3$ respectively on the graphene sheet. These three chiral vectors correspond again to the same SWCNT which is actually the nanotube by rotating the tube formed by $\vec{R}_0$ upside down, i.e. a $\hat{C}_{2x}$ operation. The net result of the operation is an exchange of A and B carbon atoms in unit cells with the sign of its chiral index $\nu = \text{mod}\{n_1 - n_2, 3\}$ also changed. There is one another operation $\hat{\sigma}_{xz}$ connecting a pair of SWCNT which is the mirror reflection onto each other with respect to xz plane. Correspondingly $\vec{R}_0$ in the sheet is changed into its countpart $\vec{R}_0'$ with $\theta \to -\theta$ but ν

kept unchanged. These vectors, as shown in Fig. 11, have the explicit forms

$$
\begin{aligned}
\vec{R}_0 &= n_1\vec{a}_1 + n_2\vec{a}_2, \\
\vec{R}_2 &= -(n_1 + n_2)\vec{a}_1 + n_1\vec{a}_2, \\
\vec{R}_4 &= n_2\vec{a}_1 - (n_1 + n_2)\vec{a}_2, \\
\vec{R}_1 &= -n_2\vec{a}_1 + (n_1 + n_2)\vec{a}_2, \\
\vec{R}_3 &= -n_1\vec{a}_1 - n_2\vec{a}_2, \\
\vec{R}_5 &= (n_1 + n_{2)}\vec{a}_1 - n_1\vec{a}_2, \\
\vec{R}'_0 &= (n_1 + n_2)\vec{a}_1 - n_2\vec{a}_2 \, .
\end{aligned}
\tag{10}
$$

Obviously, all these seven chiral vectors have the same radius.

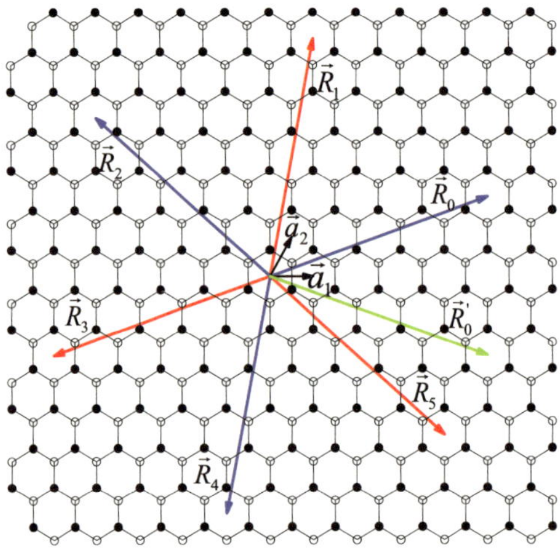

Fig. 11. Illustration of hexagonal lattice. A set of chiral vectors are shown after the symmetric operations on $\vec{R}_0$.

As one of the direct consequences of the above observations, any physical quantity $Q^{\nu}(r,\theta)$ of the SWCNT should be a periodical function of the chiral angle θ with period $\frac{2\pi}{3}$, i.e., $Q^{\nu}(r,\theta)$ can be expanded as (Ye et al., 2006)

$$
Q^{\nu}(r,\theta) = \sum_{n=0}^{\infty} a_n^{\nu}\cos(3n\theta) + b_n^{\nu}\sin(3n\theta) \, . \tag{11}
$$

This is an universal triple chiral angle (3θ) dependence of the SWCNT, which is actually the heritage of the three-fold rotational symmetry of graphene.
The permutation symmetry of the two carbon atoms in one unit cell would further give out the symmetry constrain under the operation $\theta \to \theta + \pi/3$. That is any scalars S keep unchanged, so do the radial components of the normal vectors, the azimuthal and axial components of the normal vectors change signs under the $\pi/3$ rotation in graphene plane (Jiang et al., 2008, a):

$$
\begin{cases}
S^{\nu}(\theta + \frac{\pi}{3}) = S^{-\nu}(\theta) \, , \\
\vec{v}^{\nu}(\theta + \frac{\pi}{3}) = \hat{C}_{2x}\vec{v}^{-\nu}(\theta)
\end{cases}
\tag{12}
$$

with $\hat{C}_{2x}\vec{e}_r = \vec{e}_r$, $\hat{C}_{2x}\vec{e}_\phi = -\vec{e}_\phi$, $\hat{C}_{2x}\vec{e}_z = -\vec{e}_z$. Here $\vec{e}_r$, $\vec{e}_\phi$ and $\vec{e}_z$ are unit vectors oriented towards radial, azimuthal and axial directions upon the cylindrical surface respectively.
In addition, the symmetry operation $\hat{\sigma}_{xz}$ emerged in SWCNT gives that any scalars or radial and axial components of vectors are even functions of θ, while azimuthal components are odd,

$$\begin{cases} S^\nu(-\theta) = S^\nu(\theta)\,, \\ \vec{v}^\nu(-\theta) = \hat{\sigma}_{xz}\vec{v}^\nu(\theta) \end{cases} \tag{13}$$

with $\hat{\sigma}_{xz}\vec{e}_r = \vec{e}_r$, $\hat{\sigma}_{xz}\vec{e}_\phi = -\vec{e}_\phi$, $\hat{\sigma}_{xz}\vec{e}_z = \vec{e}_z$. Combined these symmetry restrictions Eqs (12) and (13), the general expression Eq. (11) can be further reduced. Taking the physical quantities of lattice dynamics as examples, we write the final expansions with all above symmetry restrictions considered in the following (Jiang et al., 2008, a).
The frequency ω of phonon is a scalar,

$$\begin{aligned} \omega^\pm(\theta) &= a_0 \pm a_1\cos(3\theta) + a_2\cos(6\theta) \pm a_3\cos(9\theta) + \ldots; \\ \omega^0(\theta) &= a_0 + a_2\cos(6\theta) + \ldots\,. \end{aligned} \tag{14}$$

The three components of a vector can be expanded as

$$\begin{aligned} v_r^\nu(\theta) &= \sum_{n=0}^{\infty} \nu^{[\frac{1-(-1)^n}{2}]} a_n \cos 3n\theta = a_0 + \nu a_1\cos 3\theta + a_2\cos 6\theta + \ldots\,, \\ v_\phi^\nu(\theta) &= \sum_{n=0}^{\infty} \nu^{[\frac{1+(-1)^n}{2}]} b_n \sin 3n\theta = b_1\sin 3\theta + \nu b_2\sin 6\theta + \ldots\,, \\ v_z^\nu(\theta) &= \sum_{n=0}^{\infty} \nu^{[\frac{1+(-1)^n}{2}]} a_n \cos 3n\theta = \nu a_0 + a_1\cos 3\theta + \nu a_2\cos 6\theta + \ldots\,. \end{aligned} \tag{15}$$

Different from those physical quantities of normal vectors, the phonon polarization vector can be measured up to a global phase factor as ± 1. Therefore, the corresponding transformation properties with respect to the operations $\theta \to \theta + \frac{\pi}{3}$ and $\theta \to -\theta$ need to be generalized to

$$\begin{cases} v_p^{\nu(m)}(\theta + \frac{\pi}{3}) = \lambda^{(m)}(\hat{C}_{2x})\hat{C}_{2x}v_p^{-\nu(m)}(\theta)\,, \\ v_p^{\nu(m)}(-\theta) = \lambda^{(m)}(\hat{\sigma}_{xz})\hat{\sigma}_{xz}v_p^{\nu(m)}(\theta)\,, \end{cases} \tag{16}$$

respectively, where $\lambda^{(m)}(\hat{o})$ is a phase factor taking value either 1 or -1 depending on which phonon mode m we considered among $\vec{e}_r$ acoustic (AC), $\vec{e}_\phi$ optical (OP) and $\vec{e}_z$ OP modes (these three modes are non-zero frequency modes of SWCNT in long wave length limit which will be discussed in the next section) and on what kind of the symmetry operations $\hat{o}$ is applied, such as $\hat{C}_{2x}$ and $\hat{\sigma}_{xz}$. In particular,

$$\begin{aligned} \lambda^{(\vec{e}_r\mathrm{AC})}(\hat{C}_{2x}) &= \lambda^{(\vec{e}_r\mathrm{AC})}(\hat{\sigma}_{xz}) = 1\,, \\ \lambda^{(\vec{e}_\phi\mathrm{OP})}(\hat{C}_{2x}) &= \lambda^{(\vec{e}_\phi\mathrm{OP})}(\hat{\sigma}_{xz}) = -1\,, \\ \lambda^{(\vec{e}_z\mathrm{OP})}(\hat{C}_{2x}) &= -\lambda^{(\vec{e}_z\mathrm{OP})}(\hat{\sigma}_{xz}) = -1\,. \end{aligned} \tag{17}$$

The detailed expressions of polarization vectors with $\lambda^m(\sigma_{xz}) = -1$, and $\lambda^m(C_{2x}) = -1$ are

$$\begin{aligned}
v_r^\nu(\theta) &= \sum_{n=0}^{\infty} \nu^{[\frac{1+(-1)^n}{2}]} b_n \sin 3n\theta = b_1 \sin 3\theta + \nu b_2 \sin 6\theta + \dots , \\
v_\phi^\nu(\theta) &= \sum_{n=0}^{\infty} \nu^{[\frac{1-(-1)^n}{2}]} a_n \cos 3n\theta = a_0 + \nu a_1 \cos 3\theta + a_2 \cos 6\theta + \dots , \\
v_z^\nu(\theta) &= \sum_{n=0}^{\infty} \nu^{[\frac{1-(-1)^n}{2}]} b_n \sin 3n\theta = \nu b_1 \sin 3\theta + b_2 \sin 6\theta + \dots .
\end{aligned} \tag{18}$$

While for $\lambda^m(\sigma_{xz}) = 1$, and $\lambda^m(C_{2x}) = -1$, the polarization vectors can be expressed as

$$\begin{aligned}
v_r^\nu(\theta) &= \sum_{n=0}^{\infty} \nu^{[\frac{1+(-1)^n}{2}]} a_n \cos 3n\theta = \nu a_0 + a_1 \cos 3\theta + \nu a_2 \cos 6\theta + \dots , \\
v_\phi^\nu(\theta) &= \sum_{n=0}^{\infty} \nu^{[\frac{1-(-1)^n}{2}]} b_n \sin 3n\theta = \nu b_1 \sin 3\theta + b_2 \sin 6\theta + \dots , \\
v_z^\nu(\theta) &= \sum_{n=0}^{\infty} \nu^{[\frac{1-(-1)^n}{2}]} a_n \cos 3n\theta = a_0 + \nu a_1 \cos 3\theta + a_2 \cos 6\theta + \dots .
\end{aligned} \tag{19}$$

Then for $\lambda^m(\sigma_{xz}) = -1$, and $\lambda^m(C_{2x}) = 1$, they are

$$\begin{aligned}
v_r^\nu(\theta) &= \sum_{n=0}^{\infty} \nu^{[\frac{1-(-1)^n}{2}]} b_n \sin 3n\theta = \nu b_1 \sin 3\theta + b_2 \sin 6\theta + \dots , \\
v_\phi^\nu(\theta) &= \sum_{n=0}^{\infty} \nu^{[\frac{1+(-1)^n}{2}]} a_n \cos 3n\theta = \nu a_0 + a_1 \cos 3\theta + \nu a_2 \cos 6\theta + \dots , \\
v_z^\nu(\theta) &= \sum_{n=0}^{\infty} \nu^{[\frac{1+(-1)^n}{2}]} b_n \sin 3n\theta = b_1 \sin 3\theta + \nu b_2 \sin 6\theta + \dots .
\end{aligned} \tag{20}$$

For the second rank tensor ϵ, the helical symmetry of SWCNT results (Damnjanović et al., 1999)

$$\begin{aligned}
\epsilon_{xx} &= \epsilon_{yy}, \\
\epsilon_{xx} &= -\epsilon_{yy}, \\
\epsilon_{xz} &= \epsilon_{yz} = \epsilon_{zx} = \epsilon_{zy} = 0.
\end{aligned} \tag{21}$$

Similar to the analysis above, it is obtained (Ye et al., 2006) that

$$\begin{aligned}
\epsilon_{ii}^{(\nu)}(\theta + \pi/3) &= \epsilon_{ii}^{(-\nu)}(\theta), \\
\epsilon_{ii}^{(\nu)}(-\theta) &= \epsilon_{ii}^{(\nu)}(\theta), \\
\epsilon_{xy}^{(\nu)}(\theta + \pi/3) &= -\epsilon_{xy}^{(-\nu)}(\theta), \\
\epsilon_{xy}^{(\nu)}(\theta + \pi/3) &= \epsilon_{xy}^{(-\nu)}(\theta), \\
\epsilon_{xy}^{(\nu)}(-\theta) &= -\epsilon_{xy}^{(\nu)}(\theta),
\end{aligned} \tag{22}$$

where $i = x, y, z$. By noticing that the diagonal term is unchanged when the tube is reversed or reflected, and off-diagonal term vanishes, then

$$\begin{aligned} \epsilon_{ii}^{(\pm)} &= a_0 \pm a_1 \cos(3\theta) + a_2 \cos(6\theta) + \cdots , \\ \epsilon_{ii}^{(0)} &= a_0 + a_2 \cos(6\theta) + a_4 \cos(12\theta) + \cdots , \\ \epsilon_{xy}^{(\nu)} &= 0 . \end{aligned} \tag{23}$$

The coefficients $a_n's$ for different chirality have no direct relationships.

5. Coordinate systems in carbon nanotubes

As stated in the above section, an SWCNT can be identified by a chiral vector $\vec{R} = n_1\vec{a}_1 + n_2\vec{a}_2$ defined on the graphene lattice sheet with (n_1, n_2) a pair of integers. A graphene stripe is obtained by cutting the graphene sheet along the direction perpendicular to $\vec{R}$, which ensures the same edge structures of the two sides of the stripe. A nanotube is formed by wrapping the stripe, or vice verse, the stripe is the unfold plane of the nanotube. Obviously, $\vec{R}$ corresponds to the circumference of the tube and $\frac{\vec{R}}{N}$ $(N = \mathrm{GCD}(n_1, n_2))$ corresponds to a pure rotational symmetry operation $\hat{C}_N$ of the tube. Along the tube axis, i.e. z axis, a translational vector $\vec{T}$ can be defined as $\vec{T} \cdot \vec{R} = 0$ and its corresponding translation symmetry operation is $\hat{T}$. In the frame of $(\vec{a}_1, \vec{a}_2)$, it is $\vec{T} = t_1\vec{a}_1 + t_2\vec{a}_2$ with

$$t_1 = -\frac{2n_2 + n_1}{M}, \quad t_2 = \frac{2n_1 + n_2}{M}, \quad (M = \mathrm{GCD}(2n_2 + n_1, 2n_1 + n_2)).$$

It can be further obtained that

$$|\frac{\vec{R}}{N} \times \vec{T}| = \frac{n_1(2n_1 + n_2) + n_2(2n_2 + n_1)}{NM} |\vec{a}_1 \times \vec{a}_2| \equiv N_T |\vec{a}_1 \times \vec{a}_2|.$$

Because $N_T \geq 1$, the set $(\frac{\vec{R}}{N}, \vec{T})$ cannot be used as the primitive lattice vectors in unfold plane. The one-dimensional (1D) single wall carbon nanotubes belong to 1D line groups (Damnjanović et al., 1999; 2000). For chiral nanotubes, the definition of the primary chiral operation is not unique. Correspondingly, the different choices of primitive translational vectors exist on the unfold plane. Any proper pair of vectors which can map all lattice points on stripe can form the unit cell. In SWCNT, these two vectors correspond to two primitive chiral operations. There are two commonly used coordinates (Saito et al., 1998; White et al., 1993). The first one uses the rotational $\hat{C}_N$ and helical $\hat{S}_1(\alpha, h)$ symmetry descriptions for SWCNT(White et al., 1993), i.e. $\frac{\vec{R}}{N}$ and $\vec{H}$ as the primitive lattice vectors on unfold plane, which will be called White coordinate hereafter. Another chooses translational $\hat{T}$ and helical $\hat{S}_2(\alpha_0, h_b)$ symmetry descriptions with $\vec{T}$ and $\vec{Q}$ on unfold plane referred as Dresselhaus coordinate (Saito et al., 1998). The unit cells on unfold plane for two coordinate systems are shown in Fig. 12(a). Under the definition of $|\frac{\vec{R}}{N} \times \vec{H}| = |\vec{a}_1 \times \vec{a}_2|$ or $|\vec{T} \times \vec{Q}| = |\vec{a}_1 \times \vec{a}_2|$, some useful relations can be obtained (the arbitrariness of sign has be fixed),

$$\begin{cases} R = \sqrt{n_1^2 + n_2^2 + n_1 n_2}\,|\vec{a}_1|, \\ T = \frac{\sqrt{3}}{M}\sqrt{n_1^2 + n_2^2 + n_1 n_2}\,|\vec{a}_1|\,, \end{cases} \tag{24}$$

$$\begin{cases} \vec{H} = p_1\vec{a}_1 + p_2\vec{a}_2, \\ n_1 p_2 - n_2 p_1 = N\,, \end{cases} \tag{25}$$

$$\begin{cases} \vec{Q} = q_1\vec{a}_1 + q_2\vec{a}_2, \\ q_1(2n_1 + n_2) + q_2(2n_2 + n_1) = M\,. \end{cases} \tag{26}$$

The rotation angles α (α_0) as well as translations along z axis h (h_b) in helical symmetry operations can be derived by projecting $\vec{H}$ ($\vec{Q}$) to the directions of $\vec{R}$ and $\vec{T}$ respectively,

$$\begin{aligned} \vec{H}\cdot\frac{\vec{R}}{R} &\equiv \frac{R}{2\pi}\alpha \Rightarrow \alpha = 2\pi\frac{(2n_1+n_2)p_1 + (2n_2+n_1)p_2}{(2n_1+n_2)n_1 + (2n_2+n_1)n_2}, \\ \vec{H}\cdot\frac{\vec{T}}{T} &\equiv h \Rightarrow h = \frac{\sqrt{3}N}{2\sqrt{n_1^2+n_2^2+n_1n_2}}|\vec{a}_1|\,, \end{aligned} \tag{27}$$

$$\begin{aligned} \vec{Q}\cdot\frac{\vec{R}}{R} &\equiv \frac{R}{2\pi}\alpha_0 \Rightarrow \alpha_0 = 2\pi\frac{M}{(2n_1+n_2)n_1 + (2n_2+n_1)n_2}, \\ \vec{H}\cdot\frac{\vec{T}}{T} &\equiv h_b \Rightarrow h_b = \frac{\sqrt{3}(n_1q_2 - n_2q_1)}{2\sqrt{n_1^2+n_2^2+n_1n_2}}|\vec{a}_1|\,. \end{aligned} \tag{28}$$

Before go to the reciprocal space and discuss the good quantum numbers within two coordinate systems, we introduce four important integers and review their geometric meanings.

$$N_T = \frac{T}{h} = \frac{n_1(2n_1+n_2) + n_2(2n_2+n_1)}{MN}\,, \tag{29}$$

$$N_D = N_T N = \frac{2\pi}{\alpha_0} = \frac{n_1(2n_1+n_2) + n_2(2n_2+n_1)}{M}\,, \tag{30}$$

$$N_b = \frac{h_b}{h} = \frac{n_1q_2 - n_2q_1}{N}\,, \tag{31}$$

$$\Omega = \frac{\alpha}{\alpha_0} = N_D\frac{\alpha}{2\pi} = \frac{(2n_1+n_2)p_1 + (2n_2+n_1)p_2}{M}\,. \tag{32}$$

From above formulae, it shows that N_T is the number of White unit cells stacked along the tube axis to form a minimum translation period; N_D represents, in one side, the unit cell numbers in tubular section with height T called as a "giant" cell, or in the other side, the times of minimum rotation angle α_0 in one circle; N_b describes the ratio of translations along tube

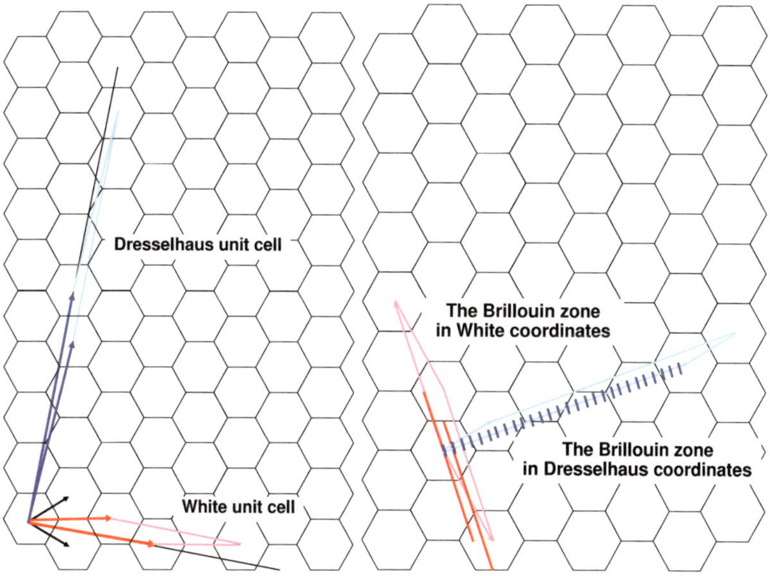

(a) Unfold plane of nanotube (b) Reciprocal lattice on unfold plane

Fig. 12. The unfold plane (a) and reciprocal lattice (b) of the nanotube (4, 2). The unit cells and primitive lattice vectors of White $\{\vec{R}/N,\ \vec{H}\} \leftrightarrow \{\hat{C}_N,\ \hat{S}_1(\alpha,h)\}$ and Dresselhaus $\{\vec{T},\ \vec{Q}\} \leftrightarrow \{\hat{T},\ \hat{S}_2(\alpha_0,h_b)\}$ coordinates are shown in (a). The Brillouin zones and quantum numbers of White $\{n,\ \kappa\}$ and Dresselhaus $\{k_z,\ n_i\}$ are shown in (b).

axis of the two helical symmetry operations; and Ω is the ratio of rotation angles of the two helical operations, which can be proved to be relatively prime with N_T.
The 1D group characters of symmetry operators in White and Dresselhaus coordinates can be written respectively as:

$$\{\hat{C}_N,\ \hat{S}_1(\alpha,h)\} \longrightarrow \{e^{i\frac{2\pi}{N}n},\ e^{i\kappa}\} \quad \text{for White coordinate,}$$
$$\{\hat{T},\ \hat{S}_2(\alpha_0,h_b)\} \longrightarrow \{e^{iN_Tk_zT},\ e^{i\kappa'}\} \quad \text{for Dresselhaus coordinate.} \tag{33}$$

The prefix N_T on the character $e^{iN_Tk_zT}$ is for taking the same length units for two coordinate systems. Using the periodical boundary conditions and notice that $\hat{S}_2^{N_D}(\alpha_0,h_b) = \hat{S}_2(2\pi,N_bNT) = \hat{T}^{N_bN} \rightarrow e^{iN_D\kappa'} = e^{iN_bNN_Tk_zT}$, it is

$$n = 0,1,\ldots,N-1\,, \tag{34}$$

$$\kappa \in [\,0,2\pi)\,, \tag{35}$$

$$k_z \in [\,0,2\pi/N_TT)\,, \tag{36}$$

$$\kappa' = N_bk_zT + \frac{2\pi}{N_D}n_i\,, \tag{37}$$

$$n_i = 0,1,\ldots,N_D-1\,. \tag{38}$$

According to symmetry operators, the good quantum numbers in the two coordinate system should be $\{n,\kappa\}$ and $\{k_z,\kappa'\}$ respectively. But as seen from Eq. (38), κ' has a complicate form,

it is commonly to use $\{k_z, n_i\}$ instead of $\{k_z, \kappa'\}$ as "good quantum number" in Dresselhaus coordinate. Since the nanotube has no $\hat{C}_{N_D}$ symmetry, which is only an isogonal group of the symmetry group of the nanotube, strictly speaking, n_i is not a good angular quantum number of the rotational symmetry as seemed to be.
For deducing the relations between two sets of quantum numbers, the following equivalence is used:

$$\begin{aligned}\hat{C}_N &= \hat{S}_2^{N_T}(\alpha_0, h_b)\hat{T}^{-N_b},\\ \hat{T} &= \hat{S}_1^{N_T}(\alpha, h)\hat{C}_N^{-\Omega},\\ \hat{S}_1(\alpha, h) &= \hat{S}_2^{\Omega}(\alpha_0, h_b)(\hat{T}^{1/N_T})^{-(\Omega N_b - 1)}.\end{aligned} \tag{39}$$

Then from the characters of the operations, it is followed that

$$\begin{aligned}\frac{2\pi}{N}n &= \frac{2\pi}{N}n_i + 2\pi\mathcal{I}_1,\\ N_T k_z T &= N_T\kappa - N_T n\alpha + 2\pi\mathcal{I}_2,\\ \kappa &= n_i\alpha + k_z T + 2\pi\mathcal{I}_3,\end{aligned} \tag{40}$$

where $\mathcal{I}_1, \mathcal{I}_2$, and $\mathcal{I}_3$ are arbitrary integers. The first two of Eq. (40) give the following relations directly,

$$n = n_i \pmod N, \quad \text{or} \quad n_i = n + NJ \quad (J \text{ an integer}), \tag{41}$$

$$k_z = \frac{2\pi}{N_T T}\mathrm{Frac}\left(N_T\frac{\kappa - n\alpha}{2\pi}\right). \tag{42}$$

The third of Eq. (40) contains fraction and integer parts of information. From its fraction part,

$$\kappa = 2\pi\mathrm{Frac}\left(\frac{k_z T + n_i\alpha}{2\pi}\right). \tag{43}$$

From the integer part,

$$J\Omega = \mathrm{Int}\left[N_T\frac{\kappa - n\alpha}{2\pi} - \mathrm{Frac}\left(N_T\frac{\kappa - n\alpha}{2\pi}\right)\right] \pmod{N_T}. \tag{44}$$

The equations (41), (42), (43), and (44) provide the complete relations between the two sets of quantum numbers (Tang et al., 2009). The Brillouin zones and quantum numbers $\{n, \kappa\}$ and $\{k_z, n_i\}$ are shown in Fig. 12(b). Actually, Eq. (44) can uniquely determine the integer J and shows the nontrivial relation between n and n_i. Interested readers can refer monographs about number theory.

6. Lattice dynamics of single-wall carbon nanotubes

6.1 Vibrational potentials

We discuss the lattice dynamics of single-wall carbon nanotubes (SWCNT) within the White coordinate system (White et al., 1993) which has been described detailed in the above section. Setting the z axis along the tube axis and the x axis across the middle point of a C-C bond, any unit cell in the (n_1, n_2) or equivalently (r, θ) tubule can be notated by a pair of integers (m, l):

$$\vec{r}_{m,l} = m\vec{H} + l\frac{\vec{R}}{N}. \tag{45}$$

In other words, any unit cell on SWCNT can be covered by m screw operations $\hat{S}_1(\alpha, h)$ and l rotations $\hat{C}_N$ successively (see the details in section 4).
Again, we use the idea of five distinctive terms (Aizawa et al., 1990) for the potential of graphene sheet, and extend them to the nanotube cases. What we stress is that the satisfying of the rigid translational and rotational invariance must be kept in the generalization, i.e., the potential energy must keep unchanged term by term when the tube is rigidly translated or rotated around any axis. Introduce $\vec{r}_i$ as the equilibrium position of atom i and $\vec{u}_i$ as its displacement vector. $\vec{r}_{ij} = \vec{r}_j - \vec{r}_i$ is the vector from atom i to j in the nanotube while the modulus r_{ij} represents the length of C-C bond between atoms i and j. The vector $\vec{r}_i$ is determined following the geometry of a warped graphene sheet. Unlike that in graphene, the geometrical curvature results to that not only the three tridental bond lengths nearest-neighbored with the atom i but also the angles between any of the two bonds are not equal to each other even in equilibrium position, especially for thin tubes. We express in the following a detailed expression of the vibrational potential for the SWCNT with the curvature effect being carefully in-cooperated. It satisfies precisely the requirements of the rigid translational and rotational invariance and realizes the corresponding general symmetry sum rules in Ref. (Popov et al., 2000).
The five terms of potential energy are basically the same as those in graphene sheet with rigorous modification of unit vectors in the present case (Jiang et al., 2006). We list them in the following for completeness.
(1) V_l is the potential of the spring force between the nearest-neighbor atom pair,

$$V_l = \frac{k_l}{2} \sum_{i=2}^{4} [(\vec{u}_i - \vec{u}_1) \cdot \vec{e}^{\,l}_{1i}]^2 , \tag{46}$$

where k_l is the first-order force constant and $\vec{e}^{\,l}_{1i} = \frac{\vec{r}_{1i}}{|\vec{r}_{1i}|}$. We'd like to point out that the component of the displacement vectors perpendicular to $\vec{e}^{\,l}_{1i}$ violates the rigid rotational invariance and is forbidden.
(2) V_{sl} is also the potential of the spring force but between the next nearest-neighbored atoms illustrated as $(1, 5 \ldots 10)$ in Fig. 1,

$$V_{sl} = \frac{k_{sl}}{2} \sum_{i=5}^{10} [(\vec{u}_i - \vec{u}_1) \cdot \vec{e}^{\,l}_{1i}]^2 \tag{47}$$

with k_{sl} the second-order force constant.
(3) The potential energy for the in-surface bond bending V_{BB} with three atoms involved is more complicated than that in graphene planar sheet due to the curvature-resulted deviation of an equilibrium angle from a constant.

$$\begin{aligned} V_{BB} &= \frac{k_{BB}}{4} \sum_{j_i} \sum_{\substack{j'_i \\ (j'_i \neq j_i)}} \left[\frac{\vec{u}_{j_i} - \vec{u}_i}{r_{ij_i}} \cdot (\vec{e}^{\,l}_{ij'_i} - \cos\theta_{j_i i j'_i} \vec{e}^{\,l}_{ij_i}) + \frac{\vec{u}_{j'_i} - \vec{u}_i}{r_{ij'}} \cdot (\vec{e}^{\,l}_{ij_i} - \cos\theta_{j'_i i j_i} \vec{e}^{\,l}_{ij'_i}) \right]^2 \\ &= \frac{k_{BB}}{4} \sum_{j_i} \sum_{\substack{j'_i \\ (j'_i \neq j_i)}} (\cos\theta'_{j_i i j'_i} - \cos\theta_{j_i i j'_i})^2 . \end{aligned} \tag{48}$$

Where i, j_i, and j'_i take the same sites as those in graphene (see Eq. (3) and Fig. 1 in section 3). The difference is here $\theta_{j_i i j'_i}$ for the equilibrium angle between the bonds $\vec{r}_{ij_i}$ and $\vec{r}_{ij'_i}$ which is no more 120° in general. $\theta'_{j_i i j'_i}$ is for the corresponding angle in vibration. The bond angles are expressed in terms of $\vec{u}_{j_i} - \vec{u}_i$ and $\vec{u}_{j'_i} - \vec{u}_i$, which can be proved that the rigid rotational invariance referred to an arbitrary axis is kept only when the differences among bond lengths and bond angles be carefully accounted.

(4) Extending the out-of-surface bond bending V_{rc} term into SWCN,

$$V_{rc} = \frac{k_{rc}}{2}[(3\vec{u}_i - \sum_{j_i}\vec{u}_{j_i}) \cdot \vec{e}_i^{rc}]^2, \tag{49}$$

$$\vec{e}_i^{rc} = -\frac{\sum_{j_i}\vec{r}_{j_i}}{|\sum_{j_i}\vec{r}_{j_i}|}, \tag{50}$$

where i takes 1 or 2 with j_i running over the three nearest neighbors of atom i. Different from graphene, an unit vector $\vec{e}_i^{rc}$ is introduced for keeping the rigid rotational invariance. We show the difference between $\vec{e}_i^{rc}$ and radial unit vector $\vec{e}_1^r$ in Fig. 13. When the radius of tube is large enough, the vector $\vec{e}_i^{rc}$ is close enough to the $\vec{e}_1^r$. However, we stress that the potential term V_{rc} with $\vec{e}_i^{rc}$ substituted by $\vec{e}_i^r$ would break the rotational symmetry.

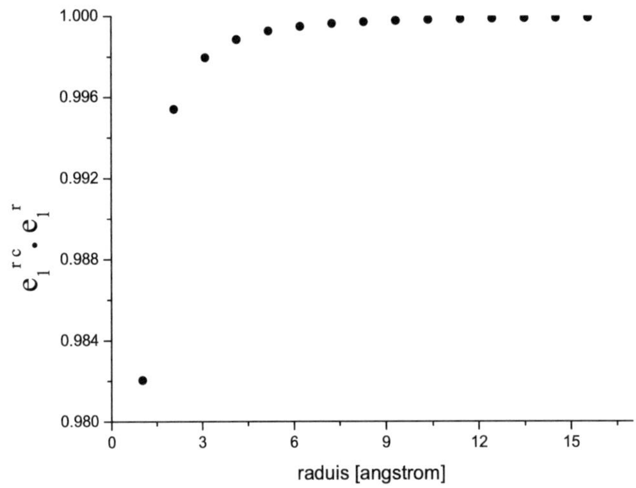

Fig. 13. The projection of $\vec{e}_1^{rc}$ on $\vec{e}_1^r$ for tubes $(2n, n)$ with $n \in [1, 15]$. It shows that $\vec{e}_1^{rc}$ only deviates about 2% from $\vec{e}_1^r$ even in the small radius (2, 1) tube.

(5) The twist potential energy for bond $\vec{r}_{1k}$ is generalized as

$$V_{tw} = \frac{k_{tw}}{2}\sum_{\langle i,j\rangle}[(\vec{u}_i - \vec{u}_j - (\vec{u}_{i'} - \vec{u}_{j'})) \cdot \vec{e}_{1k}^r]^2, \tag{51}$$

where $\vec{e}_{1k}^r$ is the unit vector along the radial direction of the middle point of $\vec{r}_{1k}$, $\langle i, j\rangle$ represents a pair of atoms nearest-neighbored with atom 1 while k the third of its nearest neighbors (see Fig. 1 for reference). Pair $\langle i', j'\rangle$ is the image of $\langle i, j\rangle$ referring to a $\hat{C}_2$ rotation around the axis in $\vec{e}_{1k}^r$.

Obviously, it can be checked that all above five potential energy terms satisfy the translational invariance (Jiang et al., 2006). When $\vec{u}_i = \vec{u}_j$, $\vec{u}_i - \vec{u}_j = 0$ leads to

$$V_l = V_{sl} = V_{BB} = V_{rc} = V_{tw} = 0 \,.$$

For the rotational invariance, we have to consider the five potentials term by term separately (Born & Huang, 1954; Madelung, 1978). When the tube rotates rigidly around an arbitrary axis for a small angle $\delta\vec{\omega}$ with its direction along the axis $\frac{\delta\vec{\omega}}{|\delta\vec{\omega}|}$, each lattice site acquires a displacement $\vec{u}_i = \delta\vec{\omega} \times \vec{r}_i$,

$$\vec{u}_i - \vec{u}_j = \delta\vec{\omega} \times (\vec{r}_i - \vec{r}_j) = \delta\vec{\omega} \times \vec{r}_{ji} \,. \tag{52}$$

Substituting Eq. (52) into the first two potential terms (46) and (47), it is straightforward to have $(\vec{u}_j - \vec{u}_i) \cdot \vec{e}^{\,l}_{ij} = r_{ij}(\delta\vec{\omega} \times \vec{e}^{\,l}_{ij}) \cdot \vec{e}^{\,l}_{ij} = 0$. Then

$$V_l = V_{sl} = 0 \,.$$

Substituting Eq. (52) into the third potential term (48), a typical representative term in summation becomes

$$V_{BB} \sim \frac{k_{BB}}{4}[\delta\vec{\omega} \cdot (\vec{e}^{\,l}_{12} \times \vec{e}^{\,l}_{13} + \vec{e}^{\,l}_{13} \times \vec{e}^{\,l}_{12})]^2 = 0 \,.$$

In which a fact has been used that r_{ij} in the denominate is canceled by that in the numerator when Eq. 52 is applied. Moreover, for each typical representative term in potentials (49) and (51), we have

$$V_{rc} \sim \tfrac{k_{rc}}{2}[\delta\vec{\omega} \times (\vec{r}_{12} + \vec{r}_{13} + \vec{r}_{14}) \cdot \vec{e}^{\,rc}_1]^2 = 0 \,,$$
$$V_{tw} \sim \tfrac{k_{tw}}{2}[\delta\vec{\omega} \times (\vec{r}_{43} - \vec{r}_{56}) \cdot \vec{e}^{\,r}_{12}]^2 = 0 \,.$$

For further clarification, we show the phonon spectrum with all the bond lengths and bond angles assumed to be equal to that of the graphene in Fig. 14(a) for SWCNT (5,2). It shows clearly that the twisting mode (TW) at $(\kappa, n) = (0, 0)$ is no longer a zero mode and there is a finite gap with the order of 0.5 cm^{-1}. Although it is a minute number and entirely negligible in practice, it is of qualitative significance. When we take the proper bond lengths with about -1.3%, -0.3% and 0.0% shorter than that of graphene 1.42 Å respectively, the correct phonon spectrum is calculated and shown in Fig. 14(b) for the same SWCNT.

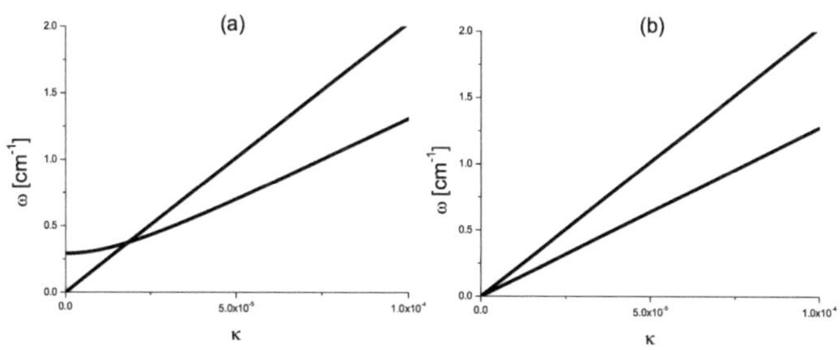

Fig. 14. The effect of bond lengths on TW mode in tube (5, 2). (a) All bonds are assumed to be the same. The frequency of TW mode at $\kappa = 0$ is nonzero. (b) The differences between bonds are considered. The frequency of TW mode is precisely zero.

As a one-dimensional system, SWCNT has a distinguished feature that the two degenerate transversal acoustic (TA) modes shown up at $(\kappa, n) = \pm(\alpha, 1)$ are flexure modes. Instead of the conventional linear behaviors, the low frequency limits of their dispersions are parabolic as $\omega^2 = \beta^2(\kappa \mp \alpha)^4$. We stress that the rigid rotational invariance around z axis itself is not a sufficient condition for the existence of the flexure modes. It can be declared more clearly by a counter example. By introducing a potential term $V_{\tau i} = \frac{k_{\tau i}}{2}[(\vec{u}_2 - \vec{u}_1) \cdot \vec{e}_{12}^{\tau i}]^2$ with $\vec{e}_{12}^{\tau i} = \vec{e}_{12}^{r} \times \vec{e}_{12}^{l}$, which satisfies $V_{\tau i} = 0$ when the tube rotates around the z axis, there will be no flexure mode in SWCNT (Dobardžić et al., 2003). This is because that $V_{\tau i}$ is not zero when the tube rotates around any axis perpendicular to the z axis.
By tuning the calculated results to the experimental data (Rao et al., 1997) for the Raman modes of (10, 10) tube (see Table 3), the corresponding force constants are fit as $k_l = 364.0\,\mathrm{Nm}^{-1}$, $k_{sl} = 62.0\,\mathrm{Nm}^{-1}$, $k_{BB} = 1.07 \times 10^{-11}$ erg, $k_{rc} = 14.8\,\mathrm{Nm}^{-1}$, and $k_{tw} = 6.24\,\mathrm{Nm}^{-1}$ (Jiang et al., 2006). For chiral SWCNT, all tubes belong to the fifth 1D line group (Popov et al., 2000) with two atoms in one unit cell and have six phonon modes at any point of reciprocal space . Looking at three specific points $(\kappa, n) = (0, 0)$, $(\kappa, n) = (\alpha, 1)$, and $(\kappa, n) = (2\alpha, 2)$, 18 phonon modes are successfully identified (Alon, 2001; Jiang et al., 2006). They are (I) 3 zero modes: two at $(\kappa, n) = (0, 0)$ as longitudinal acoustic (LA) and TW modes both belonging to ${}_0A_0^-$ representation (Reps), and one at $(\kappa, n) = (\alpha, 1)$ as the flexure mode; (II) 9 Raman active modes: three at $(\kappa, n) = (0, 0)$ belonging to ${}_0A_0^+$ Reps as $\vec{e}_r$ acoustic (AC), $\vec{e}_\theta$ optical (OP) and $\vec{e}_z$ OP, and six at $(\kappa, n) = (2\alpha, 2)$ belonging to the same ${}_{2\alpha}E_2$ Reps; (III) 1 Ir active mode: at $(\kappa, n) = (0, 0)$ as the OP mode with A and B atoms oscillating out of surface in tubular radial direction and belonging to ${}_0A_0^-$ Reps; (IV) 5 Raman and Ir active modes at $(\kappa, n) = (\alpha, 1)$ assigned to ${}_\alpha E_1$ Reps.

Reps	${}_0A_0^+$	${}_0A_0^+$	${}_\alpha E_1^-$	${}_\alpha E_1^-$	${}_{2\alpha}E_2^+$	${}_{2\alpha}E_2^+$	${}_{2\alpha}E_2^+$	${}_{2\alpha}E_2^+$
Theory	167	1588	105	1588	21	367	873	1584
Experiment (Rao et al., 1997)	186	1593	118	1567	/	377	/	1606

Table 3. Comparison between the calculated results and the experimental values for several mode frequencies (in the unit of cm^{-1}) of SWCNT (10,10).

6.2 Chiral angle and radius dependence of phonon properties

With the phenomenological vibration potential model described in above subsection, the phonon frequencies, acoustic velocities, and eigenvectors are calculated as the functions of radius and chiral angles, and further fitted following the three-fold symmetry expansion discussed in section 4. Because we are interested in SWCNT with smaller radius which should be quite different from those of graphene stripe, the fitting range is chosen as $r \in [4.0, 10.0]$ Å and $\theta \in [-\frac{\pi}{6}, \frac{\pi}{6}]$ for frequencies and velocities, $r \in [2.0, 10.0]$ Å and $\theta \in [-\frac{\pi}{3}, \frac{\pi}{3}]$ for polarization vectors. The relative errors in fitting are set less than 5×10^{-4}. The expansions of frequencies of Raman and Ir active modes, the velocities at $(\kappa, n) = (0, 0)$ and flexure parabolic at $(\kappa, n) = (\alpha, 1)$, and the polarization vectors of nonzero modes at $(\kappa, n) = (0, 0)$ are listed in Tables 4, 5, and 6 respectively (Jiang et al., 2006). In Table 4, data are represented in three parts corresponding to $(\kappa, n) = (0, 0)$, $(\alpha, 1)$, and $(2\alpha, 2)$ respectively. From the fitting results, we can see that the contributions of θ dependence are notable comparing to those of r dependence. The velocity of the twist mode (in Table 5) as well as the polarization vectors of modes with ${}_0A_0^+$ Reps (in Table 6) are typical examples shown up an evident θ dependence.
It can be checked that all the numerically fitting expressions satisfy the symmetry requirements of Eqs (12) and (13) in section 4, where the velocities (slopes of the dispersions)

Reps	Mode	$\omega(\theta)$	$f_i(r)$
${}_0A_0^+$ R	$\vec{e}_r$ AC, 1	$f_0(r)$	$f_0(r) = \frac{1133.86}{r} - \frac{139.65}{r^3}$
	$\vec{e}_\theta$ OP, 2	$f_0(r) + f_1(r)\cos 6\theta$	$f_0(r) = 1594.00 - \frac{266.98}{r^2}, f_1(r) = \frac{8.65}{r^2}$
	$\vec{e}_z$ OP, 3	$f_0(r) + f_1(r)\cos 6\theta$	$f_0(r) = 1594.00 - \frac{91.81}{r^2}, f_1(r) = -\frac{15.68}{r^2}, f_2(r) = -\frac{0.68}{r^2}$
		$+f_2(r)\cos 12\theta$	
${}_0A_0^-$ Ir	$\vec{e}_r$ OP, 4	$f_0(r) + f_1(r)\cos 6\theta$	$f_0(r) = 864.81 + \frac{990.22}{r^2} - \frac{1117.30}{r^4}, f_1(r) = \frac{9.16}{r^2}$
${}_\alpha E_1$ R Ir	1	$f_0(r) + f_1(r)\cos 6\theta$	$f_0(r) = \frac{710.16}{r} + \frac{45.07}{r^3}, f_1(r) = \frac{1.99}{r^3} - \frac{31.55}{r^4}$
	2	$f_0(r) + f_1(r)\cos 6\theta$	$f_0(r) = \frac{1603.51}{r} - \frac{746.51}{r^3}, f_1(r) = -\frac{115.54}{r^3}$
	3	$f_0(r) + f_1(r)\cos 6\theta$	$f_0(r) = 864.84 + \frac{860.00}{r^2} - \frac{1758.70}{r^4}, f_1(r) = \frac{11.63}{r^2} - \frac{206.52}{r^4}$
	$\nu = \pm 1, 4$	$f_0(r) \pm f_1(r)\cos 9\theta$	$f_0(r) = 1594.13 - \frac{316.67}{r^2}, f_1(r) = \frac{31.92}{r^3}$
	$\nu = 0, 4$	$f_0(r) + f_1(r)\cos 6\theta$	$f_0(r) = 1594.14 - \frac{318.48}{r^2}, f_1(r) = \frac{7.83}{r^2} - \frac{19.03}{r^4}$
		$+f_2(r)\cos 12\theta$	$f_2(r) = \frac{2.70}{r^2} + \frac{0.60}{r^4}$
	5	$f_0(r) + f_1(r)\cos 6\theta$	$f_0(r) = 1593.97 - \frac{277.49}{r^2}, f_1(r) = -\frac{12.45}{r^2}$
${}_{2\alpha}E_2$ R	1	$f_0(r) + f_1(r)\cos 6\theta$	$f_0(r) = \frac{959.33}{r^2} - \frac{736.60}{r^4} + \frac{779.59}{r^5}$
		$+f_2(r)\cos 12\theta$	$f_1(r) = \frac{6.19}{r^3} + \frac{73.37}{r^4}, f_2(r) = -\frac{0.06}{r^3} + \frac{9.34}{r^4}$
	2	$f_0(r) + f_1(r)\cos 6\theta$	$f_0(r) = \frac{1420.21}{r} + \frac{54.52}{r^3} - \frac{1246.29}{r^5}, f_1(r) = \frac{204.34}{r^3}$
	3	$f_0(r) + f_1(r)\cos 6\theta$	$f_0(r) = \frac{2535.48}{r} - \frac{2426.65}{r^3}, f_1(r) = -\frac{412.23}{r^3}$
	4	$f_0(r) + f_1(r)\cos 6\theta$	$f_0(r) = 864.80 + \frac{486.71}{r^2} - \frac{4711.81}{r^4} + \frac{12425.61}{r^6}$
			$f_1(r) = \frac{9.89}{r^2} - \frac{524.74}{r^4}$
	5	$f_0(r) + f_1(r)\cos 6\theta$	$f_0(r) = 1594.00 - \frac{869.19}{r^2} + \frac{978.77}{r^4}$
			$f_1(r) = -\frac{16.15}{r^2} + \frac{363.41}{r^4}$
	6	$f_0(r) + f_1(r)\cos 6\theta$	$f_0(r) = 1594.01 - \frac{392.92}{r^2} - \frac{2160.15}{r^4} + \frac{5416.26}{r^6}$
			$f_1(r) = \frac{7.88}{r^2} - \frac{297.88}{r^4}$

Table 4. Frequencies (in the unit of cm^{-1}) of 15 Raman and Ir active modes as functions of r (in Å) and θ.

	Velocity(θ)	$f_i(r)$
C_{TW}	$f_0(r) + f_1(r)\cos 6\theta$	$f_0 = 13.5 - \frac{1.63}{r^2}, f_1 = \frac{2.38}{r^2}$
C_{LA}	$f_0(r) + f_1(r)\cos 6\theta$	$f_0 = 21.0706 + \frac{0.0055}{r} - \frac{0.6860}{r^2}, f_1 = \frac{0.00091}{r} - \frac{0.01679}{r^2}$
β	$f_0(r) + f_1(r)\cos 6\theta$	$f_0 = 1.3767r - 0.00142r^2 - 5.8 \times 10^{-5}r^3, f_1 = -\frac{0.143}{r} + \frac{0.04994}{r^3}$

Table 5. Sound velocities (in kms^{-1}) of the TW and LA modes, and β (in $10^{-6}m^2s^{-1}$) of the flexure mode as functions of r (in Å) and θ.

and frequencies are scalars while polarizations are vectors. One of the ${}_\alpha E_1$ modes in Table 4 manifests different parameter dependence for different chiral index ν, i.e. it has different expressions for $\nu = \pm 1$ and $\nu = 0$ respectively. This is still consistent with the general constrains Eqs (12) and (13).

In certain extent, the planar graphene can be viewed as the $r \to \infty$ limit of the SWCNT. As pointed out in Ref. (Jiang et al., 2006), six modes at different (κ, n) points in Table 4 evolve in this limit to the two degenerate in-plane optical modes of the graphene with approximately the same frequency limit 1594.0. While three modes with $f_0(r) \to 864.8$ approach the out-of-plane optical mode of the graphene. Moreover as shown in Table 5, the sound velocities of the two zero modes, i.e. LA and TW modes belonging to ${}_0A_0^-$, have nonzero limits with different values. Therefore, it is expected that these two modes would approach two in-plane non-degenerate acoustic modes of the graphene sheet. The situation of flexure mode

		Vector(θ)	$f_i(r)$
R_1 ($\vec{e}_r$ AC)	$u_r(A)$	$f_0(r)$	$f_0(r) = 0.7071 - \frac{0.0028}{r^2}$
	$u_\phi(A)$	$f_1(r)\sin 3\theta$	$f_1(r) = \frac{0.0518}{r} + \frac{0.0468}{r^2}$
	$u_z(A)$	$f_1(r)\cos 3\theta$	$f_1(r) = \frac{0.0517}{r} + \frac{0.0749}{r^2}$
R_2 ($\vec{e}_\phi$ OP)	$u_r(A)$	$f_1(r)\sin 3\theta$	$f_1(r) = -\frac{0.0542}{r} - \frac{0.0455}{r^2}$
	$u_\phi(A)$	$f_0(r) + f_1(r)\cos 12\theta$	$f_0(r) = 0.7056 + \frac{0.0019}{r^2}$, $f_1(r) = 0.0015 - \frac{0.003}{r^2}$
	$u_z(A)$	$f_1(r)\sin 6\theta + f_2(r)\sin 12\theta$	$f_1(r) = 0.0656 - \frac{0.0801}{r^2}$, $f_2(r) = 0.0048 - \frac{0.0112}{r^2}$
R_3 ($\vec{e}_z$ OP)	$u_r(A)$	$f_1(r)\cos 3\theta$	$f_1(r) = -\frac{0.0447}{r} - \frac{0.0417}{r^2}$
	$u_\phi(A)$	$f_1(r)\sin 6\theta + f_2(r)\sin 12\theta$	$f_1(r) = -0.0656 + \frac{0.0773}{r^2}$, $f_2(r) = -0.0048 + \frac{0.0111}{r^2}$
	$u_z(A)$	$f_0(r) + f_1(r)\cos 12\theta$	$f_0(r) = 0.7056 + \frac{0.0019}{r^2}$, $f_1(r) = 0.0015 - \frac{0.0033}{r^2}$

Table 6. Polarization vectors $\vec{u} \equiv (\vec{u}(A), \vec{u}(B))$ at $(\kappa, n) = (0,0)$ as functions of r (in Å) and θ. Where $\vec{u}(A)$ and $\vec{u}(B)$ indicate the displacement vectors of atoms A and B in the $(0,0)$ unit cell respectively. For the three modes in this table, $u_r(B) = u_r(A)$, $u_\phi(B) = -u_\phi(A)$, $u_z(B) = -u_z(A)$.

is complicated. The frequency of the $\vec{e}_r$ AC mode in SWCNT (Table 4) tends towards zero with its polarization vector perpendicular to the limiting sheet. This might be interpreted as a kind of precursor of the flexure mode of graphene. However, at $(\kappa, n) = \pm(\alpha, 1)$ of SWCNT, the dispersion of the two TA branch is quadratic in κ. It should be noticed that the parameterization for the coefficient β cannot be extrapolated to $r \to \infty$. This is prohibited by a kind of symmetry argument that the rod-like tube has two flexure modes with the cylindrical symmetry while the plate-like graphene sheet breaks the symmetry so as to have only one flexure mode. There is no way to cross continuously from the former to the latter.

7. References

Aizawa, T.; Souda, R.; Otani, S.; Ishizawa, Y.; Oshima, C. (1990). Bond softening in monolayer graphene fprmed on transition-metal carbide surfaces. *PHYSICAL REVIEW B*, Vol. 42, No. 18, Dec 15, 11469–11478, ISSN 1098-0121

Aizawa, T. (1991). Correction. *PHYSICAL REVIEW B*, Vol. 43, No. 14, May 15, 12060–12061, ISSN 1098-0121

Alon, O. E. (2001). Number of Raman- and infrared-active vibrations in single-walled carbon nanotubes. *PHYSICAL REVIEW B*, Vol. 63, No. 20, May 15, 201403, ISSN 1098-0121

Berger, C.; Song, Z. M.; Li, T. B.; Li, X. B.; Ogbazghi, A. Y.; Feng, R.; Dai, Z. T.; Marchenkov, A.N.; Conrad, E. H.; First, P. N.; de Heer, W. A. (2004). Ultrathin epitaxial graphite: 2D electron gas properties and a route toward graphene-based nanoelectronics. *JOURNAL OF PHYSICAL CHEMISTRY B*, VOl. 108, No. 52, December 30, 19912–19916, ISSN 1520-6106

Born, M.; & Huang, K. (1954). *Dynamical Theory of Crystal Lattices*, Oxford University Press, Oxford

Brillson, L.; Burskin, E.; Maradudin, A. A.; Stark, T. (1971). Frequencies of the long wavelength optical vibration modes of graphite, In: *The Physics of Semimetals and Narrow-Gap Semiconductors*, Carter, D.L.; & Bate, R.T., 187–193, Pergamon Press, ISBN 008016661X, Oxford, New York

Castro Neto, A. H.; Guinea, F.; Peres, N. M. R.; Novoselov, K. S.; & Geim, A. K. (2009). The electronic properties of graphene. *REVIEWS OF MODERN PHYSICS*, Vol. 81, No. 1, Januray-March, 109–154, ISSN 0034-6861

Damnjanović, M.; Milošević, I.; Vuković, T.; Sredanović, R. (1999). Full symmetry, optical activity, and potentials of single-wall and multiwall nanotubes. *PHYSICAL REVIEW B*, Vol. 60, No. 4, July 15, 2728–2739, ISSN 1098-0121

Damnjanović, M.; Vuković, T.; & Milošević, I. (2000). Modified group projectors: tight-binding method. *JOURNAL OF PHYSICS A-MATHEMATICAL AND GENERAL*, Vol. 33, No. 37, September 22, 6561–6571, ISSN 0305-4470

Das, A.; Chakraborty, B.; Sood, A. K. (2008). Raman spectroscopy of graphene on different substrates and influence of defects. *BULLETIN OF MATERIALS SCIENCE*, Vol. 31, No. 3, June, 579–584, ISSN 0250-4707

Dobardžić, E.; Milošević, I.; Nikolić, B.; Vuković, T.; & Damnjanović, M. (2003). Single-wall carbon nanotubes phonon spectra: Symmetry-based calculations. *PHYSICAL REVIEW B*, Vol. 68, No. 4, July 15, 045408, ISSN 1098-0121

Elliott, J. P. & Dawber, P. G. (1979). *Symmetry in Physics, Vol. 1*, Macmillan Press Ltd., ISBN 0333264266, London

Eyring, H.; Walter, J. & Kimball, G. E. (1949). *Quantum Chemistry*, Fifth Pringting, John Wiley and Sons Inc., New York

Ferrari, A. C.; Meyer, J. C.; Scardaci, V.; Casiraghi, C.; Lazzeri, M.; Mauri, F.; Piscanec, S.; Jiang, D.; Novoselov, K. S.; Roth, S.; Geim, A. K. (2006). Raman spectrum of graphene and graphene layers. *PHYSICAL REVIEW LETTERS*, Vol. 97, No. 18, November 3, 187401, ISSN 0031-9007

Gunlycke, D. ; Lawler, H. M.; White, C. T. (2008). Lattice vibrations in single-wall carbon nanotubes. *PHYSICAL REVIEW B*, Vol. 77, No. 1, January, 014303-1–014303-9, ISSN 1098-0121

Gupta, A.; Chen, G.; Joshi, P.; Tadigadapa, S.; Eklund, P. C. (2006). Raman scattering from high-frequency phonons in supported n-graphene layer films. *NANO LETTERS*, Vol. 6, No. 12, December 13, 2667–2673, ISSN 1530-6984

Iijima, S. (1991). Helical microtubules of graphitic carbon. *NATURE*, Vol. 354, No. 6348, November 7, 56–58, ISSN 0028-0836

Jeon, G. S.; & Mahan, G. D. (2005). Theory of infrared-active phonons in carbon nanotubes. *PHYSICAL REVIEW B*, Vol. 72, No. 15, October, 155415, ISSN 1098-0121

Jiang, J. W.; Tang, H.; Wang, B. S.; & Su, Z. B. (2006). Chiral symmetry analysis and rigid rotational invariance for the lattice dynamics of single-wall carbon nanotubes. *PHYSICAL REVIEW B*, Vol. 73, No. 23, June, 235434, ISSN 1098-0121

Jiang, J. W.; Tang, H.; Wang, B. S.; Su, Z. B. (2008)(a). Raman and infrared properties and layer dependence of the phonon dispersions in multilayered graphene. *PHYSICAL REVIEW B*, Vol.77, No. 23, June, 235421-1–235421-8, ISSN 1098-0121

Jiang, J. W.; Tang, H.; Wang, B. S.; & Su, Z. B. (2008)(b). A lattice dynamical treatment for the total potential energy of single-walled carbon nanotubes and its applications: relaxed equilibrium structure, elastic properties, and vibrational modes of ultra-narrow tubes. *JOURNAL OF PHYSICS-CONDENSED MATTER*, Vol.20, No. 4, Jan 30, 045228, ISSN 0953-8984

Madelung, O. (1978). *Introduction to Solid-State Theory*, Chap. 3, Springer-Verlag, ISBN 3540085165, Berlin

Mahan, G. D.; & Jeon, G. S. (2004). Flexure modes in carbon nanotubes *PHYSICAL REVIEW B*, Vol. 70, No. 7, August, 075405, ISSN 1098-0121

Manes, J. L.; Guinea, F.; Vozmediano, Maria A. H. (2007). Existence and topological stability of Fermi points in multilayered graphene. *PHYSICAL REVIEW B*, Vol. 75, No. 15, April, 155424-1–155424-6, ISSN 1098-0121

Mani, K. K. & Ramani, R. (1974). Lattice dynamics of graphite. *PHYSICA STATUS SOLIDI B-BASIC RESEARCH*, Vol.61, No.2, 659–668, ISSN 0370-1972

Maultzsch, J.; Reich, S.; Thomsen, C.; Requardt, H.; & Ordejon, P. (2004). Phonon dispersion in graphite. *PHYSICAL REVIEW LETTERS*, Vol. 92, No. 7, Feb 20, 075501, ISSN 0031-9007

Mohr, M.; Maultzsch, J.; Dobardzic, E.; Reich, S.; Milošević, I.; Damnjanović, M.; Bosak, A.; Krisch, M.; & Thomsen, C. (2007). Phonon dispersion of graphite by inelastic x-ray scattering. *PHYSICAL REVIEW B*, Vol. 76, No. 3, July, 035439-1–035439-7, ISSN 1098-0121

Nicklow, R.; Wakabayashi, N.; & Smith, H. G. (1972). Lattice-dynamics of pyrolytic-graphite. *PHYSICAL REVIEW B*, Vol. 5, No. 12, June 15, 4951–4962, ISSN 1098-0121

Novoselov, K. S.; Geim, A. K.; Morozov, S. V.; Jiang, D.; Zhang, Y.; Dubonos, S. V.; Grigorieva, I. V.; Firsov, A. A. (2004). Electric Field Effect in Atomically Thin Carbon Films. *SCIENCE*, Vol. 306, No. 5696, October 22, 666–669, ISSN 0036-8075 (print), 1095-9203 (online)

Painter, G.S.; & Ellis, D.E. (1970). Electronic band structure and optical properties of graphite from a variational approach. *PHYSICAL REVIEW B*, Vol. 1, No. 12, June 15, 4747–4752, ISSN 1098-0121

Partoens, B.; & Peeters, F. M. (2006). From graphene to graphite: Electronic structure around the K point. *PHYSICAL REVIEW B*, Vol. 74, No. 7, August, 075404, ISSN 1098-0121

Partoens, B.; & Peeters, F. M. (2007). Normal and Dirac fermions in graphene multilayers: Tight-binding description of the electronic structure. *PHYSICAL REVIEW B*, Vol. 75, No. 19, May, 193402, ISSN 1098-0121

Popov, V. N.; Van Doren, V. E.; & Balkanski, M. (2000). Elastic properties of single-walled carbon nanotubes. *PHYSICAL REVIEW B*, Vol. 61, No. 4, Jan 15, 3078–3084, ISSN 1098-0121

Rao, A. M.; Richter, E.; Bandow, S.; Chase, B.; Eklund, P. C.; Williams, K. A.; Fang, S.; Subbaswamy, K.R.; Menon, M.; Thess, A.; Smalley, R. E.; Dresselhaus, G.; & Dresselhaus, M. S. (1997). Diameter-selective Raman scattering from vibrational modes in carbon nanotubes. *SCIENCE* Vol. 275, No. 5297, Jan 10, 187–191, ISSN 0036-8075

Saito, R.; Dresselhaus, G.; & Dresselhaus, M. S. (1998). *Physical Properties of Carbon Nanotubes*, Imperial College Press, ISBN 1-86094-093-5, London

Tang, H.; Jiang, J. W.; Wang, B. S.; & Su, Z. B. (2009). A full spd tight-binding treatment for electronic bands of graphitic tubes. *SOLID STATE COMMUNICATIONS*, Vol. 149, No. 1–2, Jan, 82–86, ISSN 0038-1098

White, C. T.; Robertson, D. H.; Mintmire, J. W. (1993). Helical and rotational symmetries of nanoscale graphitic tubules. *PHYSICAL REVIEW B*, Vol. 47, No. 9, March, 5485–5488, ISSN 1098-0121

Ye, F.; Wang, B. S.; Su, Z. B. (2006). Symmetry restrictions in the chirality dependence of physical properties of single-wall nanotubes. *PHYSICAL REVIEW B*, Vol. 73, No. 15, April, 155424, ISSN 1098-0121

11

Universal Quantification of Chemical Bond Strength and Its Application to Low Dimensional Materials

Bo Xu, Xiaoju Guo and Yongjun Tian
State Key Laboratory of Metastable Materials Science and Technology
Yanshan University, Qinhuangdao, Hebei 066004
China

1. Introduction

The chemical bond strength describes the ability of a chemical bond holding two constituting atoms together. Many physical and mechanical properties of a material, such as melting point, activation energy of phase transition, tensile and shear strength, and hardness, are closely related to the bond strength (Kittle, 2004). However, a universal quantification of bond strength in crystal is hard to be performed due to the lack of effective microscopic parameters to characterize the bond strength. Usually, different characteristic parameters are chosen for different materials. For simple substances and covalent compounds, bond strength is correlated to cohesive energy. For ionic crystals, breaking a chemical bond means overcoming of the electrostatic interaction between anion and cation, which is defined as lattice energy and used to characterize bond strength. Correspondingly, several theoretical definitions of chemical bond strength have been proposed, such as Pauling's definition for ionic crystals (Pauling, 1929), orbital scaling for covalent crystals (Hultgren, 1932), and two power-law expressions for a variety of materials (Brown & Shannon, 1973; Gibbs et al., 2003). These definitions of bond strength are only valid for some specific types of crystals, and a generalized model of bond strength has not been reached.

In a crystal, bond strength is an intrinsic property of chemical bond, and is regulated by the constitutional atoms as well as the crystal structure. From this viewpoint, bond strength is directly determined by the bond length and the shared bonding electrons. Obviously, greater bond strength would be expected with shorter bond length. The extent of electron sharing, related to the electronegativity difference of bond-forming atoms, is determined by the localized electron density in the binding region. It was found that the greater the localized electron density, the more the effective bonding electrons, and the stronger the bond strength (Gibbs et al., 2003). Most recently, we established a universal semi-empirical quantitative scale to describe the strength of chemical bond in crystals (Guo et al., 2009). The chemical bond strength is defined as the maximum force that a chemical bond can resist under the uniaxial tension along the bond direction which is called tensile unbinding force. We found that the bond strength only relies on two parameters, the bond length and effectively bonded valence electron number of a chemical bond.

In the following, the concept of effectively bonded valence electron number of chemical bond is introduced and the universal quantification model of chemical bond strength is established based on effectively bonded valence electron number and bond length exclusively. The correlation between ideal tensile strength and chemical bond strength is presented. This model allows a convenience determination of chemical bond strength for a variety of materials, ranging from covalent crystals to ionic crystals as well as low dimensional materials. Its application to low dimensional materials, such as graphene, h-BN sheet, and SWNT, are also presented.

2. Methodology

As mentioned in the introduction, the shared bonding electrons in the binding region of two bonded atoms plays a vital role in determining the bond strength. To establish an effective quantification model of bond strength, we must find a practical way to estimate the population of these electrons.

Considering two atoms, *A* and *B*, forming a bond in a crystal, the valence electrons are Z_A and Z_B with coordination numbers of N_A and N_B, respectively. We first consider a simple case where bonded atom pair possesses totally eight valence electrons. The nominal valence electrons contributed to the *A-B* bond are $n_A = Z_A/N_A$ and $n_B = Z_B/N_B$ from atom *A* and *B*, respectively. Larger localized electron density in the binding region would result in stronger bond. However, the bonded electrons localized in the binding region are basically smaller than n_A or n_B. The Mulliken overlap population of a bond from first-principles calculations can provide a measurement of the bonded electrons (Mulliken, 1955). For example, the calculated population is 0.75 for C-C bonds in diamond and 0.19 for Na-Cl bonds in NaCl crystal. While the determination of population greatly depends on the calculation formalisms (Segall et al., 1996), it is more convenient to find a parameter, which can be easily determined, to serve as an alternative population. Here we propose the effectively bonded valence electron (EBVE) number, n_{AB}, of *A-B* bond in terms of n_A and n_B as

$$n_{AB} = \frac{n_A n_B}{\sqrt{n_A^2 + n_B^2}} \tag{1}$$

The EBVE numbers of diamond (0.707) and NaCl (0.163) are in good agreement with the Mulliken population. Some EBVE numbers of various covalent and ionic crystals are listed in our previous publication (Guo et al., 2009) as well as in the following text.

The ideal *A-B* bond strength can essentially be defined as an unbinding force of chemical bond, and the physical feature of bond strength for solids becomes more apparent and more accessible than Pauling's definition or other above-mentioned energy scales. Previously, the power-law behavior of the bond length on bond strength are suggested (Brown & Shannon, 1973; Gibbs et al., 1998, 2003), and the exponential dependence of the resistance of a bond to indenter on the population-related ionicity is emphasized (Gao et al., 2003; He et al., 2005; Li et al., 2008; Simunek & Vackar, 2006). These studies have highlighted the role of the bond length and valance electrons on the strength of a chemical bond. The chemical bond strength, defined as the tensile unbinding force F_{AB}, can then be described quantitatively in terms of the bond length d_{AB} and EBVE number n_{AB} with the formalism,

$$F_{AB} = Cd_{AB}^{-m} \exp(kn_{AB}) \quad (2)$$

where constants C, m, and k can be deduced from first-principles calculations.

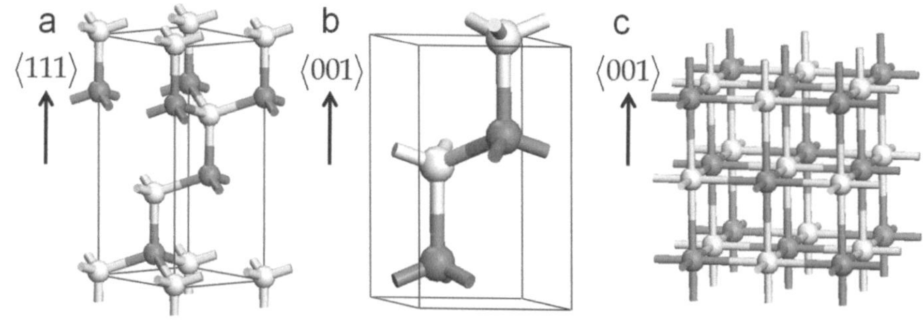

Fig. 1. Typical crystal structures of a) zincblende, b) wurtzite, and c) rock salt with the weakest tensile directions marked.

As soon as the tensile unbinding force of a bond is known, the ideal tensile strength of a crystal is easily accessible. For simple structural crystals shown in Figure 1, generally, the weakest tensile directions, such as $\langle 111 \rangle$ of zincblende (ZB), $\langle 001 \rangle$ of wurtzite (WZ), and $\langle 001 \rangle$ of rock salt (RS) structures, are parallel to the axes of the bonds to be broken. Thus, the ideal tensile strength, σ_{hkl}, of a crystal along the weakest $\langle hkl \rangle$ direction should be correlated with F_{AB} through

$$\sigma_{hkl} = S_{hkl} F_{AB} = CS_{hkl} d_{AB}^{-m} \exp(kn_{AB}) \quad (3)$$

where S_{hkl} in the unit of m^{-2} is the number of the broken bonds per unit area on the (hkl) plane, which has the lowest bond density.
Alternatively but more time-consumingly, the ideal tensile strength of a crystal can be determined from first-principles calculations (Roundy et al., 1999). We acquired the ideal tensile strength of a wide variety of covalent and ionic crystals with a single type of chemical bond with ZB, WZ, or RS structures from first-principles calculations, for which n_{AB}, d_{AB} and S_{hkl} are already known or can be calculated from the experimental values of the lattice parameters (Guo et al., 2009). Three parameters, C, m, and k, in Eqn. 3 can be deduced using the Levenberg-Marquardt method (Levenberg, 1944; Marquardt, 1963) by fitting Eqn. 3 with the crystal tensile strength values from first-principles calculations, leading to,

$$\sigma_{hkl}^{theor}(Pa) = 6.6 \times 10^{-10} S_{hkl} d_{AB}^{-1.32} \exp(3.7 n_{AB}) \quad (4)$$

and

$$F_{AB}^{theor}(N) = 6.6 \times 10^{-10} d_{AB}^{-1.32} \exp(3.7 n_{AB}) \quad (5)$$

The square of correlation coefficient R^2 is 0.996 with the mean absolute fractional deviation of ~7%, indicating that Eqn. 4 and 5 are accurate enough to estimate σ_{hkl} and F_{AB}.
Up to now, we are considering two bonding atoms, A and B, with totally 8 valence electrons. The simple formula for tensile unbinding force should be valid for complicate crystals, such as β-Si_3N_4, α-quartz, and α-Al_2O_3, where the total valence electrons of two bonding atoms

are more than 8. The chemical bonds in these crystals are typical two-electron bond. β–Si_3N_4 and α-quartz follow the *8-N* rule of structure, where *N* is the valence electron of an atom and *8-N* is the corresponding coordination number. Since there are one pair of non-bonding electrons for N in β–Si_3N_4 and two pairs of non-bonding electrons for O in β-quartz, n_N and n_O are both equal to 1. In α–Al_2O_3, The coordination number for Al and O are 6 and 4, respectively, leading to n_{Al} of 0.5 and n_O of 1.5. It is then straight forward to calculate the tensile unbinding force (bond strength) with Eqn. 5 for these crystals.

3. Results and discussions

In this section, we will start with the calculation of bond strength in chosen types of materials to understand the relationship of bond strength and crystal structure as well as to trace the relations between the macroscopic properties and bond strength. We will end this section with the bond strength calculations for some low-dimensional materials, such as graphene, h-BN sheet, and SWNT, to demonstrate the effectiveness of our semi-empirical quantification model to these systems.

3.1 IV-A semiconductors

The IV-A semiconductors belong to the family of A^NB^{8-N} semiconductors which have similar structures under normal conditions and follow similar phase transition rules under high pressure (Mujica et al., 2003). The procedure presented next can easily be applied to other members of A^NB^{8-N} semiconductors.The stable phases of the IV-A materials are graphite, diamond structured Si, Ge, and Sn, and a huge variety of polytypic forms of SiC with comparable energies such as 3C, 2H, 4H, and 6H. The phase transitions of these materials have been itemized previously with fifteen types of crystal structures for IV-A materials (Mujica et al., 2003). ZB, WZ, and RS structures are shown in Figure 1. The other 12 structures are summarized in Figure 2 with the coordination states of the atoms marked. ZB, WZ, RS, graphite, and sh (simple hexagonal) are simple with the marked directions parallel to the axes of broken bonds. Tensile strength can be calculated for materials crystallized in these structures in addition to bond strength.

Several points need to be mentioned for these complicated structures before we discuss the calculation results. Firstly, we give the hexagonal representation instead of the simple rhombohedral representation for r8 structure to show the structure more clearly. Secondly, Si atoms in Imma structure are eight-fold coordinated while Ge atoms in the same structure of are six-fold coordinated (Figure 2f). Thirdly, there are two types of coordination states in the most complicated structure Cmca, although these atoms are identical. We denote the ten-fold coordinated atoms located at 8d sites with white spheres, and the eleven-fold coordinated atoms at 8f sites with gray spheres, as shown in Figure 2g.

The lattice parameters and the calculated bond strengths are listed in Table I. Except the sh-Si and sh-Ge, the arithmetic average of the bond length are given in Table I for structures with different bond lengths. The bond strength of graphite, diamond, lonsdaleite, ZB-Si, WZ-Si, ZB-Ge, ZB-Sn, and ZB-SiC can be referred to our recent publication (Guo et al., 2009). The bond strength of the IV-A materials as a function of bond length is presented in Figure 3. The tensile unbinding forces are unambiguously grouped by distinct n_{AB} values of 0.943, 0.707, 0.471, 0.354, 0.269, and 0.236, respectively (from top to bottom). As n_{AB} reflects the coordination states, a lower value means a higher coordination number. Bond strength of sequential structures appearing along the pressure induced phase transitions will leap from the higher line to the next higher one. For instance, the experimental phase transition sequence of Si under compression is ZB-Si (n_{AB}=0.707) → β-Sn (0.471) → (Imma→sh) (0.354) → Cmca (0.269) → (hcp→fcc) (0.236). In Figure 3, the bond strength in the same line decreases from left to right in the sequence of C-C, Si-C, Si-Si, Ge-Ge, and Sn-Sn except in the line of n_{AB}=0.269 where the order for Si-Si and Ge-Ge is reversed. The bond strengths of Si-Si and Ge-Ge on other lines are not as distinguishable as those of other bonds. Such an anomaly can be explained by the experimental facts that the radii difference of Si and Ge is not so significant as other elements (r_C=0.062 Å, r_{Si}=1.068 Å, r_{Ge}=1.090 Å, r_{Sn}=1.240 Å). Alternatively, the high transition pressure makes the bond length of Ge comparable to that of Si or even shorter, reminiscent of Cmca (n_{AB}=0.269). Briefly, the bond strength of the chemical bond in this type of materials is determined by the coordination number: the lower the coordination number, the higher the bond strength.

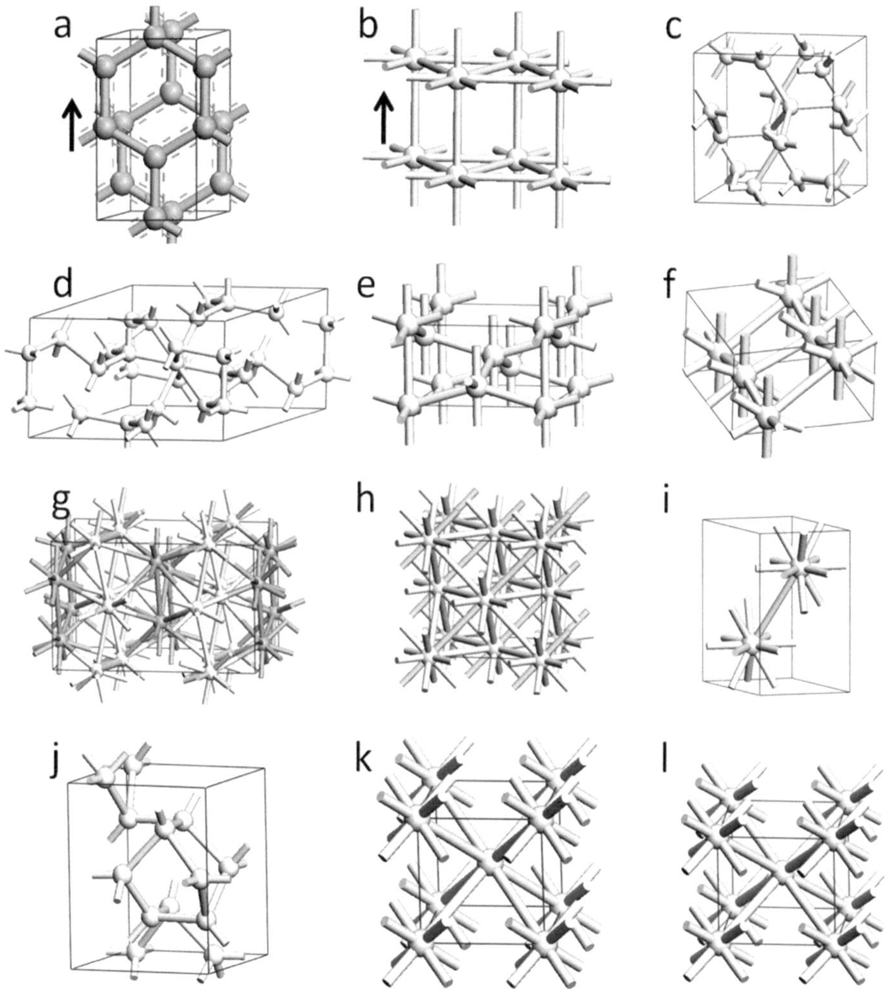

Fig. 2. Crystal structures of IV-A compounds without those in Figure 1. a) Graphite, b) sh, c) bc8, d) r8-h, e) β-Sn, f) Imma, g) Cmca, h) fcc, i) hcp, j) st12, k) bcc, and l) bct.

Bond	Crystal	Space Group	a (Å)	b (Å)	c (Å)	d (Å)	n_A	n_B	n_{AB}	F_{AB} (nN)
Si-Si	bc8	$Ia\bar{3}$	6.636			2.370	1	1	0.707	2.89
Si-Si	r8	$R\bar{3}$	9.125		5.447	2.319	1	1	0.707	2.98
Si-Si	β-Sn	$I4_1/amd$	4.686		2.585	2.475	0.667	0.667	0.471	1.14
Si-Si	sh	$P6/mmm$	2.549		2.383	2.549	0.5	0.5	0.354	0.71
						2.383	0.5	0.5	0.354	0.78
Si-Si	Imma	*Imma*	4.737	4.502	2.55	2.462	0.5	0.5	0.354	0.74
Si-Si	Cmca	*Cmca*	8.024	4.796	4.776	2.491	0.4	0.364	0.269	0.54
Si-Si	fcc	$Fm\bar{3}m$	3.34			2.362	0.333	0.333	0.236	0.51
Si-Si	hcp	$P6_3/mmc$	2.404		4.063	2.437	0.333	0.333	0.236	0.49
Ge-Ge	st12	$P4_32_12$	5.93		6.98	2.485	1	1	0.707	2.72
Ge-Ge	bc8	$Ia\bar{3}$	6.658			2.393	1	1	0.707	2.85
Ge-Ge	Imma	*Imma*	4.931	4.913	2.594	2.562	0.667	0.667	0.471	1.09
Ge-Ge	β-Sn	$I4_1/amd$	4.959		2.746	2.631	0.667	0.667	0.471	1.05
Ge-Ge	sh	$P6/mmm$	2.657		2.556	2.657	0.5	0.5	0.354	0.67
						2.556	0.5	0.5	0.354	0.71
Ge-Ge	Cmca	*Cmca*	7.886	4.656	4.667	2.430	0.4	0.364	0.269	0.55
Ge-Ge	hcp	$P6_3/mmc$	2.776		4.573	2.793	0.333	0.333	0.236	0.41
Sn-Sn	β-Sn	$I4_1/amd$	5.833		3.182	3.076	0.667	0.667	0.471	0.86
Sn-Sn	bcc	$Im\bar{3}m$	3.287			2.847	0.5	0.5	0.354	0.61
Sn-Sn	bct	$I4/mmm$	3.7		3.37	3.112	0.5	0.5	0.354	0.55
Si-C	4H	$P6_3mc$	3.079		10.073	1.894	1	1	0.707	3.89
Si-C	RS	$Fm\bar{3}m$	4.001			2.001	0.667	0.667	0.471	1.51

Table 1. Lattice parameters and bond strength of IV-A semiconductors.

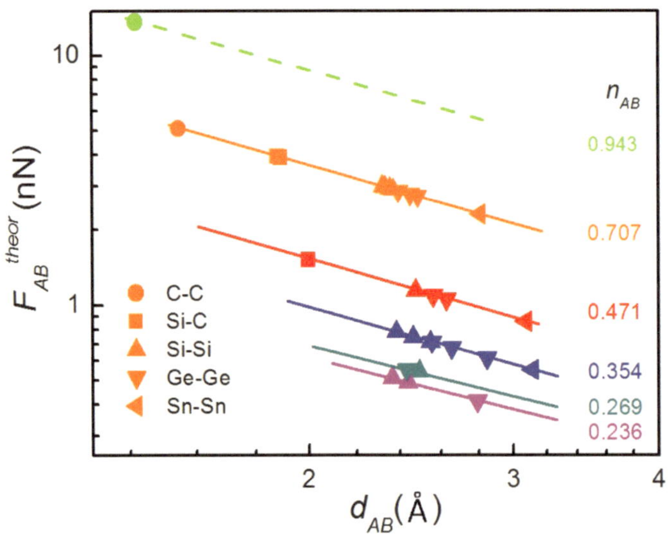

Fig. 3. Bond strength F_{AB}^{theor} vs. d_{AB} for IV-A compounds.

3.2 A^NB^{8-N} ionic crystals

Our simple model for chemical bond strength can easily be applied to ionic crystals, just like the above considered pure covalent and polar covalent A^NB^{8-N} materials (Guo et al., 2009). The elemental combinations of IA-VIIA, IB-VIIA, and IIA (except Be)-VIA tend to form ionic crystals. The typical structures of ionic A^NB^{8-N} materials are RS and CsCl. ZB and WZ structures are also founded for some IB-VIIA crystals (Shindo et al., 1965). These four structures have been presented in Figure 1 and 2. Other structures, such as $R\bar{3}m$ (H) (Hull & Keen, 1994), P3m1 (Sakuma, 1988), P4/nmm (Liu, 1971), P21/m and Cmcm (Hull & Keen, 1999), can also be found in I-VIIA and IIA-VIA compounds and are shown in Figure 4. The concrete coordination state of each atom in these five structures is clearly shown. The distance between the two neighboring iodine atoms in P3m1-structured CuI is 4.353 Å. Consequently, this compound has a lamellar structure, significantly different from other dense structures.

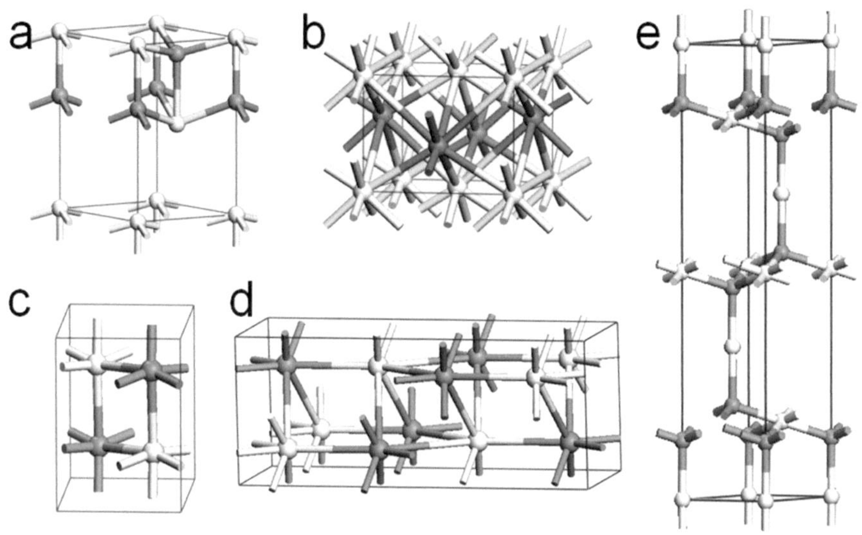

Fig. 4. Crystal structures of ionic AB compounds. a) P3m1, b) P4/nmm, c) $P2_1/m$, d) Cmcm, and e) $R\bar{3}m$.

In our previous work, the bond strength and uniaxial tensile strength of ten types of rocksalt structured compounds has been calculated (Guo et al., 2009). Crystal parameters and calculated bond strength of other ionic AB compounds are shown in Table 2. For the high pressure phases, the lattice parameters are given under compression. For the monoclinic structures of $P2_1/m$-structured AgCl, AgBr and AgI, β angles are 98.4°, 95.9°, and 98.4°, respectively. The unbinding tensile force versus bond length for I-VIIA and IIA-VIA compounds listed in Table 2 together with those given in previous work is shown in Figure 5. The bond strength of seventy six chemical bonds locates on seven parallel lines from top to the bottom with decreasing n_{AB} values.

3.3 III-VI crystals

Next step is to treat complicate crystals of A_mB_n with our semi-empirical quantification model of bond strength. We have discussed the most familiar compounds of β–Si_3N_4, α-

quartz, and α-Al_2O_3 in **Section 2**. The material members in this group are tremendous. Here only typical materials with the chemical composition of A_2B_3 (IIIA-VIA) are chosen for a demonstration. The crystal structures of the compounds as shown in Figure 6 are so complicated that sometimes identical atoms in the same structure have different coordination numbers. The nominal bonding valence electrons of atoms with lower electronegativity are equal to the numbers of outmost electrons, while the determination of the nominal bonding valence electrons of atoms with greater electronegativity is usually challenging.

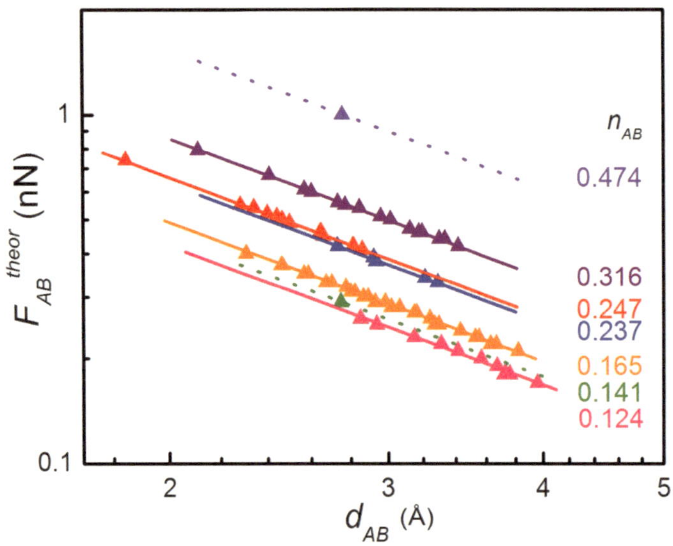

Fig. 5. Bond strength F_{AB}^{theor} vs. d_{AB} for ionic AB compounds.

For B_2O_3 of *P3₁21* and B_2S_3 of lamellar *P2₁/c*, all B atoms are three-fold coordinated with O (S) atoms which are bonded to two B atoms. Considering the sp hybridization character of O atoms, the bonding valence electrons of O atoms are set to two. The situation differs from the *Cmc2*-structured B_2O_3 where B atoms are four-fold coordinated (Prewitt & Shannon, 1968). In this structure, one third of the O atoms are two-fold coordinated and the other two thirds are three-fold coordinated. For the three-fold coordinated O atoms, the nominal bonding valence electrons number is set to four due to the sp^2 hybridized orbital. While for the two-fold coordinated O atoms, only two nominal valence electrons are considered in bonding states and the other four stay at non-bonding states. In the case of Al_2S_3 with $P6_1$ symmetry, one fourth of Al atoms are six-fold coordinated (Al_{VI}) and other three fourths are four-fold coordinated (Al_{IV}), and all S atoms are three-fold coordinated with two Al_{VI} atoms and one Al_{IV} atom (Krebs et al., 1993).

Next structure is C_c for Ga_2Se_3 where Ga atoms are four-fold coordinated, one-third of Se atoms are two-fold coordinated and the other two thirds are three-fold coordinated with the nominal bonding valence electrons of 2 and 4, respectively (Lübbers & Leute, 1982). For In_2O_3 with $R\bar{3}c$ symmetry, the coordination number of In and O atoms are six and four, respectively (Prewitt et al., 1969). The *C2/m*-structured Ga_2O_3 is even more complicated(Ahman et al., 1996). Half of the Ga atoms are four-fold coordinated (Ga_{IV}) while the others six-fold coordinated (Ga_{VI}). O atoms in this structure are rather complicated: one third of O atoms are four-fold coordinated to three Ga_{VI} atoms and one Ga_{IV} atom, another one third are three-fold coordinated to two Ga_{VI} atoms and one Ga_{IV} atom, and the last one third are three-fold coordinated to one Ga_{VI} atoms and two Ga_{IV} atom.

The lattice parameters, bond lengths, and bond strength are listed in Table 3. Most of the chemical bonds are two-electron bond. The bond strength versus bond length for these A_2B_3 and A_3B_4 compounds are shown in Figure 7 together with the data for Al_2O_3, β-C_3N_4 and β-Si_3N_4 from the previous work (Guo et al., 2009). Bond strength of the sixteen chemical bonds falls on seven parallel lines. Among all the two-electron bonds, the B-O bonds in $P3_121$-structured B_2O_3 exhibit the greatest unbinding tensile strength, even higher than that of C-C bond in diamond.

Bond	Crystal	Space group	a (Å)	b (Å)	c (Å)	d (Å)	n_A	n_B	n_{AB}	F_{AB} (nN)
Cu-F	ZB	$F\bar{4}3m$	4.255			1.842	0.25	1.75	0.247	0.74
Cu-Cl	ZB	$F\bar{4}3m$	5.406			2.341	0.25	1.75	0.247	0.54
Cu-Cl	WZ	$P63mc$	3.910		6.420	2.399	0.25	1.75	0.247	0.52
Cu-Cl	sc16	$Pa\bar{3}$	6.310			2.284	0.25	1.75	0.247	0.55
Cu-Cl	RS	$Fm\bar{3}m$	4.929			2.465	0.167	1.167	0.165	0.37
Cu-Br	ZBr	$F\bar{4}3m$	5.691			2.464	0.25	1.75	0.247	0.50
Cu-Br	WZ	$P63mc$	4.060		6.660	2.498	0.25	1.75	0.247	0.49
Cu-Br	sc16	$Pa\bar{3}$	6.738			2.442	0.25	1.75	0.247	0.51
Cu-Br	RS	$Fm\bar{3}m$	5.17			2.585	0.167	1.167	0.165	0.35
Cu-I	ZB	$F\bar{4}3m$	6.604			2.860	0.25	1.75	0.247	0.41
Cu-I	WZ	$P63mc$	4.310		7.090	2.659	0.25	1.75	0.247	0.45
Cu-I	P3m1	$P3m1$	4.279		7.17	2.644	0.25	1.75	0.247	0.46
Cu-I	$R\bar{3}m$	$R\bar{3}m$	4.155		20.48	2.648	0.25	1.75	0.247	0.46
Cu-I	RS	$Fm\bar{3}m$	6.121			3.061	0.167	1.167	0.165	0.28
Ag-F	RS	$Fm\bar{3}m$	4.920			2.460	0.167	1.167	0.165	0.37
Ag-Cl	RS	$Fm\bar{3}m$	5.546			2.773	0.167	1.167	0.165	0.32
Ag-Cl	$P2_1/m$	$P2_1/m$	3.587	3.992	5.307	2.700	0.167	1.167	0.165	0.33
Ag-Cl	Cmcm	$Cmcm$	3.399	10.124	4.023	2.747	0.143	1	0.141	0.29
Ag-Br	RS	$Fm\bar{3}m$	5.780			2.890	0.167	1.167	0.165	0.30
Ag-Br	$P2_1/m$	$P2_1/m$	3.821	3.98	5.513	2.798	0.167	1.167	0.165	0.31
Ag-I	ZB	$F\bar{4}3m$	6.499			2.814	0.25	1.75	0.247	0.42
Ag-I	WZ	$P63mc$	4.580		7.494	2.810	0.25	1.75	0.247	0.42
Ag-I	RS	$Fm\bar{3}m$	6.034			3.017	0.167	1.167	0.165	0.28
Ag-I	$P2_1/m$	$P2_1/m$	4.056	4.057	5.615	2.927	0.167	1.167	0.165	0.29
Ag-I	CsCl	$Pm\bar{3}m$	4.31			3.733	0.125	0.875	0.124	0.18
Li-Cl	RS	$Fm\bar{3}m$	5.130			2.565	0.167	1.167	0.165	0.35
Li-I	RS	$Fm\bar{3}m$	6.031			3.016	0.167	1.167	0.165	0.28
Li-I	NiAs	$P63/mmc$	4.48		7.26	3.160	0.167	1.167	0.165	0.27
K-F	RS	$Fm\bar{3}m$	5.344			2.672	0.167	1.167	0.165	0.33

K-Cl	CsCl	$Pm\bar{3}m$	3.634		3.147	0.125	0.875	0.124	0.23
K-Br	RS	$Fm\bar{3}m$	6.585		3.293	0.167	1.167	0.165	0.25
K-I	RS	$Fm\bar{3}m$	7.049		3.525	0.167	1.167	0.165	0.23
K-I	CsCl	$Pm\bar{3}m$	3.94		3.412	0.125	0.875	0.124	0.21
Rb-F	RS	$Fm\bar{3}m$	5.73		2.865	0.167	1.167	0.165	0.30
Rb-F	CsCl	$Pm\bar{3}m$	3.29		2.849	0.125	0.875	0.124	0.26
Rb-Cl	CsCl	$Pm\bar{3}m$	3.82		3.308	0.125	0.875	0.124	0.22
Rb-Br	RS	$Fm\bar{3}m$	6.855		3.428	0.167	1.167	0.165	0.24
Rb-Br	CsCl	$Pm\bar{3}m$	4.24		3.672	0.125	0.875	0.124	0.19
Rb-I	RS	$Fm\bar{3}m$	7.329		3.665	0.167	1.167	0.165	0.22
Rb-I	CsCl	$Pm\bar{3}m$	4.34		3.759	0.125	0.875	0.124	0.18
Cs-F	RS	$Fm\bar{3}m$	6.030		3.015	0.167	1.167	0.165	0.28
Cs-F	CsCl	$Pm\bar{3}m$	3.39		2.936	0.125	0.875	0.124	0.25
Cs-Cl	RS	$Fm\bar{3}m$	7.095		3.548	0.167	1.167	0.165	0.23
Cs-Cl	Cs	$Pm\bar{3}m$	4.115		3.564	0.125	0.875	0.124	0.20
Cs-Br	RS	$Fm\bar{3}m$	7.253		3.627	0.167	1.167	0.165	0.22
Cs-Br	CsCl	$Pm\bar{3}m$	4.296		3.720	0.125	0.875	0.124	0.18
Cs-I	RS	$Fm\bar{3}m$	7.631		3.816	0.167	1.167	0.165	0.21
Cs-I	CsCl	$Pm\bar{3}m$	4.568		3.956	0.125	0.875	0.124	0.17
Mg-Se	RS	$Fm\bar{3}m$	5.463		2.732	0.333	1	0.316	0.56
Mg-Te	WZ	$P63mc$	4.53	7.35	2.756	0.5	1.5	0.474	1.00
Ca-S	RS	$Fm\bar{3}m$	5.689		2.845	0.333	1	0.316	0.54
Ca-Se	RS	$Fm\bar{3}m$	5.916		2.958	0.333	1	0.316	0.51
Ca-Te	RS	$Fm\bar{3}m$	6.348		3.174	0.333	1	0.316	0.46
Ca-Te	CsCl	$Pm\bar{3}m$	3.387		2.933	0.25	0.75	0.237	0.38
Sr-O	RS	$Fm\bar{3}m$	5.134		2.567	0.333	1	0.316	0.61
Sr-S	RS	$Fm\bar{3}m$	6.023		3.012	0.333	1	0.316	0.50
Sr-S	CsCl	$Pm\bar{3}m$	3.372		2.920	0.25	0.75	0.237	0.39
Sr-Se	RS	$Fm\bar{3}m$	6.243		3.122	0.333	1	0.316	0.47
Sr-Te	RS	$Fm\bar{3}m$	6.659		3.330	0.333	1	0.316	0.44
Sr-Te	CsCl	$Pm\bar{3}m$	3.708		3.211	0.25	0.75	0.237	0.34
Ba-O	P4/nmm	$P4/nmm$	4.397	3.196	2.73	0.25	0.75	0.237	0.42
Ba-S	RS	$Fm\bar{3}m$	6.387		3.194	0.333	1	0.316	0.46
Ba-Se	RS	$Fm\bar{3}m$	6.593		3.297	0.333	1	0.316	0.44
Ba-Se	CsCl	$Pm\bar{3}m$	3.795		3.287	0.25	0.75	0.237	0.33
Ba-Te	Rs	$Fm\bar{3}m$	6.830		3.415	0.333	1	0.316	0.42

Table 2. Lattice parameters and bond strength of AB ionic compounds.

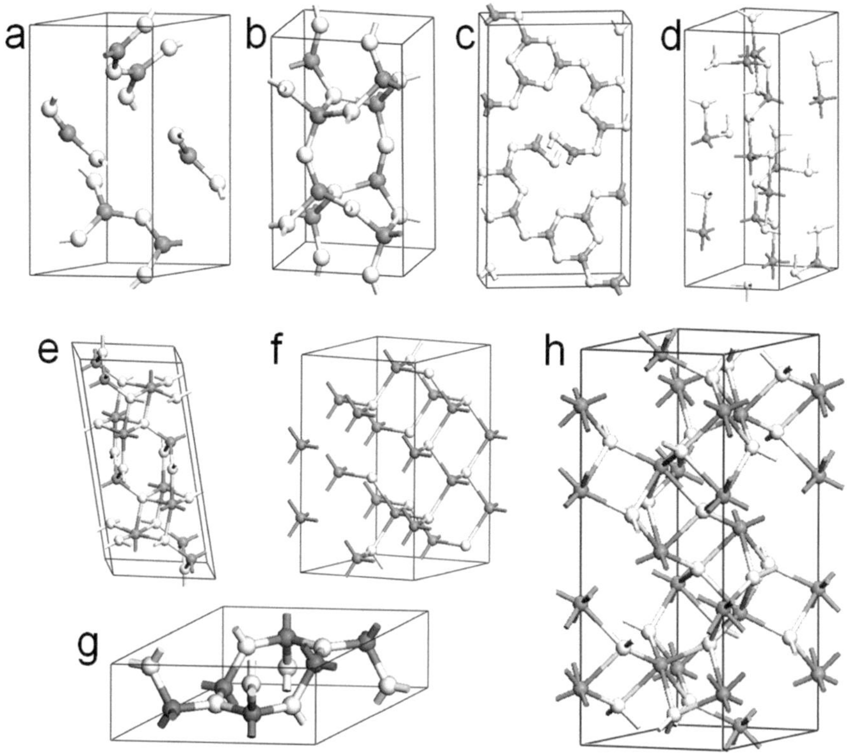

Fig. 6. Crystal structures of typical A_2B_3 and A_3B_4 compounds. White and gray spheres correspond to elements with higher and lower electronegativity, respectively. a) $P3_121$, b) $Cmc2_1$, c) $P2_1/c$, d) $P6_1$, e) C2/m, f) C_c, g) P3, and h) $R\bar{3}c$.

An extensive analysis of the A^NB^{8-N} materials will give twenty-four n_{AB} values, some of them has been shown in this work. The highest two are 0.943 and 0.856 occurring for three-fold coordinated C-C bond in graphite and three-fold coordinated B-N bond in h-BN, respectively. The common feature of these two bonds is more than two nominal valence electrons distributed on the bond, that is to say the bond order of them is higher than 1. The high bonded valence electron number makes them the strongest bonds to resist tensile force along the direction parallel to the axes of the bond. The next highest n_{AB} of 0.707 is from the four-fold coordinated two-electron bond where each atom contributes one valence electron to the bond, following by 0.643, 0.589, 0.530, 0.474, 0.471 and so on in the sequence. The lowest n_{AB} value is 0.124 corresponds to the Ag-I bond in the $Pm\bar{3}m$ -structured AgI.
Generally, lower coordination number results in higher bonded valence electron number, especially when the valence electron numbers of bonded atoms is the same. However, this argument does not hold when the valence electron numbers are different. For example, the Si atoms forming Si-Si bond in hcp- and fcc-structured Si are twelve-fold coordinated, which is the highest among the chemical bonds discussed above. The bonded valence electron number n_{AB} of these Si-Si bonds are 0.236, higher than the six-fold coordinated I-VIIA bonds (0.165), seven-fold coordinated I-VIIA bonds (0.141), and eight-fold coordinated I-VIIA bonds (0.124). It should be emphasized that the lines denoted with the same n_{AB} in different

figures are the same line, for instance, 0.707 in Figure 3 and 7, 0.474 in Figure 5 and 7. As long as the chemical bonds have the same n_{AB}, they will lie on the same line, and bond strength depends strictly on the bonded valence electron number n_{AB}. The relation of coordination number and n_{AB} therefore means that a lower coordination number corresponds to higher bond strength.

Crystal	Bond	Space group	a (Å)	b (Å)	c (Å)	d (Å)	n_A	n_B	n_{AB}	F_{AB} (nN)
B_2O_3	B-O	$P3_121$	4.336		8.340	1.368	1	1	0.707	5.97
B_2O_3	B-O_{III}	$Cmc2$	7.803	4.613	4.129	1.508	0.75	1.333	0.654	4.31
	B-O_{II}					1.373	0.75	1	0.6	4.00
B_2S_3	B-S	$P2_1/c$	4.039	10.722	18.620	1.794	1	1	0.707	4.18
Al_2S_3	Al_V-S	$P6_1$	6.491		17.169	2.376	0.6	1.333	0.547	1.59
	Al_{IV}-S					2.248	0.75	1.333	0.654	2.54
Ga_2O_3	Ga_{VI}-O_{IV}	$C2/m$	12.21	3.037	2.798	2.040	0.5	1.5	0.474	1.49
	Ga_{VI}-O_{III}					1.936	0.5	1.333	0.468	1.56
	Ga_{IV}-O_{IV}					1.863	0.75	1.5	0.671	3.47
	Ga_{IV}-O_{III}					1.833	0.75	1.333	0.654	3.33
Ga_2Se_3	Ga-Se_{III}	Cc	6.661	11.652	6.649	2.364	0.75	1.333	0.654	2.38
	Ga-Se_{II}					2.365	0.75	1	0.6	1.95
In_2O_3	In-O	$R\bar{3}c$	5.487		14.510	2.187	0.5	1.5	0.474	1.36

Table 3. Crystal parameters, bond length, and bond strength of A_2B_3 compounds.

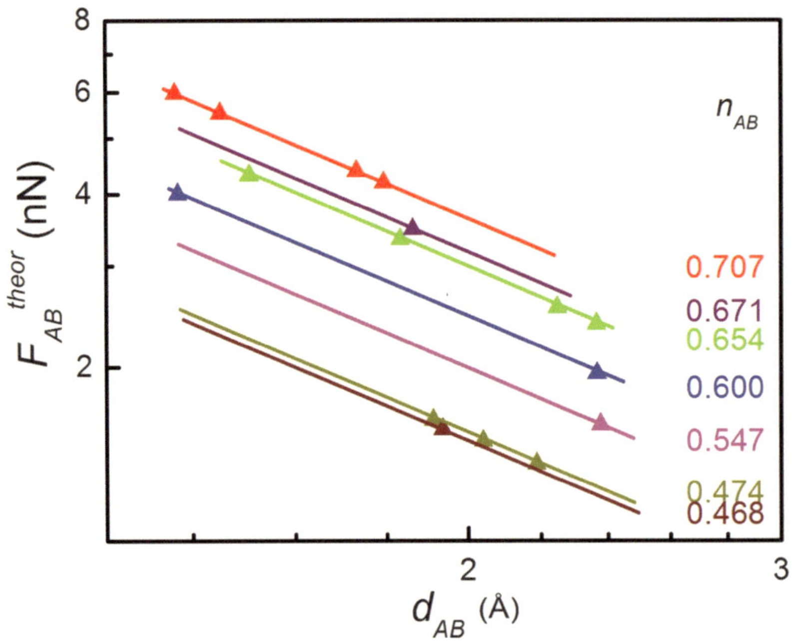

Fig. 7. Bond strength F_{AB}^{theor} vs. d_{AB} for typical A_2B_3 and A_3B_4 compounds

3.4 Low dimensional systems

We now apply our semi-empirical model to evaluate the theoretical tensile strength of the low dimensional systems, such as graphene, h-BN sheet, SWNT. As we mentioned before, the highest effective bonded valence electron numbers of 0.943 and 0.856 occur for three-fold coordinated C-C bond in graphite and three-fold coordinated B-N bond in h-BN, respectively. This is consistent with the argument that the sp^2 hybridized C-C bond in graphite is the strongest chemical bond (Coulson, 1952). Recently, graphene, a single atomic layer of graphite, has stirred enormous research interests owning to its exceptionally high crystallinity and electronic quality, as well as a fertile ground for applications (Geim & Novoselov, 2007). Experimentally, the mechanical properties of graphene have been identified with atomic force microscope (AFM) nanoindentation, giving a tensile strength of 130 GPa (Lee et al., 2008). Here we predict the theoretical tensile strength σ_{10}^{theor} of graphene in <*10*> direction using the present model as follows

$$\sigma_{10}^{theor} = S_{10} F_{cc} = \frac{F_{cc}}{\sqrt{3} d_{cc} \times \delta R} \tag{6}$$

where δR is the thickness of graphene taken as the interlayer separation 3.4 Å of graphite. F_{cc} and d_{cc} values are listed in Table 4 for graphene and other low dimensional systems. The theoretical tensile strength obtained is 162.7 GPa in the <*10*> direction, 20% higher than the experimental value.

Whilst graphene has a great application potential in microelectronics, hexagonal boron nitride (h-BN) sheets can find uses as an effective insulator in graphene based electronics. The mechanic properties of h-BN sheets have recently been investigated. With an elastic constant (E^{2D}) of 292 Nm^{-1}, a breaking strain of 0.22, and a thickness of 0.33 nm for a single atomic layer of h-BN (Song et al., 2010), the tensile strength can be deduce to be 97 GPa with the procedure suggested for graphene (Lee et al., 2008). Accordingly, the theoretical tensile strength of h-BN in the <*10*> direction can be calculated with our semi-empirical model, giving a value of 117 GPa, which is in good agreement with the experimental deduced value.

Carbon nanotube (CNT), with the same covalent sp^2 bonds formed between individual carbon atoms as graphen, is one of the strongest and stiffest materials. Direct tensile testing of individual tubes is challenging due to their small size (10 nm or less in diameter). There are several experimental efforts on the mechanical properties of CNT (Yu et al., 2000; Demczyk et al., 2002; Ding et al., 2006; Barber et al., 2005). However, the reported failure stress values display a large variance and are well below the theoretical predicted values in most cases (Ozaki et al., 2000; Mielke et al., 2004), which are attributed to the large number of defects presented on the nanotubes. Accurate measurements of tensile strength require high-quality CNT with well-defined sample parameters, as well as the elimination of measurement uncertainties. Notwithstanding, for zigzag single wall nanotube (SWNT), the theoretical tensile strength along the axial direction can be predicted with our simple model as,

$$\sigma_{axial}^{theor} = S_{axial} F_{AB} = \frac{nF_{AB}}{\pi[(\frac{D_{ep} + \delta R}{2})^2 - (\frac{D_{ep} - \delta R}{2})^2]} \tag{7}$$

where D_{ep} is the diameter of selected nanotube and n is the first index of the chiral vector (n, m) for nanotubes. The theoretical tensile strength of C (10, 0), SiC (10, 0), BN (10, 0), and AlN (10, 0) SWNTs are listed in Table 4.

Bond	Material	D_{ep} (Å)	D_{AB} (Å)	n_A	n_B	n_{AB}	F_{AB}^{theor} (nN)	σ_{theor} (GPa)
C-C	Graphene $\langle 01 \rangle$	–	1.419	1.333	1.333	0.943	13.6	162.7
B-N	h-BN sheet $\langle 01 \rangle$	–	1.446	1	1.667	0.857	9.65	117
C-C	C (10,0)	7.91	1.42	1.333	1.333	0.943	13.6	161.0
Si-C	SiC (10,0)	9.95	1.80	1.333	1.333	0.943	9.94	93.6
B-N	BN (10,0)	8.11	1.45	1	1.667	0.857	9.65	111.4
Al-N	AlN (10,0)	10.33	1.83	1	1.667	0.857	7.10	64.3

Table 4. Parameters, calculated bond strength, and tensile strength for selected low dimensional systems.

Before we end this section, there is one last point need to be mentioned regarding to the dependence of the tensile strength on the direction of the applied tensile stress. A tensile stress tilted away from the axis of a chemical bond would generate a shear component with respect to the bond, and we cannot perform the tensile strength calculation with Eqn. 3 under this circumstance. However, if the shear unbinding strength can be expressed, the ideal strength along any specific direction of a crystal will be accessible. Further studies are therefore highly expected.

4. Conclusion

The bond strength of a variety of chemical bonds are analysized with our semi-empirical unbinding tensile force model. This model proves to be valid for a wide selection of crystals, as well as low-dimensional materials such as graphene and nanotubes. In this model, the chemical bond strength, defined as the tensile unbinding force F_{AB}, can be calculated quantitatively in terms of the bond length d_{AB} and effective bonded valence electron number n_{AB}. It is demonstrated that the bond strength relies strongly on the crystal structure of a solid, in particular, the coordinated states of the bonded atoms. As a result, a chemcial bond formed by identical atom pair would have distinct bond strength in different crystal structures. For example, the the C-C bond strength in graphite is 2.67 times as high as that in diamonds. The definition of unbinding tensile strength provides a more intuitive and general representation of bond strength than those of cohensive energy for covalent crystals and lattice anergy for ionic crystals.

5. Acknowledgment

This work was supported by NSFC (Grant Nos. 50821001 and 91022029), by NBRPC (Grant No. 2011CB808205).

6. References

Ahman, J., Svensson, G. & Albertsson, J. (1996). A reinvestigation of beta-gallium oxide. *Acta Crystallographica Section C-Crystal Structure Communications*, Vol.52, No.6, pp. 1336-1338, ISSN 0108-2701

Barber, A. H., Andrews, R., Schadler, L. S. & Wagner, H. D. (2005). On the tensile strength distribution of multiwalled carbon nanotubes. *Applied Physics Letters*, Vol.87, No.20, 203106, ISSN 0003-6951

Brown, I. D. & Shannon, R. D. (1973). Empirical bond-strength bond-length curves for oxide. *Acta Crystallographica Section A*, Vol. 29, No.5, pp. 266-282, ISSN 0108-7673

Coulson, C. A. (1950). *Valence*, Oxford University Press, Oxford.

Demczyk, B. G., Wang, Y. M., Cumings, J., Hetman, M., Han, W., Zettl, A. & Ritchie, R. O. (2002). Direct mechanical measurement of the tensile strength and elastic modulus of multiwalled carbon nanotubes. *Materials Science and Engineering a-Structural Materials Properties Microstructure and Processing*, Vol.334, No.1-2, pp. 173-178, ISSN 0921-5093

Ding, W., Calabri, L., Kohlhaas, K. M., Chen, X., Dikin, D. A. & Ruoff, R. S. (2007). Modulus, fracture strength, and brittle vs. plastic response of the outer shell of arc-grown multi-walled carbon nanotubes. *Experimental Mechanics*, Vol.47, No.1, pp. 25-36, ISSN 0014-4851

Gao, F. M., He, J. L., Wu, E. D., Liu, S. M., Yu, D. L., Li, D. C., Zhang, S. Y. & Tian, Y. J. (2003). Hardness of covalent crystals. *Physical Review Letters*, Vol.91, No.1, 015502, ISSN 0031-9007

Geim, A. K. & Novoselov, K. S. (2007). The rise of graphene. *Nature Materials*, Vol.6, No.3, pp. 183-191, ISSN 1476-1122

Gibbs, G. V., Hill, F. C., Boisen, M. B. & Downs, R. T. (1998). Power law relationships between bond length, bond strength and electron density distributions. *Physics and Chemistry of Minerals*, Vol.25, No.8, pp. 585-590, ISSN 0342-1791

Gibbs, G. V., Rosso, K. M., Cox, D. F. & Boisen, M. B. (2003). A physical basis for Pauling's definition of bond strength. *Physics and Chemistry of Minerals*, Vol.30, No.5, pp. 317-320, ISSN 0342-1791

Guo, X. J., Wang, L. M., Xu, B., Liu, Z. Y., Yu, D. L., He, J. L., Wang, H. T. & Tian, Y. J. (2009). Unbinding force of chemical bonds and tensile strength in strong crystals. *Journal of Physics-Condensed Matter*, Vol.21, No.48, 485405, ISSN 0953-8984

He, J. L., Wu, E. D., Wang, H. T., Liu, R. P. & Tian, Y. J. (2005). Ionicities of boron-boron bonds in B-12 icosahedra. *Physical Review Letters*, Vol.94, No.1, 101906, ISSN 0031-9007

Hull, S. & Keen, D. A. (1994). High-pressure polymorphism of the copper(I) halides - a neutron-diffraction study to ~10 GPa. *Physical Review B*, Vol.50, No.9, pp. 5868-5885, ISSN 0163-1829

Hull, S. & Keen, D. A. (1999). Pressure-induced phase transitions in AgCl, AgBr, and AgI. *Physical Review B*, Vol.59, No.2, pp. 750-761, ISSN 0163-1829

Hultgren, R. (1932). Equivalent chemical bonds formed by s, p, and d eigenfunctions. *Physical Review*, Vol.40, No.6, pp. 0891-0907, ISSN 0031-899X

Kittle, C. (2004). *Introduction to Solid State Physics*, Wiley, ISBN 978-0471415268, New York.

Krebs, B., Schiemann, A. & Lage, M. (1993). Synthesis and crystal-structure of a novel hexagonal modification of Al_2S_3 with 5-coordinated aluninum. *Zeitschrift Fur Anorganische Und Allgemeine Chemie*, Vol.619, No.6, pp. 983-988, ISSN 0044-2313

Lee, C., Wei, X. D., Kysar, J. W. & Hone, J. (2008). Measurement of the elastic properties and intrinsic strength of monolayer graphene. *Science*, Vol.321, No.5887, pp. 385-388, ISSN 0036-8075

Levenberg, K. (1944). A method for the solution of certain non-linear problems in least squares. *Quarterly of Applied Mathematics* Vol.2, pp. 164-168, ISSN 0033-569X

Li, K. Y., Wang, X. T., Zhang, F. F. & Xue, D. F. (2008). Electronegativity identification of novel superhard materials. *Physical Review Letters*, Vol.100, No.23, 235504, ISSN 0031-9007

Liu, L. G. (1971). Dense modification of BaO and its crystal structure. *Journal of Applied Physics*, Vol.42, No.10, pp. 3702-3704, ISSN 0021-8979

Lübbers, D. & Leute, V. (1982). The crystal structure of β-Ga_2Se_3. *Journal of Solid State Chemistry*, Vol.43, No.3, pp. 339-345, ISSN 0022-4596

Marquardt, D. W. (1963). An algorithm for least-square estimation of nonlinear parameters. *Journal of the Society for Industrial and Applied Mathematics*, Vol.11, No.2, pp. 431-441, ISSN 0368-4245

Mielke, S. L., Troya, D., Zhang, S., Li, J. L., Xiao, S. P., Car, R., Ruoff, R. S., Schatz, G. C. & Belytschko, T. (2004). The role of vacancy defects and holes in the fracture of carbon nanotubes. *Chemical Physics Letters*, Vol.390, No.4-6, pp. 413-420, ISSN 0009-2614

Mujica, A., Rubio, A., Munoz, A. & Needs, R. J. (2003). High-pressure phases of group-IV, III-V, and II-VI compounds. *Reviews of Modern Physics*, Vol.75, No.3, pp. 863-912, ISSN 0034-6861

Mulliken, R. S. (1955). Electronic population analysis on LCAO-MO molecular wave functions. *Journal of Chemical Physics*, Vol.23, No.10, pp. 1833-1840, ISSN 0021-9606

Ozaki, T., Iwasa, Y. & Mitani, T. (2000). Stiffness of single-walled carbon nanotubes under large strain. *Physical Review Letters*, Vol.84, No.8, pp. 1712-1715, ISSN 0031-9007

Pauling, L. (1929). The principles determining the structure of complex ionic crystals. *Journal of the American Chemical Society*, Vol.51, No.4, pp. 1010-1026, ISSN 0002-7863

Prewitt, C. T. & Shannon, R. D. (1968). Crystal structure of a high-pressure form of B_2O_3. *Acta Crystallographica Section B-Structural Crystallography and Crystal Chemistry*, Vol.24, No.6, pp. 869-874, ISSN 0108-7681

Prewitt, C. T., Shannon, R. D., Rogers, D. B. & Sleight, A. W. (1969). C rare earth oxide-corundum transition and crystal chemisty of oxides having the corundum structure. *Inorganic Chemistry*, Vol.8, No.9, pp. 1985-1993, ISSN 0020-1669

Roundy, D., Krenn, C. R., Cohen, M. L. & Morris, J. W. (1999). Ideal shear strengths of fcc aluminum and copper. *Physical Review Letters*, Vol.82, No.13, pp. 2713-2716, ISSN 0031-9007

Sakuma, T. (1988). Crystal-structure of β-CuI. *Journal of the Physical Society of Japan*, Vol.57, No.2, pp. 565-569, ISSN 0031-9015

Segall, M. D., Shah, R., Pickard, C. J. & Payne, M. C. (1996). Population analysis of plane-wave electronic structure calculations of bulk materials. *Physical Review B*, Vol.54, No.23, pp. 16317-16320, ISSN 0163-1829

Shindo, K., Morita, A. & Kamimura, H. (1965). Spin-orbit coupling in ionic crystal with zincblende and wurtzite structures. *Journal of the Physical Society of Japan*, Vol.20, No.11, pp. 2054-2059, ISSN 0031-9015

Simunek, A. & Vackar, J. (2006). Hardness of covalent and ionic crystals: First-principle calculations. *Physical Review Letters*, Vol.96, No.8, 085501, ISSN 0031-9007

Song, L., Ci, L. J., Lu, H., Sorokin, P. B., Jin, C. H., Ni, J., Kvashnin, A. G., Kvashnin, D. G., Lou, J., Yakobson, B. I. & Ajayan, P. M. (2010). Large scale growth and characterization of atomic hexagonal boron nitride layers. *Nano Letters*, Vol.10, No.8, pp. 3209-3215, ISSN 1530-6984

Yu, M. F., Lourie, O., Dyer, M. J., Moloni, K., Kelly, T. F. & Ruoff, R. S. (2000). Strength and breaking mechanism of multiwalled carbon nanotubes under tensile load. *Science*, Vol.287, No.5453, pp. 637-640, ISSN 0036-8075

Part 3

Development of Quantum Theory on Graphene

12

The Photoeffect on Graphene and Axion Detection by Graphene

Miroslav Pardy
Department of Physical Electronics
Masaryk University, Kotlářská , Brno
Czech Republic

1. Introduction

The photoelectric effect is a quantum electronic phenomenon in which electrons are emitted from matter after the absorption of energy from electromagnetic radiation. Frequency of radiation must be above a threshold frequency, which is specific to the type of surface and material. No electrons are emitted for radiation with a frequency below that of the threshold. These emitted electrons are also known as photoelectrons in this context. The photoelectric effect was theoretically explained by Einstein in his paper (Einstein, 1905), and the term "light quanta" called "photons" was introduced by chemist G. N. Lewis, in 1926. Einstein writes: *In accordance with the assumption to be considered here, the energy of light ray spreading out from point source is not continuously distributed over an increasing space but consists of a finite number of energy quanta which are localized at points in space, which move without dividing, and which can only be produced and absorbed as complete units.*

There is a textbook proof that the free electron in vacuum cannot absorb photon. It follows from the special theory of relativity. If p_1, p_2 are the initial and final 4-momenta of electron with rest mass m and k is the 4-momentum of photon, then after absorption of photon by electron we write $k + p_1 = p_2$, which gives when squared $k^2 + 2kp_1 + p_1^2 = p_2^2$. Then, with $p_1^2 = p_2^2 = -m^2$ and $k^2 = 0$, we get for the rest electron with $\mathbf{p}_1 = 0$ the elementary relation $m\omega = 0$, Q.E.D..

The linear dependence on the frequency was experimentally determined in 1915 when Robert Andrews Millikan showed that Einstein formula

$$\hbar\omega = \frac{mv^2}{2} + A \tag{1}$$

was correct. Here $\hbar\omega$ is the energy of the impinging photon, v is electron velocity measured by the magnetic spectrometer and A is the work function of concrete material. The work function for Aluminium is 4.3 eV, for Beryllium 5.0 eV, for Lead 4.3 eV, for Iron 4.5 eV, and so on (Rohlf, 1994). The work function concerns the surface photoelectric effect where the photon is absorbed by an electron in a band. The theoretical determination of the work function is the problem of the solid state physics. On the other hand, there is the so called atomic photoeffect (Amusia, 1987), where the ionization energy plays the role of the work function.

The system of the ionization energies is involved in the tables of the solid state physics. The work fuction of graphene, or, work fuction of the Wigner crystal in graphene was never determined, and it is the one of the prestige problem of the contemporary experimental and theoretical graphene physics.
In case of the volume photoeffect, the ionization work function is defined in many textbooks on quantum mechanics. Or,

$$W = \int_{x_1}^{x_2} \left(\frac{dE}{dx}\right) dx, \tag{2}$$

where E is the energy loss of moving electron.
The formula (1) is the law of conservation of energy. The classical analogue of the equation (1) is the motion of the Robins ballistic pendulum in the resistive medium.
The Einstein ballistic principle is not valid inside of the blackbody. The Brownian motion of electrons in this cavity is caused by the repeating Compton process $\gamma + e \to \gamma + e$ and not by the ballistic collisions. The diffusion constant for electrons must be calculated from the Compton process and not from the ballistic process. The same is valid for electrons immersed into the cosmic relic photon sea.
The idea of the existence of the Compton effect is also involved in the Einstein article. He writes (Einstein, 1905): *The possibility should not be excluded, however, that electrons might receive their energy only in part from the light quantum.* However, Einstein was not sure, a priori, that his idea of such process is realistic. Only Compton proved the reality of the Einstein statement. The time lag between the incidence of radiation and the emission of a photoelectron is very small, less than 10^{-9} seconds.
At energies $\hbar\omega < W$, the photoeffect is not realized. However, the photo-conductivity is the real process for such energies. The photoeffect is realized only in medium and with low energy photons, but with energies $\hbar\omega > W$, which gives the Compton effect negligible. Compton effect can be realized with electrons in medium and also with electrons in vacuum. For $\hbar\omega \gg W$ the photoeffect is negligible in comparison with the Compton effect. At the same time it is necessary to say that the Feynman diagram of the Compton effect cannot be reduced to the Feynman diagram for photoeffect. In case of the high energy gamma rays, it is possible to consider the process called photoproduction of elementary particles on protons in LHC, or photo-nuclear reactions in nuclear physics. Such processes are energetically far from the photoelectric effect in solid state physics.
Eq. (1) represents so called one-photon photoelectric effect, which is valid for very weak electromagnetic waves. At present time of the petawatt laser physics, where the high electromagnetic intensity is possible, we know that so called multiphoton photoelectric effect is possible. Then, instead of equation (1) we can write

$$\hbar\omega_1 + \hbar\omega_2 + ...\hbar\omega_n = \frac{mv^2}{2} + A. \tag{3}$$

As na analogue of the equation (3), the multiphoton Compton effect is also possible: $\gamma_1 + \gamma_2 + ...\gamma_n + e \to \gamma + e$. To our knowledge the Compton process with the entangled photons was still not discovered and elaborated. On the other hand, there is the deep inelastic Compton effect in the high energy particle physics.
More than 70 years ago, Peierls (1934, 1935) and Landau (1937) performed a proof that the 2-dimensional crystal is not thermodynamically stable and cannot exist. They argued that the

thermodynamical fluctuations of such crystal lead to such displacements of atoms that these displacements are of the same size as the interatomic distances at the any finite temperature. The argument was extended by Mermin (1968) and it seemed that many experimental observations supported the Landau-Peierls-Mermin theory. So, the "impossibility" of the existence of graphene was established.
In 2004, Andre Geim, Kostia Novoselov (Novoselov et al. 2004; 2005; Kane, 2005) and co-workers at the University of Manchester in the UK by delicately cleaving a sample of graphite with sticky tape produced something that was long considered impossible: a sheet of crystalline carbon just one atom thick, known as graphene. Geim group was able to isolate graphene, and was able to visualize the new crystal using a simple optical microscope. Nevertheless, Landau-Peierls-Mermin proof is of the permanent historical and pedagogical meaning.
At present time, there are novel methods how to create graphene sheet. For instance, Dato et al. (2008) used the plasma reactor, where the graphene sheets were synthesized by passing liquid ethanol droplets into an argon plasma.
Graphene is the benzene ring (C_6H_6) stripped out from their H-atoms. It is allotrope of carbon because carbon can be in the crystalline form of graphite, diamond, fullerene (C_{60}), carbon nanotube and glassy carbon (also called vitreous carbon).
Graphene unique properties arise from the collective behaviour of electrons. The electrons in graphene are governed by the Dirac equation. The Dirac equation in graphene physics is used for the description of so called pseudoelectrons with pseudospins formed by the hexagonal lattice.
The Dirac fermions in graphene carry one unit of electric charge and so can be manipulated using electromagnetic fields. Strong interactions between the electrons and the honeycomb lattice of carbon atoms mean that the dispersion relation is linear and given by $E = vp$, v is called the Fermi-Dirac velocity, p is momentum of a pseudoelectron.
The linear dispersion relation follows from the relativistic energy relation for small mass together with approximation that the Fermi velocity is approximately only about 300 times less than the speed of light.
The pseudospin of the pseudoelectron is constructed in the graphene physics as follows. The graphene is composed of the system of hexagonal cells and it geometrically means that graphene is composed from the systems of two equalateral triangles. If the wave function of the first triangle sublattice system is φ_1 and the wave function of the second triangle sublattice system is φ_2, then the total wave function of the electron noving in the hexagonal system is superposition $\psi = c_1\varphi_1 + c_2\varphi_2$, where c_1 and are c_2 appropriate functions of coordinate $\mathbf{x}$ and functions φ_1, φ_2 are functions of wave vector $\mathbf{k}$ and coordinate $\mathbf{x}$. The next crucial step in the graphene physics is the definition of the new spinor function discussed by Lozovik et al.(2008).

$$\chi = \left\{ \begin{matrix} \varphi_1 \\ \varphi_2 \end{matrix} \right\} \tag{4}$$

and it is possible to prove that this spinor function is solution of the Pauli equation in the nonrelativistic situation and Dirac equation of the generalized case. The corresponding mass of such effective electron is proved to be zero.
The introduction of the Dirac relativistic Hamiltonian in gaphene physics is the description of the graphene physics by means of electron-hole medium. It is the analogue of the description

of the electron-positron vacuum by the Dirac theory of quantum electrodynamics. The pseudoelectron is not an electron of QED and pseudospin is not the spin of QED, because QED is the quantum theory of the interaction of real electrons and photons where the mass of electron is the mass defined by classical mechanics and not by collective behavior in hexagonal sheet called graphene. The interaction of photons with pseudoelectrons was still not published by Geim and Novoselov and others, however, there is an analogue with the photodesintegration of deuteron being the bound state of neutron and proton. Or,

$$\gamma + D \rightarrow n + p, \tag{5}$$

where dissociation energy D of the process is 2.225 MeV. We use this analogy in the photoelectric effect in graphene and we must accept the idea that the initial process in the photoelectric effect in graphene is the photodesintegration of pseudoelectron and then the interaction of the photon with the real electron. So, we generalize the Einstein equation to the novel form:

$$\hbar\omega = \frac{mv^2}{2} + A_d + A, \tag{6}$$

where the desintegration energy necessary for the desintegration of pseudoelectron must be determined experimentally. The today graphenic theory does not solve this problem and the corresponding experiment was not performed. We suppose that this missing experiment is crucial in the graphene physics.

After dissociation of pseudoelectron, we can use the Schrödinger equation in order to establish the photoeffect in graphene in the magnetic field. Our calculation of the photoelectric effect is applicable also for the Wigner crystal, being the crystalline phase of electrons first predicted by Eugene Wigner (1934).

We derive here the differential probability of the photoelectric effect realized at the very low temperature of graphene in the very strong magnetic field. So, the term magnetic photoeffect. We deal also with graphene on the Riemann surfaces and consider the possibility of axion detection by the graphene detectors.

2. The quantum theory of the photoelectric effect in the 2D electron gas at zero temperature and strong magnetic field

The photoelectric effect in graphene is presented here 100 years after the publication of well known Einstein article (Einstein, 2005). While the experimental investigation of the photoelectric effect was performed in past many times, the photoelectric effect in graphene, is still the missing experiment in the graphene physics. Many discoveries in physics were performed when the physical, or, chemical object was immersed into the magnetic field.

The photoelectric effect in semiconductor by onset of magnetic field is discussed by Kleinert et al. (2008), where the Bloch functions in a magnetic field are considered as the adequate for the solution of the problem. The calculation of photoeffect of graphene including both electron-electron and electron-photon interactions on the same footing is performed for instance by Park et al. (2010). The crucial object is the spectral function with the self-energy which accounts for the many-body interactions going beyond the mean-field picture.

Ratnikov et al. (2008a) define graphene as a two-dimensional zero gap semiconductor with zero conduction and valence band overlap to determine ground state of energy of current carriers. Then he et al. calculate the transition of graphene on a substrate to a semi-metallic state (Ratnikov et al. 2008b). The Ratnikov at al. theory can be applied to graphene in magnetic field and bilayer graphene in magnetic field. However, it is not immediately possible to apply the Ratnikov et al. theory to calculate the photoeffect in magnetic field. So, we use the Landau approach applied in Landau diamagnetism.

The rigorous theory of the photoeffect in the metallic thin films was given by Tamm and Schubin (1931). However, graphene is the one-atom sheet and cannot be identified with the thin film situation of Tamm and Schubin. In case of the thin metallic film, the light penetrates to the distance 1000 - 10 000 atomic layers under the surface of the film, while graphene is the thin film with only one atomic layer. So, the original idea is to consider the photoeffect on the 2D electron system immersed into the constant magnetic field and graphene enables to realize this project.

The quantum mechanical description of the photoeffect is based on the appropriate S-matrix element involving the interaction of atom with the impinging photon with the simultaneous generation of the electron, the motion of which can be described approximately by the plane wave

$$\psi_{\mathbf{q}} = \frac{1}{\sqrt{V}} e^{i\mathbf{q}\cdot\mathbf{x}}, \quad \mathbf{q} = \frac{\mathbf{p}}{\hbar}, \tag{7}$$

where $\mathbf{p}$ is the momentum of the ejected electron.

The standard approach consists in the definition of the cross-section by the quantum mechanical equation

$$d\sigma = \frac{2\pi}{\hbar} |V_{fi}| \delta(-I + \hbar\omega - E_f) \frac{d^3p}{(2\pi)^3}, \tag{8}$$

where I is the ionization energy of an atom and E_f is the the final energy of the emitted electron, $|V_{fi}|$ is the matrix element of the transition of electron from the initial bound state to the final state. The matrix element follows from the perturbative theory and it involves the first order term of the interaction between electron and photon. We use here the Davydov elementary approach (Davydov, 1976).

We suppose here that magnetic field is applied locally to the carbon sheet, so, in a sufficient distance from it the wave function is of the form of the plane wave (7). This situation has an analog in the classical atomic effect discussed by Davydov (1976).

The probability of the emission of electron by the electromagnetic wave is of the well-known form (Davydov, 1976):

$$dP = \frac{e^2 p}{8\pi^2 \varepsilon_0 \hbar m \omega} \left| \int e^{i(\mathbf{k}-\mathbf{q})\cdot\mathbf{x}} (\mathbf{e}\cdot\nabla)\psi_0 dxdydz \right|^2 d\Omega = C|J|^2 d\Omega, \tag{9}$$

where the interaction for absorption of the electromagnetic wave is normalized to *one photon in the unit volume*, $\mathbf{e}$ is the polarization of the impinging photon, ε_0 is the dielectric constant of vacuum, ψ_0 is the basic state of and atom. We have denoted the integral in $||$ by J and the constant before $||$ by C.

We consider the case with electrons in magnetic field as an analog of the Landau diamagnetism. So, we take the basic function ψ_0 for one electron in the lowest Landau level, as

$$\psi_0 = \left(\frac{m\omega_c}{2\pi\hbar}\right)^{1/2} \exp\left(-\frac{m\omega_c}{4\hbar}(x^2+y^2)\right), \quad \omega_c = \frac{|e|H}{mc}, \tag{10}$$

which is solution of the Schrödinger equation in the magnetic field with potentials $\mathbf{A} = (-Hy/2, -Hx/2, 0)$, $A_0 = 0$ (Drukarev, 1988):

$$\left[\frac{p_x^2}{2m} + \frac{p_y^2}{2m} - \frac{m}{2}\left(\frac{\omega_c}{2}\right)^2 (x^2+y^2)\right]\psi = E\psi. \tag{11}$$

We have supposed that the motion in the z-direction is zero and it means that the wave function $\exp[(i/\hbar)p_z z] = 1$.
So, The main problem is to calculate the integral

$$J = \int e^{i(\mathbf{K}\cdot\mathbf{x})}(\mathbf{e}\cdot\nabla)\psi_0 dxdydz; \quad \mathbf{K} = \mathbf{k} - \mathbf{q}. \tag{12}$$

with the basic Landau function ψ_0 given by the equation (10).
Operator $(\hbar/i)\nabla$ is Hermitean and it means we can rewrite the last integrals as follows:

$$J = \frac{i}{\hbar}\mathbf{e}\cdot\int\left[\left(\frac{\hbar}{i}\nabla\right)e^{i(\mathbf{K}\cdot\mathbf{x})}\right]^* \psi_0 dxdydz, \tag{13}$$

which gives

$$J = i\mathbf{e}\cdot\mathbf{K}\int e^{-i(\mathbf{K}\cdot\mathbf{x})}\psi_0 dxdydz, \tag{14}$$

The integral in eq. (14) can be transformed using the cylindrical coordinates with

$$dxdydz = \varrho d\varrho d\varphi dz, \quad \varrho^2 = x^2 + y^2 \tag{15}$$

which gives for vector $\mathbf{K}$ fixed on the axis z with $\mathbf{K}\cdot\mathbf{x} = Kz$ and with physical condition $\mathbf{e}\cdot\mathbf{k} = 0$, expressing the physical situation where polarization is perpendicular to the direction of the wave propagation. So,

$$J = (i)(\mathbf{e}\cdot\mathbf{q})\int_0^\infty \varrho d\varrho \int_{-\infty}^{\infty} dz \int_0^{2\pi} d\varphi e^{-iKz}\psi_0. \tag{16}$$

Using

$$\psi_0 = A\exp\left(-B\varrho^2\right); \quad A = \left(\frac{m\omega_c}{2\pi\hbar}\right)^{1/2}; \quad B = \frac{m\omega_c}{4\hbar}. \tag{17}$$

The integral (16) is then

$$J = (-\pi i)\frac{A}{B}(\mathbf{e}\cdot\mathbf{q})\int_{-\infty}^{\infty} e^{-iKz}dz = (-\pi i)\frac{A}{B}(\mathbf{e}\cdot\mathbf{q})(2\pi)\delta(K). \tag{18}$$

Then,

$$dP = C|J|^2 d\Omega = 4\pi^4\frac{A^2}{B^2}C(\mathbf{e}\cdot\mathbf{q})^2\delta^2(K)d\Omega. \tag{19}$$

Now, let be the angle Θ between direction $\mathbf{k}$ and direction $\mathbf{q}$, and let be the angle Φ between planes $(\mathbf{k}, \mathbf{q})$ and $(\mathbf{e}, \mathbf{k})$. Then,

$$(\mathbf{e}\cdot\mathbf{q})^2 = q^2\sin^2\Theta\cos^2\Phi. \quad (20)$$

So, the differential probability of the emission of photons from the graphene in the strong magnetic field is as follows:

$$dP = \frac{4e^2 p}{\pi\varepsilon_0 m^2\omega\omega_c}\left[q^2\cos^2\Theta\sin^2\Phi\right]\delta^2(K)d\Omega; \quad \omega_c = \frac{|e|H}{mc}. \quad (21)$$

We can see that our result differs form the result for the original photoelectric effect which involves the term

$$\frac{1}{(1-\frac{v}{c}\cos\Theta)^4}, \quad (22)$$

which means that the most intensity of the classical photoeffect is in the direction of the electric vector of the electromagnetic wave ($\Phi = \pi/2, \Theta = 0$). While the non-relativistic solution of the photoeffect in case of the Coulomb potential was performed by Stobbe (1930) and the relativistic calculation by Sauter (1931), the general magnetic photoeffect (with electrons moving in the magnetic field and forming atom) was not still performed in a such simple form. The delta term $\delta\cdot\delta$ represents the conservation law $|\mathbf{k}-\mathbf{q}| = 0$ in our approximation.

3. Photoeffect with Volkov solution

It is valuable from the pedagogical point of view (Berestetzkii et al., 1989) to remember the Volkov solution, where the motion of the Dirac electron is considered in the following four potential

$$A_\mu = a_\mu\varphi; \quad \varphi = kx; \quad k^2 = 0. \quad (23)$$

From equation (23), it follows that $F_{\mu\nu} = \partial_\mu A_\nu - \partial_\nu A_\mu = a_\nu k_\mu - a_\mu k_\nu = const.$, which means that electron moves in the constant electromagnetic field with the components **E** and **H**. The parameters a and k can be chosen in a such a way that $\mathbf{E} = 0$. So, the motion of electron is performed in the constant magnetic field.
The Volkov (1935) solution of the Dirac equation for an electron moving in a field of a plane wave was derived in the form (Berestetzkii et al., 1989; Pardy, 2003; Pardy, 2004; Pardy, 2007):

$$\psi_p = \frac{u(p)}{\sqrt{2p_0}}\left[1 + e\frac{(\gamma k)(\gamma A(\varphi))}{2kp}\right]\exp\left[(i/\hbar)S\right] \quad (24)$$

and S is an classical action of an electron moving in the potential $A(\varphi)$:

$$S = -px - \int_0^{kx}\frac{e}{(kp)}\left[(pA) - \frac{e}{2}(A)^2\right]d\varphi. \quad (25)$$

It was shown that for the potential (23) the Volkov wave function is (Berestetzkii et al., 1989):

$$\psi_p = \frac{u(p)}{\sqrt{2p_0}}\left[1 + e\frac{(\gamma k)(\gamma a)}{2kp}\varphi\right]\exp\left[(i/\hbar)S\right] \quad (26)$$

with

$$S = -e\frac{ap}{2kp}\varphi^2 + e^2\frac{a^2}{6kp}\varphi^3 - px. \tag{27}$$

We used $c = \hbar = 1$.
However, the relativistic wave function can be obtained by solving the Dirac equation in magnetic field. It was derived in the form (Sokolov et all. 1983).

$$\Psi(\mathbf{x}, t) = \frac{1}{L}\exp\{-\frac{i}{\hbar}\epsilon Et + ik_2 y + ik_3 z\}\psi; \quad \psi = \begin{pmatrix} C_1 u_{n-1}(\eta) \\ iC_2 u_n(\eta) \\ C_3 u_{n-1}(\eta) \\ iC_4 u_n(\eta) \end{pmatrix}, \tag{28}$$

where $\epsilon = \pm 1$ and the spinor components are given by the following formulas:

$$u_n(\eta) = \sqrt{\frac{\sqrt{2\gamma}}{2^n n!}\sqrt{\pi}\, e^{-\eta^2/2} H_n(\eta)} \tag{29}$$

with

$$H_n(\eta) = (-1)^n e^{\eta^2}\left(\frac{d}{d\eta}\right)^n e^{-\eta^2}, \tag{30}$$

$$\eta = \sqrt{2\gamma}\, x + k_2/\sqrt{2\gamma}; \quad \gamma = eH/2c\hbar. \tag{31}$$

The coefficients C_i are defined in the Sokolov et al. monograph (Sokolov et al., 1983). So, our approach can be generalized.
It is evident that the model which we have used is the one-pseudoelectron model of the 2D hexagon crystal. However the photoelectric effect should be considered in the many pseudoelectron model which is substantially more complicated than the one-pseudoelectron electron model. The investigation of such approach is not at present time published.

4. Green function in magnetic field for photoeffect

The electron propagator is a building stone in the mass operator from which can be computed the energy shift, the power spectrum of the synchrotron radiation, the anomalous magnetic moment and so on. It was intensively discussed for instance by Dittrich et al. and Schwinger et al.. Here we will follow the Dittrich treatment in order to derive the special representation of the Dirac propagator of a particle in a constant external magnetic field (Dittrich et al., 1985). The application of the formalism to graphene physics is evident.
If we write the Green function as

$$G_+(x, x') = \langle x|G_+|x'\rangle, \tag{32}$$

then, from the Green function equation for spin 1/2 particles

$$(\gamma\Pi + m)G_+(x, x') = \delta(x - x') = \langle x|x'\rangle \tag{33}$$

we have

$$G_+ = \frac{1}{\gamma\Pi + m - i\varepsilon}; \quad \varepsilon > 0, \tag{34}$$

where

$$\Pi_\mu = p_\mu - eA_\mu; \quad p_\mu = \frac{1}{i}\partial_\mu \tag{35}$$

with simultaneous omitting the charge matrix q in eqs. (34) and (35).
The equivalent form of G_+ in eq. (34) is obviously given by relation

$$G_+ = \frac{\gamma\Pi - m}{(\gamma\Pi)^2 - m^2 - i\varepsilon}, \tag{36}$$

where

$$(\gamma\Pi)^2 = -\Pi^2 - \frac{i}{2}\sigma_{\mu\nu}[\Pi_\mu, \Pi_\nu] \tag{37}$$

with

$$\sigma^{\mu\nu} = \frac{i}{2}[\gamma^\mu, \gamma^\nu], \quad \{\gamma^\mu, \gamma^\nu\} = -2g^{\mu\nu} \tag{38}$$

and

$$[\Pi^\mu, \Pi^\nu] = ieF^{\mu\nu}, \tag{39}$$

which gives

$$(\gamma\Pi)^2 = -\Pi^2 + \frac{e}{2}\sigma_{\mu\nu}F^{\mu\nu}. \tag{40}$$

For the constant magnetic field chosen in the z-direction we have $F_{12} = -F_{21} = B = const$ and

$$(\gamma\Pi)^2 = -\Pi^2 - eB\sigma^{12}, \tag{41}$$

where we designated by σ^{12} the following matrix:

$$\sigma^{12} = \begin{pmatrix} \sigma^3 & 0 \\ 0 & \sigma^3 \end{pmatrix} = \begin{pmatrix} 1 & 0 & 0 & 0 \\ 0 & -1 & 0 & 0 \\ 0 & 0 & 1 & 0 \\ 0 & 0 & 0 & -1 \end{pmatrix} \stackrel{d}{=} \sigma^3, \tag{42}$$

where σ^3 is 4×4 matrix. Using the last formulas, we write

$$G_+ = \frac{m - \gamma\Pi}{\Pi^2 - \kappa^2 - i\varepsilon}, \tag{43}$$

where

$$\kappa^2 \stackrel{d}{=} m^2 - eB\sigma^3. \tag{44}$$

Space-time representation of the Green function with potential A is then

$$G_+(x', x''|A) = \langle x'| \frac{m - \gamma\Pi}{\Pi^2 - \kappa^2 - i\varepsilon} |x''\rangle. \tag{45}$$

Now, we can evaluate the formula (45). We use the ansatz (Dittrich et al., 1985)

$$G_{+}(x',x''|A) = \Phi(x',x'')\left[m - \gamma^{\nu}\left(\frac{1}{i}\partial'_{\nu} - eA'_{\nu}(x')\right)\right]\Delta_{+}(x',x''|A') \qquad (46)$$

with

$$A'_{\mu}(x')] = -\frac{1}{2}F_{\mu\nu}(x'-x'')^{\nu} \qquad (47)$$

and

$$\Phi(x',x'') = \exp\left\{ie\int_{x''}^{x'} dx_{\mu}[A^{\mu}(x) + \frac{1}{2}F^{\mu\nu}(x'_{\nu} - x''_{\nu})]\right\}. \qquad (48)$$

where the integral is not dependent of the choice of the integration path because of vanishing rot of the integrand. If we choose the integration path in the form of the straight line

$$x(t) = x'' + t(x'-x''); \quad t \in <0,1>, \qquad (49)$$

we find that the second term of the integrand gives no contribution and we have instead of eq. (48):

$$\Phi(x',x'') = \exp\left\{ie\int_{x''}^{x'} dx_{\mu}A^{\mu}(x)\right\} \qquad (50)$$

and the derivatives are as follows:

$$\frac{\partial\Phi(x',x'')}{\partial x'^{\mu}} = ie[A_{\mu}(x') - A'_{\mu}(x')]\Phi(x',x''), \qquad (51)$$

where after substitution of the ansatz (46) into eq. (33) we get

$$\left((\frac{1}{i}\partial' - eA')^2 + \kappa^2\right)\Delta_{+}(x',x''|A) = \delta(x'-x''), \qquad (52)$$

where A' is defined by eq. (47). After modification of eq. (52) we have

$$(-\partial^2 + \kappa^2 - i\varepsilon - \frac{e^2}{4}x_{\mu}F^{2\mu\nu}x_{\nu})\Delta_{+}(x|A') = \delta(x) \qquad (53)$$

with

$$F^{2\mu\nu} = F^{\mu}_{\alpha}F^{\alpha\mu}. \qquad (54)$$

We can solve the equation (53) by using the Fourier transform (with the notation $(dk) = d^4k$)

$$\Delta_{+}(x|A') = \int\frac{(dk)}{(2\pi)^4}\Delta_{+}(k|A'), \qquad (55)$$

which gives

$$\left(k^2 + \frac{e^2}{4}\frac{\partial}{\partial k^{\mu}}F^{2\mu\nu}\frac{\partial}{\partial k^{\nu}} + \kappa^2 - i\varepsilon\right)\Delta_{+}(k|A') = 1. \qquad (56)$$

Using the ansatz

$$\Delta_+(x|A') = i\int_0^\infty dse^{-M(is)}e^{-is(\kappa^2 - is)} \tag{57}$$

with

$$M(is) = k^\alpha X_{\alpha\beta}(is)k^\beta + Y(is); \quad X_{\alpha\beta} = X_{\beta\alpha}, \tag{58}$$

we get from eq. (56)

$$i\int_0^\infty ds\left\{k\left[1 + e^2X(is)F^2X(is)\right]k - \frac{e^2}{2}\text{tr}\,[F^2X] + \kappa^2 - i\varepsilon\right\}e^{-M(is)}e^{-is(\kappa^2 - is)} = 1, \tag{59}$$

or, in the equivalent form

$$i\int_0^\infty dsg(is)e^{-f(is)} = 1 \tag{60}$$

which is equation for X and Y.
Let us try to put

$$g(is) = f'(is). \tag{61}$$

Then,

$$i\int_0^\infty dsg(is)e^{-f(is)} = e^{-f(0)} - e^{-f(i\infty)}. \tag{62}$$

From the comparison of eq. (62) with eq. (60) the requirement follows

$$f(0) = 0; \quad \text{Re}\, f(i\infty) = \infty. \tag{63}$$

The relation $g = f'$ reads in our case

$$k\left[1 + e^2X(is)F^2X(is)\right]k - \frac{e^2}{2}\text{tr}\,[F^2X(is)] = M'(is). \tag{64}$$

It enables to write after rotation of the integration path according to $is \to s$, the following equations:

$$1 + e^2X(s)F^2(s) = \dot{X}(s), \tag{65}$$

$$-\frac{e^2}{2}\text{tr}\,[F^2X(s)] = \dot{Y}(s). \tag{66}$$

The solution of eqs. (65) and (66) are

$$X(s) = (eF)^{-1}\tan(eFs), \tag{67}$$

$$Y(s) = \frac{1}{2}\text{tr}\,\ln\cos(eFs). \tag{68}$$

which can be verified by differentiation. At the same time it can be verified the first condition in (63) is fulfilled because of $X(0) = Y(0) = 0$.

To write $X(s)$ and $Y(s)$ in the explicit form, we use the advantage of the special form of the strength tensor

$$F^{\mu\nu} = \begin{pmatrix} 0 & 0 & 0 & 0 \\ 0 & 0 & B & 0 \\ 0 & -B & 0 & 0 \\ 0 & 0 & 0 & 0 \end{pmatrix} \tag{69a}$$

with

$$(iF)^2 = B^2 \begin{pmatrix} 0 & 0 & 0 & 0 \\ 0 & 1 & 0 & 0 \\ 0 & 0 & 1 & 0 \\ 0 & 0 & 0 & 0 \end{pmatrix} \tag{69b}$$

Then, we have:

$$X(is) = (eF)^{-1} \tan(ieFs) \quad =$$

$$is \left[\begin{pmatrix} 1 & 0 & 0 & 0 \\ 0 & 0 & 0 & 0 \\ 0 & 0 & 0 & 0 \\ 0 & 0 & 0 & 1 \end{pmatrix} + \begin{pmatrix} 0 & 0 & 0 & 0 \\ 0 & 1 & 0 & 0 \\ 0 & 0 & 1 & 0 \\ 0 & 0 & 0 & 0 \end{pmatrix} \frac{\tan(eBs)}{(eBs)} \right]. \tag{70}$$

Now, let us introduce the notation for the specific vectors as follows

$$a_{\|} \stackrel{d}{=} (a^0, 0, 0, a^3); \quad a_{\perp} \stackrel{d}{=} (0, a^1, a^2, 0). \tag{71}$$

$$(ab)_{\|} \stackrel{d}{=} -a^0 b^0 + a^3 b^3; \quad (ab)_{\perp} \stackrel{d}{=} a^1 b^1 + a^2 b^2. \tag{72}$$

Then, $(X = X_{\alpha}^{\beta})$

$$kX(is)k \; = is \left[k_{\|}^2 + \frac{\tan(eBs)}{eBs} k_{\perp}^2 \right]. \tag{73}$$

For $Y(is)$ we get by the similar way

$$Y(is) = \frac{1}{2} \text{tr } \ln \cos(ieFs) \quad =$$

$$\frac{1}{2} \ln \det \left[\begin{pmatrix} 1 & 0 & 0 & 0 \\ 0 & 0 & 0 & 0 \\ 0 & 0 & 0 & 0 \\ 0 & 0 & 0 & 1 \end{pmatrix} + \begin{pmatrix} 0 & 0 & 0 & 0 \\ 0 & 1 & 0 & 0 \\ 0 & 0 & 1 & 0 \\ 0 & 0 & 0 & 0 \end{pmatrix} \cos(eBs) \right] \quad =$$

$$\ln \cos(eBs). \tag{74}$$

At this stage we are prepared to compute $\exp(-f(i\infty))$. Let us recall that

$$f(is) = M(is) + is(\kappa^2 - is) = kX(is)k + Y(is) + is(\kappa^2 - is). \tag{75}$$

Then,

$$e^{-f(i\infty)} = \lim_{s\to\infty} e^{-f(is)} =$$

$$\lim_{s\to\infty} \exp\left\{-is\left(k_\parallel^2 + k_\perp^2\,\frac{\tan(eBs)}{eBs}\right)\right\}\frac{e^{-is(\kappa^2-is)}}{\cos(eBs)} = 0. \tag{76}$$

If we introduce

$$z = eBs, \tag{77}$$

we get from eqs. (55), (57), and (75)

$$\Delta(x',x''|A) = i\int_0^\infty ds\Phi(x',x'')\int \frac{(dk)}{(2\pi)^4}e^{ik(x'-x'')} \quad\times$$

$$\frac{1}{\cos z}\exp\left\{-is\left(k_\parallel^2 + k_\perp^2\,\frac{\tan(z)}{z}\right)\right\}e^{-is(\kappa^2-is)}. \tag{78}$$

However, because of

$$\Delta_+(x',x''|A) = \langle x'|\frac{1}{\Pi^2+\kappa^2-i\varepsilon}|x''\rangle = i\int_0^\infty ds\langle x'|e^{-is\Pi^2}|x''\rangle e^{-is(\kappa^2-is)} \tag{79}$$

we get after comparison of eq. (79) with eq. (78)

$$\langle x'|e^{-is\Pi^2}|x''\rangle = \Phi(x',x'')\int \frac{(dk)}{(2\pi)^4}e^{ik(x'-x'')}\frac{1}{\cos z}e^{-is(k_\parallel^2+k_\perp^2\,\frac{\tan(z)}{z})}. \tag{80}$$

The formula (80) may be used as a starting point in derivation of the further formulas. For instance, putting $s \to a_\perp s$ in eq. (80) we get

$$\langle x'|e^{-isa_\perp\Pi^2}|x''\rangle =$$

$$\Phi(x',x'')\int \frac{(dk)}{(2\pi)^4}e^{ik(x'-x'')}\frac{1}{\cos(a_\perp z}\exp\left\{-is\left(k_\parallel^2 + k_\perp^2\,\frac{\tan(a_\perp z)}{a_\perp z}\right)a_\perp\right\}. \tag{81}$$

Further, because of

$$\langle x'|e^{-isa_\perp\Pi^2}\Pi_\mu|x''\rangle = (i\partial'' - eA(x''))_\mu\langle x'|e^{-isa_\perp\Pi^2}|x''\rangle \tag{82}$$

we get

$$\langle x'|\exp\left\{-is\left(a^0\Pi_0\Pi^0 + a^3\Pi_3\Pi^3 + a_\perp\Pi_\perp^2\right)\right\}|x''\rangle =$$

$$\langle x'|e^{-isa_\perp\Pi^2}\exp\left\{-is\left((a^0-a_\perp)\Pi_0\Pi^0 + (a^3-a_\perp)\Pi_3\Pi^3\right)\right\}|x''\rangle =$$

$$\exp\left\{-is\left(-(a^0-a_\perp)\left(i\partial''^0 - eA^0(x'')\right)^2 + (a^3-a_\perp)\left(i\partial''^3 - eA^3(x'')\right)^2\right)\right\}$$

$$\times\quad \langle x'|e^{-isa_\perp\Pi^2}|x''\rangle. \tag{83}$$

Using

$$\partial''^{\mu}\Phi(x',x'') = -ie\left(A^{\mu}(x'') - \frac{1}{2}F^{\mu\nu}(x'-x'')_{\nu}\right)\Phi(x',x'') \tag{84}$$

and

$$\begin{aligned}\exp\left\{-is\left(-(a^0-a_\perp)\left(i\partial''^0\right)^2+(a^3-a_\perp)\left(i\partial''^3\right)^2\right)\right\}e^{ik(x'-x'')} &= \\ \exp\left\{-is\left(-(a^0-a_\perp)\left(k^0\right)^2+(a^3-a_\perp)\left(k^3\right)^2\right)\right\}e^{ik(x'-x'')}, &\end{aligned} \tag{85}$$

we get from eqs. (82) and (83)

$$\begin{aligned}\langle x'|\exp\left\{-is\left(a^0\Pi_0\Pi^0+a^3\Pi_3\Pi^3+a_\perp\Pi_\perp^2\right)\right\}\Pi^{\mu}|x''\rangle &= \\ \Phi(x',x'')\int\frac{(dk)}{(2\pi)^4}e^{ik(x'-x'')} &\times \\ \frac{1}{\cos(a_\perp z)}\exp\left\{-is\left(a^0k_0k^0+a^3k_3k^3+a_\perp k_\perp^2\frac{\tan(a_\perp)z}{a_\perp z}\right)\right\}. &\end{aligned} \tag{86}$$

Similarly, we can derive

$$\begin{aligned}\langle x'|\exp\left\{-is\left(a^0\Pi_0\Pi^0+a^3\Pi_3\Pi^3+a_\perp\Pi_\perp^2\right)\right\}|x''\rangle &= \\ \Phi(x',x'')\int\frac{(dk)}{(2\pi)^4}e^{ik(x'-x'')}\frac{1}{\cos(a_\perp z)}\left[k^{\mu}-ea_\perp sF^{\mu\nu}k_\nu\frac{\tan(a_\perp z)}{a_\perp z}\right] &\times \\ \exp\left\{-is\left(a^0k_0k^0+a^3k_3k^3+a_\perp k_\perp^2\frac{\tan(a_\perp)z}{a_\perp z}\right)\right\}. &\end{aligned} \tag{87}$$

From this equation and using

$$\gamma_\mu F^{\mu\nu}k_\nu\tan(a_\perp z) = (\gamma_1k_2-\gamma_2k_1)B\tan(a_\perp z) = \frac{i}{\cos(a_\perp z)}B\sin(a_\perp z\sigma^3)(\gamma k)_\perp \tag{88}$$

and

$$(k\gamma)_\perp - esa_\perp(\gamma_\mu F^{\mu\nu}k_\nu)\frac{\tan(a_\perp z)}{a_\perp z} = \frac{1}{\cos(a_\perp z)}e^{-ia_\perp z\sigma^3}(\gamma k)_\perp \tag{89}$$

we get the following relations

$$\begin{aligned}\langle x'|\exp\left\{-is\left(a^0\Pi_0\Pi^0+a^3\Pi_3\Pi^3+a_\perp\Pi_\perp^2\right)\right\}\left(1,\gamma^0\Pi^0,\gamma^3\Pi^3,(\gamma\Pi)_\perp\right)|x''\rangle &= \\ \Phi(x',x'')\int\frac{(dk)}{(2\pi)^4}e^{ik(x'-x'')}\frac{1}{\cos(eBsa_\perp z)} &\times \\ \exp\left\{-is\left(a^0k_0k^0+a^3k_3k^3+a_\perp k_\perp^2\frac{\tan(eBsa_\perp)}{eBsa_\perp}k_\perp^2\right)\right\} &\times\end{aligned}$$

$$\left(1, \gamma^0 k^0, \gamma_3 k^3, \frac{1}{\cos(eBsa_\perp)} e^{-ieBsa_\perp \sigma^3} (\gamma k)_\perp\right). \tag{90}$$

For the propagator $G_+(x', x'')$ which we write in the form

$$G_+(x', x') = \langle x' | \frac{m - \gamma\Pi}{\Pi^2 - \kappa^2 - i\varepsilon} | x'' \rangle \quad =$$

$$i \int_0^\infty ds e^{-is(\kappa^2 - is)} \langle x' | e^{-is\Pi^2} (m - \gamma\Pi) | x'' \rangle, \tag{91}$$

we get

$$G_+(x', x') = \Phi(x', x'') \int \frac{(dk)}{(2\pi)^4} e^{ik(x' - x'')} G_+(k), \tag{92}$$

where

$$G(k) = i \int_0^\infty ds \exp\left\{-is\left(m^2 - i\varepsilon + k_\parallel^2 + \frac{\tan z}{z} k_\perp^2\right)\right\} \quad \times$$

$$\frac{e^{i\sigma^3 z}}{\cos z}\left(m - \gamma k_\parallel - \frac{e^{-i\sigma^3 z}}{\cos z} \gamma k_\perp\right). \tag{93}$$

This representation of the Green function of electron in the constant magnetic field will be used in calculation of the mass operator and the polarization tensor in the constant magnetic field.

The mass operator tin the x-representation is usually denoted by the symbol $M(x', x'')$ and in the past literature as $\Sigma(x', x'')$. It is defined by the relation (Schvinger, 1988):

$$M(x', x'') = ie^2 \gamma^\mu G_+(x', x'') D_+(x' - x'') \gamma_\mu + C.T., \tag{94}$$

where $D_+(x' - x'')$ is the photon propagator

$$D_+(x) = \int \frac{(dk)}{(2\pi)^4} \frac{e^{ikx}}{k^2 - i\varepsilon} \tag{95}$$

and $G_+(x', x'')$ is the electron propagator

$$G_+(x', x'') = \Phi(x', x'') \int \frac{(dp)}{(2\pi)^4} e^{ip(x' - x'')} G(p), \tag{96}$$

where $\Phi(x', x'')$ and $G(p)$ are function which were specified.

The contact terms C.T. in eq. (94) can be determined by the physical normalization condition such that for $\gamma\Pi \to -m$, both M and its first derivative with respect to $\gamma\Pi$ are zero. The motivation for such definition of the contact terms can be found in the monography of Dittrich and Reuter (1985).

The considered formalism can be reduced to the two-dimensional situation which is the situation of the graphene physics. The two-dimensional Green function applied to the dynamical mass generation was elaborated by Shpagin (1996) and can be also applied to the graphene physics.

5. Axion detection by graphene

Axions were proposed as an extension to the Standard Model of particle physics to explain why CP violation, it is charge-parity symmetry violation, is observed in weak but not strong interactions - the so-called strong-CP problem.
The CP violation is measured in terms of a parameter denoted usually by symbol θ in QCD and also in Glashow-Weinberg-Salam electroweak model (GWS) and experimentally is observed that

$$|\theta| = |\theta_{QCD} + \theta_{GWS}| < 10)^{-9} \approx 0, \tag{97}$$

which means that the strong interactions are invariant with regard to the CP transformation. The question is why θ is so small?
One of the most striking consequences of CP-violation is the calculated neutron electric dipole moment (EDM) to be ten orders of magnitude larger than its measured upper limit.
The total EDM is expected to receive contributions from both the TeV electroweak scale, via the quark spin, and from the GeV QCD scale, via the spatial distribution of the quark wave function within the neutron. It is difficult to understand how this two contributions could cancel to such precision to produce a CP conserving QCD ground state without fine tuning of parameters. Or, in other words, why A + B = 0, if A and B have different physical origin.
The problem was solved by Peccei, Quinn, Weinberg and Wilczek by introducing a new scalar field which rolls within its potential into a state of minimum action, a CP conserving QCD vacuum state. Any imbalance between the contributions to the EDM of neutron from TeV and GeV scales is absorbed by an axion. Axion is a postulated particle which is generated by the Higgs mechanism.
The Higgs mechanism as the mechanism of the spontaneous broken symmetry can be demonstrated easily by the Lagrangian for the scalar field φ (Kane, 1987):

$$L = T - V = \frac{1}{2}\partial_\nu \varphi \partial^\nu \varphi - \left(\frac{1}{2}\mu^2\varphi^2 + \frac{1}{4}\lambda\varphi^4\right) ..., \tag{98}$$

where μ, λ are some constants.
The potential energy has its minimum for $\mu^2 < 0$ at points

$$\varphi_{min} = \pm\sqrt{-\frac{\mu^2}{\lambda}} \equiv \pm v. \tag{99}$$

We can write around the minimal point that

$$\varphi(x) = v + \eta(x). \tag{100}$$

Then after insertion of eq. (100) into the original Lagrangian (98) we get approximately

$$L \approx \frac{1}{2}\partial_\nu \varphi \partial^\nu \varphi - \left(\lambda v^2\eta^2 + \lambda v\eta^2 + \frac{1}{2}\mu^2\varphi^2 + \frac{1}{4}\lambda\eta^4\right) + ... \quad . \tag{101}$$

It is elementary to see that the mass term in the last Lagrangian is the mass term of the field η and it is:

$$m_\eta^2 = 2\lambda v^2 = -2\mu^2 > 0. \tag{102}$$

In case of the axionic situation the mass which is generated by he so called Higgs mechanism just described, is the mass of axion.
The interaction of axionic field a with the electromagnetic field is postulate in the literature as

$$L_{inter} \approx ga\mathbf{E} \cdot \mathbf{B}, \tag{103}$$

where g is some interaction constant of axionic field with the electric and magnetic field $\mathbf{E}, \mathbf{B}$. In case of the interaction of axion with the electron via the Compton process, the interaction Lagrangian is postulate as follows:

$$L_{ae} = \frac{const}{2m_e}(\bar{\psi}\gamma^{\mu}\gamma^5\psi)\partial_{\mu}a. \tag{104}$$

In case that the Compton process is considered in the external electromagnetic field, the ψ functions must be determined from the Dirac equation involving the external electromagnetic field.
In case of the interaction of the axion with the graphene, the interaction is the axion-pseudoelectron interaction via the Compton process, then the wave function ψ is the wave function of the graphene pseudoelectron. Graphene is considered in the presence of no external field, or in the presence of the external electromagnetic field. This process was still not solved in graphene physics, or in the physics of elementary particles.
The candidate for source of axions is considered in the last years the Sun. A decade ago US Brookhaven Laboratory first pointed an axion telescope at the Sun - a highly useful source of weakly interacting particles for fundamental research.
Axions would be produced in the Sun through the scattering of photons from electric charges - the Primakoff effect - and their numbers could equal those of solar neutrinos. The idea behind the Brookhaven experiment, first proposed by Pierre Sikivie, was to put the Primakoff effect to work in reverse, using a magnetic field to catalyze conversion of solar axions back into X-ray photons of a few kilo-electronvolts.
The CERN Solar Axion Telescope, CAST, tries the detection of axions originating from the 15 million degree plasma in the Sun's core.
There is also the alternative look on axion as an physical object existing only inside of a medium and not as a free elementary particle. Then, the detection of free axion is not possible. The analogue of this situation is for instance the Cooper pair which is present only inside of superconductor and not outside of superconductor.

6. Photoeffect on graphene wrapped on the geometrical surface

Carbon hexagonal structures can be formed also on surfaces and not only on the Euclidean sheet. We first determine the density of states of graphene from QFT. The density of states is the imaginary part $\Im G$ of the Green function integrated over all positions, in the limit $\mathbf{x} \to \mathbf{x}'$ (Cortijo, et al., 2007a).

$$\varrho(\omega) = \int \Im G(\omega, \mathbf{x}, \mathbf{x})d\mathbf{x}. \tag{105}$$

In terms of the Green function in momentum representation, $\varrho(\omega)$ can be written as:

$$\varrho(\omega) = \Im \int\int \frac{d\mathbf{k}}{(2\pi)^2}\frac{d\mathbf{k}'}{(2\pi)^2} e^{i\mathbf{k}\mathbf{x}} e^{i\mathbf{k}'\mathbf{x}} G(\omega, \mathbf{k}, \mathbf{k}')d\mathbf{x}. \tag{106}$$

The integration over $\mathbf{x}$ gives delta function $4\pi^2\delta(\mathbf{k}-\mathbf{k}')$, and $\varrho(\omega)$ then reads:

$$\varrho(\omega) = \int \frac{d\mathbf{k}}{(2\pi)^2} \Im G(\omega, \mathbf{k}, \mathbf{k}). \tag{107}$$

Generalization to the curvature is possible by including the local curvature coupling of the Dirac equation to a curved space. This approach was historically introduced to study the electronic spectrum of closed fullerenes and has been used for fullerenes of different shapes. In case of a two dimensional space in presence of a single cosmic string in polar coordinates, the metric is (Cortijo et al., 2007a):

$$-ds^2 = -dt^2 + dr^2 + c^2r^2d\theta^2, \tag{108}$$

where the parameter c is a constant related to the deficit angle by $c = 1 - b$.
The dynamics of a massless Dirac spinor in a curved spacetime is governed by the Dirac equation:

$$i\gamma^\mu \nabla_\mu \psi = 0. \tag{109}$$

The difference with the flat space lies in the definition of the γ matrices that satisfy generalized anti-commutation relations

$$\{\gamma^\mu, \gamma^\nu\} = 2g^{\mu\nu}, \tag{110}$$

where $g^{\mu\nu}$ is given by (108), and in the covariant derivative operator, defined as

$$\nabla_\mu = \partial_\mu - \Gamma_\mu, \tag{111}$$

where Γ_μ is the spin connection of the spinor field that can be calculated using the tetrad formalism (Cortijo et al., 2007a).
From equation (109), we can write down the Dirac equation for the electron propagator, $S_F(x, x')$:

$$i\gamma^\mu(x)(\partial_\mu - \Gamma_\mu)S_F(x, x') = \frac{1}{\sqrt{-g}}\delta^3(x - x'), \tag{112}$$

where x = (t, $\mathbf{x}$). The local density of states $N(\omega, \mathbf{x})$ is obtained by Fourier transforming the time component and taking the limit $\mathbf{x} \to \mathbf{x}'$:

$$N(\omega, \mathbf{x}) = \Im \mathrm{Tr} S_F(\omega, \mathbf{x}, \mathbf{x}). \tag{113}$$

The solution was performed considering the curvature induced by the defects as a perturbation of the flat graphene layer. The details of the calculation is given elsewhere (Cortijo et al., 2007b)
The ideas considered here can be extended to the general Riemann metric. However, the Riemann metric is also the metric of general relativity. So, there is the analogy between metric generated by deformation of the Euclidean surface and Einstein gravity, which is the Tartaglia et al. (2009) and author idea (Pardy, 2005). The deformed graphene obviously leads to the modification of the photoeffect in graphene and it can be used as the introduction to the photoelectric effect influenced by the gravitational field.

The generalization of our formalism to the Lobačcevskii geometry is also possible if we consider the Lobačevskii geometry as the partial sum of the local pseudospheres, where pseudosphere is the two dimensional manifold with the constant negative curvature. It was proved by Beltrami (Mc Cleary, 1994), that the Lobačevskii geometry is the geometry of the sphere with the imaginary radius (iR), where R is the radius of sphere.

7. Discussion

The article is in a some sense the preamble to the any conferences of ideas related to the photoeffect on graphene, bi-layer graphite, n-layer graphite and on the Wigner crystals which are spontaneously formed in graphite structures, or, in other structures. We have considered the photoeffect on the planar crystal at zero temperature and in the very strong magnetic field. We calculated only the process which can be approximated by the Schrödinger equation for an electron orbiting in magnetic field.

At present time, the most attention in graphene physics is devoted to the conductivity of a graphene with the goal to invent new MOSFETs and new transistors for new computers forming the basis for the artificial intelligence. However, we do not know, a priori, how many discoveries are involved in the investigation of he photo-electric effects in graphene.

An pseudoelectron in the graphene is considered as the particle with all attributes of electron as an elementary particle and it is the consequence of the collective quantum motion of crystal electrons described by the solid state physics.

The existence of the pseudoelectron in vacuum is not possible. The photoeffect on graphene is the process where the photon interacts with the pseudoelectron and after some time the pseudoelectron decays in such a way that the decay product is electron. At present time the theory of such process was not elaborated. The new experiments are necessary in order to verify the photoelectric equation in graphene.

The ballistic interaction of photon with electron is not possible in vacuum for point-like electrons. The experiments in CERN, Hamburg, Orsay, (The L3 collaboration, CELLO collaboration, ALEPH collaboration, ...) and other laboratories confirmed that there are no excited states of electron in vacuum. In other words, ballistic process in vacuum with electrons was not confirmed. It means that electron in the standard model of elementary particles is a point particle. It seems that the ballistic process is possible only in a medium. However, not only in the photoelectric media. We know, that electron accelerated in PASER by the ballistic method is accelerated for instance in CO_2. On the other hand, the photo-desintegrations of nuclei involves the ballistic process (Levinger, 1960), or, the interaction of light with carbon C_{60}, C_{70}, C_{80}... involves also the ballistic process with photons. The interaction of photon with pseudolectron in graphene is also the ballistic process.

The photoelectric effect at zero temperature can be realized only by very short laser pulses, because in case of the continual laser irradiation the zero temperature state is not stable. Only very short pulses can conserve the zero temperature 2D of the system. So, the experiment needs the laser with very short laser pulses.

The interaction of axion with pseudoelectron is also considered here, however this process is substantially new in the physics of elementary particles. The axion detection by graphene is based on the interaction of axion with pseudoelectron.

The information on the photoelectric effect in graphene and also the elementary particle interaction with graphene is necessary not only in the solid state physics, but also in the elementary particle physics in the big laboratories where graphene can form the substantial components of the particle detectors. The graphene can be probably used as the appropriate components in the solar elements, the cathode surface in the electron microscope, or, the medium of the memory hard disks in the computers. We hope that these possibilities will be consider in the physical laboratories.

The Volkov solution of the Dirac equation in the magnetic field was also mentioned. Such solution was used by Ritus (1979), Nikishov (1979) and others. Author (Pardy, 2003, 2004, 2007), used it for the electron in the laser field, synchrotron radiation, or, in case of the massive photons leading to the Riccati equation (Pardy, 2004).

We did not consider the photoeffect in LHC where the orbiting protons can be irradiated by the laser gun if installed in the tube of LHC. The elaboration of such difficult problem needs specific approach to the problem.

The graphene can be deformed in such a way that the metric of the deformed sheet is the Riemann one (Cortijo, et al., 2007b). However, the Riemann metric of general relativity is the metric of the deformed 4D sheet as was proposed by Tartaglia et al. (2009). So, in other words, there is the analogy between deformation and Einstein gravity (Pardy, 2005). The deformed graphene obviously leads to the modification of the photoeffect in graphene leading to the theory of the photoelectric effect influenced by the gravitational field.

The monolithic structures can be also built into graphene by addition and re-arrangement of carbon atoms (Lusk et al., 2008). The repeating patterns can be created to form new carbon allotropes called haeckelites. The introducing such architectonic defects modifies mechanical, electrical and chemical properties of graphene. The unconventional materials can be prepared by such technique in order to do revolution in the solid state physics.

While the last century economy growth was based on the inventions in the Edison-Tesla electricity, the economy growth in this century will be obviously based on the graphene physics. We hope that these perspective ideas will be considered at the universities and in the physical laboratories.

8. References

Amusia, M. Y. (1987). Atomic photoeffect, (Moscow, Nauka). (in Russian).

Berestetzkii, V. B.; Lifshitz, E. M. and Pitaevskii, L. P. (1989). Quantum electrodynamics (Moscow, Nauka). (in Russian).

Cortijo, A. and Vozmediano, M. A. H. (2007a). Electronic properties of curved graphene sheet, *Europhys. Lett.* (EPL) 77,47002.

Cortijo, A. and Vozmediano, M. A. H, (2007b). *Nucl. Phys.* B 763, 293 .

Dato, A.; Radmilovic, V.; Lee, Z.; Philips, J. and Frenklach, M. (2008). Substrate-free gas phase synthesis of graphene sheets, *Nano Lett.*, 8 (7), 2012.

Davydov, A. S. (1976). Quantum mechanics (Pergamon press, Oxford, ...).

Dittrich, W. and Reuter, M. (1985). Effective Lagrangians in Quantum Electrodynamics, Lecture Notes in Physics 220, (Springer-Verlag, Berlin, Heidelberg, New York, Tokyo).

Drukarev, G. F. (1988). Quantum mechanics (St. Petersburgh University). (in Russsian).

Einstein, A. (1905). Über einen die Erzeugung und Verwandlung des Lichtes betreffenden heuristischen Gesichtspunkt, *Annalen der Physik*, 17, 132. (in German); There is an

English translation: On the Heuristic Viewpoint Concerning the Production and Transformation of Light, in *AJP*, 33, No. 5, May (1965).
Kane, Ch. L. (2005). Erasing electron mass, *Nature*, 438, November 205, 168.
Kane, G. (1987). Modern Elementary particle physics (Addison-Wesley Publishing Company, Inc.).
Kleinert, H. and Xue She-S. (2008). Photoproduction in semiconductors by onset of magnetic field, arXiv:0810.2367v1 [cond-mat.other].
Landau, L. D. (1937). Zur Teorie der Phasenumwandlungen II., *Phys. Z. Sowjetunion* 11, 26.
Levinger, J. S. (1960). Nuclear photo-desintegration, (Oxford University Press). (in Russian).
Lozovik, Yu. E.; Merkulov, S. P. and Sokolik, A. A. (2008). Collective electron phenomena in graphene, *Uspekhi Fiz. Nauk*, 178, (7), 758.
Lusk, M. T.; Carr, L. D. (2008). cond-mat.mtrl-sci/0809.3160v2. Creation of graphene allotropes using patterned defects.
Mc Cleary, J. (1994).Geometry from a differentiable viewpoint, (Cambridge University Press).
Mermin, N. D. (1968). Crystalline order in two dimensions, *Phys. Rev.*, 176.
Nikishov, A. I. (1979). The problem of the intensive external field in quantum electrodynamics *Trudy FIAN* 111, 152. (in Russian).
Novoselov, K. S.; Geim, A.K.; Morozov, S. V.; Jiang, D.; Zhang. Y.; Dubonos, S. V.; Grigorieva, I.V. and Firsov, A.A. (2004). Electric field effect in atomically thin Carbon films, *Science*, 306, 666-669.
Novoselov, K.S.; Geim, A. K.; Morozov, S.V. et al. (2005). Two-dimensional gas of massless Dirac fermions in graphene, *Nature*, 438, 197.
Pardy, M. (2003). Electron in the ultrashort laser pulse, Int. J. Theor. Phys., 42(1), 99, 127.
Pardy, M. (2004). Massive photons and the Volkov solution, International Journal of Theoretical Physics, 43(1), 127.
Pardy, M. (2005). Light in metric space-time and its deflection by the screw dislocation, in: Spacetime Physics Research Trends. Horizons in World Physics, Volume 248, ISBN 1-59454-322-4, 111-125.
Pardy, M. (2007). The synchrotron radiation from the Volkov solution of the Dirac equation, arXiv: hep-ph/0703102v1.
Park, C.; Gustino, F.; Spataru, C. D.; Cohen, M.L. and Louie, S. G. (2010). Angle resolved photoemission spectra of graphene from first-principles calculations, arXiv: 1001.2803v1 [cond-mat.mtrl-sci].
Peierls, R. E. (1934). Bemerkungen über Umwandlungstemperaturen. *Helvetica Physica Acta* 7, 81; ibid: (1935). Quelques proprietes typiques des corpses solides. Ann. Inst. Henri Poincaré 5, 177.
Ratnikov, P. V. and Silin, A. P. (2008a). Ground state energy of current carriers in graphene, arXiv: 0801. 4642v1. arXiv: 0801.4642v1 [condmat. other].
Ratnikov. P. V. and Silin, A. P. (2008b). Transition of graphene on a substrate to a semimetallic state, arXiv: 0805.4510v1 [cond-mat.other].
Ritus, V. I. (1979). The quantum effects of the interaction of elementary particles with the intense electromagnetic field, *Trudy FIAN* 111, 5. (in Russian).
Rohlf, J. W. (1994). Modern Physics from α to Z^0 (John Willey & Sons, Inc., New York).
Sauter, F. (1931). Über atomaren Photoeffekt bei Grosser Härte der Anwegenden Strahlung, *Ann. der Phys.* 9, 217.

Sokolov, A. A. and Ternov, I. M. (1983). Relativistic electron, 2nd ed., (Moscow, Nauka). (in Russian).
Schwinger, J. (1988). Particles, Sources and Fields, (Addison-Wesley Publishing Company, Inc.).
Shpagin, A. V. (1996). Dynamical mass generation in 2+1-dimensional electrodynamics in an external magnetic field, arXiv: hep-ph/9611412.
Stobbe, M. (1930). Zur Quantenmechanik photoelektrischer Prozesse, *Ann. der Phys.* 7, 661.
Tamm, I. E.; Schubin, S. P. (1931). Zur Theorie des Photoeffektes an Metallen, *Zeitschrift für Physik* 131, 68, 97.
Volkov, D. M. (1935). Über eine Klasse von Lösungen der Diracschen Gleichung, *Zeitschrift für Physik* 94, 250.
Tartaglia, A. and Radicella, N. (2009). From Elastic Continua to Space-time, arXiv: 0911.3362v1 [gr-qc].
Wigner, E. P. (1935). On the Interaction of Electrons in Metals, *Phys. Rev.* 46, 1002.

13

Planar Dirac Fermions in External Electromagnetic Fields

Gabriela Murguía[1], Alfredo Raya[2] and Ángel Sánchez[3]
[1]*Facultad de Ciencias, Universidad Nacional Autónoma de México*
[2]*Instituto de Física y Matemáticas, Universidad Michoacana de San Nicolás de Hidalgo*
[3]*Department of Physics, University of Texas at El Paso*
[1,2]*Mexico*
[3]*USA*

1. Introduction

We study the electron propagator in two spatial dimensions in the presence of external electromagnetic fields, this is, we focus in (2+1)-dimensional quantum electrodynamics (QED), where a third spatial dimension is suppressed. This is not a mere theoretical simplification, and we explain ourselves: back in time, some twenty years ago, it was shown that the low-energy effective theory of graphene in a tight-binding approach is the theory of two species of massless Dirac electrons in a (2+1)-dimensional Minkowski spacetime, each on a different irreducible representation of the Clifford algebra. The isolation of graphene samples in 2004 and 2005, has given rise to the new paradigm of relativistic condensed matter, bringing a new boost, both theoretical and experimental, to the matching of common interests of the condensed matter and high energy physics communities. Thus, the massless limit of our findings is of direct relevance in this subject. We assume the electrons moving in a magnetic field alone pointing perpendicularly to their plane of motion. We first develop the general case and then, we present a couple of examples: the motion of electrons in a uniform magnetic field, which is a canonical example to present the Ritus method and the case of a static magnetic field which decays exponentially along the x-axis (Murguía et al, 2010; Raya & Reyes, 2010).
There are many problems relating electrons in non-uniform magnetic fields of relevance in graphene. In particular, it has been established the possibility to confine quasiparticles in magnetic barriers (De Martino et al, 2007; Ramezani et al, 2009). This could be feasible creating spatially inhomogeneous, but constant in time, magnetic fields depositing ferromagnetic layers over the substrate of a graphene sample layer (Reijniers et al, 2001). The physical properties of graphene make it a promising novel material to control the transport properties in nanodevices. It has been considered to be used in electronics and spintronics applications, like in single-electron transistors (Ponomarenko et al, 2008; Wu et al, 2008), in the so called magnetic edge states (Park & Sim, 2008), which may play an important role in the spin-polarized currents along magnetic domains, and in quantum dots and antidots magnetically confined. Moreover, the quantum Hall effect in graphene has been observed at room temperature (Novoselov et al, 2007), evidence which confirms the great potential of graphene as the material to be used in carbon-based electronic devices. The effects of the exponentially decaying magnetic field can hardly be considered with other approaches,

but it can be straightforwardly studied within the Ritus method, which consist in the diagonalization of the electron propagator in external electromagnetic fields in the basis of the operator $(\not{\Pi})^2$ with $\Pi^\mu = p^\mu - eA^\mu$. Exploiting the Ritus formalism, we also derive the exact Foldy-Wouthuysen (FW) transformation for Dirac fermions in a time independent external electromagnetic field, where the transformation acquires a free form involving the dynamical quantum numbers induced by the field (Murguía & Raya, 2010). This is related with the fact that for some class of Hamiltonians of Dirac particles in presence of external static electromagnetic fields, it is possible to show a supersymmetric character, in the quantum mechanical sense, and in these cases the Ritus method provide a direct calculation of the exact FW transformation in arbitrary dimensions. Powerful applications of FW transformation in semiclassical calculations can be exploited in systems of other than (3+1) dimensions. The FW transformation has proven to be a favorite way to obtain the nonrelativistic limit of the Dirac equation, because it provides a block diagonal form representation of quantum operators and hence of the Dirac Hamiltonian itself. It has been widely used in both, gravitational and electromagnetic backgrounds, including the case of different stationary metrics. Going further in the applications of the Ritus formalism, we also explore other kind of useful transformations, like the Cini-Touschek transformation for the ultrarelativistic case.

2. Irreducible Dirac fermions

In order to describe planar Dirac electrons, let us start by defining the Dirac γ^μ-matrices, which satisfy the Clifford algebra

$$\{\gamma^\mu, \gamma^\nu\} = 2g^{\mu\nu}. \tag{1}$$

The lowest dimensional representation of these matrices is 2×2. We can choose, for instance

$$\gamma^0 = \sigma^3, \qquad \gamma^1 = i\sigma^1, \qquad \gamma^2 = i\sigma^2. \tag{2}$$

It is straightforward to verify that these matrices satisfy the relations

$$[\gamma^\mu, \gamma^\nu] = -2i\epsilon^{\mu\nu\rho}\gamma_\rho, \qquad \text{and} \qquad \gamma^\mu\gamma^\nu = g^{\mu\nu} - i\epsilon^{\mu\nu\rho}\gamma_\rho. \tag{3}$$

This is the starting point to build up the planar Dirac theory. The free Dirac Lagrangian takes the form of its 4D counterpart, namely

$$\mathcal{L} = \bar{\psi}(i\not{\partial} - m)\psi\,, \tag{4}$$

where m is the mass of the electrons and we have used the γ matrices given in Eq. (2). The spectrum of solutions of the Dirac equation is

$$\begin{aligned} \psi^P(x) &= \begin{pmatrix} 1 \\ \frac{p_x - ip_y}{E+m} \end{pmatrix} e^{-ix\cdot p} \equiv u(p)e^{-ix\cdot p}\,, \\ \psi^N(x) &= \begin{pmatrix} \frac{p_x + ip_y}{E+m} \\ 1 \end{pmatrix} e^{ix\cdot p} \equiv v(p)e^{ix\cdot p}\,. \end{aligned} \tag{5}$$

These solutions verify the completeness relations

$$\sum u\bar{u} = \not{p} + m, \qquad \text{and} \qquad \sum v\bar{v} = \not{p} - m, \tag{6}$$

and the positive energy solution (labeled by P), describes a particle with spin up, whereas the negative energy one (labeled by N) describes an antiparticle with spin down (Anguiano

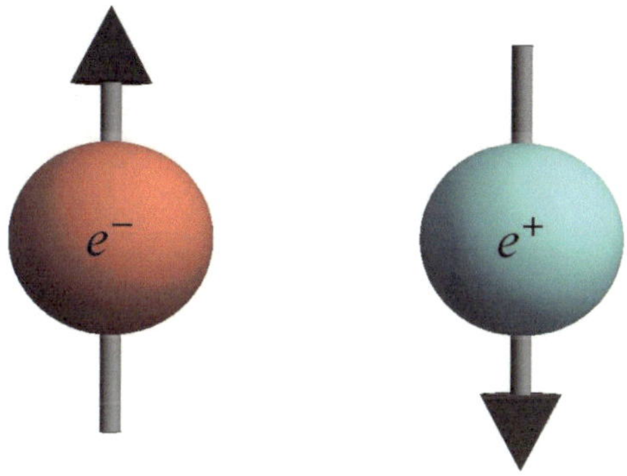

Fig. 1. The particle spectrum of solutions to the Dirac equation in Eq. (5). Adapted from Ref. (Hernández Ortíz, 2011).

& Bashir, 2005), as shown in Fig. 1. This fact is better seen from the definition of the spin operator

$$\Sigma_3 = \frac{\gamma_0}{2} = \frac{1}{2}\begin{pmatrix} 1 & 0 \\ 0 & -1 \end{pmatrix}. \tag{7}$$

Thus *the Dirac particles have spin* $\pm 1/2$. These solutions, however, fail to incorporate important features of the ordinary Dirac spectrum familiar in relativistic quantum mechanics. For example, the solutions are not invariant under a Parity transformation $\mathcal{P}$, which for consistency with Lorentz symmetry in (2+1)-dimensions corresponds to the operation [1]

$$(t, x, y) \to (t, -x, y)^{\mathcal{P}}, \tag{8}$$

nor under a time reversal transformation $\mathcal{T}$. This is due to the fact that under $\mathcal{P}$, $\bar{\psi}\psi \to -\bar{\psi}\psi$. Furthermore, only one out of the two spin states of the physical electrons is present. A point of caution has to be raised here, in Condensed Matter Physics spin plays the role of flavor in High Energy Physics, thus one might be tempted to put by hand the spin factors of 2 whenever it is required. However, one cannot simply push this argument to the fully relativistic theory. The two spin states and symmetry features of the familiar spectrum of solutions to the Dirac equation can be recovered owing to the fact that there exists a second irreducible representation of the Dirac matrices. In graphene, the two representations describe two different electron species in each of the two triangular sub-lattices of the honeycomb lattice. The origin of the second irreducible representation is in direct connection with the fact that there is no chiral symmetry to be defined in (2+1)-dimensions with an irreducible representation of the Clifford algebra. Let γ be the product of all Dirac matrices, i.e, the would-be "γ_5" on the plane. From the properties of the σ matrices,

$$\gamma = \gamma^0\gamma^1\gamma^2 = \pm iI, \tag{9}$$

I being the 2×2 unit matrix. This allows us to define one of the matrices in terms of the other two. For instance $\gamma^2 = \pm i\gamma^0\gamma^1$. The second representation can then be chosen as

$$\gamma^0 = \sigma^3, \qquad \gamma^1 = i\sigma^1, \qquad \gamma^2 = -i\sigma^2, \tag{10}$$

[1] Parity is an improper Lorentz transformation, which should have determinant -1. If the usual definition of parity is employed, that would correspond to a rotation by an angle π of the plane.

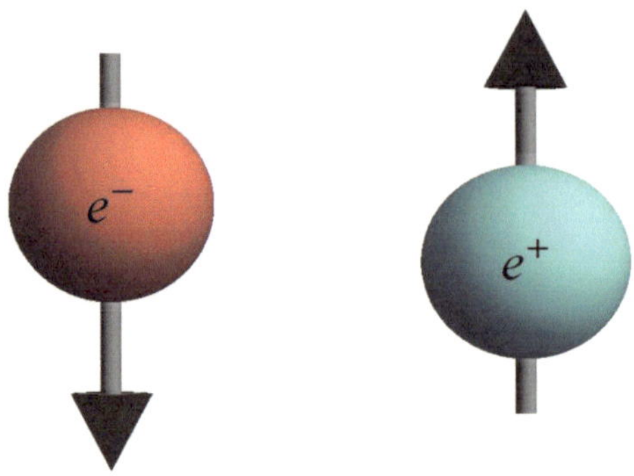

Fig. 2. The particle spectrum of solutions to the Dirac equation in Eq. (11). Adapted from Ref. (Hernández Ortíz, 2011).

with the property $\gamma^\mu\gamma^\nu = g^{\mu\nu} + i\epsilon^{\mu\nu\lambda}\gamma_\lambda$. Representations (2) and (10) are inequivalent, and hence correspond to physically different situations. This can be seen from the solutions of the Dirac equation in representation (10), which expanded in the representation (2) are

$$\psi^P(x) = \begin{pmatrix} \frac{p_x+ip_y}{E+m} \\ 1 \end{pmatrix} e^{-ix\cdot p} \equiv u(p)e^{-ix\cdot p}\,,$$
$$\psi^N(x) = \begin{pmatrix} 1 \\ \frac{p_x-ip_y}{E+m} \end{pmatrix} e^{ix\cdot p} \equiv v(p)e^{ix\cdot p}\,, \tag{11}$$

and correspond to particle spinor with spin down and antiparticle with spin up (Anguiano & Bashir, 2005), as shown in Fig. 2. These solutions fulfill the completeness relations (6), but present also only one spin state for electron and yield a $\mathcal{P}$ and $\mathcal{T}$ non-invariant Lagrangian. Nevertheless, taking into account solutions for both representations, (2) and (10), labeled A and B, respectively, we recover two spin states for the electrons and their corresponding Lorentz conjugated positron states. The two "irreducible" fermion fields can be cast into the following *extended* form of the free Dirac Lagrangian (Anguiano & Bashir, 2005; Shimizu, 1985) [2]:

$$\mathcal{L} = \bar{\psi}_A(i\partial\!\!\!/ - m)\psi_A + \bar{\psi}_B(i\partial\!\!\!/ + m)\psi_B\,. \tag{12}$$

As we noticed before, neither under $\mathcal{P}$ nor under $\mathcal{T}$, the fields ψ_A and ψ_B transform onto themselves. In fact, under $\mathcal{C}, \mathcal{P}$ and $\mathcal{T}$ transformations, these fields transform as

$$\begin{aligned}
(\psi_A)^{\mathcal{C}} &= \gamma^2 e^{i\eta_1}(\bar{\psi}_A)^T\,, & (\psi_B)^{\mathcal{C}} &= \gamma^2 e^{i\eta_2}(\bar{\psi}_B)^T\,,\\
(\psi_A)^{\mathcal{P}} &= -i\gamma^1 e^{i\phi_1}(\psi_B)\,, & (\psi_B)^{\mathcal{P}} &= -i\gamma^1 e^{i\phi_2}(\psi_A)\,,\\
(\psi_A)^{\mathcal{T}} &= i\gamma^0 e^{i\varphi_1}(\bar{\psi}_B)^T\,, & (\psi_B)^{\mathcal{T}} &= i\gamma^0 e^{i\varphi_2}(\bar{\psi}_A)^T\,.
\end{aligned} \tag{13}$$

where η_i, ϕ_i and φ_i, $i = 1,2$ are constant phases. This shows that the extended Lagrangian (12) is $\mathcal{CPT}$ invariant (Shimizu, 1985). There are two chiral symmetries which can be defined. In infinitesimal form, these are

[2] Notice that only one irreducible representation of the Dirac matrices, say (2) is used.

Set I

$$\psi_A \to \psi'_A = \psi_A + \alpha\psi_B\,, \qquad \psi_B \to \psi'_B = \psi_B - \alpha\psi_A\,, \tag{14}$$

leading to the conserved current

$$j_I^\mu = \bar{\psi}_A\gamma^\mu\psi_B - \bar{\psi}_B\gamma^\mu\psi_A. \tag{15}$$

Set II

$$\psi_A \to \psi'_A = \psi_A + i\alpha\psi_B\,, \qquad \psi_B \to \psi'_B = \psi_B + i\alpha\psi_A\,, \tag{16}$$

leading to the conserved current

$$j_{II}^\mu = \bar{\psi}_A\gamma^\mu\psi_B + \bar{\psi}_B\gamma^\mu\psi_A. \tag{17}$$

The presence of two irreducible fermion fields in (12) naturally suggest that these can be merged into one reducible four-component spinor and hence we can make use of the ordinary 4×4 Dirac matrices. Such an issue is discussed below.

3. Reducible Dirac fermions

Planar Dirac fermions can also be described with the ordinary 4×4 matrices. Nevertheless, only three of them are required to describe the Dirac equation, for example $\{\gamma^0,\ \gamma^1,\ \gamma^2\}$, which in Euclidean space can be represented as

$$\gamma_E^0 = \begin{pmatrix} -i\sigma^3 & 0 \\ 0 & i\sigma^3 \end{pmatrix}, \quad \gamma_E^1 = \begin{pmatrix} i\sigma^1 & 0 \\ 0 & -i\sigma^1 \end{pmatrix}, \quad \gamma_E^2 = \begin{pmatrix} i\sigma^2 & 0 \\ 0 & -i\sigma^2 \end{pmatrix}. \tag{18}$$

In such a case, we have two other γ matrices (from now onwards we omit de subscript E for the Euclidean matrices) which commute with all the three matrices above, in such a fashion that the corresponding massless Dirac Lagrangian is invariant under the chiral-like transformations $\psi \to e^{i\alpha\gamma^3}\psi$, and $\psi \to e^{i\beta\gamma^5}\psi$, that is, it is invariant under a global $U(2)$ symmetry with generators $1, \gamma^3, \gamma^5$ and $[\gamma^3, \gamma^5]$. Here

$$\gamma^3 = \begin{pmatrix} 0 & I \\ I & 0 \end{pmatrix}, \quad \gamma^5 = \begin{pmatrix} 0 & -I \\ I & 0 \end{pmatrix}. \tag{19}$$

This symmetry is broken by an ordinary mass term $m_e\bar{\psi}\psi$. But there exists a second mass term, referred to as Haldane mass term (Haldane, 1988), which is invariant under the "chiral" transformations

$$m_o\bar{\psi}\frac{1}{2}[\gamma^3,\gamma^5]\psi \equiv m_o\bar{\psi}\tau\psi, \tag{20}$$

and has to be included in the complete Lagrangian when parity is allowed to be broken. If we write the 4-spinor as

$$\psi = \begin{pmatrix} \psi_A \\ \psi_B \end{pmatrix}, \tag{21}$$

we observe that under $\mathcal{P}$ and $\mathcal{T}$, the components of spinors transform, up to a phase, as (Jackiw & Templeton, 1981)

$$\begin{aligned} (\psi_A(t,x,y))^{\mathcal{P}} &\to \sigma^1\psi_B(t,-x,y), \qquad & (\psi_B(t,x,y))^{\mathcal{P}} &\to \sigma^1\psi_A(t,-x,y), \\ (\psi_A(t,x,y))^{\mathcal{T}} &\to \sigma^2\psi_B(-t,x,y), \qquad & (\psi_B(t,x,y))^{\mathcal{T}} &\to \sigma^2\psi_A(-t,x,y). \end{aligned} \tag{22}$$

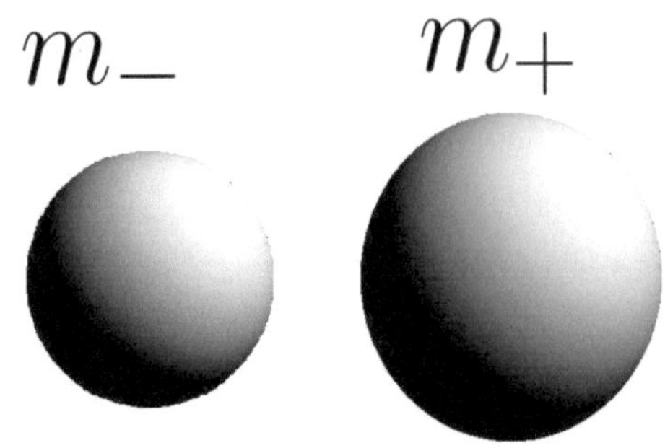

Fig. 3. Light and heavy fermion species of Lagrangian 25. Adapted from Ref. (Hernández Ortíz, 2011).

Thus, the term $m_e\bar{\psi}\psi$ is even under each of these transformations, but $m_o\bar{\psi}\tau\psi$ is not, although it is $\mathcal{PT}$ and thus $\mathcal{CPT}$ symmetric. The Euclidean space free *reducible* Dirac Lagrangian in this case has the form

$$\mathcal{L} = \bar{\psi}(i\not\partial - m_e - m_o\tau)\psi \,. \tag{23}$$

Written in this form, neither m_e nor m_o represent "physical" masses for electrons. In order to disentangle the species described by this Lagrangian, it is convenient to introduce the chiral-like projectors

$$\chi_\pm = \frac{1}{2}(1 \pm \tau)\,, \tag{24}$$

which verify (Kondo, 1996) $\chi_\pm^2 = \chi_\pm$, $\chi_+\chi_- = 0$, $\chi_+ + \chi_- = I$. The "right handed" ψ_+ and "left handed" ψ_- fermion fields in this case are given by $\psi_\pm = \chi_\pm\psi$. The $\chi_\pm$ project the upper and lower two component spinors, i.e, fermion species, out of the four-component ψ. In terms of the chiral fields, the Dirac Lagrangian acquires the form (see for instance (Jackiw & Pi, 2007))

$$\mathcal{L}_F = \bar{\psi}_+(i\not\partial - m_+)\psi_+ + \bar{\psi}_-(i\not\partial - m_-)\psi_-, \tag{25}$$

with $m_\pm = m_e \pm m_o$. This Lagrangian explicitly describes two fermion species of physical masses m_+ and m_-, respectively. For each species, the mass terms breaks chiral symmetry and parity at the same time. Moreover, the effect of the parity-violating mass is seen to remove the mass degeneracy between species. There is a light species and a heavy species, as illustrated in Fig. 3. Below we introduce interactions of fermions with a classical electromagnetic field within the Ritus formalism.

4. Propagator in magnetic fields

We start from the free Dirac equation derived from the Lagrangian (4)

$$(i\not\partial - m)\psi = 0\,, \tag{26}$$

and work with representation (2). The extension to other representations is straightforward. In a background electromagnetic field, the Dirac equation takes the form

$$(\not\Pi - m)\Psi = 0, \tag{27}$$

where $\Pi_\mu = i\partial_\mu + eA_\mu$ and A_μ is the electromagnetic potential defining the external field. From now onwards, let us consider a magnetic field alone pointing perpendicularly to the plane of motion of the electrons. Moreover, let us work in a Landau-like gauge by choosing $A^\mu = (0, 0, W(x))$, where $W(x)$ is some function for which, in the general case, its derivative $W'(x) = \partial_x W(x)$ defines the profile of the field. We are interested in finding the Green's function or propagator for this equation, namely, the function $G(x, x')$ which satisfies

$$(\not{\Pi} - m)\, G(x, x') = \delta(x - x'). \tag{28}$$

Since $\not{\Pi}$ does not commute with the momentum operator, neither the wave function nor $G(x, x')$ can be expanded in plane-waves, and this does not allow to have a diagonal propagator in momentum space. The scheme we choose to deal with the external fields was developed by Ritus (Ritus, 1972; 1974; 1978). The crucial observation is that the Green's function above should be a combination of all scalar structures obtained by contracting the γ^μ-matrices, the canonical momentum Π_μ and the electromagnetic field strength tensor $F_{\mu\nu} = [\Pi_\mu, \Pi_\nu]/e \equiv \partial_\mu A_\nu - \partial_\nu A_\mu$, which are compatible with Lorentz symmetry, gauge invariance and charge conjugation, namely,

$$G(x, x') = G(\not{\Pi}, \sigma^{\mu\nu} F_{\mu\nu}, (\tilde{F}^\nu \Pi_\nu)^2), \tag{29}$$

where $\sigma_{\mu\nu} = i[\gamma_\mu, \gamma_\nu]/2$ and $\tilde{F}_\mu \equiv (1/2)\epsilon_{\mu\nu\alpha} F^{\nu\alpha}$ is the dual field strength tensor, which in (2+1)-dimensions is simply a vector. The key observation is that all the above structures commute with $(\not{\Pi})^2$, and thus

$$\left[(\gamma \cdot \Pi)^2, G(x, x')\right] = 0\,. \tag{30}$$

This fact allows us to expand the Green's function $G(x, x')$ in the same basis of eigenfunctions of $(\not{\Pi})^2$. Furthermore, if we perform a similarity transformation on $(\not{\Pi})^2$ in which it acquires a diagonal form in momentum space, then the same transformation makes the Green's function diagonal too. Such a similarity transformation is

$$\mathbb{E}_p^{-1} (\not{\Pi})^2 \mathbb{E}_p = p^2 \mathbb{I}\,, \tag{31}$$

where $\mathbb{E}_p$ are the transformation matrices, $\mathbb{I}$ is the unit matrix and p^2 can be any real number. Therefore, when we apply $\mathbb{E}_p$ functions to the propagator, it will become diagonal in momentum space. It is important to notice that in the fermionic case, the spin operator is realized in terms of the γ^μ-matrices, and thus the $\mathbb{E}_p$ functions inherit its matrix form. For different charged particles, the spin operator is realized in a different ways. For example, for scalar particles, the $\mathbb{E}_p$ functions are simply scalars (Ginzburg, 1995), whereas in the case of charged gauge bosons, the spin structure is embedded in a Lorentz tensor, and therefore the $\mathbb{E}_p$ functions also comply a Lorentz tensor structure (Elizalde et al, 2002). Our goal in this work is to study the structure of the $\mathbb{E}_p$ matrices for the case of Dirac fermions in (2+1)-dimensions. The similarity transformation (31) can be written as

$$(\not{\Pi})^2 \mathbb{E}_p = p^2 \mathbb{E}_p\,, \tag{32}$$

which is an eigenvalue equation for the matrices $\mathbb{E}_p$, which are referred to as the Ritus eigenfunctions in the specialized literature. Now,

$$(\not{\Pi})^2 = \Pi^2 + \frac{e}{2} \sigma^{\mu\nu} F_{\mu\nu}\,. \tag{33}$$

The only non-vanishing elements of the field strength tensor are $F_{12} = -F_{21} = W'(x)$, and because $\sigma^{12} = \sigma_3$, the $\mathbb{E}_p$ functions satisfy

$$\left(\Pi^2 + e\sigma_3 W'(x)\right) \mathbb{E}_p = p^2\, \mathbb{E}_p\,. \tag{34}$$

Moreover, because

$$[\not\Pi, i\partial_t] = \left[\not\Pi, -i\partial_y\right] = [\not\Pi, \mathcal{H}] = 0\,, \tag{35}$$

with

$$\mathcal{H} = -\not\Pi^2 + \Pi_0^2, \tag{36}$$

the functions $\mathbb{E}_p$ are eigenfunctions of these operators with eigenvalues

$$i\partial_t \mathbb{E}_p = p_0 \mathbb{E}_p\,, \qquad i\partial_y \mathbb{E}_p = -p_2 \mathbb{E}_p\,, \qquad \mathcal{H}\mathbb{E}_p = k\mathbb{E}_p\,, \tag{37}$$

which label the solutions to the massless Dirac equation in the background field. Notice that $p^2 = p_0^2 - k$, and hence, the $\mathbb{E}_p$ functions verify

$$\left(-\Pi_1^2 - \Pi_2^2 + e\sigma_3 W'(x)\right) \mathbb{E}_p = -k\mathbb{E}_p\,. \tag{38}$$

The first two terms of the operator on the l.h.s. of this equation act on the orbital degrees of freedom of the eigenfunctions $\mathbb{E}_p$, whereas the last term acts only in its spin degrees of freedom. Hence we can make the ansatz

$$\mathbb{E}_p = E_{p,\sigma}\, \omega_\sigma\,, \tag{39}$$

where ω_σ is the matrix of eigenvectors of σ_3 with eigenvalues $\sigma = \pm 1$, respectively, and

$$E_{p,\sigma} = N_\sigma e^{-i(p_o t - p_2 y)} F_{k,p_2,\sigma}\,, \tag{40}$$

with N_σ being the corresponding normalization constant. Substituting the ansätze (39) and (40) into Eq. (38), we arrive at

$$\left(\frac{d^2}{dx^2} - (-p_2 + eW(x))^2 + e\sigma W'(x)\right) F_{k,p_2,\sigma} = -kF_{k,p_2,\sigma}\,. \tag{41}$$

For the squared-integrability of the solutions, the eigenvalue k must be discrete. The above expression has the form of the Pauli Hamiltonian with the constrained vector potential, mass $m = 1/2$ and gyromagnetic factor $g = 2$, and turns out to be supersymmetric in the Quantum Mechanical sense (SUSY-QM) (Cooper et al, 1995; 2001). From the solutions to the above equation, we can construct the Ritus eigenfunctions $\mathbb{E}_p$ as

$$\mathbb{E}_p = \begin{pmatrix} E_{p,1}(z) & 0 \\ 0 & E_{p,-1}(z) \end{pmatrix}, \tag{42}$$

where the label $p = (p_0, p_2, k)$ and $z = (t, x, y)$. Being a complete set, the eigenfunctions $\mathbb{E}_p$ given in Eq. (42), satisfy

$$\begin{aligned} &\int dz\, \overline{\mathbb{E}}_{p'}(z)\mathbb{E}_p(z) = \mathbb{I}\delta(p-p')\,, \\ &\int dp\, \mathbb{E}_p(z)\overline{\mathbb{E}}_p(z') = \mathbb{I}\delta(z-z')\,, \end{aligned} \tag{43}$$

with $\overline{\mathbb{E}}_p(z) = \gamma^0 \mathbb{E}_p^*(z)\gamma^0$ and $\mathbb{I}$ is the 2×2 unit matrix. Let us introduce the bar-momentum $\overline{p}_\mu = (p_0, 0, \sqrt{k})$, which plays an important role in the method. Its definition involves the dynamical quantum numbers p_0 and k, but not p_2, which merely fixes the origin of the x coordinate. In other words, in the Ritus method the propagator is written only in terms of the eigenvalues of the dynamical operators commuting with $\not{\Pi}$. Notice that the bar-momentum verifies $\overline{p}^2 = p_0^2 - k = p^2$, and it is defined through the relation

$$\not{\Pi}\mathbb{E}_p = \mathbb{E}_p \not{\overline{p}}\,. \tag{44}$$

This relation will become important in the discussion of unitary transformations of the Dirac Hamiltonian. With the $\mathbb{E}_p$ functions, we can consider the Green's function method to obtain the propagator in the presence of the field. From Eq. (28), we define the Green's function in momentum space as

$$G(z,z') = \int dp\; \mathbb{E}_p(z) S_F(p) \overline{\mathbb{E}}_p(z')\,. \tag{45}$$

Here, the integral sign might as well represent a sum, depending upon the continuous or discrete nature of the components of the momentum. Applying the Dirac operator $(\not{\Pi} - m)$ to $G(z,z')$, we have that

$$\begin{aligned}(\not{\Pi} - m)\, G(z,z') &= \int dp\; \mathbb{E}_p(z)(\not{\overline{p}} - m) S_F(p) \overline{\mathbb{E}}_p(z') \\ &= \int dp\; \mathbb{E}_p(z) \overline{\mathbb{E}}_p(z')\,, \end{aligned} \tag{46}$$

where in the last step we have used the representation of the δ-function in the $\mathbb{E}_p$ basis. Hence we notice that, in this basis, the Ritus propagator takes the form of a free propagator, namely,

$$S_F(p) = \frac{1}{\not{\overline{p}} - m}\,, \tag{47}$$

with $\overline{p}_\mu$ defined through Eq. (44). On physical grounds, the $\mathbb{E}_p$ functions correspond to the states of electrons with momentum $\overline{p}$ in the background of the external field under consideration. With the help of these functions and the property (44), we can find the solutions of the Dirac equation (27) in a straightforward manner. To this end, we propose

$$\Psi = \mathbb{E}_p u_{\overline{p}}, \tag{48}$$

where $u_{\overline{p}}$ is a spinor of momentum $\overline{p}$. Then

$$(\not{\Pi} - m)\mathbb{E}_p u_{\overline{p}} = \mathbb{E}_p(\not{\overline{p}} - m) u_{\overline{p}} = 0, \tag{49}$$

and thus we see that $u_{\overline{p}}$ is simply a free spinor describing an electron with momentum $\overline{p}$. Notice that with this form of Ψ, the information concerning the interaction with the background magnetic field has been factorized into the $\mathbb{E}_p$ functions and throughout the $\overline{p}$ dependence of $u_{\overline{p}}$. Several relevant physical observables can then be found immediately, such as the probability density, the transmission and reflection coefficients between magnetic domains, and the density matrix, which are all useful, for example, in graphene applications as those which were mentioned in the Introduction, we discuss below, where the Ritus method plays useful.

5. Examples

5.1 Uniform magnetic field

Consider in the first place the case of a uniform magnetic field (Khalilov, 1999). This corresponds to the choice $W(x) = B_0 x$. To simplify the calculations, we rename the quantum number $k \to 2|eB_0|k$ in Eq. (37). In this case, Eq. (41) simplifies to

$$\left[\frac{d^2}{dx^2} - (-p_2 + eB_0 x)^2 + \sigma e B_0\right] F_{k,p_2,\sigma}(x) = -2|eB_0|kF_{k,p_2,\sigma}(x)\,. \tag{50}$$

Letting $\eta = \sqrt{2|eB_0|}[x - p_2/(eB_0)]$, the above expression acquires the form

$$\left[\frac{d^2}{d\eta^2} + k + \frac{\sigma}{2}\mathrm{sgn}(eB_0) - \frac{\eta^2}{4}\right] F_{k,p_2,\sigma}(\eta) = 0\,, \tag{51}$$

that is, the equation for a quantum harmonic oscillator, with center of oscillation in $x_0 = p_2/(eB_0)$ and cyclotron frequency $w_c = 2eB_0$. Thus, the normalized functions $E_{p,\sigma}$ acquire the form

$$\begin{aligned} E_{p,1} &= \frac{(\pi|eB_0|)^{1/4}}{2\pi^{3/2}k!^{1/2}} e^{-ip_0 t + ip_2 y} D_k(\eta)\,, \\ E_{p,-1} &= \frac{(\pi|eB_0|)^{1/4}}{2\pi^{3/2}(k-1)!^{1/2}} e^{-ip_0 t + ip_2 y} D_{k-1}(\eta)\,, \end{aligned} \tag{52}$$

where

$$D_n(x) = 2^{-n/2} e^{-x^2/4} H_n\left(x/\sqrt{2}\right) \tag{53}$$

is the parabolic cylinder function of order

$$n = k + \frac{\sigma}{2}\mathrm{sgn}(eB_0) - \frac{1}{2}\,, \tag{54}$$

and $H_n(x)$ are the Hermite's polynomials. Expectedly, the uniform magnetic field renders the $(n-1)$-th state with spin down with the same energy of the n-th state with spin up. Inserting these functions into Eq. (42), we obtain the Ritus eigenfunctions which render the propagator diagonal in momentum space. Alternative forms of this propagator were recently reported (Rusin & Zawadzki, 2011).

Observe that physical observables like probability densities are linear combinations of $|E_p|^2$. These functions have the profile shown Fig. 4. The dashed curve enveloping these solutions corresponds to the potential

$$y = \bar{W}^2(x) - \bar{W}'(x)\,, \tag{55}$$

where $\bar{W} = eW - p_2$ is referred to as the superpotential in the SUSY-QM literature (Cooper et al, 1995; 2001).

5.2 Exponential magnetic field

In this section we study the electron propagator in a background static magnetic field which has an exponentially decaying spatial profile along one direction, described through the function $W(x) = -B_0[\exp\{-\alpha x\} - 1]/\alpha$. In this case, Eq. (41) simplifies to

$$\left[\frac{d^2}{dx^2} - \left(-p_2 - \frac{eB_0}{\alpha}(\exp\{-\alpha x\} - 1)\right)^2 + \sigma e B_0 \exp\{-\alpha x\}\right] F_{k,p_2,\sigma}(x) = -kF_{k,p_2,\sigma}(x)\,. \tag{56}$$

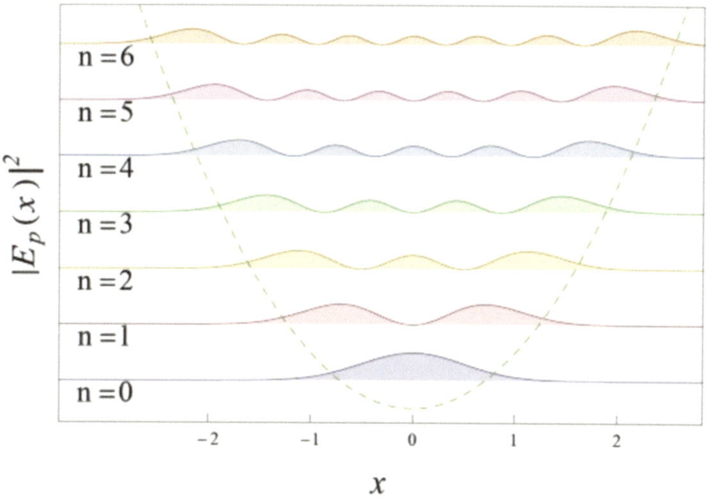

Fig. 4. The $|E_p|^2$ function from Eq. (52) in arbitrary units along the dynamical direction for various Landau levels. The dashed curve $y = x^2 - 1$ corresponds to the potential (55) for this field configuration. The scale of the graphic is set by $eB_0 = 1$ and $p_2 = 0$.

Let $\varrho = (2eB_0/\alpha^2)\exp\{-\alpha x\}$ and $s = -(p_2 - eB/\alpha)/\alpha$, then, the above expression is equivalent to

$$\left[\varrho^2 \frac{d^2}{d\varrho^2} + \varrho\frac{d}{d\varrho} - \left(s - \frac{1}{2}\varrho\right)^2 + \frac{\sigma}{2}\varrho + \frac{k}{\alpha^2}\right] F_{k,p_2,\sigma}(\varrho) = 0\,. \tag{57}$$

This equation has the normalized solutions $E_{p,\sigma}$ given as

$$\begin{aligned} E_{p,1} &= \frac{1}{2\pi}\left(\frac{2\alpha n!(s-n)}{\Gamma(2s-n+1)}\right)^{1/2} e^{-ip_0 t + ip_2 y} e^{-\varrho/2} \varrho^{(s-n)} L_n^{2(s-n)}(\varrho), \\ E_{p,-1} &= \frac{1}{2\pi}\left(\frac{2\alpha(n-1)!(s-n)}{\Gamma(2s-n)}\right)^{1/2} e^{-ip_0 t + ip_2 y} e^{-\varrho/2} \varrho^{(s-n)} L_{n-1}^{2(s-n)}(\varrho), \end{aligned} \tag{58}$$

where $L_a^b(x)$ are the associate Laguerre polynomials with

$$n = s - \sqrt{-\frac{k}{\alpha^2} + s^2}. \tag{59}$$

The quantum number n is the principal quantum number, whereas s a center of oscillation weighted by the damping factor α. Fig. 5 we show $|E_p|^2$ for various values of n at fixed $s = 8$. Notice that in this case the potential (55) also envelops the squares of the solutions.

6. Non relativistic and ultrarelativistic forms of the Dirac equation

The study of semiclassical and nonrelativistic limits of the Dirac equation is a useful method to understand some effects on fermions coupled to external fields. In both, gravitational (Goncalves et al, 2007; Obukhov et al, 2009) and electromagnetic backgrounds (Barducci et al, 2009; Silenko, 2008), the Foldy-Wouthuysen (FW) transformation (Foldy & Wouthuysen, 1950) has proven to be a favorite way to obtain

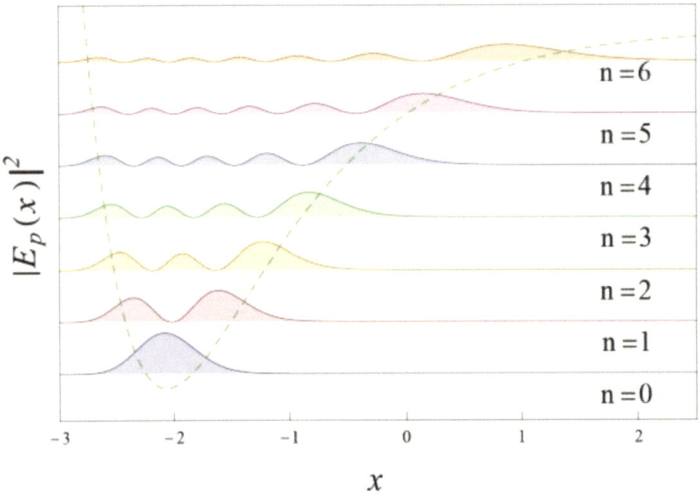

Fig. 5. The $|E_p|^2$ functions from Eq. (58) in arbitrary units along the dynamical direction for various values of n at fixed $s = 8$. The dashed curve $y = (8 - e^{-x})^2 + e^{-x}$ corresponds to the potential (55) for this field configuration. The scale of the graphic is set by $eB_0 = \alpha = 1$.

the nonrelativistic limit of the Dirac equation, because it provides a block diagonal form representation of quantum operators and hence of the Dirac Hamiltonian itself. Powerful applications of FW transformation in semiclassical calculations can be exploited in systems of other than (3+1) dimensions (Moreno & Méndez-Moreno, 1992) due to its relation with the supersymmetric character, in the quantum mechanical sense (Cooper et al, 1995; 2001), of some class of Hamiltonians, as well as in different stationary metrics (Buhl et al, 2008; Heidenreich et al, 2006). In the non-relativistic domain, the components of the Dirac spinors in either eq. (5) or eq. (11) are such that $u_1 \gg u_2$ and $v_1 \ll v_2$, thus the Dirac equation reduces to the *Pauli equation*, which is a first non-relativistic approximation of the Dirac equation for an electron in an external electromagnetic field,

$$i\frac{\partial}{\partial t}\psi = e\Phi\psi + \alpha^j\Pi^j\psi + \beta m\psi, \tag{60}$$

where $\alpha^j = \gamma^0\gamma^j$, $\beta = \gamma^0$ and $A^\mu = (\Phi, \mathbf{A})$. Now we write $\psi = \begin{pmatrix}\phi' \\ \chi'\end{pmatrix}$ and substitute it into (60):

$$\begin{aligned} i\frac{\partial}{\partial t}\phi' &= e\Phi\phi' + (\Pi^1 - i\Pi^2)\chi' + m\phi', \\ i\frac{\partial}{\partial t}\chi' &= e\Phi\chi' + (\Pi^1 + i\Pi^2)\phi' - m\chi'. \end{aligned} \tag{61}$$

Writing

$$\psi = \begin{pmatrix}\phi \\ \chi\end{pmatrix} e^{-imt}$$

in order to subtract the relativistic rest energy, the coupled system of equations (61) now takes the form:

$$i\frac{\partial}{\partial t}\phi = e\Phi\phi + (\Pi^1 - i\Pi^2)\chi,$$
$$i\frac{\partial}{\partial t}\chi = e\Phi\chi + (\Pi^1 + i\Pi^2)\phi - 2m\chi. \quad (62)$$

In the weak external electromagnetic field, $m \gg e\Phi$ and $m \gg e|\mathbf{A}|$, and so, from the second equation in (62):

$$\chi \approx \frac{1}{2m}(\Pi^1 + i\Pi^2)\phi. \quad (63)$$

Substituting the last expression into the first equation (62), we obtain

$$i\frac{\partial}{\partial t}\phi = e\Phi\phi + \frac{1}{2m}(\Pi^1 - i\Pi^2)(\Pi^1 + i\Pi^2)\phi. \quad (64)$$

The second term on the r.h.s. of the last equation may be rewritten as

$$\frac{1}{2m}\left((\Pi^1)^2 + (\Pi^2)^2 + i[\Pi^1, \Pi^2]\right) = \frac{1}{2m}\left((\mathbf{p} - e\mathbf{A})^2 - e\left(\frac{\partial A^2}{\partial x^1} - \frac{\partial A^1}{\partial x^2}\right)\right),$$

and recalling that $\mathbf{B} = \nabla \times \mathbf{A}$, we reach at the well-known Pauli equation

$$i\frac{\partial}{\partial t}\phi = e\Phi\phi + \frac{(\mathbf{p} - e\mathbf{A})^2}{2m}\phi + \frac{e}{2m}B\phi, \quad (65)$$

which is a *one component* equation for an electron in an external electromagnetic field. We can obtain the non-relativistic form of the Dirac equation through the FW transformation. Notice that in the free Dirac equation,

$$i\frac{\partial \Psi}{\partial t} = (\boldsymbol{\alpha} \cdot p + \beta m)\Psi, \quad (66)$$

the large and small components of the Dirac spinor Ψ, labeled by the momentum $\mathbf{p}$, get mixed by the odd operator $\boldsymbol{\alpha} \cdot \mathbf{p}$ involving off-diagonal elements. The FW is a canonical transformation which, by removing such an operator from the Dirac Hamiltonian,

$$H_{\text{free}} = \boldsymbol{\alpha} \cdot \mathbf{p} + \beta m, \quad (67)$$

allows to decouple these large and small components of Ψ. The free FW transformation,

$$U_{\text{free}}(p) = e^{iS_{\text{free}}(\mathbf{p})} = \cos|\mathbf{p}|\theta + \frac{\boldsymbol{\gamma} \cdot \mathbf{p}}{|\mathbf{p}|}\sin|\mathbf{p}|\theta, \quad (68)$$

with θ given through

$$\tan(2|\mathbf{p}|\theta) = \frac{|\mathbf{p}|}{m}, \quad (69)$$

is exact, and renders the free Hamiltonian in the form

$$H_{\text{free}}^{FW} = \gamma^0\sqrt{\mathbf{p}^2 + m^2}. \quad (70)$$

In the presence of an external electromagnetic field, the FW transformation can be ordinarily obtained through successive approximations as an expansion in powers of $1/m$ (Bjorken & Drell, 1984). For example, at $O(1/m^3)$,

$$S(\Pi) = -i\left(\frac{\gamma^0}{2m}\right)^2 \left([\mathcal{O}', \mathcal{E}'] + i\dot{\mathcal{O}}'\right), \tag{71}$$

with

$$\mathcal{O}' = \frac{\gamma^0}{2m}[\mathcal{O}, \mathcal{E}] - \frac{\mathcal{O}^3}{3m^2} + i\frac{\gamma^0 \dot{\mathcal{O}}}{2m}, \tag{72}$$

$$\mathcal{E}' = \mathcal{E} + \gamma^0\left(\frac{\mathcal{O}}{2m} - \frac{\mathcal{O}^4}{8m^3}\right) - \frac{1}{8m^2}[\mathcal{O}, [\mathcal{O}, \mathcal{E}]] - i\frac{1}{8m^2}[\mathcal{O}, \dot{\mathcal{O}}]. \tag{73}$$

In the expressions above, the even (diagonal) and odd (off-diagonal) operators correspond to $\mathcal{E} = e\Phi$ and $\mathcal{O} = \gamma^0\boldsymbol{\gamma}\cdot\boldsymbol{\Pi} \equiv \tilde{\boldsymbol{\Pi}}$, respectively, and the dot represents the time derivative. To this order, the FW transformation renders the Dirac Hamiltonian to its leading non-relativistic form. For an external static inhomogeneous magnetic field the series can be written in closed form as

$$H_{\Pi}^{FW} = \gamma^0\sqrt{(\tilde{\boldsymbol{\Pi}})^2 + m^2}, \tag{74}$$

where the transformation matrix for this case is

$$S(\Pi) = -i(\boldsymbol{\gamma}\cdot\boldsymbol{\Pi})\theta, \tag{75}$$

with θ given through

$$\tan(2|\tilde{\boldsymbol{\Pi}}|\theta) = \frac{|\tilde{\boldsymbol{\Pi}}|}{m}. \tag{76}$$

Here, $|\tilde{\boldsymbol{\Pi}}| = \sqrt{(\gamma^0\boldsymbol{\gamma}\cdot\boldsymbol{\Pi})^2}$ plays the role of the momentum $|\mathbf{p}|$ in the free case and $(\tilde{\boldsymbol{\Pi}})^2 = \mathcal{H}$ as it was defined by Eq. (36). With the decomposition (48) of the Dirac wave function, the stationary Schrödinger form of the Dirac equation becomes

$$E_D \mathbb{E}_p u_{\overline{p}} = \gamma^0(\boldsymbol{\gamma}\cdot\boldsymbol{\Pi} + m)\mathbb{E}_p u_{\overline{p}} \tag{77}$$

which with the aid of property (44), simplifies to

$$E_D \mathbb{E}_p u_{\overline{p}} = \mathbb{E}_p \gamma^0(\boldsymbol{\gamma}\cdot\overline{\boldsymbol{p}} + m)u_{\overline{p}}. \tag{78}$$

In the above expressions E_D represent the eigenenergies of the Dirac equation. Moreover, the Hamiltonian on the r.h.s. of Eq. (78) acquires a free form involving $\overline{p}$ alone. Thus, it is straightforward to convince oneself that the Ritus eigenfunctions map the FW transformation in external fields to a free transformation which depends on $\overline{p}$, namely

$$e^{iS(\Pi)}\mathbb{E}_p = \mathbb{E}_p e^{iS_{\text{free}}(\overline{p})}. \tag{79}$$

So, the $\mathbb{E}_p$ functions not only render the fermion propagator in external fields diagonal in momentum space, with a free form involving the quantum numbers induced by the field. These also allow to express the exact FW transformation in the presence of the fields in a free form. To see the usefulness of Eq. (79), we first apply the Hamiltonian in Eq. (74) to the Ritus eigenfunctions $\mathbb{E}_p$,

$$H_{\Pi}^{FW}\mathbb{E}_p = \left(\gamma^0\sqrt{(\tilde{\boldsymbol{\Pi}})^2 + m^2}\right)\mathbb{E}_p, \tag{80}$$

which has to be evaluated expanding the square-root operator in a power series of $(\tilde{\mathbf{\Pi}}/m)^2$. This procedure leads to an expression in terms of the eigenvalues k of the operator $(\tilde{\mathbf{\Pi}})^2$ given through Eqs. (31) and (37) with $p^2 = \overline{p}^2$. Since $p_0 = E_D$ are the eigenvalues of $\Pi_0 = i\partial_t$, from the Dirac equation $p_0^2 = k + m^2$, thus $\sqrt{k}$ correspond to the energy eigenvalues of a particle on-shell. From Eq. (36), $\overline{p}^2 = p_0^2 - k$, and it can be fulfilled with the choice of $\overline{p}^\mu = (p_0, 0, \sqrt{k})$, in accordance to our selection of gauge. Hence, Eq. (80) simplifies to

$$H_{\Pi}^{FW}\mathbb{E}_p = \mathbb{E}_p\left(\gamma^0\sqrt{E_D^2 + m^2}\right). \tag{81}$$

On the other hand, notice that under the FW transformation, the Hamiltonian $H = \gamma^0(\gamma \cdot \overline{p} + m)$ on the r.h.s. of Eq. (78) transforms in a free form, as in Eq. (70), but involving $\overline{\mathbf{p}}^2$ alone. Thus the FW transformed Hamiltonian (78) can be written directly:

$$H_{\text{free}}^{FW}\mathbb{E}_p = \mathbb{E}_p\left(\gamma^0\sqrt{E_D^2 + m^2}\right). \tag{82}$$

The r.h.s. of this last equation precisely corresponds to the r.h.s. of Eq. (81). This last was obtained transforming the Dirac Hamiltonian of Eq. (77) with a magnetic filed in the *usual* way. As comparison, with the aid of Eq. (48), the corresponding FW transformed Hamiltonian was obtained directly from a free one, Eq. (78), given in terms of the tri-momentum $\overline{p}^\mu$ which contains all the dynamics induced by the external magnetic field. It is then straightforward to prove the relationship between the FW transformations $S(\Pi)$ and $S_{\text{free}}(\overline{p})$ established by Eq. (79) in terms of the Ritus eigenfunctions $\mathbb{E}_p$. The same idea can be generalized to the case on which the Dirac Hamiltonian (66) is expressed in its ultra-relativistic form through the Cini-Touschek (CT) transformation (Cini & Touschek, 1958). In the free case, the parameter (69) for the CT transformation acquires the form

$$\tan(2|\mathbf{p}|\theta) = -\frac{m}{|\mathbf{p}|}, \tag{83}$$

and correspondingly the hamiltonian takes the form

$$H_{\text{free}}^{CT} = \frac{\sqrt{\mathbf{p}^2 + m^2}}{|\mathbf{p}|}\alpha \cdot \mathbf{p}, \tag{84}$$

which is precisely the ultra-relativistic form of the Dirac Hamiltonian.

7. Graphene hamiltonian in diagonal form

Let us consider the "free" Hamiltonian of graphene

$$H_g = \frac{E_F}{|\mathbf{p}_F|}\gamma \cdot \mathbf{p}. \tag{85}$$

Now, the free Dirac hamiltanian can be written in its non-relativistic and ultrarelativistic forms through the FW and CT transformations, respectively:

$$\begin{aligned} U_{FW}(m, \mathbf{p}) &= \cos\left(\frac{1}{2}\arctan\frac{1}{x}\right) + \frac{\gamma \cdot \mathbf{p}}{|\mathbf{p}|}\sin\left(\frac{1}{2}\arctan\frac{1}{x}\right), \\ U_{CT}(m, \mathbf{p}) &= \cos\left(\frac{1}{2}\arctan x\right) + \frac{\gamma \cdot \mathbf{p}}{|\mathbf{p}|}\sin\left(\frac{1}{2}\arctan x\right), \end{aligned} \tag{86}$$

with $x = m/|\mathbf{p}|$. This means that graphene can be described through a CT-transformed Dirac Hamiltonian,

$$H_g = U_{CT}(\mu, \mathbf{p}_F) H_D U^{\dagger}_{CT}(\mu, \mathbf{p}_F) , \tag{87}$$

with μ some mass parameter of the transformation. In the free case, the CT transformation is exact, thus

$$H_D = U^{\dagger}_{CT}(\mu, \mathbf{p}_F) H_g U_{CT}(\mu, \mathbf{p}_F) . \tag{88}$$

Moreover, we know that

$$H^{FW}_{\Pi} = \gamma^0 E_k \equiv U_{FW}(\mu, \mathbf{p}_F) H_D U^{\dagger}_{FW}(\mu, \mathbf{p}_F) , \tag{89}$$

with $E_k = \sqrt{E_D^2 + \mu^2} = E_F$, therefore

$$H_{FW} = U_{FW}(\mu, \mathbf{p}_F) U^{\dagger}_{CT}(\mu, \mathbf{p}_F) H_g U_{CT}(\mu, \mathbf{p}_F) U^{\dagger}_{FW}(\mu, \mathbf{p}_F) . \tag{90}$$

Now, since

$$\arctan(z) \, \arctan\left(\frac{1}{z}\right) = \mathrm{sgn}(z)\frac{\pi}{2} , \tag{91}$$

we have that

$$U_{FW}(\mu, \mathbf{p}_F) U^{\dagger}_{CT}(\mu, \mathbf{p}_F) = \frac{1}{\sqrt{2}}\left(1 + \mathrm{sgn}(\mu)\frac{\boldsymbol{\gamma}\cdot\mathbf{p}}{|\mathbf{p}|}\right) . \tag{92}$$

Now, consider the Hamiltonian of graphene in a static magnetic field. We learnt that for the FW and CT transformations

$$U(\Pi)\mathbb{E}_p = \mathbb{E}_p U_{\mathrm{free}}(\overline{p}) , \tag{93}$$

in such a manner that

$$H_g = \gamma^0 E_F(\mu) \tag{94}$$

in the Ritus basis, with $E_F(\mu) = \sqrt{\mathbf{p}^2 + \mu^2}$.

8. Concluding remarks

Summarizing, we studied the electron propagator in (2+1)-dimensions in the presence of external electromagnetic fields under the Ritus formalism. We have seen that the Ritus method offers an alternative way to study the electron propagator in the presence of external magnetic fields. Within the framework of this method the electron propagator acquires a free form involving only the dynamical quantum numbers induced by the external field when it is spanned in the Ritus functions E_p, the eigenfunctions of the operator $(\gamma \cdot \Pi)^2$. We have also shown that the Ritus eigenfunctions provide a direct connection with the non-relativistic and the ultra-relativistic limit of the Dirac equation. In the non-relativistic limit case we showed that, in the Ritus basis, the exact Foldy-Wouthuysen (FW) transformation of the Dirac Hamiltonian in presence of an external and time independent electromagnetic field can be expressed in a closed form in terms of a free transformation which only depends on the dynamical quantities induced by the field. In the ultra-relativistic limit, we have shown that the Cini-Touschek (CT) transformed Dirac Hamiltonian leads into the corresponding for graphene. We have shown the relationship between the Ritus eigenfunctions and the FW and CT transformations which let us write down the solutions of the graphene Hamiltonian only in terms of dynamical quantities induced by the external fields, namely, only on terms of $\overline{p}$.

9. Acknowledgments

We acknowledge S. Hernández Ortiz for allowing us to adapt Figs. 1, 2 and 3 from his undergraduate thesis. Support has been received from CIC-UMSNH under project 4.22. AR acknowledges support from SNI and CONACYT grants under project 82230. GM acknowledges support from DGAPA-UNAM grant under project PAPIIT IN118610. AS is supported by postdoctoral fellowship in part by the Office of Nuclear Physics of the Department of Energy under contract de-sc0002179.

10. References

Anguiano, Ma. de J. & Bashir, A. (2005). Fermions in Odd Space-Time Dimensions: Back to Basics, *Few Body Systems* 37, pp. 71-78.

Barducci, A; Giachetti, R. & Pettini, G. (2009). Foldy-Wouthuysen transformation for a spinning particle with anomalous magnetic moment, *J. Phys.* A 42, pp. 355302-1–9.

Bjorken, J. D. & Drell, S. D. (1964). *Relativistic Quantum Mechanics* (New York: McGraw-Hill).

Buhl, M.; Borzeszkowski, H.-H. v. & Chrobok, T. (2008). Approximate Foldy-Wouthuysen transformation of the Dirac Hamiltonian in general spacetimes, *Phys. Rev.* D 78, pp. 104022-1–5, and references therein.

Cini, M. & Touschek, B. (1958). The relativistic limit of the theory of spin 1/2 particles, *Nuovo Cimento* 7, pp. 422-423.

Cooper, F.; Khare, A. & Shukhatme, U. (1995). Supersymmetry and quantum mechanics, *Phys. Rep.* 251, pp. 267-385.

Cooper, F.; Khare, A. & Shukhatme, U. (2001). *Supersymmetry in Quantum Mechanics* (Singapore: World Scientific), pp. 61-80.

De Martino, A.; Dell'Anna, L. & Egger, R. (2007). Magnetic confinement of massless Dirac fermions in Graphene, *Phys. Rev. Lett.* 98, 066802-1–4.

Elizalde, E.; Ferrer E. J. & de la Incera V (2002). Neutrino Self-Energy and Index of Refraction in Strong Magnetic Field: A New Approach, *Ann. of Phys.* 295, pp. 33-49.

Foldy, L. L. & Wouthuysen, S. A. (1950). On the Dirac Theory of Spin 1/2 Particles and Its Non-Relativistic Limit, *Phys. Rev.* 78, pp. 29-36.

Ginzburg, V. L. ed, *Quantum Electrodynamics with Unstable Vacuum*, Nova Science, New York, 1995, pp. 155-161.

Goncalves, B; Obukhov, Y. N. & Shapiro, I. L. (2007). Exact Foldy-Wouthuysen transformation for gravitational waves and magnetic field background, *Phys. Rev.* D 75, pp. 124023-1–7.

Haldane, F. D. M. (1988). Model for a Quantum Hall Effect without Landau Levels: Condensed-Matter Realization of the "Parity Anomaly", *Phys. Rev. Lett.* 61, pp. 2015-2018.

Heidenreich, S.; Chrobok, T. & Borzeszkowski, H.-H. v. (2006). Supersymmetry, exact Foldy-Wouthuysen transformation, and gravity, *Phys. Rev.* D 73, pp. 044026-1–10.

Hernández Ortiz, S. (2011). Mapping to a Supersymmetric Quantum Mechanics of the Dirac Equation on a plane, *Undergraduate Theses*, University of Michoacan. pp. 1-56.

Jackiw, R. & Templeton, S. (1981). How super-renormalizable interactions cure their infrared divergences, *Phys. Rev.* D 23, pp. 2291-2304.

Jackiw, R. & Pi, S.-Y. (2007). Chiral Gauge Theory for Graphene, *Phys. Rev. Lett.* 98, pp. 266402-1–4.

Khalilov, V. R. (1999). QED_{2+1} radiation effects in a strong magnetic field, *Theor. Math. Phys.* 121, pp. 1606-1616.

Kondo, K.-I. (1996). First and Second Order Phase Transitions in Maxwell-Chern-Simons Theory Coupled to Fermions, *Int. J. Mod. Phys.* A 11, pp. 777-822.

Moreno, M. & Méndez-Moreno R. M. (1992). *Proceedings of the Workshop on High-energy Phenomenology* (Singapore: World Scientific, editors Pérez M A and Huerta R), pp. 365-375.

Murguía, G.; Raya, A.; Sánchez, A. & Reyes, E. (2010). The electron propagator in external electromagnetic fields in low dimensions, *Am. J. Phys.* 78, pp. 700-707.

Murguía, G. & Raya, A. (2010). Free form of the Foldy-Wouthuysen transformation in external electromagnetic fields, *J. Phys. A: Math. Theor.* 43, pp. 402005-1–5.

Novoselov, K. S.; Jiang, Z.; Zhang, Y.; Morozov, S. V.; Stormer, H. L.; Zeitler, U.; Maan, J. C.; Boebinger, G. S.; Kim, P. & Geim, A. K. (2007). Room temperature quantum Hall effect in graphene, *Science* 315, pp. 1379-1379.

Obukhov, Y. N.; Silenko, A. J. & Teryaev, O. V. (2009). Spin dynamics in gravitational fields of rotating bodies and the equivalence principle, *Phys. Rev.* D 80, pp. 064044-1–10.

Park, S. & Sim, H.-S. (2008). Magnetic edge states in graphene in nonuniform magnetic fields, *Phys. Rev.* B 77, pp. 075433-1–8.

Ponomarenko, L. A.; Schedin, F.; Katsnelson, M. I.; Yang, R.; Hill, E. W.; Novoselov, K. S. & Geim, A. K. (2008). Chaotic Dirac billiard in graphene quantum dots, *Science 320*, pp. 356-358.

Ramezani Masir, M; Matulis, A. & Peeters, F. M. (2009). Quasibound states of Schrödinger and Dirac electrons in a magnetic quantum dot, *Phys. Rev.* B 79, 155451-1–8.

Raya, A. & Reyes, E. (2010). Fermion condensate and vacuum current density induced by homogeneous and inhomogeneous magnetic fields in (2+1) dimensions, *Phys. Rev.* D 82, pp. 016004-1–8.

Reijniers, J.; Peeters, F. M. & Matulis, A. (2001). Electron scattering on circular symmetric magnetic profiles in a two-dimensional electron gas, *Phys. Rev.* B 64, pp. 245314-1–8.

Ritus, V. I. (1972). Radiative corrections in quantum electrodynamics with intense field and their analytical properties, *Annals Phys.* 69, pp. 555-582.

Ritus, V. I. (1974). On diagonality of the electron mass operator in the constant field, *Pizma Zh. E. T. F.* 20, pp 135-138, in Russian.

Ritus V. I. (1978). The eigenfunction method and the mass operator in quantum electrodynamics of the constant field, *Zh. E. T. F.* 75, pp. 1560-1583, in Russian.

Rusin, T. M. & Zawadzki, W. (2011). Green functions of electrons in monolayer and bilayer graphene in a magnetic field, *J. Phys. A: Math. Theor.* 44, pp. 105201-1–7.

Shimizu, K. (1985). C, P and T Transformations in Higher Dimensions, *Prog. Theor. Phys.* 74, pp. 610-613.

Silenko, A. J. (2008). Foldy-Wouthyusen transformation and semiclassical limit for relativistic particles in strong external fields, *Phys. Rev.* A 77, pp. 012116-1–7.

Wu, X.; Sprinkle, M.; Li, X.; Ming, F.; Berger, C. & de Heer, W. A. (2008). Epitaxial-graphene/graphene-oxide junction: An essential step towards epitaxial graphene electronics, *Phys. Rev. Lett.* 101, pp. 026801-1–4.

14

Nonlinear Plasmonics Near the Dirac Point in Negative-Zero-Positive Index Metamaterials –Optical Simulations of Electron in Graphene

Ming Shen and Linxu Ruan
Physics Department, Shanghai University
People's Republic of China

1. Introduction

In 2004, graphene, a single layer of carbon atoms arranged in a hexagonal lattice, has been experimentally realized by A. K. Geim and K. S. Novoselov (Novoselov et al., 2004). In graphene, the conduction and valence bands touch each other at Dirac point (DP) with a double-cone structure (Novoselov et al., 2004). Near DP, the dispersion of electron is linear with two branches (Novoselov et al., 2004). The valence electron dynamics in such a truly two-dimensional (2D) material is governed by a massless Dirac equation. So graphene exhibits many unique electronic properties (Beenakker, 2008; Castro Neto et al., 2009), including half-integer and unconventional quantum Hall effect (Zhang et al., 2005), observation of minimum conductivity (Novoselov et al., 2005), and Klein tunneling (Katsnelson et al., 2006). The optical-like behaviors of electron waves in graphene have also drawn considerable attention recently, such as focusing (Cheianov et al., 2007), collimation (Park et al., 2008a), subwavelength optics (Darancet et al., 2009), Bragg reflection (Ghosh et al., 2009), and Goos-Hänchen effect (Beenakker et al., 2009; Zhao et al., 2010). In this regard, one of the recent work is to investigate the guided modes in monolayer graphene waveguide, by analogy of optical waveguides (Zhang et al., 2009). The exotic properties of the graphene waveguide are found in two different cases of classical motion and Klein tunneing (Zhang et al., 2009). Similar behaviors also happened to the transmission of Dirac-like electron in 2D monolayer graphene barrier at nonzero angle of incidence (Chen and Tao, 2009). The modulation of the transmission gap by the incidence angle, the height, and width of potential barrier may lead to potential applications in graphene-based electronic devices (Chen and Tao, 2009). The electronics waves in graphene can also be treated as the guided waves in an optical fibre, i.e., graphene based electronic fibre (Wu, 2011). The graphene analog of the optical device, the fibre optic, has also been demonstrated both experimentally and numerically in a $p-n$ junction (Williams et al., 2011).

On the other hand, the DP with double-cone structure in photonic crystals (PCs) for the Bloch states is found from the similarity of the photonic bands of the 2D PCs with the electronic bands of solids. Several novel optical transport properties near the DP have been investigated, such as conical diffraction (Peleg et al., 2007), "pseudodiffusive" scaling (Sepkhanov et al., 2007), photon's Zitterbewegung (Zhang, 2008), and perfect Klein tunneling (Bahat-Treidel et al., 2010a). Up to now, the dynamics of Dirac-like excitations in 2D PCs (honeycomb lattices) have been well studied when the propagation equation is linear.

However, the nonlinear dynamics has drawn little attention in the 2D PCs by present. The first study of nonlinear dynamics in honeycomb lattices was conducted in (Peleg et al., 2007), demonstrating gap solitons, which had no overlap with Bloch modes residing in the vicinity of the Dirac points. Subsequently, the nonlinear version of the massless Dirac equation in honeycomb lattices were studied with the Dirac approximation (Ablowitz et al., 2009; Haddad et al., 2009). The nonlinear interactions can also break down the Dirac dynamics in honeycomb photonic lattices (Bahat-Treidel et al., 2010b), open a gap between the first two bands which can support a gap soliton (Bahat-Treidel et al., 2008).

In optical system, Wang *et al.* have shown that, the analogy phenomenon of Dirac point with double-cone structure can be realized in the negative-zero-positive index metamaterial (NZPIM) (Wang et al., 2009a). It is further found that the light field near DP possesses of pseudodiffusive property obeying the $1/L$ scaling law (Wang et al., 2009a). Subsequently, they study the Zitterbewegung of optical pulses near the Dirac point inside a NZPIM (Wang et al., 2009b). The transmission gap, Bragg-like reflection, and Goos-Hänchen shifts near the DP inside a NZPIM slab was further studied comprehensively by Chen *et al.* (Chen et al., 2009). In addition, when the thermal emission frequency is close to the DP, the spectral hemispherical power of thermal emission in layered structures containing NZPIM is strongly suppressed and the emission can become a high directional source with large spatial coherence (Wang et al., 2010). Shen et al. have studied the guided modes in NZPIM waveguide (Shen et al., 2010a), in which the properties of the guided modes are analogous to the propagation of electron waves in graphene waveguide (Zhang et al., 2009). These unique results suggest that many exotic phenomena in graphene can be simulated by the relatively simple optical NZPIM.

Recently, the nonlinear optical response of graphene has been another interesting subject. It has been predicted that the graphene exhibits a strongly nonlinear optical behavior at microwave and terahertz frequencies (Mikhailov, 2007). At higher, optical frequencies one can also expect an enhanced optical nonlinearity as, due to graphene's band structure, interband optical transitions occur at all photon energies. Hendry et al. have performed the first measurements of the coherent nonlinear optical response of single- and few-layer graphene using four-wave mixing (Hendry et al., 2010). Their results demonstrate that graphene exhibits a very strong nonlinear optical response in the near-infrared spectral region. All the results will stimulate much research on the studies of the nonlinear optical response of graphene. Based on the facts of optics-like phenomena of electron wave in graphene and the similarly nonlinear optical dynamics of Dirac equation in 2D PCs, Shen et al. have studied the nonlinear surface waves (Shen et al., 2010b) and the nonlinear guided modes (Shen et al., 2011) near the DP in NZPIM. For the nonlinear NZPIM waveguide, when the nonlinearity is self-focusing, there exists an asymmetric forbidden band near DP which can be modulated by the strength of the nonlinearity. However, the self-defocusing nonlinearity can completely eliminate the asymmetric band gap (Shen et al., 2011).

This chapter presents a review on the propagation of nonlinear plasmonics in NZPIM. The chapter is organized as follow. In Sec. 2, the nonlinear surface wave is discussed at the interface between a nonlinear conventional dielectric media and a linear NZPIM. By analogy of electron wave in monolayer graphene waveguide, the guided modes in NZPIM waveguide is studied in Sec. 3 by the graphic method. In Sec. 4, the nonlinear guided modes are investigated in NZPIM waveguide with a nonlinear dielectric media substrate. Finally, we make brief prospects of the research and conclusion of this chapter in Sec. 5 and 6, respectively.

2. Nonlinear surface waves near the DP in NZPIM

Surface waves, which propagate along the interface between two media and decay in transverse direction, were previously studied at the interface between metal and dielectric medium (Kivshar, 2008). In such systems, the permittivity of the metal is negative and then only the TM-mode surface waves can exist. The TM-mode surface waves can also exist in nonlinear ferromagnetic and antiferromagnetic materials (Wang and Awai, 1998). Surface waves can also propagate in negative refractive metamaterial with simultaneous negative permittivity and negative permeability (Smith et al., 2000). Following the seminal work on the linear TE- and TM-modes surface waves and the nonlinear surface modes in metamaterial (Ruppin, 2000; Shadrivov et al., 2003; 2004), some very recent studies have illustrated that the nonlinear surface wave becomes a more interesting topic from the physical point of view, due to a host of new phenomena in comparison with the linear surface wave. For example, Xu *et. al.* (Xu et al., 2009) have investigated the nonlinear surface polaritons in anisotropic Kerr-type metamaterials. Very lately, we also discovered the bistable and negative lateral shifts in Kretschmann configuration (Chen et al., 2010), where the nonlinear surface waves can be excited at a Kerr nonlinear metamaterial-metal interface.
In this section, we present a comprehensive study of the properties of the nonlinear surface waves at the interface between semi-infinite media of two types, nonlinear conventional dielectric and linear NZPIM, and demonstrate a number of unique properties of surface waves near the DP in a NZPIM.

2.1 Dispersion equation of the nonlinear surface waves

We consider an interface between the nonlinear conventional medium in the $x > 0$ region and the linear NZPIM in the $x < 0$ region. The permittivity and permeability are ϵ_1^{NL} and μ_1 for the nonlinear conventional medium and

$$\epsilon_2(\omega) = 1 - \frac{\omega_{ep}^2}{\omega^2 + i\gamma_e\omega}, \tag{1}$$

$$\mu_2(\omega) = 1 - \frac{\omega_{mp}^2}{\omega^2 + i\gamma_m\omega}, \tag{2}$$

for the NZPIM (Wang et al., 2009a; Ziolkowski, 2004), respectively. ω_{ep}^2 and ω_{mp}^2 are the electronic and magnetic plasma frequencies, and γ_e and γ_m are the damping rates relating to the absorption of the material. Here we assume $\gamma_e = \gamma_m = \gamma \ll \omega_{ep}^2, \omega_{em}^2$. It is important that when $\omega_{ep} = \omega_{em} = \omega_D$ and $\gamma = 0$ (no loss), then $\epsilon_2(\omega) = \mu_2(\omega) = 1 - \omega_D^2/\omega^2$, which indicates both $\epsilon_2(\omega_D)$ and $\mu_2(\omega_D)$ may be zero simultaneously. ω_D is the frequency of the optical DP (corresponding wavelength is $\lambda_D = 2\pi c/\omega_D$), where two bands touch each other forming a double cone structure. In this case, the linear dispersion near the DP, $\omega \approx \omega_D$, can be written as

$$\kappa(\omega) = \frac{\omega - \omega_D}{v_D} = \frac{2(\omega - \omega_D)}{c}, \tag{3}$$

due to the fact that $\kappa(\omega_D) \approx 0$ and $v_D \simeq c/2$ at the DP, where c is the light speed in vacuum. Near the DP, the light transport obeys the massless Dirac equation as follows (Wang et al., 2009a):

$$\begin{bmatrix} 0 & -i(\frac{\partial}{\partial x} - i\frac{\partial}{\partial y}) \\ -i(\frac{\partial}{\partial x} + i\frac{\partial}{\partial y}) & 0 \end{bmatrix} \Psi = \left(\frac{\omega - \omega_D}{v_D}\right) \Psi, \tag{4}$$

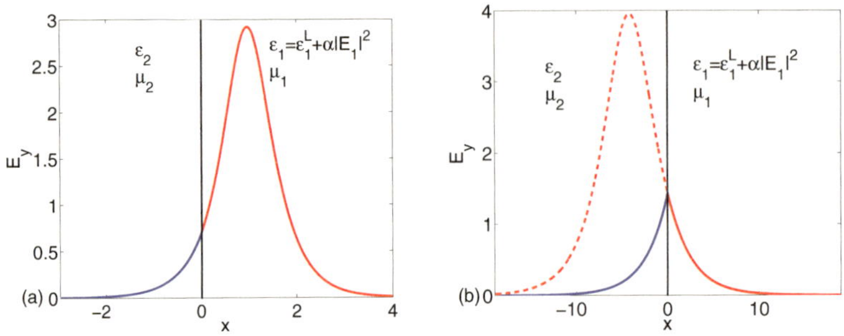

Fig. 1. (Color online). The transverse profile of the nonlinear surface waves: (a) $\mu_2 > 0$ and (b) $\mu_2 < 0$.

where $\Psi = \begin{pmatrix} E_{z1}(x,y,\omega) \\ E_{z2}(x,y,\omega) \end{pmatrix}$ are the eigenfunctions of the electric fields with the same $k(\omega)$. We consider the TE-mode electric field in this paper, and the propagation of the surface waves obey the following nonlinear differential equations

$$\frac{\partial^2 E_y}{\partial x^2} - (\kappa^2 - \kappa_0^2 \epsilon_1^{NL} \mu_1) E_y = 0, x > 0, \tag{5}$$

$$\frac{\partial^2 E_y}{\partial x^2} - \kappa_2^2 E_y = 0, x < 0, \tag{6}$$

where $\kappa_0 = \omega / c$ is the wave vector in vacuum, $\epsilon_1^{NL} = \epsilon_1^L + \alpha |E_1|^2, \epsilon_1^L$ is the linear permittivity and α is the nonlinear index of medium 1, κ is the propagation constant of the nonlinear surface wave, and $\kappa_1^2 = \kappa^2 - \kappa_0^2 \epsilon_1^L \mu_1$ and $\kappa_2^2 = \kappa^2 - [2(\omega - \omega_D)/c]^2$ are the decay constants in nonlinear medium and NZPIM, respectively. We only consider a self-focusing nonlinearity with $\alpha > 0$. It should be strengthened that for the self-defocusing nonlinearity $\alpha < 0$, the transverse electronic field will very different from the case of self-focusing nonlinearity. The solution of Eqs. (5) and (6) have the following form

$$E_{1y} = \frac{\kappa_1}{\kappa_0} \sqrt{\frac{2}{\alpha \mu_1}} sech[\kappa_1 (x - x_0)], x > 0, \tag{7}$$

$$E_{2y} = E_2 \exp(\kappa_2 x), x < 0. \tag{8}$$

Applying the continuity of wave function at the interface $x = 0$, we obtain two equations as follow

$$E_2 = \frac{\kappa_1}{\kappa_0} \sqrt{\frac{2}{\alpha \mu_1}} sech(\kappa_1 x_0), \tag{9}$$

$$\frac{\kappa_2}{\mu_2} E_2 = \frac{\kappa_1^2}{\mu_1 \kappa_0} \sqrt{\frac{2}{\alpha \mu_1}} \frac{\tanh(\kappa_1 x_0)}{\cosh(\kappa_1 x_0)}, \tag{10}$$

which yields

$$\tanh(\kappa_1 x_0) = \frac{\mu_1 \kappa_2}{\mu_2 \kappa_1}, \tag{11}$$

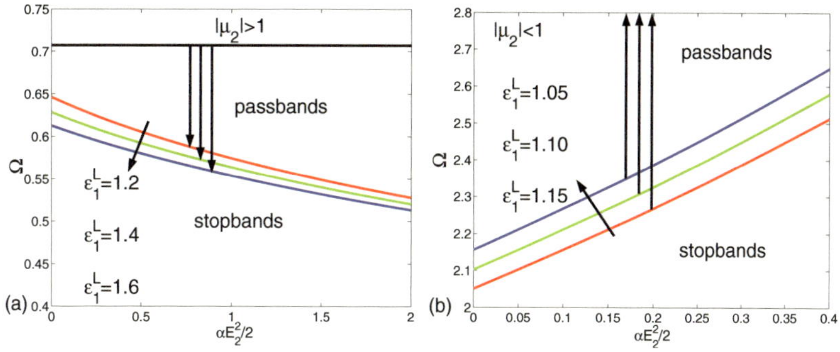

Fig. 2. (Color online). Variation of Ω with $\alpha E_2^2/2$, for three different value ϵ_1^L when (a) $|\mu_2| > 1$ ($\mu_2 < 0$) and (b) $|\mu_2| < 1$ ($\mu_2 > 0$).

and the dispersion equation

$$\frac{\kappa^2}{\kappa_0^2} = \frac{\mu_1}{\mu_2^2 - \mu_1^2}\left[\mu_2^2\epsilon_1^L + \frac{\mu_2^2\alpha E_2^2}{2} - 4\mu_1(1-\frac{\omega_D}{\omega})^2\right]. \tag{12}$$

From Eq. (11), we know that the transverse profile of the nonlinear surface waves depend on the sign of μ_2. The maximum of the dielectric fields amplitude are located at the interface when $\mu_2 < 0$ and located inside the nonlinear dielectric medium when $\mu_2 > 0$, as shown in Fig. 1.

The power fluxes is described by the Poynting vector $P = 1/2\int(\vec{E}\times\vec{H}^*)_z dx = P^{NL} + P^L$ with

$$P^{NL} = \frac{\kappa\kappa_1}{\mu_0\mu_1^2\kappa_0^2\omega\alpha}\left(1 + \frac{\mu_1\kappa_2}{\mu_2\kappa_1}\right), \tag{13}$$

$$P^L = \frac{\kappa\kappa_1^2}{2\mu_0\mu_1\mu_2\kappa_2\kappa_0^2\omega\alpha}\left[1 - \left(\frac{\mu_1\kappa_2}{\mu_2\kappa_1}\right)^2\right], \tag{14}$$

are the power fluxes in the nonlinear medium and NZPIM, respectively.

The dispersion equation Eq. (12) tells us that increasing αE_2^2 will reduce or increase the effective wave index κ/κ_0. For the surface waves, the propagation constant should be larger than wave vector in the nonlinear medium and NZPIM

$$\kappa^2 > \kappa_0^2\epsilon_1^L\mu_1, \kappa^2 > \left[\frac{2(\omega-\omega_D)}{c}\right]^2 = 4\kappa_0^2(1-\frac{\omega_D}{\omega})^2. \tag{15}$$

Define a parameter $\Omega = \omega/\omega_D$, and then the permittivity and the permeability of NZPIM are $\epsilon_2 = \mu_2 = 1 - 1/\Omega^2$. Eq. (12) also gives the following necessary conditions for the nonlinear surface wave existence, since κ^2/κ_0^2 should be positive:

$$\mu_2^2\epsilon_1^L + \frac{\mu_2^2\alpha E_2^2}{2} - 4\mu_1(1-\frac{\omega_D}{\omega})^2 < 0, (|\mu_2| < \mu_1), \tag{16}$$

$$\mu_2^2\epsilon_1^L + \frac{\mu_2^2\alpha E_2^2}{2} - 4\mu_1(1-\frac{\omega_D}{\omega})^2 > 0, (|\mu_2| > \mu_1). \tag{17}$$

In the next, we will study the nonlinear surface waves in NZPIM when $|\mu_2| < \mu_1$ and $|\mu_2| > \mu_1$, respectively.

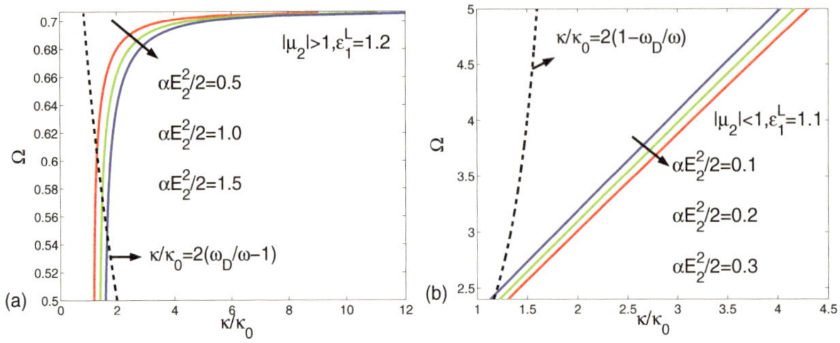

Fig. 3. (Color online). Dispersion relation of Ω versus normalized propagation constant, for three different $\alpha E_2^2/2$ when (a) $|\mu_2| > 1$ and (b) $|\mu_2| < 1$.

2.2 Pass and stop bands, power fluxes, and group velocity of the nonlinear surface waves

Case A: $|\mu_2| < \mu_1$. From the Eqs. (15) and (16), we obtain the following condition for the surface propagation:

$$\epsilon_1^L < 4(1 - \frac{1}{\Omega})^2 - \frac{\alpha E_2^2}{2}. \tag{18}$$

For simplicity, we assume that $\mu_1 = 1$ in this article. Since $|\mu_2| < 1$, Ω obeys $\sqrt{2}/2 < \Omega < 1$ or $\Omega > 1$. We rewrite Eq. (18) as

$$A\Omega^2 + B\Omega + C > 0, \tag{19}$$

where $A = 4 - (\epsilon_1^L + \frac{\alpha E_2^2}{2})$, $B = -8$, $C = 4$, and $a = \epsilon_1^L + \frac{\alpha E_2^2}{2} > 1$ for general dielectric medium. The solution of Eq. (19) is

$$\Omega > \frac{4 + 2\sqrt{a}}{4 - a}, 1 < a < 4. \tag{20}$$

Case B: $|\mu_2| > \mu_1$. From the Eqs. (15) and (17), we obtain that

$$\epsilon_1^L > 4(1 - \frac{1}{\Omega})^2 - \frac{\alpha E_2^2}{2}. \tag{21}$$

Since $|\mu_2| > 1$, Ω obeys $0 < \Omega < \sqrt{2}/2$. Similarly, we can get the solution of Eq. (21)

$$\frac{4 - 2\sqrt{a}}{4 - a} < \Omega < \frac{\sqrt{2}}{2}, a > 1. \tag{22}$$

From Eqs. (20) and (22), we know that there is a forbidden band near the optical dirac point for the nonliear surface waves. For $\sqrt{2}/2 < \Omega < 2$, the nonlinear surface wave do not exist. The existence regions of the nonlinear surface waves also depend on the frequency properties and the nonlinear value $\alpha E_2^2/2$, i.e., the passbands and stopbands, as shown in Fig. 2. It is shown that when $|\mu_2| > 1$ ($\mu_2 < 0$) the passbands of the surface waves have a maximum frequency limit $\sqrt{2}/2\omega_D$. The increase of the nonlinear part $\alpha E_2^2/2$ reduces the lower frequency limit Ω and widens the passband, as shown in Fig. 2 (a). From Fig. 2 (b), we know that, the passbands have a minimum frequency limit, and the increase of the nonlinear part $\alpha E_2^2/2$ increases the minimum frequency limit and reduces the passbands when $|\mu_2| < 1$ ($\mu_2 > 0$). For that the value of the nonlinear part $\alpha E_2^2/2$ can be modulated by increasing or decreasing the power, so

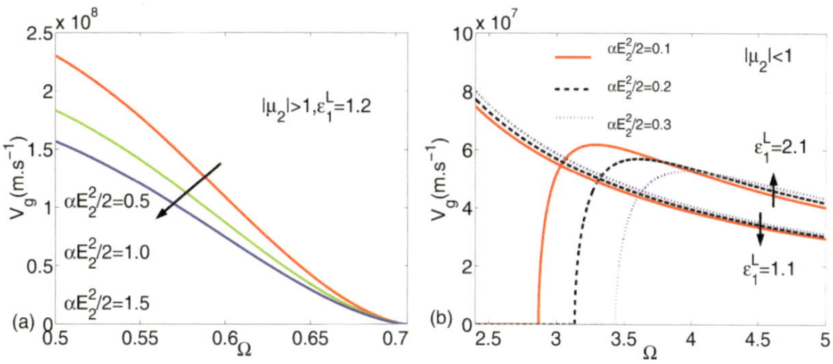

Fig. 4. (Color online). Group velocity V_g versus normalized frequency, for three different $\alpha E_2^2/2$ when (a) $|\mu_2| > 1$ and (b) $|\mu_2| < 1$.

it maybe have potential applications in optical devices and guided wave optics.
In Fig. 3, we plot the frequency dispersion of the nonlinear surface waves. We find that the normalized propagation constants will increase when the normalized frequency increase for both $\mu_2 > 0$ and $\mu_2 < 0$. The dispersion curves are always positive in the existence regions which means the nonlinear surface waves are always forward with a positive power fluxes. This is very different from the frequency dispersion in left-hand materials, whereas the dispersion curves may be negative under some conditions (Shadrivov et al., 2004). Fig. 3 also shows that the dispersion curves have a maximum (minimum) limit for the parameter Ω when $|\mu_2| > 1$ ($|\mu_2| < 1$), which corresponding to the forbidden bands described in Fig. 2. As to the case $|\mu_2| > 1$ (negative index), we can see the upper limit for the parameter Ω is $\Omega = \sqrt{2}/2$.
Group velocity is an important parameter for the propagation of the surface waves. It can describe the direction of the power fluxes. We rewrite dispersion relation Eq. (12) as

$$\kappa = \kappa_0 \left\{ \frac{\mu_1}{\mu_2^2 - \mu_1^2} \left[\mu_2^2 \epsilon_1^L + \frac{\mu_2^2 \alpha E_2^2}{2} - 4\mu_1 (1 - \frac{\omega_D}{\omega})^2 \right] \right\}^{1/2}, \tag{23}$$

which yields that

$$\begin{aligned} \frac{d\kappa}{d\omega} &= \frac{\kappa}{\omega} + \frac{\kappa_0^2}{2\kappa} \{ (\frac{\mu_1}{\mu_2^2 - \mu_1^2})' [\mu_2^2 \epsilon_1^L + \frac{\mu_2^2 \alpha E_2^2}{2} - 4\mu_1 (1 - \frac{\omega_D}{\omega})^2] \\ &+ \frac{\mu_1}{\mu_2^2 - \mu_1^2} [(2\epsilon_1^L + \alpha E_2^2)\mu_2 \mu_2' - 8\mu_1 (1 - \frac{\omega_D}{\omega})(1 - \frac{\omega_D}{\omega})'] \}, \end{aligned} \tag{24}$$

where $(\frac{\mu_1}{\mu_2^2 - \mu_1^2})' = -\frac{2\mu_1 \mu_2 \mu_2'}{(\mu_2^2 - \mu_1^2)^2}$, $\mu_2' = 2\omega_D^2/\omega^3$, and $(1 - \frac{\omega_D}{\omega})' = \omega_D/\omega^2$. The group velocity can be represented as

$$V_g = d\omega/d\kappa = (d\kappa/d\omega)^{-1}. \tag{25}$$

In Fig. 4, we plot the group velocity of the nonlinear surface wave in two different conditions $|\mu_2| > 1$ and $|\mu_2| < 1$. We know that the group velocity of the nonlinear surface waves is always positive which means the power fluxes is always positive and the surface waves are always forward. When $\mu_2 < 0$, the group velocity will decrease when the frequency increase.

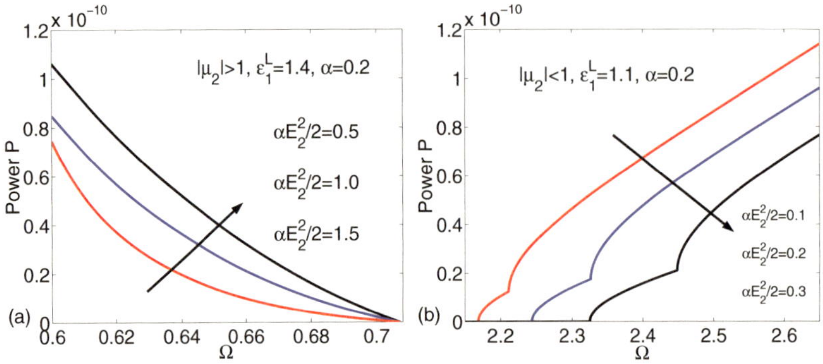

Fig. 5. (Color online). Total power versus normalized frequency, for three different $\alpha E_2^2/2$ when (a) $|\mu_2| > 1$ and (b) $|\mu_2| < 1$.

It will reach zero at the point $\omega = \sqrt{2}/2\omega_D$, which means the nonlinear surface waves is forbidden at this point. When $\omega = \sqrt{2}/2\omega_D$, $|\mu_2| = \mu_1 = 1$ ($\mu_2 = -1$), we have $\kappa \to \infty$ and the nonlinear surface wave stops. The group velocity will decrease when the nonlinear part $\alpha E_2^2/2$ increase. However, they will all stop at the point $\omega = \sqrt{2}/2\omega_D$, as shown in Fig. 4 (a). Fig. 4 (b) shows that the group velocity has a lower frequency limit for a given ϵ_1^L and the nonlinear part $\alpha E_2^2/2$ when $|\mu_2| < 1$. This lower frequency limit imply a power threshold of the surface waves. The lower frequency limit will increase when the linear permittivity of ϵ_1^L or the nonlinear part $\alpha E_2^2/2$ increase. This result can also be obtained from Fig. 2 (b), the passbands have a minimum frequency limit, and the increase of the nonlinear part $\alpha E_2^2/2$ increases the minimum frequency limit and reduces the passbands. We also find the group velocity will increase firstly and then decrease when it reaches the maximum. The group velocity is always positive when $\mu_2 < 0$ which means the nonlinear surface waves are always forward with positive power fluxes.

We also plot the power fluxes of the nonlinear surface waves in Fig. 5 by calculating Eqs. (13) and (14). It is shown that the power flux will approximately linear decrease near the Dirac point when $\mu_2 < 0$, as shown in Fig. 5 (a). And the power flux will be zero when $\omega = \sqrt{2}/2\omega_D$ for that the nonlinear surface waves stops at this point. We also find that increase the nonlinear part $\alpha E_2^2/2$ can effectively increase the power fluxes of the nonlinear surface waves when the frequency is near the Dirac point. When $\mu_2 > 0$, from Fig. 5 (b) we can see that the power fluxes have a frequency threshold for a given ϵ_1^L and the nonlinear part $\alpha E_2^2/2$. This threshold will increase when the nonlinear part $\alpha E_2^2/2$ increase, and this result also shown in Fig. 4 (b). The power flux will also increase when the frequency increase. However, the power fluxes will decrease when the nonlinear part $\alpha E_2^2/2$ increase for the same frequency.

3. Guide modes in NZPIM waveguide

In 2009, Zhang et al. have investigated the guided modes in monolayer graphene waveguide, by analogy of optical waveguides (Zhang et al., 2009). In this section, we will investigate systemically the guided modes in NZPIM waveguide by using the graphic method. For the fast wave guided modes, it is shown that the fundamental mode is absent when the angular frequency is smaller than the DP. Whereas the NZPIM waveguide behaves like conventional dielectric waveguide, when the angular frequency is larger than the DP. The unique properties of the guided modes are very similar as the propagation of electronic wave in graphene waveguide, corresponding to the classical motion and the Klein tunneling Zhang et al. (2009).

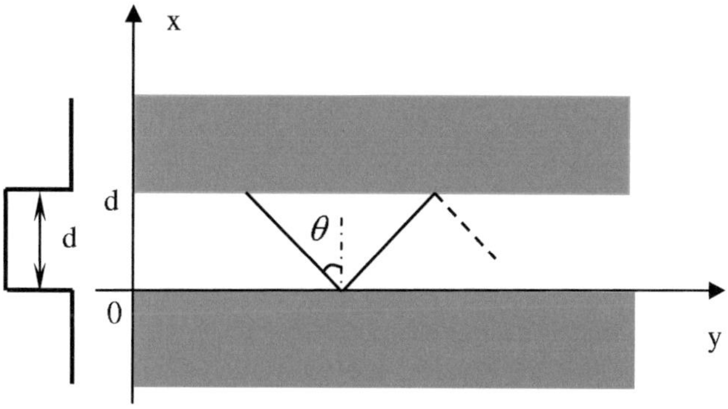

Fig. 6. Schematic structure of negative-zero-positive index metamaterial (NZPIM) waveguide, where the core is the air with the thickness is d and the cladding is the so-called NZPIM.

3.1 Model and basic equation

We consider a waveguide structure of NZPIM, as shown in Fig. 6, where the core is the air with the thickness is d and the cladding is the so-called NZPIM, the optical wave with angle θ is incident upon the waveguide, the direction of the guide modes is z axis, and there are two types of situations: (a) when the incident angle is less than the total internal reflection (TIR) angle, the modes become radiation modes; (b) if the incident angle is more than the critical angle, there will exist oscillating guided modes. What as follows we will focus on the latter case. The TIR angle is defined by $\sin\theta_c = \kappa_2/\kappa_1$, where $\kappa_1 = \omega/c$ is the wavevector in the air, and $\kappa_2 = (\omega - \omega_D)/v_D$ is the wave vector of the NZPIM near the DP (Wang et al., 2009a).
We consider the transverse electric (TE) guided modes [TM modes can be obtained in the same way], the electric fields in the three regions can be written as

$$\psi_A(x) = \begin{cases} Ae^{\alpha x}e^{i\beta y}, & x < 0, \\ [B\cos(\kappa_x x) + C\sin(\kappa_x x)]\, e^{i\beta y}, & 0 < x < d, \\ De^{-\alpha(x-d)}e^{i\beta y}, & x > d, \end{cases} \tag{26}$$

where $\kappa_x = \kappa_1\cos\theta$, $\beta = \kappa_1\sin\theta$ is the propagation constant of the guide modes, and $\alpha = \sqrt{\beta^2 - \kappa_2^2}$ is the decay constant in the cladding region.
Applying the continuity of wave function at the interface $x = 0$ and $x = d$, we obtain the corresponding dispersion equation as follow:

$$\tan(\kappa_x d) = \frac{2\mu_1\mu_2\alpha\kappa_x}{\mu_2^2\kappa_x^2 - \mu_1^2\alpha^2}. \tag{27}$$

We make Eq. (27) in dimensionless form

$$F(\kappa_x d) = \frac{2\mu_1\mu_2(\kappa_x d)\sqrt{(\kappa_1 d)^2 - (\kappa_x d)^2 - (\kappa_2 d)^2}}{\mu_2^2(\kappa_x d)^2 - \mu_1^2[(\kappa_1 d)^2 - (\kappa_x d)^2 - (\kappa_2 d)^2]}. \tag{28}$$

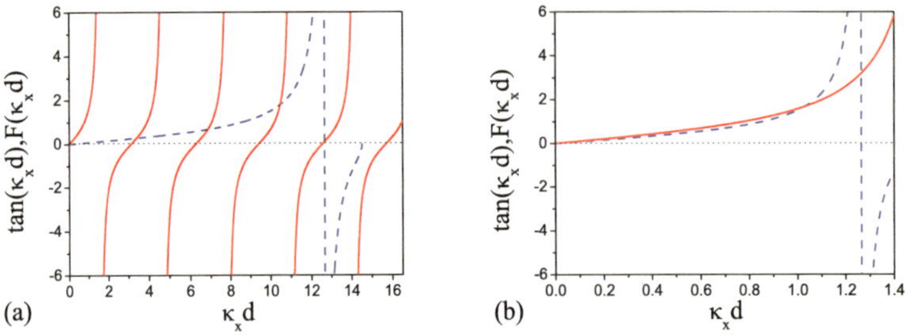

Fig. 7. (Color online). Graphical determination of $\kappa_x d$ for fast wave guided modes when $w < w_D$. The solid and dashed curves correspond to tan $(\kappa_x d)$ and $F(\kappa_x d)$, respectively. The initial parameters are $\omega_D = 2\pi \times 10GHz$, $\omega = 0.8\omega_D$ which means the total reflection angle is $\theta_c = 30°$, the thickness of the core are (a) $d = 10cm$ and (b) $d = 1cm$.

The dispersion Eq. (28) is a transcendental one and cannot be solved analytically, so we propose a graphical method to determine the solution of $\kappa_x d$ for the guided modes. We will discuss the properties of the guided modes in two cases $\omega < \omega_D$ and $\omega > \omega_D$, respectively.

3.2 Fast wave guided modes

Case 1: $\omega < \omega_D$. The critical angle is defined as

$$\theta_c = \sin^{-1}\left[2\left(\frac{\omega_D}{\omega} - 1\right)\right] \tag{29}$$

with the necessary condition $\frac{2}{3}\omega_D < \omega < \omega_D$ (Chen et al., 2009).

As shown in Fig. 7 (a), we plot the dependencies of $\tan(\kappa_x d)$ and $F(\kappa_x d)$ on $\kappa_x d$. The intersections show the existence of the guided modes, as shown in Fig. 8 (a), (b), and (c), corresponding to the TE_2, TE_3, and TE_4 modes, respectively. We find that for some waveguide parameters, the lower-order mode TE_1 can not coexist with higher-order guided modes. So we can not solve TE_1 mode in the same graph Fig 7 (a). This is because that the waveguide parameters used in Fig. 7 (a) does not satisfy the dispersion relation of Eq. (30) when $m = 1$. We can reduce the thickness of the waveguide to obtain the TE_1 mode in the NZPIM waveguide, as shown in Fig. 8 (d), which corresponds to the dispersion relation graphic of Fig 7 (b) with the waveguide thickness is $d = 1cm$.

Another interesting property of the guided modes is that the absence of fundamental TE_0 mode for any parameters of the NZPIM waveguide, which is a novel property different from that in conventional waveguide. The unique property is similar as the guide modes of electron waves in graphene waveguide, where the fundamental mode is absent in the Klein tunneling case (Zhang et al., 2009). For the TE modes, we can write the dispersion relation Eq. (27) as

$$\kappa_x d = m\pi + 2\phi, m = 0, 1, 2, ... \tag{30}$$

where

$$\phi = \arctan\left(\frac{\mu_1 \alpha}{\mu_2 \kappa_x}\right), \tag{31}$$

is negative (angular frequency is smaller than the Dirac point, $\mu_2 < 0$, corresponding Klein tunneling in graphene), which represents the phase retardation upon the total internal reflection at the interface between air and the NZPIM. From Eq. (30), we know that for the

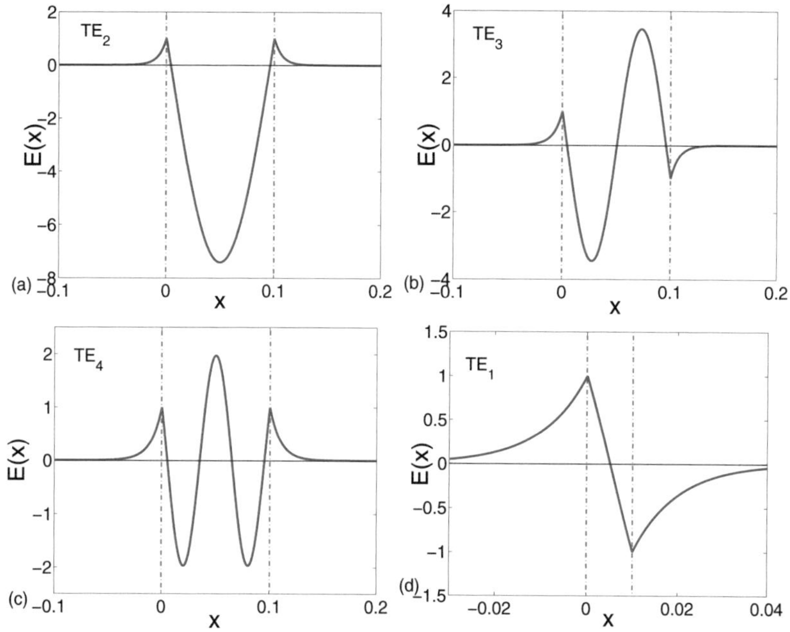

Fig. 8. (Color online) The wave function of guided modes as a function of distance of NZPIM waveguide corresponding to the intersection in Fig. 7 when $\omega < \omega_D$. (a) TE_2: $\kappa_x d = 3.41$; (b) TE_3: $\kappa_x d = 6.87$; (c) TE_4: $\kappa_x d = 10.49$; (d) TE_1: $\kappa_x d = 1.03$.

fundamental mode ($m = 0$), it does not meet with the required dispersion relation. In fact, the condition for the guided waves to exist in a slab waveguide, has a simple physical meaning: the round-trip accumulation of phase due to wave propagation across the layer, $2\phi_{prop}$, including the phase retardation upon the total internal reflection, $2\phi_{refl}$, should be equal to a multiple of 2π (Shadrivov et al., 2005). When the angular frequency is smaller than the Dirac point (the permittivity and the permeability are both negative, NZPIM can be treated as left-handed material), the total phase change does not satisfy the required dispersion relation of Eq. (30), and no fundamental guided modes exist (Shadrivov et al., 2003; 2005). This result is also shown in Fig. (11), where we plot the propagation constant of the guided modes as a function of incident frequency near the DP. It is obviously that the dispersion of TE_0 mode only exist when $\omega > \omega_D$.

Case 2: $\omega > \omega_D$. The critical angle is defined as

$$\theta_c = \sin^{-1}\left[2\left(1 - \frac{\omega_D}{\omega}\right)\right] \tag{32}$$

with the necessary condition $\omega_D < \omega < 2\omega_D$ (Chen et al., 2009). Similarly, we obtain the guided modes of the NZPIM waveguide by using the graphical method, as shown in Fig. 9. It is shown that when $\omega > \omega_D$, the properties of the NZPIM waveguide can be treated as a conventional dielectric waveguide. From Fig. 10, we can see that the fundamental odd and even guided modes can coexist with higher-order modes within the same waveguide for general parameters, which is very different from the case when $\omega < \omega_D$. Under this condition, it corresponds to the guided modes in graphene waveguide in classical motion (Zhang et al., 2009).

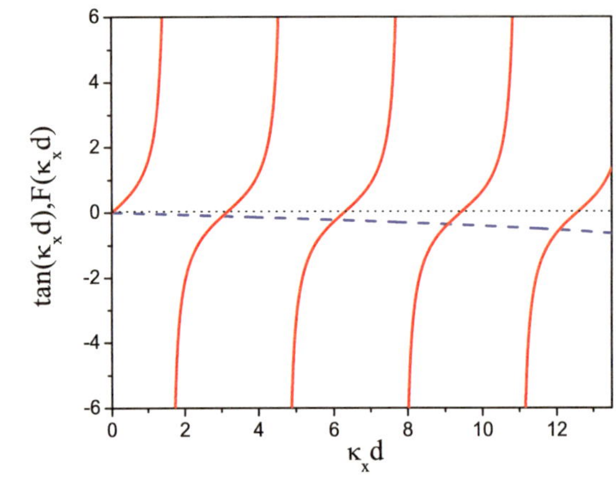

Fig. 9. (Color online) Graphical determination of $\kappa_x d$ for fast wave guided modes when $\omega > \omega_D$. The solid and dashed curves correspond to tan $(\kappa_x d)$ and $F(\kappa_x d)$, respectively. The initial parameters are $\omega_D = 2\pi \times 10GHz$, $\omega = 4\omega_D/3$ which means the total reflection angle is $\theta_c = 30°$, the thickness of the core is $d = 10cm$.

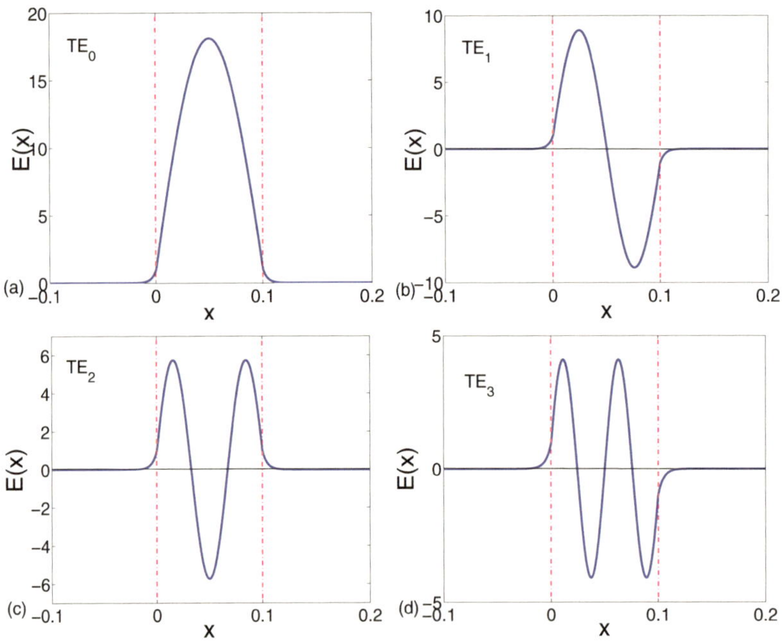

Fig. 10. (Color online) The wave function of guided modes as a function of distance of NZPIM waveguide corresponding to the intersection in Fig. 9 when $\omega > \omega_D$. (a) TE_0: $\kappa_x d = 3.03$; (b) TE_1: $\kappa_x d = 6.06$; (c) TE_2: $\kappa_x d = 9.07$; (d) TE_3: $\kappa_x d = 12.07$.

In order to show further the unique properties of guides modes near the DP in NZPIM waveguides, we plot the dispersion of the guided modes when the incident frequency varies from $\omega < \omega_D$ to $\omega > \omega_D$ in Fig. 11. As discussed above, we can see that TE_0 mode only exist when $\omega > \omega_D$. In addition, another important and interesting property of the guided modes is that there exists an asymmetric forbidden band for the dispersion. The band will also become wider when the order of the guided modes increases with increasing the incidence angle. The result indicates that the modes are not continuous near the DP. This behavior on the forbidden band discussed here is very similar to the transmission gap in the NZPIM slab (Chen et al., 2009). It seems that the guided modes near the DP are quite different from the negative refractive index metamaterial waveguides discussed in Ref. (Shadrivov et al., 2003), though one can divide NZPIM two parts with positive index and negative index respectively by DP, which corresponds to $\omega > \omega_D$ and $\omega < \omega_D$.

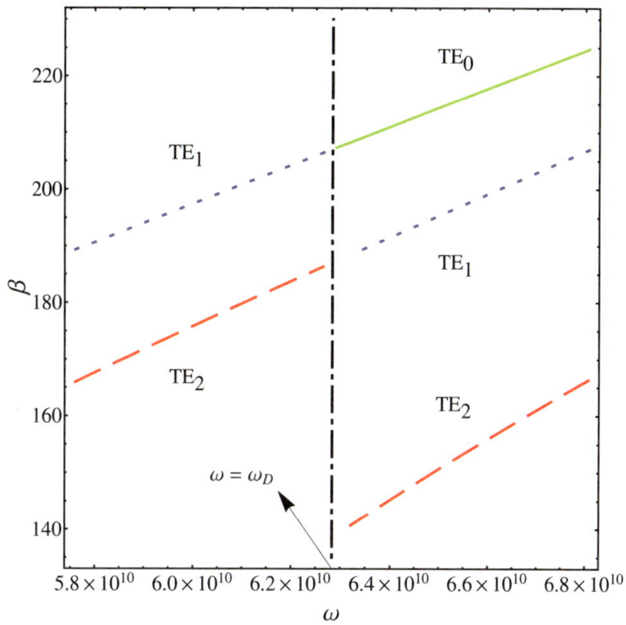

Fig. 11. (Color online) The propagation constant β versus the incident frequency ω near the DP in the NZPIM waveguide.

3.3 Slow wave guided modes

We also find that when $\omega < \omega_D$, the NZPIM waveguide can propagate surface guided modes-slow wave. In this case, the function of the modes in core become sinh and cosh with the imaginary transverse κ_x, and the electric fields in three regions can be written as

$$\psi_A(x) = \begin{cases} Ae^{\alpha x}e^{i\beta y}, & x < 0, \\ [B\cosh(\kappa_x x) + C\sinh(\kappa_x x)]\, e^{i\beta y}, & 0 < x < d, \\ De^{-\alpha(x-d)}e^{i\beta y}, & x > d, \end{cases} \tag{33}$$

where κ_x is the transverse decay constant in the core region, and $\beta^2 = \kappa_1^2 + \kappa_x^2$ is the propagation constant of the slow wave guided modes, and $\alpha = \sqrt{\beta^2 - \kappa_2^2}$ is the decay constant in the cladding region.

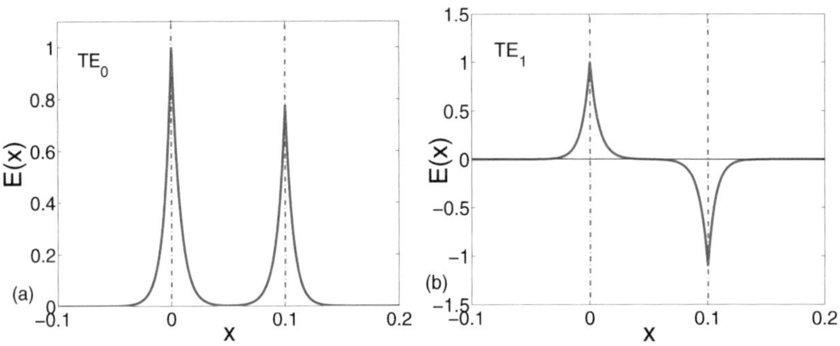

Fig. 12. (Color online). The wave function of guided modes as a function of distance of NZPIM waveguide. The initial parameters are $\omega_D = 2\pi \times 10GHz$, $\omega = 0.69\omega_D$, and the thickness of the core is $d = 10cm$.(a) TE_0: $\kappa_x d = 13.8109$; (b) TE_1: $\kappa_x d = 13.8112$.

Similarly, we obtain the corresponding dispersion equation of this system

$$\tanh(\kappa_x d) = -\frac{2\mu_1\mu_2\alpha\kappa_x}{\mu_2^2\kappa_x^2 + \mu_1^2\alpha^2}. \tag{34}$$

Write Eq. (34) in dimensionless form as follow

$$F(\kappa_x d) = -\frac{2\mu_1\mu_2(\kappa_x d)\sqrt{(\kappa_1 d)^2 + (\kappa_x d)^2 - (\kappa_2 d)^2}}{\mu_2^2(\kappa_x d)^2 + \mu_1^2[(\kappa_1 d)^2 + (\kappa_x d)^2 - (\kappa_2 d)^2]}. \tag{35}$$

As discussed above, we also propose a graphical method to solve the surface guided modes. We find that only fundamental odd and even surface guided modes can exist in the waveguide for some parameters. As shown in Fig. 12, higher-order surface modes are forbidden except the TE_0 and TE_1 surface guided modes. These results obtained here also predict the surface mode of electrons and holes in graphene waveguide.

We emphasize that these results discussed here do extend the investigations (Shadrivov et al., 2003; 2005) and applications (Tsakmakidis et al., 2007) of the waveguide containing only left-handed material. On one hand, we can control the properties of guides modes for the potential applications by adjusting the angular frequency with respect to the DP. On the other hand, our work will also motivate the further work to simulate many exotic phenomena in graphene with relatively simple optical benchtop experiments, based on the links between Klein paradox and negative refraction (Güney and Meyer, 2009).

4. Tunable band gap near the Dirac point in nonlinear negative-zero-positive index metamaterial waveguide

The optical-like behaviors of electron waves in graphene have also drawn considerable attention recently, for example, graphene based electronic fibre and waveguide. In Sec. 3, we have studied the guided modes in NZPIM waveguide (Shen et al., 2010a), in which the properties of the guided modes are analogous to the propagation of electron waves in graphene waveguide (Zhang et al., 2009). However, the nonlinearity may affect the properties of guided modes near the DP in a special manner. In this section, we investigate systemically the guided modes in nonlinear NZPIM waveguide. When the nonlinearity is self-focusing, there exists an asymmetric forbidden band near DP which can be modulated by the strength of the nonlinearity. However, the self-defocusing nonlinearity can completely eliminate the asymmetric band gap.

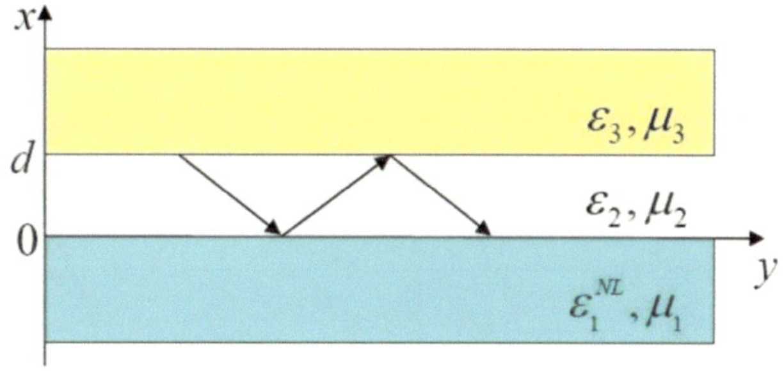

Fig. 13. (Color online) Schematic structure of the NZPIM waveguide with a nonlinear substrate, the core thickness is d.

4.1 Model and basic equation

Considering a nonlinear NZPIM waveguide structure, as shown in Fig. 13. The core of the waveguide is conventional dielectric medium with the thickness d, the permittivity and the permeability are ϵ_2 and μ_2. The substrate is the nonlinear medium with the permittivity and the permeability are $\epsilon_1^{NL} = \epsilon_1^L + \alpha|E_1|^2$ and μ_1, where ϵ_1^L is the linear refractive index and α is the nonlinear coefficient index, $\alpha > 0$ ($\alpha < 0$) corresponding to self-focusing (self-defocusing) nonlinearity. The cladding is NZPIM with the permittivity and the permeability are $\epsilon_3 = \mu_3 = 1 - \omega_D^2/\omega^2$ (Shen et al., 2010a; Wang et al., 2009a).

We only consider the transverse electric (TE) nonlinear guided modes. When the substrate has a self-focusing nonlinearity $\alpha > 0$, the electric fields in three regions can be written as

$$\psi(x) = \begin{cases} \frac{k_1}{k_0}\sqrt{\frac{2}{\alpha\mu_1}}\operatorname{sech}[k_1(x - x_0)], & x < 0, \\ Ae^{ik_2(x-d)} + Be^{-ik_2(x-d)}, & 0 < x < d, \\ Ce^{-k_3(x-d)}, & x > d, \end{cases} \tag{36}$$

where $k_1^2 = \beta^2 - k_0^2\epsilon_1^L\mu_1$ and $k_3^2 = \beta^2 - [2(\omega - \omega_D)/c]^2$ are the transverse decay indexes in substrate and cladding, $k_2^2 = k_0^2\epsilon_2\mu_2 - \beta^2$ is the transverse wave vector of the guided modes in core and it is real, β is the propagation constant of the nonlinear fast wave guide modes. For convenience, we assume the nonlinear substrate and the core are non-magnetic medium with $\mu_1 = \mu_2 = 1$.

Applying the continuity of wave function at the interfaces $x = 0$ and $x = d$, we obtain the corresponding dispersion equation as follow:

$$\tan(k_2 d) = \frac{\mu_1\mu_2 k_2 k_3 - \mu_2\mu_3 k_1 k_2 \tanh(-k_1 x_0)}{\mu_1\mu_3 k_2^2 + \mu_2^2 k_1 k_3 \tanh(-k_1 x_0)}, \tag{37}$$

where $x_0 = -\frac{1}{k_1}\operatorname{sech}^{-1}\left[\frac{k_0}{k_1}\sqrt{\frac{\alpha\mu_1}{2}}(\cos k_2 d + \frac{\mu_2 k_3}{\mu_3 k_2}\sin k_2 d)C\right]$ is the position of the maximum of the amplitude in nonlinear substrate, C being the amplitude of the electric field at the interface $x = d$. In the whole paper, we make the assumption $C = 1$. Next, we will discuss the properties of the nonlinear guided modes by using the graphic method (Shen et al., 2010a; Zhang et al., 2009) near DP in two cases $\omega < \omega_D$ and $\omega > \omega_D$, respectively.

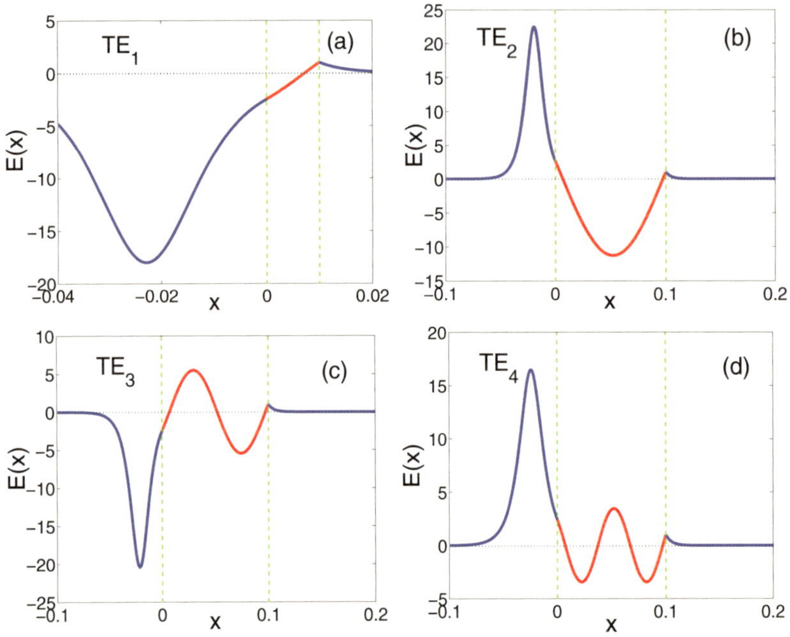

Fig. 14. (Color online) Guided modes in nonlinear NZPIM waveguide when $\omega = 0.8\omega_D < \omega_D$, $\alpha = 0.003$, $\epsilon_1^L = 1.2$ and $\epsilon_2 = 2$. (a) TE_1: $\kappa_2 d = 0.93864$, $d = 1cm$; (b) TE_2: $\kappa_2 d = 3.46529$, $d = 10cm$; (c) TE_3: $\kappa_2 d = 6.95257$, $d = 10cm$; (d) TE_4: $\kappa_2 d = 10.5005$, $d = 10cm$.

4.2 Nonlinear fast wave guided modes

Case 1: $\omega < \omega_D$, and $\epsilon_3 = \mu_3 < 0$. As shown in Fig. 14, we find that the fundamental TE_0 mode is absent for any parameters of the nonlinear NZPIM waveguide. The unique property is very different from the case of conventional nonlinear waveguide where the lowest-order TE_0 guided mode is always exist (Stegeman et al., 1984). The lowest-order guided mode of the nonlinear NZPIM waveguide is TE_1 mode, which can not coexist with higher-order guided modes for some waveguide parameters. For the TE_0 guided modes, it should satisfy the condition $k_2 d = 2\phi_{refl1} + 2\phi_{refl2}$ (Shadrivov et al., 2003; 2005; Shen et al., 2010a), where $2\phi_{refl1}$ and $2\phi_{refl2}$ are the phase retardation upon the total internal reflection at the interface between core and cladding and at the interface between core and substrate. In previous works (Shadrivov et al., 2003; 2005; Shen et al., 2010a), both $2\phi_{refl1}$ and $2\phi_{refl2}$ are negative, then the TE_0 mode does not exist. In a linear waveguide with left handed material cover, $2\phi_{refl1}$ is negative but $2\phi_{refl2}$ is positive, which may support the TE_0 mode with appropriate physical parameters. However, the nonlinear physical mechanism in this work is different from the linear dynamics (Wang et al., 2008). We emphasize that the nonlinear NZPIM waveguide can not support the TE_0 mode due to the nonlinear dispersion when the angular frequency is smaller than DP.

Case 2: $\omega > \omega_D$, and $\epsilon_3 = \mu_3 > 0$. In Fig. 15, we plot the nonlinear guided modes when the angular frequency is larger than DP. It is shown that the properties of the nonlinear NZPIM waveguide can be treated as a conventional nonlinear dielectric waveguide. From Fig. 15, we can see that the nonlinear NZPIM can support the fundamental guided mode [Fig. 15 (a)], though it can not coexist with higher-order modes within the same waveguide for general parameters. This result is different from the case that the nonlinear NZPIM waveguide can not

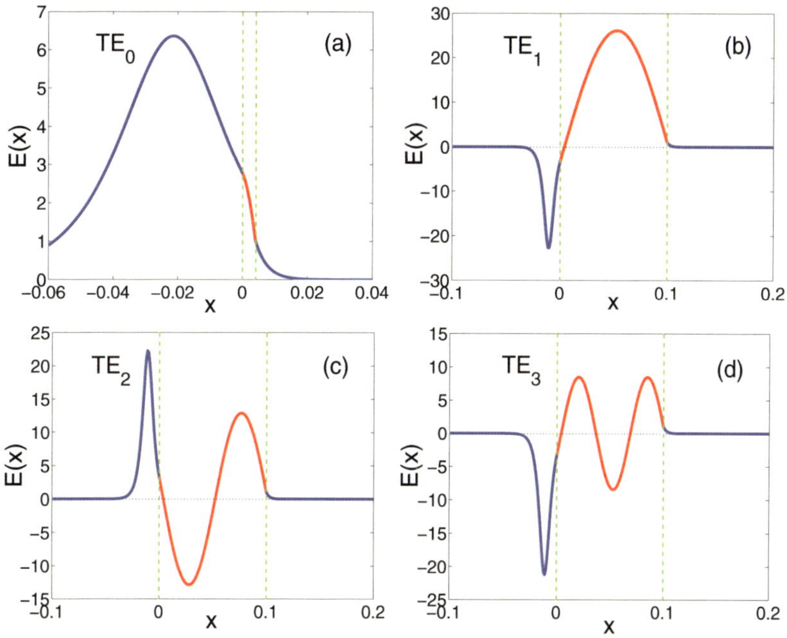

Fig. 15. (Color online) Guided modes in nonlinear NZPIM waveguide when $\omega = 4/3\omega_D > \omega_D$, $\alpha = 0.003$, $\epsilon_1^L = 1.2$ and $\epsilon_2 = 2$. (a) TE_0: $\kappa_2 d = 0.960548$, $d = 0.4cm$; (b) TE_1: $\kappa_2 d = 3.23447$, $d = 10cm$; (c) TE_2: $\kappa_2 d = 6.47046$, $d = 10cm$; (d) TE_3: $\kappa_2 d = 9.70986$, $d = 10cm$.

support the fundamental mode when the angular frequency is smaller than DP. Although the physical mechanism of nonlinear dispersion is different from the linear case, we can see that the nonlinear guided modes can also be treated as electronic wave in graphene waveguide (Zhang et al., 2009), corresponding to the Kelin tunneling (lack of fundamental mode) and classical motion (support fundamental mode), respectively.

Since the maximum of the magnitude at the interface between a nonlinear media and a linear media locates inside the nonlinear media (Shadrivov et al., 2004), the guided modes not only have mode energy in the core, but also a peak mode energy in the nonlinear substrate region, as shown in both Fig. 14 and Fig. 15. This result is identical with the conventional nonlinear waveguide (Stegeman et al., 1984).

We further show the unique properties of nonlinear guides modes near DP in the nonlinear NZPIM waveguide with the angular frequency varying from $\omega < \omega_D$ to $\omega > \omega_D$ in Fig. 16 when the nonlinearity is self-focusing [Fig. 16 (a) and (b)] and self-defocusing [Fig. 16 (c) and (d)], respectively. When the nonlinearity is self-focusing, like the behaviors of light in linear NZPIM (Chen et al., 2009; Shen et al., 2010a), there also exists an asymmetric forbidden band for the dispersion [Fig. 16 (a)] which means the nonlinear guided modes are not continuous near DP. The band will become narrower when the order of the guided modes increases which is opposite to the case that the band will become wider when the order of the guided modes increases in linear NZPIM waveguide (Shen et al., 2010a). Another important and interesting phenomenon is that the bang gap can be modulated by the strength of the nonlinearity. It is obviously that the band gap will become wider when the nonlinear index α increases [Fig. 16 (b)]. The band gap will also become wider with the increase of the wave intensity when the nonlinear index α is fixed (not shown). However, the self-defocusing nonlinearity can eliminate the asymmetric band gap, leading to the continuation of the guided modes near DP, as shown in Fig. 16 (c) and (d). The field in the core of the NZPIM waveguide with self-defocusing nonlinearity is represented in the form of **csch** function, and its dispersion can be obtained in the same way. The tunable gap may have potential application in electron wave filters in nonlinear graphene fibre optics (Chen and Tao, 2009).

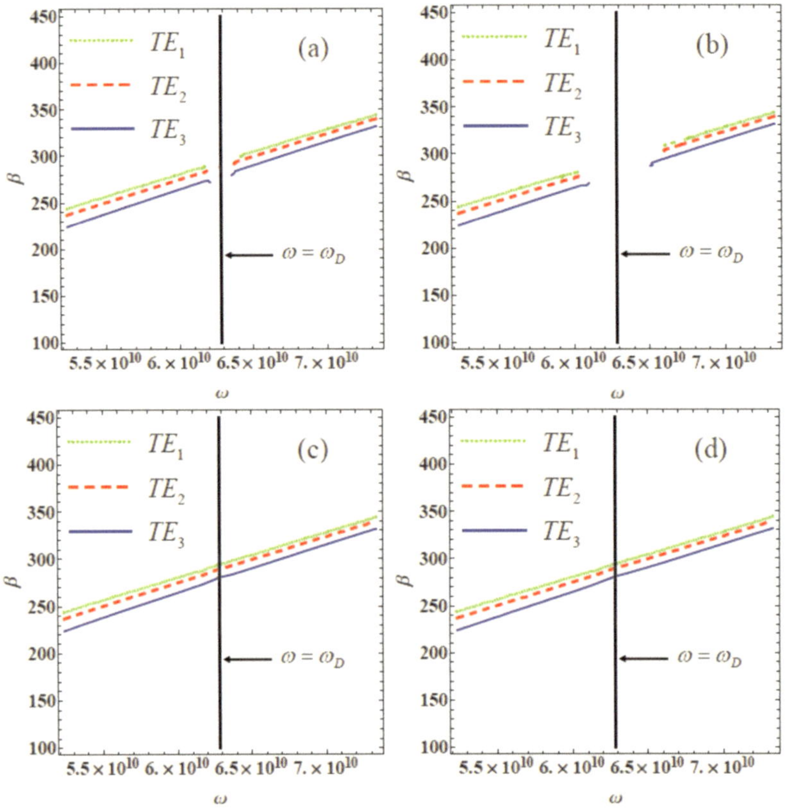

Fig. 16. (Color online) The propagation constant β versus the incident frequency ω near the DP in the nonlinear NZPIM waveguide, the core thickness is $d = 0.1$, $\epsilon_1^L = 1.2$, $\epsilon_2 = 2$, and the nonlinearity are (a) $\alpha = 0.0001$, (b) $\alpha = 0.0008$, (c) $\alpha = -0.0001$, and (d) $\alpha = -0.0008$.

4.3 Nonlinear surface guided modes

We also find that when $\omega < \omega_D$, the nonlinear NZPIM waveguide can propagate nonlinear surface guided modes-slow wave. In this case, the wave vectors in core, substrate and cladding are all imaginary. The electric fields in three regions can be written as

$$\psi(x) = \begin{cases} \frac{k_1}{k_0}\sqrt{\frac{2}{\alpha\mu_1}}\operatorname{sech}[k_1(x - x_0)], & x < 0, \\ A\cosh[k_2(x-d)] + B\sinh[k_2(x-d)], & 0 < x < d, \\ Ce^{-k_3(x-d)}, & x > d, \end{cases} \tag{38}$$

where $k_1^2 = \beta^2 - k_0^2\epsilon_1^L\mu_1$, $k_2^2 = \beta^2 - k_0^2\epsilon_2\mu_2$, and $k_3^2 = \beta^2 - [2(\omega - \omega_D)/c]^2$ are the transverse decay wave vectors in the substrate, core, and cladding, β is the propagation constant of the slow wave guided modes.
Similarly, we obtain the following dispersion relation

$$\tanh(\kappa_2 d) = -\frac{\mu_1\mu_2 k_2 k_3 - \mu_2\mu_3 k_1 k_2 \tanh(-k_1 x_0)}{\mu_1\mu_3 k_2^2 - \mu_2^2 k_1 k_3 \tanh(-k_1 x_0)}. \quad (39)$$

where $x_0 = -\frac{1}{k_1}\text{sech}^{-1}\left[\frac{k_0}{k_1}\sqrt{\frac{\alpha\mu_1}{2}}(\cosh k_2 d + \frac{\mu_2 k_3}{\mu_3 k_2}\sinh k_2 d)C\right]$ is the position of the maximum of the amplitude in nonlinear substrate, C being the amplitude of the electric field at the interface $x = d$.

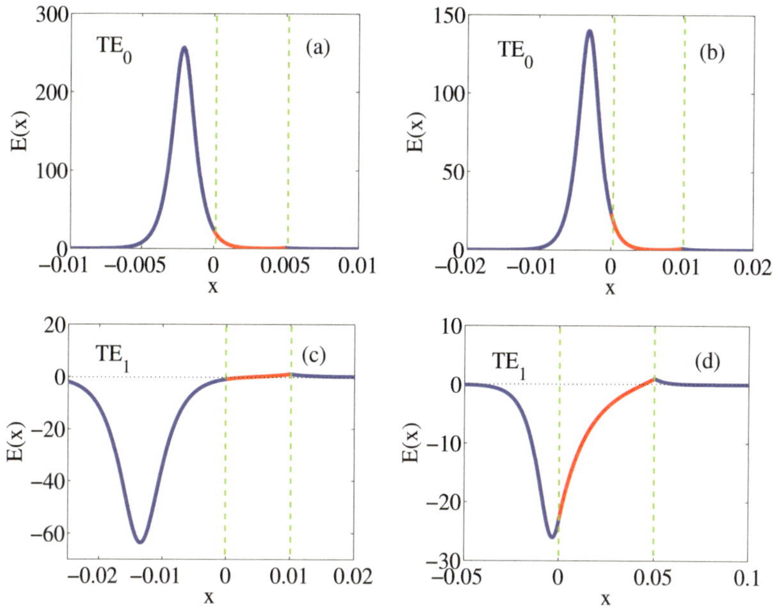

Fig. 17. (Color online) Slow wave guided modes in nonlinear NZPIM waveguide when $\omega = 0.7\omega_D < \omega_D$, the nonlinearity is $\alpha = 0.003$, $\epsilon_1^L = 1.2$, $\epsilon_2 = 2$. (a) TE_0: $\kappa_2 d = 7.25054$, $d = 0.5cm$; (b) TE_0: $\kappa_2 d = 7.8406$, $d = 1cm$; (c) TE_1: $\kappa_2 d = 3.36844$, $d = 1cm$; (d) TE_1: $\kappa_2 d = 3.42543$, $d = 5cm$.

We plot the nonlinear slow waves guided modes in Fig. 17. We find that only the lowest order even (TE_0) or odd (TE_1) surface guided modes can exist in the nonlinear NZPIM waveguide which crucially depend on the physical parameters. When the core thickness is smaller, the waveguide can only support TE_0 mode surface wave [Fig. 17 (a)]. When the core thickness getting bigger, the waveguide can support both the TE_0 mode and the TE_1 mode surface waves [Fig. 17 (b) and (c)]. However, when the core thickness is larger, the waveguide can only support the TE_1 mode surface wave[Fig. 17 (d)]. Recent research showed that the optically discrete and surface solitons in honeycomb photonic lattices can be regarded as an optical analog of graphene nanoribbons (Molina and Kivshar, 2010). Surface solitons (Savin and Kivshar, 2010a) and vibrational Tamm states (Savin and Kivshar, 2010b) at the edges of graphene nanoribbons have also been reported recently. We hope that our results obtained here may also predict the nonlinear surface mode of electrons and holes in nonlinear graphene waveguide.

5. Further work on optical DP in NZPIM

In the recent years, graphene supperlatices with periodic potential structures has drawn considerable attention due to the fact that superlattices are very successful in controlling the electronic structures of many conventional semiconducting materials (Tsu, 2005). Many theoretical (Park et al., 2008a;b) and experimental (Sutter et al., 2008) works have been focus on the devices of graphene-based superlattices. Wang *et al.* have presented the result on a new DP which is exactly located at the energy which corresponds to the zero-averaged wavenumber inside the one-dimensional (1D) periodic potentials (Wang and Zhu, 2010a; Wang and Chen, 2011). The gap for the zero-averaged wavenumber is quite different from the Bragg gap, which is analogous to the case of the one-dimensional PCs containing left-handed and right-handed materials (Bliokh et al., 2009; Wang and Zhu, 2010b).

Based on the rapid developments in both theoretical and experimental works on graphene-based superlattices, and the analogy phenomena between electron waves in graphene and optics in NZPIM, the optical propagation in one-dimensional PCs containing NZPIM will be an interesting and challenge task in the future. We will study the transmission of optics through an one-dimensional PCs containing NZPIM, and predict some novel properties, such as Goos-Hänchen shifts (Chen et al., 2009), zero-averaged index gap (Wang and Zhu, 2010b), new Dirac gap, and Bragg gap etc. The propagation of one-dimensional NZPIM PCs containing a nonlinear defect will be another significative question as well.

6. Conclusion

In summary, we have investigated the nonlinear plasmonics in NZPIM and shown that the dynamics of electron wave in graphene can be simulated by the analogy of optics in NZPIM. The unique propagation of optics near the DP in NZPIM, such as frequency threshold of nonlinear surface waves, and tunable band gap of the nonlinear guided modes, will lead to the potential applications in guided wave optics, integral optics and optical-based devices. Our results will also give the deeper understanding of several exotic phenomena in graphene. We hope our work will motivate the further work to simulate and predict many exotic phenomena in graphene with relatively simple optical experiments.

7. Acknowledgement

The authors appreciate L.-G. Wang and X. Chen for their pioneering and enlightening works on optical DP in NZPIM. This work is supported by the Chinese National Natural Science Foundation (Grant No. 60808002) and the Shanghai Leading Academic Discipline Project (Grant No. S30105).

8. References

Ablowitz, M. J.; Nixon, S. D. & Zhu, Y. (2009). Conical diffraction in honeycomb lattices. *Phys. Rev. A*, Vol. 79, No. 5, May -2009, 053830 (1-6), ISSN 1050-2947

Bahat-Treidel, O.; Peleg, O. & Segev, M. (2008). Symmetry breaking in honeycomb photonic lattices. *Opt. Lett.*, Vol. 33, No. 19, Sep -2008, 2251-2253, ISSN 0146-9592

Bahat-Treidel, O.; Peleg, O.; Grobman, M.; Shapira, N. & Segev, M. (2010). Klein tunneling in deformed honeycomb lattices. *Phys. Rev. Lett.*, Vol. 104, No. 6, Feb -2010, 063901 (1-4), ISSN 0031-9007

Bahat-Treidel, O.; Peleg, O.; Segev, M. & Buljan, H. (2010). Breakdown of Dirac dynamics in honeycomb lattices due to nonlinear interactions. *Phys. Rev. A*, Vol. 82, No. 1, Jul -2010, 013830 (1-6), ISSN 1050-2947

Beenakker, C. W. (2008). Colloquium: Andreev reflection and Klein tunneling in graphene.*Rev. Mod. Phys.*, Vol. 80, No. 4, Oct -2008, 1337-1354, ISSN 0034-6861

Beenakker, C. W. J.; Sepkhanov, R. A.; Akhmerov, A. R. & Tworzydlo, J. (2009). Quantum Goos-Hänchen effect in graphene. *Phys. Rev. Lett.*, Vol. 102, No. 14, Jan -2009, 146804 (1-4), ISSN 0031-9007

Bliokh, Y. P.; Freilikher, V.; Savel'ev, S. & Nori, F. (2009). Transport and localization in periodic and disordered graphene superlattices. *Phys. Rev. B*, Vol. 79, No. 7, Feb -2009, 075123 (1-4), ISSN 1098-0121

Castro Neto, A. H.; Guinea, F.; Peres, N. M. R.; Novoselov, K. S. & Geim, A. K. (2009). The electronic properties of graphene. *Rev. Mod. Phys.*, Vol. 81, No. 1, Jan -2009, 109ÍC162, ISSN 0034-6861

Cheianov, V. V.; Fal'ko, V. & Altshuler, B. L. (2007). The focusing of electron flow and a veselago lens in graphene p-n junctions. *Science*, Vol. 315, No. 5816, Mar -2007, 1252-1255, ISSN 0036-8075

Chen, X. & Tao, J.-W. (2009). Design of electron wave filters in monolayer graphene by tunable transmission gap. *Appl. Phys. Lett.*, Vol. 94, No. 26, Jun -2009, 262102 (1-3), ISSN 0003-6951

Chen, X.; Wang, L.-G. & Li, C.-F. (2009). Transmission gap, Bragg-like reflection, and Goos-Hänchen shifts near the Dirac point inside a negative-zero-positive index metamaterial slab. *Phys. Rev. A*, Vol. 80, No. 4, Oct -2009, 043839 (1-5), ISSN 1050-2947

Chen, X.; Wei, R.-R.; Shen, M.; Zhang Z.-F. & Li, C.-F. (2010). Bistable and negative lateral shifts of the reflected light beam from Kretschmann configuration with nonlinear left-handed metamaterials. *Appl. Phys. B*, Vol. 101, No. 1-2, May -2010, 283-289, ISSN 0946-2171

Darancet, P.; Olevano, V. & Mayou, D. (2009). Coherent electronic transport through graphene constrictions: Subwavelength regime and optical analogy. *Phys. Rev. Lett.*, Vol. 102, No. 13, Mar -2009, 136803 (1-4), ISSN 0031-9007

Ghosh, S. & Sharma, M. (2009). Electron optics with magnetic vector potential barriers in graphene. *J. Phys.: Condens. Matter*, Vol. 21, No. 29, Jul -2009, 292204 (1-8), ISSN 0953-8984

Güney D. Ö. & Meyer, D. A. (2009). Negative refraction gives rise to the Klein paradox. *Phys. Rev. A*, Vol. 79, No. 6, Jun -2009, 063834 (1-4), ISSN 1050-2947

Haddad L. H. & Carr, L. D. (2009). The nonlinear Dirac equation in Bose-Einstein condensates: Foundation and symmetries. *Phys. D*, Vol. 238, No. 15, Jul -2009, 1413-1421, ISSN 0167-2789

Hendry, E.; Hale, P. J.; Moger, J.; Savchenko, A. K. & Mikhailov, S. A. (2010). Coherent Nonlinear Optical Response of Graphene. *Phys. Rev. Lett.*, Vol. 105, No. 9, Aug -2010, 097401 (1-4), ISSN 0031-9007

Katsnelson, M. I.; Novoselov, K. S. & Geim, A. K. (2006). Chiral tunnelling and the Klein paradox in graphene. *Nature Physics*, Vol. 2, Feb -2006, 620-625, ISSN 1745-2473

Kivshar, Y. S. (2008). Nonlinear optics: The next decade. *Opt. Exp.*, Vol.16, No. 26, Dec-2008, 22126-22128, ISSN 1094-4087

Mikhailov, S. A. (2007). Non-linear electromagnetic response of graphene. *Europhys. Lett.*, Vol. 79, No. 2, Jul -2007, 27002 (1-4), ISSN 0295-5075

Molina, M. I. & Kivshar, Y. S. (2010). Discrete and surface solitons in photonic graphene nanoribbons. *Opt. Lett.*, Vol. 35, No. 17, Sep -2010, 2895-2897, ISSN 0146-9592

Novoselov, K. S.; Geim, A. K.; Morozov, S. V.; Jiang, D.; Zhang, Y.; Dubonos, S. V.; Grigorieva, I. V. & Firsov, A. A.(2004). Electric field effect in atomically thin carbon films. *Science*, Vol. 306, No. 5696, Oct -2004, 666-669, ISSN 0036-8075

Novoselov, K. S.; Geim, A. K.; Morozov, S. V.; Grigorieva, I. V.; Dubonos, S. V. & Firsov, A. A. (2005). Two-dimensional gas of massless Dirac fermions in graphene. *Nature (London)*, Vol. 438, Nov -2005, 197-200, ISSN 0028-0836

Park, C.-H.; Yang, L.; Son, Y.-W.; Cohen, M. L. & Louie, S. G. (2008). Electron beam supercollimation in graphene superlattices. *Nano Lett.*, Vol. 8, No. 9, Aug -2008, 2920-2924, ISSN 1530-6984

Park, C.-H.; Yang, L.; Son, Y.-W.; Cohen, M. L. & Louie, S. G. (2008). Anisotropic behaviours of massless Dirac fermions in graphene under periodic potentials. *Nature Physics*, Vol. 4, No. 3, Feb -2008, 213-217, ISSN 1745- 2473

Peleg, O.; Bartal, G.; Freedman,B.; Manela, O.; Segev, M. & Christodoulides, D. N. (2007). Conical diffraction and gap solitons in honeycomb photonic lattices. *Phys. Rev. Lett.*, Vol. 98, No. 10, Mar -2007, 103901 (1-4), ISSN 0031-9007

Ruppin, R. (2000). Surface polaritons of a left-handed medium. *Phys. Lett. A*, Vol. 277, No. 1, Nov -2000, 61-64, ISSN 0375-9601

Savin, A. V. & Kivshar, Y. S. (2010). Surface solitons at the edges of graphene nanoribbons. *Europhys. Lett.*, Vol. 89, No. 4, Mar -2010, 46001 (1-6), ISSN 0295-5075

Savin, A. V. & Kivshar, Y. S. (2010). Vibrational Tamm states at the edges of graphene nanoribbons. *Phys. Rev. B*, Vol. 81, No. 16, Apr -2010, 165418 (1-9) , ISSN 1098-0121

Shadrivov, I. V.; Sukhorukov, A. A. & Kivshar, Y. S. (2003). Guided modes in negative-refractive-index waveguides. *Phys. Rev. E*, Vol. 67, No. 5, May -2003, 057602 (1-4), ISSN 1539-3755

Shadrivov, I. V.; Sukhorukov, A. A.; Kivshar, Y. S.; Zharov, A. A.; Boardman, A. D. & Egan, P. (2004). Nonlinear surface waves in left-handed materials. *Phys. Rev. E*, Vol. 69, No. 1, Jan -2004, 016617 (1-9), ISSN 1539-3755

Shadrivov, I. V.; Sukhorukov, A. A. & Kivshar, Y. S. (2005). Complete band gaps in one-dimensional left-handed periodic structures. *Phys. Rev. Lett.*, Vol. 95, No. 19, Nov -2005, 193903 (1-4), ISSN 0031-9007

Shen, M.; Ruan, L.-X. & Chen, X. (2010). Guided modes near the Dirac point in negative-zero-positive index metamaterial waveguide. *Opt. Exp.*, Vol. 18, No. 12, May -2010, 12779-12787, ISSN 1094-4087

Shen, M.; Ruan, L.-X.; Chen, X.; Shi, J.-L.; Ding, H.-X.; Xi, N. & Wang Q. (2010). Nonlinear surface waves near the Dirac point in negative-zero-positive index metamaterial. *J. of Opt.*, Vol. 12, N, 8, Aug- 2010, 085201 (1-5), ISSN 2040-8978

Shen, M.; Ruan, L.-X.; Wang, X.-L; Shi, J.-L. & Wang Q. (2011). Tunable band gap near the Dirac point in nonlinear negative-zero-positive index metamaterial waveguide. *Phys. Rev. A*, (accepted), ISSN 1050-2947

Smith, D. R.; Padilla W. J., Vier, D. C.; Nemat-Nasser, S. C. & Schultz, S. (2000). Composite Medium with Simultaneously Negative Permeability and Permittivity. *Phys. Rev. Lett.*, Vol. 84, No. 18, 4184-4187, May-2000, ISSN 0031-9007

Sepkhanov, R. A.; Bazaliy, Y. B. & Beenakker, C. W. J. (2007). Extremal transmission at the Dirac point of a photonic band structure. *Phys. Rev. A*, Vol. 75, No. 6, Jun -2007, 063813 (1-5), ISSN 1050-2947

Stegeman, G. I.; Seaton, C. T.; Chilwell, J. & Smith, S. D. (1984). Nonlinear waves guided by thin films. *Appl. Phys. Lett.*, Vol. 44, No. 9, May -1984, 830-832, ISSN 0003-6951

Sutter, P. W.; Flege, J. I. & Sutter, E. A. (2008). Epitaxial graphene on ruthenium. *Nature Materials*, Vol. 7, No. 5, May -2008, 406-411, ISSN 1476-1122

Tsakmakidis, K. L.; Boardman, A. D. & Hess, O. (2007). 'Trapped rainbow' storage of light in metamaterials. *Nature (London)*, Vol. 440, Nov -2007, 397-401, ISSN 0028-0836

Tsu, R. (2005), Superlattice to Nanoelectronics, Elsevier, ISBN: 978-0-08-044377-5, Oxford.

Wang, L.-G.; Wang, Z.-G.; Zhang, J.-X. & Zhu, S.-Y. (2009). Realization of Dirac point with double cones in optics. *Opt. Lett.*, Vol. 34, No. 10, May -2009, 1510-1512, ISSN 0146-9592

Wang, L.-G.; Wang, Z.-G. & Zhu, S.-Y. (2009). Zitterbewegung of optical pulses near the Dirac point inside a negative-zero-positive index metamaterial. *Europhys. Lett.*, Vol. 86, No. 4, Jun -2009, 47008 (1-5), ISSN 0295-5075

Wang, L.-G.; Li, G.-X. & Zhu, S.-Y. (2010). Thermal emission from layered structures containing a negative-zero-positive index metamaterial. *Phys. Rev. B*, Vol. 81, No. 7, Jan -2010, 073105 (1-4) , ISSN 1098-0121

Wang, L.-G. & Zhu, S.-Y. (2010). Electronic band gaps and transport properties in graphene superlattices with one-dimensional periodic potentials of square barriers. *Phys. Rev. B*, Vol. 81, No. 20, May -2010, 205444 (1-9), ISSN 1098-0121

Wang, L.-G. & Zhu, S.-Y. (2010), The reversibility of the Goos-Hänchen shift near the band- crossing structure of one-dimensional photonic crystals containing left-handed metamaterials. *Appl. Phys. B: Lasers and Optics*, Vol.98, No. 2-3, Feb.-2010, 459-463, ISSN 0946-2171

Wang, L.-G. & Chen, X. (2011). Robust zero-averaged wave-number gap inside gapped graphene superlattices. *J. Appl. Phys.*, Vol.109, No.3, Feb.-2011, 033710 (1-8), ISSN 0021-8979

Wang, Q. & Awai, I. (1998). Frequency characteristics of the magnetic spatial solitons on the surface of an antiferromagnet. *J. Appl. Phys.*, Vol.83, No.1, Jan.-1998, 382-387, ISSN 0021-8979

Wang, Z.-H.; Xiao, Z.-Y. & Li, S. P. (2008). Guided modes in slab waveguides with a left handed material cover or substrate. *Opt. Commun.*, Vol.281, No.4, Feb.-2008, 607-613, ISSN 0030-4018

Williams, J. R.; Low, Tony; Lundstrom, M. S. & Marcus C. M. (2011). Gate-controlled guiding of electrons in graphene. *Nature Nanotechnology*, Vol.281, No.4, Apr-2011, 222-225, ISSN 1748-3387

Wu, X. (2011). Electronic fiber in graphene. *Appl. Phys. Lett.*, Vol. 98, No. 8, Feb -2011, 082117 (1-3), ISSN 0003-6951

Xu, G.-D.; Pan, T.; Zang, T.-C. & Sun, J. (2009). Nonlinear surface polaritons in anisotropic Kerr-type metamaterials. *J. Phys. D: Appl. Phys.*, Vol. 42, No. 4, Jan -2009, 045303 (1-7), ISSN 1361-6463

Zhao, L. & Yelin, S. F. (2010). Proposal for graphene-based coherent buffers and memories. *Phys. Rev. B*, Vol. 81, No. 11, Mar -2010, 115441 (1-4), ISSN 1098-0121

Zhang, Y.; Tan, Y.-W.; Stormer, H. L. & Kim, P. (2005). Experimental observation of the quantum Hall effect and Berry's phase in graphene. *Nature (London)*, Vol. 438, Nov -2005, 201-204, ISSN 0028-0836

Zhang, X. (2008). Observing Zitterbewegung for photons near the Dirac point of a two-dimensional photonic crystal. *Phys. Rev. Lett.*, Vol. 100, No. 11, Mar -2008, 113903 (1-4), ISSN 0031-9007

Zhang, F.-M.; He, Y. & Chen, X. (2009). Guided modes in graphene waveguides. *Appl. Phys. Lett.*, Vol. 94, No. 21, May -2009, 212105 (1-3), ISSN 0003-6951

Ziolkowski, R. W. (2004). Propagation in and scattering from a matched metamaterial having a zero index of refraction. *Phys. Rev. E*, Vol. 70, No. 4, Oct -2004, 046608 (1-12), ISSN 1539-3755

15

Zitterbewegung (Trembling Motion) of Electrons in Graphene

Tomasz M. Rusin[1] and Wlodek Zawadzki[2]
[1]*Orange Customer Service sp. z o. o., Warsaw*
[2]*Institute of Physics, Polish Academy of Sciences, Al. Lotników, Warsaw*
Poland

1. Introduction

This chapter describes a somewhat mysterious phenomenon known in the literature under the German name of "Zitterbewegung" (trembling motion). Both the phenomenon and its name were conceived by Erwin Schrodinger who, in 1930, published the paper *Ueber die kraeftefreie Bewegung in der relativistischen Quantenmechanik* in which he observed that in the Dirac equation, describing relativistic electrons in a vacuum, the 4×4 operators corresponding to components of relativistic velocity do not commute with the free-electron Hamiltonian (Barut & Bracken, 1981; Schroedinger, 1930). In consequence, the electron velocity is not a constant of the motion also in absence of external fields. Such an effect in a vacuum must be of a quantum nature as it does not obey Newton's first law of classical motion. Schrodinger calculated the resulting time dependence of the electron velocity and position concluding that, in addition to classical motion, they experience very fast periodic oscillations which he called Zitterbewegung (ZB). Schrodinger's idea stimulated numerous theoretical investigations but no experiments since the predicted frequency $\hbar\omega_Z \simeq 2m_0c^2 \simeq 1$ MeV and the amplitude of about $\lambda_c = \hbar/mc \simeq 3.86 \times 10^{-3}$Å are not accessible to current experimental techniques. It was recognized that the ZB is due to an interference of states corresponding to the positive and negative electron energies resulting from the Dirac equation (Bjorken & Drell, 1964; Greiner, 1994; Sakurai, 1997). Lock (1979) showed that, if an electron is represented by a wave packet, its ZB has a transient character, i.e. it disappears with time.

It was conceived years later that the trembling electron motion should occur also in crystalline solids if their band structure could be represented by a two-band model reminiscent of the Dirac equation (Cannata et al., 1990; Lurie & Cremer, 1970; Vonsovskii et al., 1990; Zawadzki, 1997). An intense interest in ZB of electrons in semiconductors was launched only in 2005 (Schliemann et al., 2005; Zawadzki, 2005). There followed a wave of theoretical considerations for various semiconductor and other systems. It was recognized that the phenomenon of ZB occurs every time one deals with two or more interacting energy bands (Cserti & David, 2006; Rusin & Zawadzki, 2007a; Winkler et al., 2007).

As mentioned above, in a vacuum the ZB characteristics are not favorable. In solids, the ZB characteristics are much better but it is difficult to observe the motion of a single electron. However, recently Gerritsma et al. (2010) simulated experimentally the Dirac equation and the resulting electron Zitterbewegung with the use of trapped ions and laser excitations. The power of the simulation method is that one can adjust experimentally the essential parameters of the Dirac equation: mc^2 and c, and thus achieve more favorable values of the ZB frequency and amplitude. Interestingly, it turned out that analogues of ZB can occur also in classical wave propagation phenomena. Several predictions were made, but in two systems, namely macroscopic sonic crystals (Zhang & Liu, 2008), and photonic superlattices (Dreisow et al., 2010), the ZB-like effects were actually observed.

2. ZB of electrons in graphene

Here we consider the ZB in three modern materials: monolayer graphene, bilayer graphene and carbon nanotubes. Monolayer graphene is currently by far the most studied material in the world (Novoselov et al., 2004). The description is based mostly on our theoretical works (Rusin & Zawadzki, 2007b; 2008; 2009), but papers of other authors concerned with ZB in graphene are also mentioned. It is emphasized that the trembling electrons should emit electromagnetic radiation which would provide a direct evidence for the existence of this phenomenon. We discuss the origin of ZB in crystalline solids not going into detailed mathematical derivations. An exact formulation of the electron dynamics in monolayer graphene in the presence of a magnetic field is presented in Appendix A. The electron classical velocity and its mass for a linear energy band of monolayer graphene are considered in Appendix B. It is hoped that our presentation will simulate experimental investigations of the electron trembling motion in graphene.

2.1 Bilayer graphene

We begin our description considering bilayer graphene since the results can be obtained in the analytical form, which allows one to see directly important features of the trembling motion. Two-dimensional Hamiltonian for bilayer graphene in the K point of Brillouin zone is (McCann & Fal'ko, 2006)

$$\hat{H}_B = -\frac{1}{2m^*}\begin{pmatrix} 0 & (\hat{p}_x - i\hat{p}_y)^2 \\ (\hat{p}_x + i\hat{p}_y)^2 & 0 \end{pmatrix}, \tag{1}$$

where $m^* = 0.054m_0$. The energy spectrum is $\mathcal{E}_k = \pm E_k$, where $E_k = \hbar^2k^2/2m^*$, i.e. there is no energy gap between the parabolic conduction and valence bands. The position operator in the Heisenberg picture is a 2×2 matrix $\hat{x}(t) = \exp(i\hat{H}_Bt/\hbar)\hat{x}\exp(-i\hat{H}_Bt/\hbar)$. One calculates

$$x_{11}(t) = x(0) + \frac{k_y}{k^2}\left[1 - \cos\left(\frac{\hbar k^2 t}{m^*}\right)\right], \tag{2}$$

where $k^2 = k_x^2 + k_y^2$. The third term represents the Zitterbewegung with the frequency $\hbar\omega_Z = 2\hbar^2k^2/2m^*$, corresponding to the energy difference between the upper and lower energy branches for a given value of k. We want to calculate the ZB of a charge carrier represented by a two-dimensional wave packet

$$\psi(\boldsymbol{r},0) = \frac{1}{2\pi}\frac{d}{\sqrt{\pi}}\int d^2\boldsymbol{k}e^{-\frac{1}{2}d^2k_x^2 - \frac{1}{2}d^2(k_y-k_{0y})^2}e^{i\boldsymbol{kr}}\begin{pmatrix}1\\0\end{pmatrix}. \tag{3}$$

The packet is centered at $\boldsymbol{k}_0 = (0, k_{0y})$ and is characterized by a width d. The unit vector $(1, 0)$ is a convenient choice, selecting the (11) component of $\hat{x}(t)$, see Eq. (2). An average of $\hat{x}_{11}(t)$ is a two-dimensional integral which can be calculated analytically in the form

$$\bar{x}_{11}(t) = \langle\psi(\boldsymbol{r})|\hat{x}(t)|\psi(\boldsymbol{r})\rangle = \bar{x}_c + \bar{x}_Z(t) \tag{4}$$

where $\bar{x}_c = (1/k_{0y})\left[1 - \exp(-d^2k_{0y}^2)\right]$, and

$$\bar{x}_Z(t) = \frac{1}{k_{0y}}\left[\exp\left(-\frac{\delta^4 d^2 k_{0y}^2}{d^4+\delta^4}\right)\cos\left(\frac{\delta^2 d^4 k_{0y}^2}{d^4+\delta^4}\right) - \exp(-d^2k_{0y}^2)\right], \tag{5}$$

in which $\delta = \sqrt{\hbar t/m^*}$ contains the time dependence. In Figure 1a we show the ZB of the electron position $\bar{x}_{11}$ as given in Eqs. (4) and (5).

Here are the main features of ZB following from Eqs. (4) and (5). First, in order to have ZB in the direction x one needs an initial transverse momentum $\hbar k_{0y}$. Second, the ZB frequency depends only weakly on the packet width: $\omega_Z = (\hbar k_{0y}^2/m^*)(d^4/(d^4+\delta^4))$, while its amplitude is strongly dependent on the width d. Third, the ZB has a transient character since it is attenuated by the exponential term. For small t the amplitude of $\bar{x}_Z(t)$ diminishes as $\exp(-\Gamma_Z^2 t^2)$ with

$$\Gamma_Z = \frac{\hbar k_{0y}}{m^* d}. \tag{6}$$

Fourth, as t (or δ) increases, the cosine term tends to unity and the first term in Eq. (5) cancels out with the second term. After the oscillations disappear, the charge carrier is displaced by the amount $\bar{x}_c$, which is a "remnant" of ZB. Fifth, for very wide packets ($d \to \infty$) the exponent in Eq. (5) tends to unity, the oscillatory term is $\cos(\delta^2 k_{0y}^2)$ and the last term vanishes. In this limit one recovers undamped ZB oscillations.

Next, we consider other quantities related to ZB, beginning by the current. The latter is given by the velocity multiplied by charge. The velocity is simply $\bar{v}_x = \partial\bar{x}_Z/\partial t$, where $\bar{x}_Z$ is given by Eq. (5). The calculated current is plotted in Figure 1b, its oscillations are a direct manifestation of ZB. The transient character of ZB is accompanied by a temporal spreading of the wave packet. The question arises whether the attenuation of ZB is not simply caused by the spreading of the packet. The calculated packet width ΔR is plotted versus time in Figure 1c. It is seen that during the initial 80 femtoseconds the packet's width increases only twice compared to its initial value, while the ZB disappears almost completely. We conclude that the spreading of the packet is *not* the main cause of the transient character of ZB. Looking for physical reasons behind the transient character of ZB we decompose the total wave function $\psi(\boldsymbol{r}, t)$ into the positive (p) and negative (n) components $\psi^p(\boldsymbol{r}, t)$ and $\psi^n(\boldsymbol{r}, t)$. We have

$$\begin{aligned}|\psi(t)\rangle &= e^{-i\hat{H}t/\hbar}|\psi(0)\rangle \\ &= e^{-iEt/\hbar}\langle p|\psi(0)\rangle|p\rangle + e^{iEt/\hbar}\langle n|\psi(0)\rangle|n\rangle,\end{aligned} \tag{7}$$

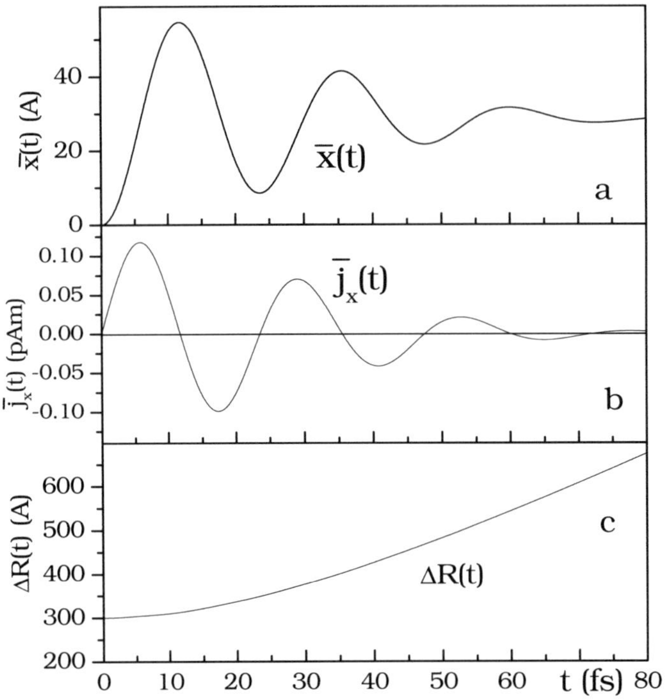

Fig. 1. Zitterbewegung of a charge carrier in bilayer graphene *versus* time, calculated for a gaussian wave packet width $d = 300$Å and $k_{0y} = 3.5 \times 10^8\text{m}^{-1}$: a) position, b) electric current, c) dispersion $\Delta R(t)$. After the ZB disappears a constant shift of the position remains. After Rusin & Zawadzki (2007b).

where $|p\rangle$ and $|n\rangle$ are the eigen-functions of the Hamiltonian (1) in $\boldsymbol{k}$ space corresponding to positive and negative energies, respectively. Further

$$\langle \boldsymbol{k}|p\rangle = \frac{1}{\sqrt{2}}\begin{pmatrix} 1 \\ k_+^2/k^2 \end{pmatrix}\delta(\boldsymbol{k}-\boldsymbol{k}'), \tag{8}$$

$$\langle \boldsymbol{k}|n\rangle = \frac{1}{\sqrt{2}}\begin{pmatrix} 1 \\ -k_+^2/k^2 \end{pmatrix}\delta(\boldsymbol{k}-\boldsymbol{k}'). \tag{9}$$

After some manipulations we obtain

$$\psi^p(\boldsymbol{r},t) = \frac{1}{4\pi}\frac{d}{\sqrt{\pi}}\int d^2\boldsymbol{k} e^{-\frac{1}{2}d^2(k_x^2+(k_y-k_{0y})^2)} e^{i\boldsymbol{k}\boldsymbol{r}} e^{-iEt/\hbar} \times \begin{pmatrix} 1 \\ k_+^2/k^2 \end{pmatrix}. \tag{10}$$

The function $\psi^n(\boldsymbol{r},t)$ is given by the identical expression with the changed signs in front of E and k_+^2/k^2 terms. There is $\psi(\boldsymbol{r},t) = \psi^p(\boldsymbol{r},t) + \psi^n(\boldsymbol{r},t)$ and $\langle\psi^n|\psi^p\rangle = 0$. Now, one can calculate the average values of $\bar{x}$ and $\bar{y}$ using the positive and negative components in the above sense. We have

$$\bar{x}(t) = \int (\psi^n + \psi^p)^\dagger x(\psi^n + \psi^p)d^2\boldsymbol{r}, \tag{11}$$

so that one deals with four integrals. A direct calculation gives

$$\int |\psi^p|^2 x d^2\boldsymbol{r} + \int |\psi^n|^2 x d^2\boldsymbol{r} = \bar{x}_c, \quad (12)$$

$$\int \psi^{n+} x \psi^p d^2\boldsymbol{r} + \int \psi^{p+} x \psi^n d^2\boldsymbol{r} = \bar{x}_Z(t), \quad (13)$$

where $\bar{x}_c$ and $\bar{x}_Z(t)$ have been defined in Eq. (4). Thus the integrals involving only the positive and only the negative components give the constant shift due to ZB, while the mixed terms lead to the ZB oscillations. All terms together reconstruct the result from Eq. (4). Next we calculate the average value $\bar{y}$. There is no symmetry between $\bar{x}$ and $\bar{y}$ because the wave packet is centered around $k_x = 0$ and $k_y = k_{0y}$. The average value $\bar{y}$ is again given by four integrals. However, now the mixed terms vanish, while the integrals involving the positive and negative components give

$$\int |\psi^p|^2 y d^2\boldsymbol{r} = \frac{\hbar k_{0y}}{2m^*} t, \quad (14)$$

$$\int |\psi^n|^2 y d^2\boldsymbol{r} = -\frac{\hbar k_{0y}}{2m^*} t. \quad (15)$$

This means that the "positive" and "negative" subpackets move in the opposite directions with the velocity $v = \hbar k_{0y} t/2m^*$. The relative velocity is $v^{rel} = \hbar k_{0y} t/m^*$. Each of these packets has the initial width d and it (slowly) spreads in time. After the time $\Gamma_Z^{-1} = d/v^{rel}$ the distance between the two packets equals d, so the integrals in Eq. (12) are small, resulting in the diminishing Zitterbewegung amplitude. This reasoning gives the decay constant $\Gamma_Z = \hbar k_{0y}/m^* d$, which is exactly what we determined above from the analytical results (see Eq. (6)). Thus, *the transient character of the ZB oscillations is due to the increasing spatial separation of the subpackets corresponding to the positive and negative energy states.* This confirms our previous conclusion that it is not the packet's slow spreading that is responsible for the attenuation. The separation of subpackets with the resulting decay of ZB turns out to be *a general feature* of this phenomenon.

2.2 Monolayer graphene

Now we consider monolayer graphene. The two-dimensional band Hamiltonian describing its band structure at the K point of the Brillouin zone is (Semenoff, 1984; Slonczewski & Weiss, 1958; Wallace, 1947)

$$\hat{H}_M = u \begin{pmatrix} 0 & \hat{p}_x - i\hat{p}_y \\ \hat{p}_x + i\hat{p}_y & 0 \end{pmatrix}, \quad (16)$$

where $u \approx 1 \times 10^8$cm/s. The resulting energy dispersion is linear in momentum: $\mathcal{E} = \pm u\hbar k$, where $k = \sqrt{k_x^2 + k_y^2}$. The quantum velocity in the Schrodinger picture is $\hat{v}_i = \partial H_M/\partial \hat{p}_i$, it does not commute with the Hamiltonian (16). In the Heisenberg picture we have $\hat{\boldsymbol{v}}(t) = \exp(i\hat{H}_M t/\hbar)\hat{v}\exp(-i\hat{H}_M t/\hbar)$. Using Eq. (16) one obtains

$$v_x^{(11)} = u\frac{k_y}{k}\sin(2ukt). \quad (17)$$

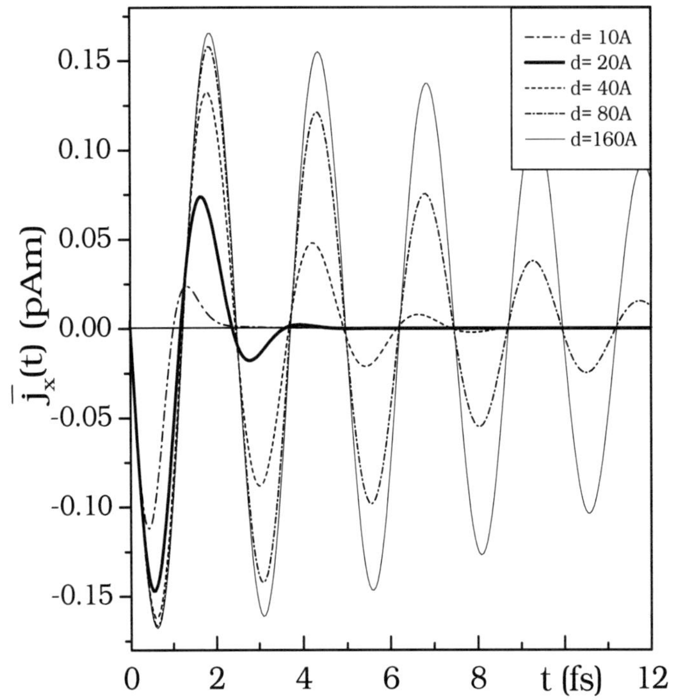

Fig. 2. Oscillatory electric current in the x direction caused by the ZB in monolayer graphene *versus* time, calculated for a gaussian wave packet with $k_{0y} = 1.2 \times 10^9 \mathrm{m}^{-1}$ and various packet widths d. Transient character of ZB is clearly seen. After Rusin & Zawadzki (2007b).

The above equation describes the trembling motion with the frequency $\omega_Z = 2uk$, determined by the energy difference between the upper and lower energy branches for a given value of k. As before, ZB in the direction x occurs only if there is a non-vanishing momentum $\hbar k_y$. One calculates an average velocity (or current) taken over a two-dimensional wave packet with nonzero initial momentum k_{0x}. The results for the current $\bar{j}_x = e\bar{v}_x$ are plotted in Figure 2 for different realistic packet widths d. It is seen that the ZB frequency does not depend on d and is nearly equal to ω_Z given above for the plane wave. On the other hand, the amplitude of ZB depends on d and we deal with decay times of the order of femtoseconds. For small d there are almost no oscillations, for very large d the ZB oscillations are undamped. These conclusions agree with the analytical results for bilayer graphene. The behavior of ZB depends quite critically on the values of k_{0y} and d, which is reminiscent of the damped harmonic oscillator. In the limit $d \to \infty$ the above results for the electric current resemble those of Katsnelson (2006) for ZB in graphene obtained with the use of plane wave representation.

Maksimova et al. (2008) investigated dynamics of wave packets in monolayer graphene for different pseudo-spin polarizations with the resulting ZB. For specific packet components and their relative phases a "longitudinal ZB" can take place, but its intensity is weak. Martinez et al. (2010) considered a creation of electron-hole pairs by a constant electric field in the plane of a monolayer graphene sheet. They showed that, as the pairs undergo the ZB in opposite directions, a Hall-like separation of the charge occurs giving a measurable dipole moment. We note that it is not the time-dependent motion but the ZB shift at large times which is responsible for the charge separation, see Eq. (2) and Figure 1. Englman &

Vertesi (2008) calculated a ZB-related electron current in monolayer graphene in the adiabatic approximation and related it to the Berry phase.

2.3 Carbon nanotubes

Finally we consider monolayer graphene sheets rolled into single semiconducting carbon nanotubes (CNT) (Rusin & Zawadzki, 2007b; Zawadzki, 2006). The band Hamiltonian in the vicinity of K point is (Ajiki & Ando, 1993)

$$\hat{H}_{CNT} = u\begin{pmatrix} 0 & \hbar k_{n\nu} - i\hat{p}_y \\ \hbar k_{n\nu} + i\hat{p}_y & 0 \end{pmatrix}. \tag{18}$$

This Hamiltonian is similar to that given by Eq. (16) except that, because of the periodic boundary conditions, the momentum p_x is quantized and takes discrete values $\hbar k_x = \hbar k_{n\nu}$, where $k_{n\nu} = (2\pi/L)(n - \nu/3)$, $n = 0, \pm 1, \ldots$, $\nu = \pm 1$, and L is the length of circumference of CNT. As a result, the free electron motion can occur only in the direction y, parallel to the tube axis. The geometry of CNT has important consequences. There exists an energy gap $E_g = 2u\hbar|k_{n\nu}|$ and the effective mass at the band edge $m_0^* = \hbar|k_{n\nu}|/u$. For $\nu = \pm 1$ there *always* exists a non-vanishing value of the quantized momentum $\hbar k_{n\nu}$. Finally, for each value of $k_{n\nu}$ there exists $k_{-n,-\nu} = -k_{n\nu}$ resulting in the same subband energy $\mathcal{E} = \pm E$, where

$$E = \hbar u\sqrt{k_{n\nu}^2 + k_y^2}. \tag{19}$$

The time dependent velocity $\hat{v}_y(t)$ and the displacement $\hat{y}(t)$ can be calculated for the plane electron wave in the usual way and they exhibit the ZB oscillations [see Zawadzki (2006)]. For small momenta k_y the ZB frequency is $\hbar\omega_Z = E_g$ and the ZB length is $\lambda_Z \approx 1/|k_{n\nu}|$. We are again interested in the displacement $\bar{y}(t)$ of a charge carrier represented by a one-dimensional wave packet analogous to that described in Eq. (3) The average displacement is $\bar{y}(t) = \bar{y}_Z(t) - \bar{y}_{sh}$, where

$$\bar{y}_Z(t) = \frac{\hbar^2 du^2 k_{n\nu}}{2\sqrt{\pi}} \int_{-\infty}^{\infty} \frac{dk_y}{E^2} \cos\left(\frac{2Et}{\hbar}\right) e^{-d^2k_y^2} \tag{20}$$

and $\bar{y}_{sh} = 1/2\sqrt{\pi}d\,\mathrm{sgn}(b)[1 - \Phi(|b|)]\exp(b^2)$, where $b = k_{n\nu}d$ and $\Phi(x)$ is the error function. The ZB oscillations of $\bar{y}(t)$ are plotted in Figure 3. It is seen that, after the transient ZB oscillations disappear, there remains a shift $\bar{y}_{sh}$. Thus the ZB separates spatially the charge carriers that are degenerate in energy but characterized by n, ν and $-n, -\nu$ quantum numbers. The current is proportional to $\bar{v}_y = \partial\bar{y}/\partial t$, so that the currents related to $\nu = 1$ and $\nu = -1$ cancel each other. To have a non-vanishing current one needs to break the above symmetry, which can be achieved by applying an external magnetic field parallel to the tube axis.

It is seen from Figure 3 that the decay time of ZB in CNT is much larger than that in bilayer and monolayer graphene. The oscillations decrease proportionally to $t^{-1/2}$. The reason is that we consider the situation with $k_{0y} = 0$, so that the ZB oscillations occur due to "built in" momentum $k_x = k_{n\nu}$, arising from the tube's topology. In other words, the long decay time is due to the one-dimensionality of the system. If the circumference of a CNT is increased, the energy gap (and, correspondingly, the ZB frequency) decreases, the amplitude of ZB is larger, but the decay time remains almost unchanged.

One can show that we again deal here with two sub-packets which, however, for $k_{0y} = 0$ do not run away from each other. Thus, the slow damping of ZB is due only to the slow

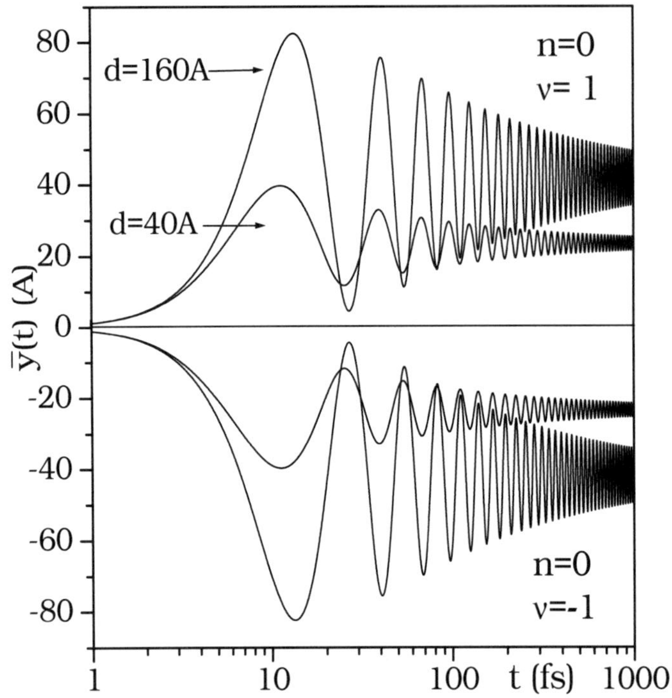

Fig. 3. Zitterbewegung of two charge carriers in the ground subband of a single carbon nanotube of $L = 200$ Å *versus* time (logarithmic scale), calculated for gaussian wave packets of two different widths d and $k_{0y} = 0$. After the ZB disappears a constant shift remains. The two carriers are described by different quantum numbers ν. At higher times the amplitude of ZB oscillations decays as $t^{-1/2}$. After Rusin & Zawadzki (2007b).

broadening of the sub-packets. We emphasize the slow decay, as illustrated in Figure 3, because it is confirmed experimentally, see Gerritsma et al. (2010). We add that for $k_{0y} \neq 0$ the sub-packets run away from each other and the decay time is much faster.

3. ZB in graphene in a magnetic field

The trembling motion of charge carriers in graphene and CNT has been described above for no external potentials. Now we consider the trembling motion of electrons in the presence of an external magnetic field (Rusin & Zawadzki, 2008). The magnetic field is known to cause no interband electron transitions, so the essential features of ZB are expected not to be destroyed. On the other hand, introduction of an external field provides an important parameter affecting the ZB behavior. This case is special because the electron spectrum is fully quantized, so we deal with discrete levels, not bands. We consider a graphene monolayer in an external magnetic field parallel to the z axis. The Hamiltonian for electrons and holes at the K_1 point of the Brillouin zone is (Slonczewski & Weiss, 1958; Wallace, 1947)

$$\hat{H} = u \begin{pmatrix} 0 & \hat{\pi}_x - i\hat{\pi}_y \\ \hat{\pi}_x + i\hat{\pi}_y & 0 \end{pmatrix}, \tag{21}$$

where $u \approx 1\times10^8$ cm/s is the characteristic velocity, $\hat{\boldsymbol{\pi}} = \hat{\boldsymbol{p}} - q\hat{\boldsymbol{A}}$ is the generalized momentum, in which $\hat{\boldsymbol{A}}$ is the vector potential and q is the electron charge. Using the Landau gauge, we take $\hat{\boldsymbol{A}} = (-By, 0, 0)$, and for an electron $q = -e$ with $e > 0$. We take the wave function in the form $\Psi(x,y) = e^{ik_x x}\Phi(y)$. Introducing the magnetic radius $L = \sqrt{\hbar/eB}$, the variable $\xi = y/L - k_x L$, and defining the standard raising and lowering operators for the harmonic oscillator $\hat{a} = (\xi + \partial/\partial\xi)/\sqrt{2}$ and $\hat{a}^\dagger = (\xi - \partial/\partial\xi)/\sqrt{2}$, the Hamiltonian is rewritten in the form

$$\hat{H} = -\hbar\Omega\begin{pmatrix} 0 & \hat{a} \\ \hat{a}^\dagger & 0 \end{pmatrix}, \tag{22}$$

where the frequency is $\Omega = \sqrt{2}u/L$. Next one determines the eigenstates and eigenenergies of the Hamiltonian $\hat{H}$. The energy is $E_{ns} = s\hbar\Omega\sqrt{n}$. Here $n = 0, 1, \ldots$, and $s = \pm 1$ for the conduction and valence bands, respectively. The above energies were confirmed experimentally. The complete wave function is

$$|\mathrm{n}\rangle \equiv |nk_x s\rangle = \frac{e^{ik_x x}}{\sqrt{4\pi}}\begin{pmatrix} -s|n-1\rangle \\ |n\rangle \end{pmatrix} \tag{23}$$

where $|n\rangle$ are the harmonic oscillator functions.

We want to calculate the velocity of a charge carrier described by a wave packet. We first calculate matrix elements $\langle f|\mathrm{n}\rangle$ between an arbitrary two-component function $f = (f^u, f^l)$ and eigenstates from Eq. (23). A straightforward manipulation gives $\langle f|\mathrm{n}\rangle = -sF^u_{n-1} + F^l_n$, where

$$F^j_n(k_x) = \frac{1}{\sqrt{2LC_n}}\int g^j(k_x, y)e^{-\frac{1}{2}\xi^2}\mathrm{H}_n(\xi)dy, \tag{24}$$

in which

$$g^j(k_x, y) = \frac{1}{\sqrt{2\pi}}\int f^j(x,y)e^{ik_x x}dx. \tag{25}$$

The superscript $j = u, l$ stands for the upper and lower components of the function f. The Hamilton equations give the velocity components: $\hat{v}_i(0) = \partial\hat{H}/\partial\hat{p}_i$, with $i = x, y$. We want to calculate averages of the time-dependent velocity operators $\hat{v}_i(t)$ in the Heisenberg picture taken on the function f. The averages are

$$\bar{v}_i(t) = \sum_{\mathrm{n,n'}} e^{iE_{\mathrm{n'}}t/\hbar}\langle f|\mathrm{n'}\rangle\langle\mathrm{n'}|v_i(0)|\mathrm{n}\rangle\langle\mathrm{n|f}\rangle e^{-\mathrm{i}E_{\mathrm{n}}t/\hbar}, \tag{26}$$

where the energies and eigenstates are given in Eq. (23). The summation in Eq. (26) goes over all the quantum numbers: n, n', s, s', k_x, k'_x. The only non-vanishing matrix elements of the velocity components are for the states states $n' = n \pm 1$. One finally obtains

$$\overline{v}_y(t) = u\sum_{n=0}^{\infty} V_n^+ \sin(\omega_n^c t) + u\sum_{n=0}^{\infty} V_n^- \sin(\omega_n^Z t) + iu\sum_{n=0}^{\infty} A_n^+ \cos(\omega_n^c t) + iu\sum_{n=0}^{\infty} A_n^- \cos(\omega_n^Z t), \tag{27}$$

$$\overline{v}_x(t) = u\sum_{n=0}^{\infty} B_n^+ \cos(\omega_n^c t) + u\sum_{n=0}^{\infty} B_n^- \cos(\omega_n^Z t) + iu\sum_{n=0}^{\infty} T_n^+ \sin(\omega_n^c t) + iu\sum_{n=0}^{\infty} T_n^- \sin(\omega_n^Z t), \tag{28}$$

where $V_n^{\pm}$, $T_n^{\pm}$, $A_n^{\pm}$ and $B_n^{\pm}$ are given by combinations of $U_{m,n}^{\alpha,\beta}$ integrals

$$U_{m,n}^{\alpha,\beta} = \int F_m^{\alpha*}(k_x)F_n^{\beta}(k_x)dk_x. \tag{29}$$

The superscripts α and β refer to the upper and lower components, see Rusin & Zawadzki (2008). In the above model the coefficients $U_{m,n}^{\alpha,\beta}$ are real numbers, so that $A_n^{\pm}$ and $T_n^{\pm}$ in Eqs. (27) and (28) vanish. The time dependent sine and cosine functions come from the exponential terms in Eq. (26). The frequencies in Eqs. (27) and (28) are $\omega_n^c = \Omega(\sqrt{n+1} - \sqrt{n})$, $\omega_n^Z = \Omega(\sqrt{n+1} + \sqrt{n})$, where Ω is given in Eq. (22). The frequencies ω_n^c correspond to the intraband energies while frequencies ω_n^Z correspond to the interband energies, see Figure 4. The interband frequencies are characteristic of the Zitterbewegung. The intraband (cyclotron) energies are due to the band quantization by the magnetic field and they do not appear for non quantized spectra.

Final calculations are carried out for a two-dimensional Gaussian wave packet centered around the wave vector $\mathbf{k_0} = (k_{0x}, 0)$ and having two non-vanishing components. In this case one can obtain analytical expressions for $U_{m,n}^{\alpha,\beta}$. The main frequency of oscillations is $\omega_0 = \Omega$, which can be interpreted either as $\omega_0^c = \Omega(\sqrt{n+1} - \sqrt{n})$ or $\omega_0^Z = \Omega(\sqrt{n+1} + \sqrt{n})$ for $n = 0$. Frequency ω_0^c belongs to the intraband (cyclotron) set, while ω_0^Z belongs to the interband set (see Figure 4). The striking feature is, that ZB is manifested by *several frequencies simultaneously*. This is a consequence of the fact that in graphene the energy distances between the Landau levels diminish with n, which results in different values of frequencies ω_n^c and ω_n^Z for different n. It follows that it is the presence of an external quantizing magnetic field that introduces various frequencies into ZB. It turns out that, after the ZB oscillations seemingly die out, they actually reappear at higher times. Thus, for all k_{0x} values (including $k_{0x} = 0$), *the ZB oscillations have a permanent character*, that is they do not disappear in time. This feature is due to the discrete character of the electron spectrum caused by a magnetic field. The above property is in sharp contrast to the no-field cases considered above, in which the spectrum is not quantized and the ZB of a wave packet has a transient character. In mathematical terms, due to the discrete character of the spectrum, averages of operator quantities taken over a wave packet are sums and not integrals. The sums do not obey the Riemann-Lebesgues theorem for integrals which guarantee the damping of ZB in time for a continuous spectrum see Lock (1979).

Finally, one calculates the displacements $\bar{x}(t)$ and $\bar{y}(t)$ of the wave packet. To this end we integrate Eqs. (27) and (28) with respect to time using the initial conditions $x_0 = \bar{x}(0) = 0$ and $y_0 = \bar{y}(0) = k_x L^2$. The results are plotted in Figure 5 in the form of $x - y$ trajectories for different initial wave vectors k_{0x}. The direction of movement is clockwise and the trajectories span early times (1ps) after the creation of a wave packet.

All in all, the presence of a quantizing magnetic field has the following important effects on the trembling motion. (1) For $B \neq 0$ the ZB oscillation are permanent, while for $B = 0$ they are transient. The reason is that for $B \neq 0$ the electron spectrum is discrete. (2) For $B \neq 0$ many ZB frequencies appear, whereas for $B = 0$ only one ZB frequency exists. (3) For $B \neq 0$ both interband and intraband (cyclotron) frequencies appear in ZB; for $B = 0$ there are no intraband frequencies. (4) Magnetic field intensity changes not only the ZB frequencies but the entire character of ZB spectrum.

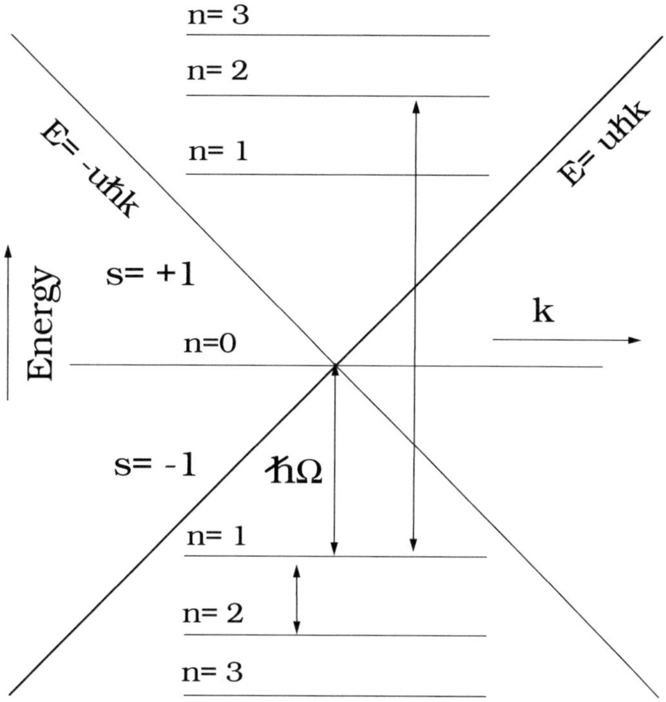

Fig. 4. The energy dispersion $E(k)$ and the Landau levels for monolayer graphene in a magnetic field (schematically). Intraband (cyclotron) and interband (ZB) energies for $n' = n \pm 1$ are indicated. The basic energy is $\hbar\Omega = \sqrt{2}\hbar u/L$. After Rusin & Zawadzki (2008).

The Zitterbewegung should be accompanied by electromagnetic dipole radiation emitted by the trembling electrons. The oscillations $\langle \boldsymbol{r}(t) \rangle$ are related to the dipole moment $-e\langle \boldsymbol{r}(t) \rangle$, which couples to the electromagnetic radiation. One can calculate the emitted electric field from the electron acceleration $\langle \ddot{\boldsymbol{r}}(t) \rangle$ and takes its Fourier transform to determine the emitted frequencies. In Figure 6 we plot the calculated intensities of various emitted lines. The strong peak corresponds to oscillations with the basic frequency $\omega = \Omega$. The peaks on the high-frequency side correspond to the interband excitations and are characteristic of ZB. The peaks on the lower frequency side correspond to the intraband (cyclotron) excitations. In absence of ZB the emission spectrum would contain only the intraband (cyclotron) frequencies. Thus the interband frequencies ω_n^Z shown in Figure 6 are a direct signature of the trembling motion. It can be seen that the ω_z^Z peaks are not much weaker than the central peak at $\omega = \Omega$, which means that there exists a reasonable chance to observe them. Generally speaking, the excitation of the system is due to the nonzero momentum $\hbar k_{0x}$ given to the electron. It can be provided by accelerating the electron in the band or by exciting the electron with a nonzero momentum by light from the valence band to the conduction band. The electron can emit light because the Gaussian wave packet is not an eigenstate of the system described by the Hamiltonian (21). The energy of the emitted light is provided by the initial kinetic energy related to the momentum $\hbar k_{0x}$. Once this energy is completely used, the emission will cease.

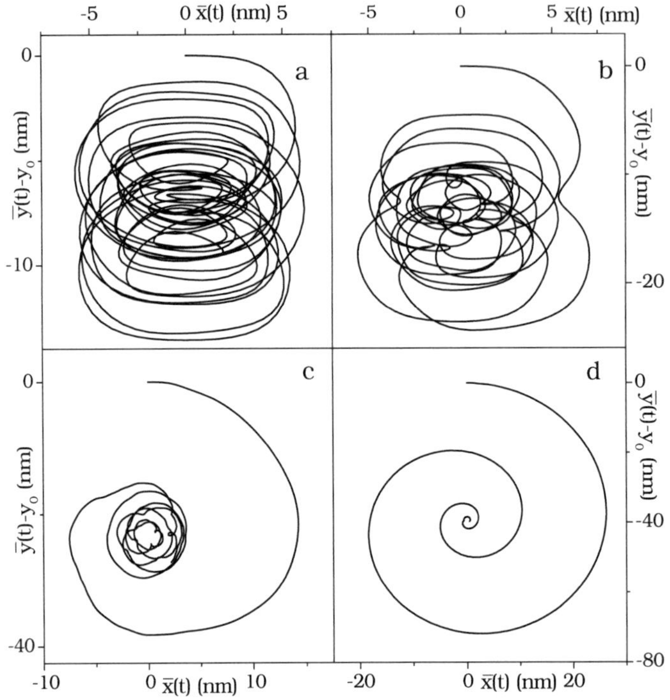

Fig. 5. Zitterbewegung trajectories of electron at the K_1 point of the Brillouin zone in monolayer graphene at B=20T during the first picosecond for various values of k_{0x}. After Rusin & Zawadzki (2008).

Schliemann (2008) described time dependence of the cyclotron motion in monolayer graphene in the presence of a magnetic field using the semiclassical approximation for high carrier energies. He showed that the cyclotron motion is perturbed by interband ZB contributions of higher frequencies. Krueckl & Kramer (2009) described time propagation of an initially concentrated wave packet in monolayer graphene in a perpendicular magnetic field. A collapse-revival pattern of ZB was investigated and an effect of impurities (disorder) on the packet dynamics was analyzed. It turned out that ZB "survives" the perturbation by impurities. Romera & de los Santos (2009) studied monolayer graphene in a magnetic field concentrating on collapse-revival pattern of ZB oscillations.

4. ZB probed by laser pulse

All the recent theoretical work on ZB assumed that initially the electrons are represented by Gaussian wave packets. While this assumption represents a real progress compared to the work that had treated electrons as plane waves, it is obviously an idealization since it is not quite clear how to prepare an electron in this form. Here we calculate a reaction of an electron in graphene excited by a laser pulse, not assuming anything about initial form of the electron wave packet. In our description we take into account currently available experimental possibilities.

The following conditions should be met for a successful observation of ZB: a) The ZB frequency must be in the range of currently detectable regimes, i.e. of the order of $\omega_Z \approx 1\ \mathrm{fs}^{-1}$,

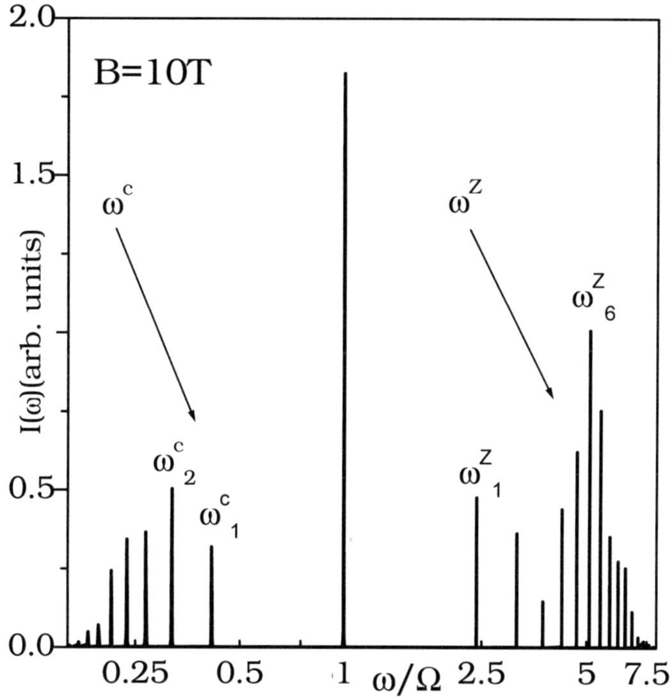

Fig. 6. Intensity spectrum versus frequency during the first 20 ps of motion of an electron described by a Gaussian wave packet having $k_{0x} = 0.035\text{Å}^{-1}$ in monolayer graphene. After Rusin & Zawadzki (2008).

and the size of oscillations should be of the order of a few Å; b) The ZB oscillations should be persistent or slowly transient; c) Both positive and negative electron energies must be excited with a sufficient probability; d) The wave packet should be created in a one-electron regime. A system that fulfills the above criteria is p-type monolayer graphene in a constant magnetic field. The wave packet should be created by an ultra short monocycle or sub-monocycle laser pulse. Because of a very wide frequency spectrum of such a pulse, the resulting wave packet will have both positive and negative energies. The electron oscillations give rise to a time-dependent dipole moment which will be a source of electric field and it will emit or absorb radiation in the far infrared range.

In the description we use the above results for the band structure of monolayer graphene in a magnetic field. The Fermi level is assumed to coincide with the Landau level (LL) $n = -2$ and we consider the initial electron to be in the $n = -1$ state. The wavelength of the laser light is assumed to be much larger that the size L of the $n = -1$ state, so we can neglect spatial variation of the electric field in the laser pulse. The total Hamiltonian, including the perturbation due to the laser light, is $\hat{H}_T = \hat{H} + \hat{W}(t)$, where $\hat{H}$ is given by Eq. (21) and the perturbing potential due laser light is

$$\hat{W}(t) = -ey\mathcal{E}_0 e^{-(2\ln 2)t^2/\tau^2}\cos(\omega_L t), \tag{30}$$

in which e is the electron charge, τ is the pulse duration (FWHM), $\omega_L = 2\pi c/\lambda_L$ is the laser frequency (being of the order of $3\times 10^{15}\text{s}^{-1}$), and $\mathcal{E}_0$ is the amplitude of electric field. A

Gaussian shape of the laser pulse is widely used in optical experiments and it parameterizes effectively a profile of electric field in the laser beam.
As a result of a laser shot, the initial state of the system $\Phi_k(t) = \psi_k e^{-iE_k t/\hbar}$ evolves into the final state $\Psi_k(t) = \sum_j c_j(t)\psi_k e^{-iE_j t/\hbar}$, which is a combination of the eigenstates of $\hat{H}$ with suitably chosen coefficients $c_j(t)$. The resulting time-dependent dipole moment is $\boldsymbol{D}(t) = e\langle\Psi_k(t)|\boldsymbol{r}|\Psi_k(t)\rangle$. The corresponding time-dependent wave functions are $\Psi_k(t) = e^{-i\hat{H}t/\hbar}\Psi_k(0)$, and the dipole moment is

$$\begin{aligned}\boldsymbol{D}(t) &= e\langle\Psi_k(0)e^{i\hat{H}t/\hbar}|\hat{\boldsymbol{r}}|e^{-i\hat{H}t/\hbar}\Psi_k(0)\rangle \\ &= e\langle\Psi_k(0)|\hat{\boldsymbol{r}}(t)|\Psi_k(0)\rangle = e\langle\boldsymbol{r}(t)\rangle. \end{aligned} \tag{31}$$

Here $\boldsymbol{r}(t)$ is the electron position in the Heisenberg picture. Thus the dipole moment $\boldsymbol{D}(t)$ is proportional to the time-dependent position averaged over the electron wave packet.
A time-dependent dipole moment is a source of electromagnetic radiation. We treat the radiation classically (Bohm, 1952) and take the radiated transverse electric field to be (Jackson, 1975)

$$\mathcal{E}_{\perp}(\boldsymbol{r},t) = \frac{\ddot{\boldsymbol{D}}(t)}{4\pi\epsilon_0 c^2}\frac{\sin(\theta)}{R}, \tag{32}$$

where ϵ_0 is the vacuum permittivity, θ is an angle between the direction of electron motion and a position of the observer $\boldsymbol{R}$. Since $\ddot{\boldsymbol{D}}(t) = e\langle\ddot{\boldsymbol{r}}(t)\rangle$, Eq. (32) relates the electric field of the dipole with the average acceleration of the packet. If the electric field is measured directly by an antenna, one measures the trembling motion of the wave packet. If the square of electric field is measured in emission or absorption experiments, the signature of ZB is the existence of peaks corresponding to interband frequencies and their dependence on packet's parameters. Accordingly, in the time resolved luminescence experiments it should be possible to detect directly the motion of the packet with interband frequencies.
The electric field emitted by the trembling electron is calculated by the time-dependent perturbation theory. The induced dipole moment and the corresponding electric field oscillate with three frequencies. The frequency $\omega(|1-\rangle, |2+\rangle) = (\sqrt{2}+1)\ \Omega$ corresponds to the Zitterbewegung, i.e., to the motion of the packet with an interband frequency. Here $|1-\rangle$ denotes the valence state $n = 1$ and $|2+\rangle$ the conduction state $n = 2$. The interband frequency is characteristic of ZB because the trembling motion occurs due to an interference of electron states related to positive and negative electron energies. The second frequency $\omega(|1-\rangle, |2-\rangle) = (\sqrt{2}-1)\ \Omega$ describes the intraband cyclotron motion of the packet. The third frequency $\omega(|1-\rangle, |0\rangle) = \Omega$ has both interband and intraband character (see Figure 4). In zero-gap materials like graphene the interband ZB frequency is not much larger than the cyclotron frequency.
In Figure 7 we plot the oscillating dipole moment within the first 1000 fs of motion after the laser shot for two magnetic fields $B = 1$ T and $B = 10$ T, and two laser pulses. In Figure 8 we plot the corresponding electric field for the same parameters during the first 250 fs of oscillations. We assume the laser intensity to be $I = 1.0\times 10^9$ W/cm^2, the emitted electromagnetic wave detected at the angle $\theta = 45^{\circ}$, and the distance $R = 1$ cm. All the quantities in Figures 7 and 8 are calculated per one electron. Since the frequencies are incommensurable, the electron trajectories $\boldsymbol{r}(t)$ are not closed and there is no repeated pattern of oscillations. The motion of the wave packet is permanent in the time scale of femtoseconds

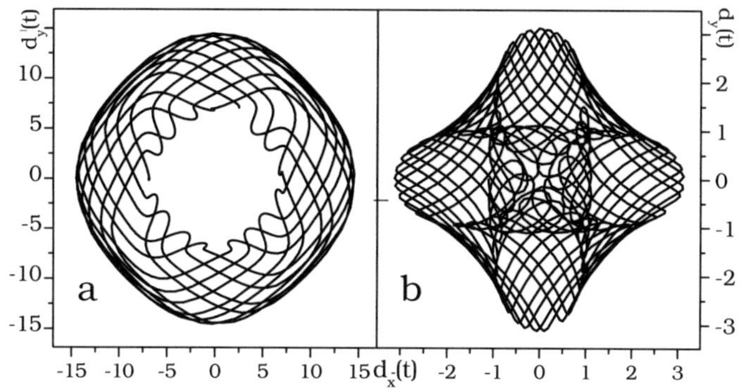

Fig. 7. Oscillations of dipole moment during the first 1000 fs of electron motion after the laser pulse. Experimental characteristics: pulse intensity 1×10^9 W/cm^2, a) $\tau = 1.6$ fs, $B = 10$ T, b) $\tau = 3.0$ fs, $B = 10$ T. Dipole moments in a) are in 10^{-28} [Cm] units, while in b) they are in 10^{-31} [Cm] units. The above results refer to very narrow Landau levels, disregarding broadening. After Rusin & Zawadzki (2009).

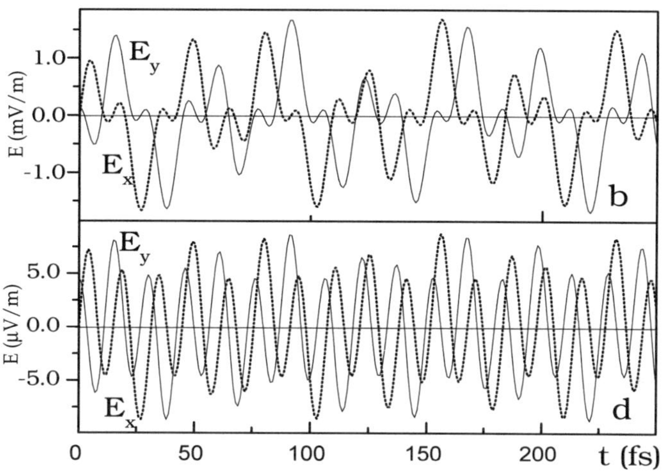

Fig. 8. Electric field emitted by one electron during the first 250 fs of electron motion after the laser pulse. Experimental characteristics: pulse intensity 1×10^9 W/cm^2, a) $\tau = 1.6$ fs, $B = 10$ T, b) $\tau = 3.0$ fs, $B = 10$ T. The above results refer to very narrow Landau levels, disregarding broadening. After Rusin & Zawadzki (2009).

or picoseconds but there is damping of the motion due to the light emission in a long time scale. The results shown in Figures 7 and 8 refer to very narrow Landau levels, disregarding broadening.

We can draw the following qualitative conclusions from Figure 7 and 8. First, for small magnetic fields B the period of oscillations is longer than for large fields, which is related to the basic frequency Ω. Second, irrespective of the variation of Ω with B, for small fields the oscillations are dominated by the low (cyclotron) frequency, while at stronger B the high (ZB) frequency dominates. Finally, comparing the magnitudes of dipole moment or emitted electric field for $\tau = 1.6$ fs with the corresponding values for $\tau = 3.0$ fs one observes that the amplitude of oscillations depends very strongly on the duration τ of the pulse.

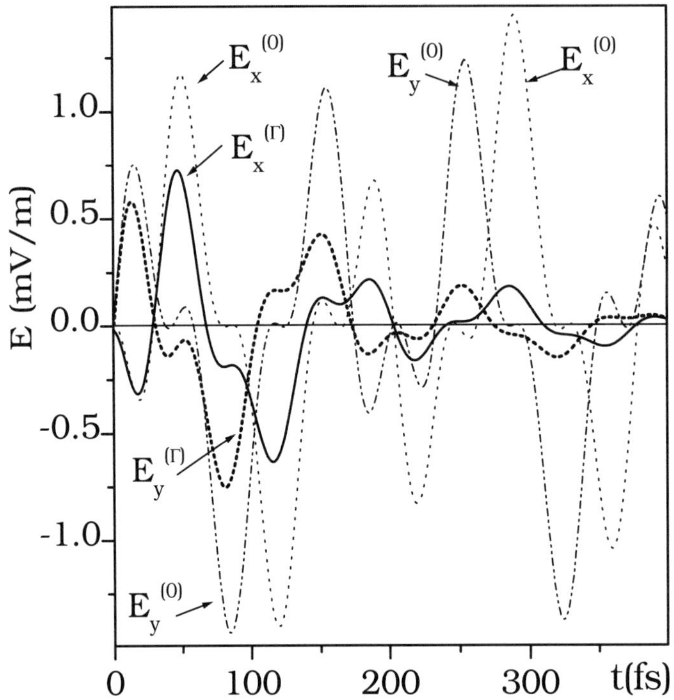

Fig. 9. Calculated electric fields $E_x(t)$ and $E_y(t)$ emitted by one trembling electron during the first 400 fs after the laser pulse. Pulse parameters: intensity 1×10^9 W/cm^2, $\tau = 1.6$ fs, magnetic field $B = 1$ T. Bold lines - electric fields for broadened Landau levels described by Lorentzian line-shapes with experimental values of Γ_n. Dotted lines - electric fields for delta-like Landau levels. After Rusin & Zawadzki (2009).

The broadening of the Landau levels may strongly influence the trembling motion of the wave packet. To analyze the overall impact of all the effects leading to the level broadening: disorder, e-e interaction, electron-phonon scattering, etc., one assumes finite widths of all energy levels, characterized by broadening parameters Γ_n. These are treated as phenomenological quantities determined experimentally and including all scattering mechanisms existing in real samples. We approximate the broadening of the density of states by a Lorentzian line-shape irrespective of the detailed scattering mechanism.
The electric field $\boldsymbol{E}(t)$ is calculated, as before, as a second time derivative of $\boldsymbol{D}(t)$. In Figure 9 we plot the electric field emitted by an oscillating electron within the first 400 fs of motion after the laser shot of the width $\tau = 1.6$ fs in a magnetic field $B = 1$ T. The two bold lines describe calculated electric fields $E_x(t)$ and $E_y(t)$ for the Landau levels having the broadening parameters Γ_n indicated above. The two dotted lines show the electric fields calculated without damping ($\Gamma_n = 0$). Within the first 50 fs of motion the electric fields emitted in the two cases are similar, but later the damping of the emitted fields for broadened levels is visible. After around 400 fs the trembling motion in real case disappears. It can be seen that the maxima of oscillations for the damped ZB motion coincide with the undamped ones. The general conclusion from Figure 9 is that the existence of disorder, many-body effects or other scattering mechanisms changes the persistent ZB motion to a decaying one, within the characteristic lifetimes for these processes: $\tau_n = 1/\Gamma_n \approx 130$ fs. Nevertheless, since the

parameters Γ_n used in the calculations correspond to the measured lifetimes in real graphene samples, it follows that the broadening of the Landau levels does not prevent the existence of ZB. Clearly, a lower disorder in better samples will result in longer decay times for ZB.
Wang et al. (2009) carried out a study similar to the one described above, but for bilayer graphene in a magnetic field. This system is somewhat different from monolayer graphene since the Landau levels are nearly uniformly spaced due to quadratic dependence of positive and negative energies on momentum, see Eq. (1). Also, the laser pulse was assumed to contain only one frequency ω_L. The authors estimated that in high quality bilayer graphene samples the stimulated ZB electric field can be of the order of volts per meter and the corresponding coherence times of tens of femtoseconds.

5. Discussion and conclusions

In spite of the great interest in the phenomenon of ZB its physical origin remained mysterious. As mentioned above, it was recognized that the ZB in a vacuum is due to an interference of states corresponding to positive and negative electron energies. Since the ZB in solids was treated by two-band Hamiltonians similar to the Dirac equation, its interpretation was also similar. This did not explain its origin, it only provided a way to describe it. For this reason we considered recently the fundamentals of electron propagation in a periodic potential (Zawadzki & Rusin, 2010). The physical origin of ZB is essential because it resolves the question of its observability. The second purpose was to decide whether the two-band **k.p** model of the band structure, used to describe the ZB in solids, was adequate.
It is often stated that an electron moving in a periodic potential behaves like a free particle characterized by an effective mass m^*. The above picture suggests that, if there are no external forces, the electron moves in a crystal with a constant velocity. This, however, is clearly untrue because the electron velocity operator $\hat{v}_i = \hat{p}_i/m_0$ does not commute with the Hamiltonian $\hat{H} = \hat{\mathbf{p}}^2/2m_0 + V(\mathbf{r})$, so that $\hat{v}_i$ is not a constant of the motion. In reality, as the electron moves in a periodic potential, it accelerates or slows down keeping its total energy constant. This situation is analogous to that of a roller-coaster: as it goes down losing its potential energy, its velocity (i.e. its kinetic energy) increases, and when it goes up its velocity decreases. We first considered the trembling frequency ω_Z. The latter is easy to determine if we assume, in the first approximation, that the electron moves with a constant average velocity $\bar{v}$ and the period of the potential is a, so $\omega_Z = 2\pi\bar{v}/a$. Putting typical values for GaAs: $a = 5.66$Å, $\bar{v} = 2.3 \times 10^7$cm/s, one obtains $\hbar\omega_Z = 1.68$eV, i.e. the interband frequency since the energy gap is $E_g \simeq 1.5$eV. The interband frequency is in fact typical for the ZB in solids. One can go further calculating classical velocity oscillations for a periodic potential using a condition of the energy conservation: $mv_z^2/2 + eV(z) = E$, which for a periodic potential $V(z)$ gives velocity oscillations. Finally, we used a quantum approach employing the Kronig-Penney delta-like potential since it allows one to calculate explicitly the eigenenergies and eigenfunctions (Kronig & Penney, 1931; Smith, 1961). The oscillations of the packet velocity calculated directly from the periodic potential have many similarities to those computed on the basis of the two-band **k.p** model. The question arises: does one deal with *the same* phenomenon in the two cases? To answer this question we calculated ZB using the two methods for the same periodic potential. It turned out that the two-band **k.p** model gives an excellent description of ZB for instantaneous velocities. This agreement demonstrates that the theories based on: (a) the periodic potential and (b) the band structure, describe *the*

same trembling motion of the electron. The procedure based on the energy band structure is more universal since it also includes cases which do not seem to have a direct classical interpretation. The main conclusion is that the electron Zitterbewegung in crystalline solids is the basic way of electron propagation in a periodic potential. The established nature of ZB indicates that the latter should certainly be observable. Another important conclusion is that, in spite of similarities between electron ZB in a vacuum and in crystalline solids, the origin of the phenomenon in the two situations is completely different.

Recently there appeared many propositions to *simulate* the Dirac equation and the resulting phenomena with the use of other systems. Since in narrow-gap semiconductors and in graphene one deals with the Hamiltonians resembling the Dirac equation, one can apply simulations also to electrons in these materials. Many simulations make use of trapped atoms or ions interacting with laser light. There are two essential advantages of such methods. First, it is possible to follow the interaction of laser light with few or even single atoms or ions. Second, when simulating the Hamiltomians it is possible to modify their basic parameters in order to make the ZB frequency much lower and its amplitude much larger. In consequence, they become measurable with current experimental techniques. Simulations of important Hamiltonians are carried out with the use of Jaynes-Cummings model (Johanning et al., 2009; Lamata et al., 2007; Leibfried et al., 2003) known from quantum and atomic optics. In particular, Gerritsma et al. (2010) simulated for the first time the 1+1 Dirac equation with the resulting one-dimensional Zitterbewegung using $^{40}Ca^{+}$ trapped ions. The results agree very well with the predictions of Zawadzki & Rusin (2010). The reason of this agreement is that the theory of Zawadzki & Rusin (2010), while concerned with semiconductors, also uses an effective Dirac equation. Gerritsma et al. (2010) showed that, if the wave packet does not have the initial momentum, the decay time of ZB is much slower. This agrees with theoretical results for carbon nanotubes, as shown in our Figure 3.

An important recognition won after the considerable effort of the last years is, that the Zitterbewegung is not a marginal, obscure and probably unobservable effect, but a real and universal phenomenon that often occurs in both quantum and non-quantum systems. Clearly, the ZB in a vacuum proposed by Schroedinger (1930) stands out as an exception since it is supposed to occur without any external force. However, in its original form it is probably not directly observable for years to come and one has to recourse to its simulations. On the other hand, manifestations of ZB in crystalline solids and other periodic systems turned out to be quite common and they are certainly observable. A universal background underlying the phenomenon of ZB in any system (including a vacuum) is an interference of states belonging to positive and negative energies (in a generalized sense, see below). The positive and negative energies belong usually to bands but they can also be discrete levels, as shown for electrons in graphene in a magnetic field, see Figure 4.

A question arises: what should be called "Zitterbewegung"? It seems that the signature of ZB phenomenon is its *interband frequency*, in which the term *interband* has the meaning "between interacting bands". An instructive example is provided by graphene in a magnetic field (see Figure 4), where the electron motion contains both intraband and interband frequencies. We believe that only the interband contributions should be called ZB, while the intraband ones are simply the cyclotron components. It appears that the second signature of ZB is the actual *motion* which, for instance, distinguishes it from the Rabi oscillations.

If an electron is prepared in the form of a wave packet, and if the electron spectrum is not completely quantized, the ZB has a *transient* character, i.e. it decays in time. One can

show that the decay time is inversely proportional to the momentum spread Δk of the wave packet, see Eq. (6). Physically, the transient character of ZB comes about as a result of waning interference of the two sub-packets belonging to positive and negative energies as they go apart because of different speeds, see Eqs. (14) and (15). The decay time is usually much longer in one-dimensional systems, see Figure 3 and Rusin & Zawadzki (2007b). On the other hand, if the electron spectrum is discrete, ZB persists in time, sometimes in the form of collapse-revival patterns. In general, the wave packet should have a non-vanishing initial momentum in one direction to exhibit the ZB in the perpendicular direction, but this is not always the case.
One should ask the question about possible observation of Zitterbewegung in graphene. Two different ways were proposed to observe the trembling electrons. The first is to detect an ac current related to the ZB velocity, see Figure 2. One needs a current meter sensitive to the ZB frequency. Then, even if the electrons do not move in phase so that the net current averages to zero, the meter should detect a clear increase of noise at the frequency ω_Z. The second possible way to observe the ZB is to detect electromagnetic radiation emitted by the trembling electrons, see Figure 6 and Rusin & Zawadzki (2008). The emission is possible because, if the electrons are prepared in form of wave packets or they respond to light wave packets, they are not in their eigenstates. Castro Neto et al. (2009) describing electronic properties of graphene considered possible manifestations of ZB in electron transport due to confinement of electron motion. The proposed ZB should not be confused with the Bloch oscillations of charge carriers in superlattices. The Bloch oscillations are basically a one-band phenomenon and they require an external electric field driving electrons all the way to the Brillouin zone boundary. On the other hand, the ZB needs at least two bands and it is a no-field phenomenon. Narrow gap superlattices and graphene can provide a suitable system for its observation.

6. Appendix A

In this appendix we describe the time evolution of electrons in monolayer graphene in the presence of a magnetic field using operator techniques rather than averages. The initial Hamiltonian for the problem reads

$$\hat{H} = u\sigma_x\hat{\pi}_x + u\sigma_y\hat{\pi}_y, \tag{33}$$

where $\hat{\pi} = \hat{p} - q\boldsymbol{A}$ is the generalized momentum, q is the electron charge, σ_i are Pauli matrices. Taking the magnetic field $\boldsymbol{B}\|\boldsymbol{z}$ we choose the vector potential $\boldsymbol{A} = (-By, 0)$. For an electron there is $q = -e$ with $e > 0$. One can look for solutions of the Schrodinger equation in the form

$$\Psi(\boldsymbol{r}) = e^{ik_x x}\Phi(y), \tag{34}$$

which gives an effective Hamiltonian $\hat{H}$

$$\hat{H} = c\hbar\left[(k_x - eBy/\hbar)\sigma_x + (\partial/i\partial y)\sigma_y\right]. \tag{35}$$

Introducing the magnetic radius $L = \sqrt{\hbar/eB}$ and $\xi = y/L - k_x L$ one has $y = \xi L + k_x L^2$, $eB/\hbar = 1/L^2$, and $\partial/\partial y = (1/L)\partial/\partial\xi$. Defining the standard raising and lowering operators for the harmonic oscillator

$$\begin{cases} \hat{a} &= (\xi + \partial/\partial\xi)/\sqrt{2}, \\ \hat{a}^+ &= (\xi - \partial/\partial\xi)/\sqrt{2}, \end{cases} \tag{36}$$

one has $[\hat{a}, \hat{a}^+] = 1$ and $\xi = (\hat{a} + \hat{a}^+)/\sqrt{2}$. The rewritten Hamiltonian $\hat{H}$ is

$$\hat{H} = -\hbar\omega \begin{pmatrix} 0 & \hat{a} \\ \hat{a}^+ & 0 \end{pmatrix}, \tag{37}$$

with $\omega = \sqrt{2}c/L$. The frequency ω is often used in our considerations.
Now we introduce an important two-component operator $\hat{A} = \mathrm{diag}(\hat{a}, \hat{a})$ and its adjoint operator $\hat{A}^+ = \mathrm{diag}(\hat{a}^+, \hat{a}^+)$. Next we define the two-component position operators

$$\hat{Y} = \frac{L}{\sqrt{2}} (\hat{A} + \hat{A}^+), \tag{38}$$

$$\hat{X} = \frac{L}{i\sqrt{2}} (\hat{A} - \hat{A}^+), \tag{39}$$

in analogy to the position operators $\hat{y}$ and $\hat{x}$. We intend to calculate the time dependence of $\hat{A}$ and $\hat{A}^+$ and then the time dependence of $\hat{Y}$ and $\hat{X}$.
To find the dynamics of $\hat{A}$ one calculates the first and second time derivatives of $\hat{A}$ using the equation of motion: $\hat{A}_t \equiv d\hat{A}/dt = (i/\hbar)[\hat{H}, \hat{A}]$. We have

$$\hat{A}_t = i\omega \begin{pmatrix} 0 & 0 \\ 1 & 0 \end{pmatrix} \tag{40}$$

$$\hat{A}_t^+ = -i\omega \begin{pmatrix} 0 & 1 \\ 0 & 0 \end{pmatrix}. \tag{41}$$

The second time derivatives of $\hat{A}$ and $\hat{A}^+$ are calculated following the trick proposed by Schrodinger. We use two versions of this trick

$$\hat{A}_{tt} = (i/\hbar)[\hat{H}, \hat{A}_t] = \frac{2i}{\hbar}\hat{H}\hat{A}_t - \frac{i}{\hbar}\{\hat{H}, \hat{A}_t\}, \tag{42}$$

$$\hat{A}_{tt}^+ = (i/\hbar)[\hat{H}, \hat{A}_t^+] = -\frac{2i}{\hbar}\hat{A}_t^+\hat{H} + \frac{i}{\hbar}\{\hat{H}, \hat{A}_t^+\}. \tag{43}$$

There is $(i/\hbar)\{\hat{H}, \hat{A}_t\} = \omega^2\hat{A}$ and $(i/\hbar)\{\hat{H}, \hat{A}_t^+\} = -\omega^2\hat{A}^+$. In consequence we obtain second order equations for $\hat{A}$ and $\hat{A}^+$

$$\hat{A}_{tt} = (2i/\hbar)\hat{H}\hat{A}_t - \omega^2\hat{A} \tag{44}$$

$$\hat{A}_{tt}^+ = -(2i/\hbar)\hat{A}_t^+\hat{H} - \hat{A}^+\omega^2. \tag{45}$$

To solve the above equations one eliminates the terms with the first derivative using the substitutions $\hat{A} = \exp(+i\hat{H}t/\hbar)\hat{B}$ and $\hat{A}^+ = \hat{B}^+\exp(-i\hat{H}t/\hbar)$, which gives

$$\hat{B}_{tt} = -(1/\hbar^2)\hat{H}^2\hat{B} - \omega^2\hat{B},. \tag{46}$$

$$\hat{B}_{tt}^+ = -(1/\hbar^2)\hat{B}^+\hat{H}^2 - \hat{B}^+\omega^2. \tag{47}$$

Finally

$$\hat{B}_{tt} = -(\hat{\Omega}^2 + \omega^2)\hat{B}, \tag{48}$$

$$\hat{B}^+_{tt} = -\hat{B}^+(\hat{\Omega}^2 + \omega^2), \tag{49}$$

where $\hat{\Omega} = \hat{H}/\hbar$. The solutions of the above equations are

$$\hat{B} = e^{-i\hat{M}t}\hat{C}_1 + e^{i\hat{M}t}\hat{C}_2, \tag{50}$$

$$\hat{B}^+ = \hat{C}_1^+ e^{-i\hat{M}t} + \hat{C}_2^+ e^{i\hat{M}t}, \tag{51}$$

where $\hat{M} = +\sqrt{\hat{\Omega}^2 + \omega^2}$ is the positive root of $\hat{M}^2 = \hat{\Omega}^2 + \omega^2$. The operator $\hat{M}$ is an important quantity in our considerations. Both $\hat{C}_1$ and $\hat{C}_2^+$ are time-independent operators. Coming back to $\hat{A}(t)$ and $\hat{A}^+(t)$ one has

$$\hat{A}(t) = e^{i\hat{\Omega}t}e^{-i\hat{M}t}\hat{C}_1 + e^{i\hat{\Omega}t}e^{+i\hat{M}t}C_2, \tag{52}$$

$$\hat{A}^+(t) = \hat{C}_1^+ e^{+i\hat{M}t}e^{-i\hat{\Omega}t} + \hat{C}_2^+ e^{-i\hat{M}t}e^{-i\hat{\Omega}t}. \tag{53}$$

In order to find the final forms of $\hat{A}(t)$ and $\hat{A}^+(t)$ one has to use the initial conditions. They are

$$\hat{A}(0) = \hat{C}_1 + \hat{C}_2,$$

$$\hat{A}^+(0) = \hat{C}_1^+ + \hat{C}_2^+,$$

$$\hat{A}_t(0) = i(\hat{\Omega} - \hat{M})\hat{C}_1 + i(\hat{\Omega} + \hat{M})\hat{C}_2,$$

$$\hat{A}_t^+(0) = -i\hat{C}_1^+(\hat{\Omega} - \hat{M}) - i\hat{C}_2^+(\hat{\Omega} + \hat{M}).$$

Simple manipulations give

$$\hat{C}_1 = \frac{i}{2}\hat{M}^{-1}\hat{A}_t(0) + \frac{1}{2}\hat{M}^{-1}\hat{\Omega}\hat{A}(0) + \frac{1}{2}\hat{A}(0), \tag{54}$$

$$\hat{C}_2 = -\frac{i}{2}\hat{M}^{-1}\hat{A}_t(0) - \frac{1}{2}\hat{M}^{-1}\hat{\Omega}\hat{A}(0) + \frac{1}{2}\hat{A}(0). \tag{55}$$

Similarly

$$\hat{C}_1^+ = -\frac{i}{2}\hat{A}_t^+(0)\hat{M}^{-1} + \frac{1}{2}\hat{A}^+(0)\hat{\Omega}\hat{M}^{-1} + \frac{1}{2}\hat{A}^+(0), \tag{56}$$

$$\hat{C}_2^+ = \frac{i}{2}\hat{A}_t^+(0)\hat{M}^{-1} - \frac{1}{2}\hat{A}^+(0)\hat{\Omega}\hat{M}^{-1} + \frac{1}{2}\hat{A}^+(0). \tag{57}$$

One can see by inspection that the initial conditions for $\hat{A}(0)$ and $\hat{A}_t(0)$ are satisfied. It is convenient to express $\hat{A}_t$ in terms of $\hat{A}$ and $\hat{\Omega}$ using the equation of motion $i\hat{A}_t = \hat{A}\hat{\Omega} - \hat{\Omega}\hat{A}$. Then the first and second terms in Eqs. (54) and (56) partially cancel out and the operator $\hat{A}(t)$

can be expressed as a sum $\hat{A}(t) = \hat{A}_1(t) + \hat{A}_2(t)$, where

$$\hat{A}_1(t) = \frac{1}{2} e^{i\hat{\Omega}t} e^{-i\hat{M}t} \left[\hat{A}(0) + \hat{M}^{-1}\hat{A}(0)\hat{\Omega}\right], \tag{58}$$

$$\hat{A}_2(t) = \frac{1}{2} e^{i\hat{\Omega}t} e^{+i\hat{M}t} \left[\hat{A}(0) - \hat{M}^{-1}\hat{A}(0)\hat{\Omega}\right]. \tag{59}$$

Similarly, one can break $\hat{A}^+(t) = \hat{A}_1^+(t) + \hat{A}_2^+(t)$, where

$$\hat{A}_1^+(t) = \frac{1}{2}\left[\hat{A}^+(0) + \hat{\Omega}\hat{A}^+(0)M^{-1}\right] e^{+i\hat{M}t} e^{-i\hat{\Omega}t}, \tag{60}$$

$$\hat{A}_2^+(t) = \frac{1}{2}\left[\hat{A}^+(0) - \hat{\Omega}\hat{A}^+(0)M^{-1}\right] e^{-i\hat{M}t} e^{-i\hat{\Omega}t}. \tag{61}$$

Using Eqs. (38) and (39) we obtain

$$\hat{Y}(t) = \frac{L}{\sqrt{2}}\left(\hat{A}_1(t) + \hat{A}_2(t) + \hat{A}_1^+(t) + \hat{A}_2^+(t)\right), \tag{62}$$

$$\hat{X}(t) = \frac{L}{i\sqrt{2}}\left(\hat{A}_1(t) + \hat{A}_2(t) - \hat{A}_1^+(t) - \hat{A}_2^+(t)\right). \tag{63}$$

The above compact equations are final expressions for the time dependence of $\hat{A}(t)$ and $\hat{A}^+(t)$ operators and, by means of Eqs. (62) and (63), for the time dependence of the position operators $\hat{Y}(t)$ and $\hat{X}(t)$. These exact results are given in terms of operators $\hat{\Omega}$ and $\hat{M}$. To finalize this description one needs to specify the physical sense of functions of these operators appearing in Eqs. (58)-(63).
As it will be seen below, operators $\hat{\Omega}$ and $\hat{M}$ have the same eigenfunctions, so they commute. Then the product of two exponential functions in Eqs. (58)-(61) is given by the exponential function with the sum of two exponents. In consequence, there appear *two sets of frequencies* ω^+ and ω^- corresponding to the sum and the difference: $\omega^- \sim \hat{M} - \hat{\Omega}$, and $\omega^+ \sim \hat{M} + \hat{\Omega}$, respectively. The first frequencies ω^-, being of the intraband type, lead in the non-relativistic limit to the cyclotron frequency ω_c. The interband frequencies ω^+ correspond to the Zitterbewegung. The electron motion is a sum of different frequency components. In absence of a magnetic field *there are no intraband frequencies and only one interband frequency*. Each of the operators $\hat{A}(t)$ or $\hat{A}^+(t)$ contains both intraband and interband terms. One could infer from Eqs. (59) and (61) that the amplitudes of interband and intraband terms are similar. However, when the explicit forms of the matrix elements of $\hat{A}(t)$ and $\hat{A}^+(t)$ are calculated, it turns out that the ZB terms are much smaller than the cyclotron terms, except at high magnetic fields. The operators $\hat{\Omega}$ and $\hat{M}$ do not commute with $\hat{A}$ or $\hat{A}^+$. In Eq. (59) the operator $\hat{A}$ acts on the exponential terms from the right-hand side, while in Eq. (61) the operator $\hat{A}^+$ acts from the left-hand side.
Let us consider the operator $\hat{M}^2 = \hat{\Omega}^2 + \omega^2$. Let $E_n/\hbar$ and $|n\rangle$ be the eigenvalue and eigenvector of $\hat{\Omega}$, respectively. Then

$$\hat{M}^2|n\rangle = (\hat{\Omega}^2 + \omega^2)|n\rangle = \frac{1}{\hbar^2}\left(E_n^2 + \hbar^2\omega^2\right)|n\rangle. \tag{64}$$

Thus, every state $|\mathrm{n}\rangle$ is also an eigenstate of the operator $\hat{M}^2$ with the eigenvalue $\lambda_\mathrm{n}^2 = E_\mathrm{n}^2/\hbar^2 + \omega^2$. To find a more convenient form of λ_n we must find an explicit form of E_n. To do this we choose again the Landau gauge $\boldsymbol{A} = (-By, 0)$. Then, the eigenstate $|\mathrm{n}\rangle$ is characterized by three quantum numbers: n, k_x, s, where n is the harmonic oscillator number, k_x is the wave vector in x direction and $s = \pm 1$ labels the positive and negative energy branches. The state $|\mathrm{n}\rangle$ in a magnetic field is

$$|\mathrm{n}\rangle = \frac{e^{ik_x x}}{N_n} \begin{pmatrix} -s|n-1\rangle \\ |n\rangle \end{pmatrix}, \tag{65}$$

The energies of electrons in graphene are $E_{ns} = s\hbar\omega\sqrt{n}$ and the norm is $N_n = 4\pi$ for $n > 0$ and $N_0 = 2\pi$. The harmonic oscillator states are

$$\langle \boldsymbol{r}|n\rangle = \frac{1}{2\pi\sqrt{L C_n}} \mathrm{H}_n(\xi) e^{-1/2\xi^2}, \tag{66}$$

where $\mathrm{H}_n(\xi)$ are the Hermite polynomials and $C_n = \sqrt{2^n n! \sqrt{\pi}}$. Using the above forms for $|\mathrm{n}\rangle$ and E_{ns} we obtain from Eq. (64)

$$\hat{M}^2|\mathrm{n}\rangle = \omega(n+1)|\mathrm{n}\rangle, \tag{67}$$

i.e. $\lambda_n = \pm\omega\sqrt{n+1} = \pm\omega_{n+1}$, where $\omega_n = \omega\sqrt{n}$. In further calculations we assume λ_n to be positive. The operator $\hat{M}^2$ is diagonal and its explicit form is $\hat{M}^2 = \mathrm{diag}[\hat{a}\,\hat{a}^+, \hat{a}^+\hat{a}]$, see Eq. (64). Because $\hat{M}^2 = \Omega^2 + \omega^2$, the eigenstates of $\hat{M}^2$ do not depend on the energy branch index s.

To calculate the functions of operators $\hat{\Omega}$ and $\hat{M}$ we use the fact that, for every reasonable function f of operators $\hat{\Omega}$ or $\hat{M}^2$ there is $f(\hat{\Omega}) = \sum_\mathrm{n} f(s\omega_n)|\mathrm{n}\rangle\langle\mathrm{n}|$, and$f(\hat{M}^2) = \sum_\mathrm{n} f(\lambda_n^2)|\mathrm{n}\rangle\langle\mathrm{n}|$, see e.g Feynmann (1972). Thus

$$e^{\pm i\hat{\Omega}t} = \sum_\mathrm{n} e^{\pm ist\omega_n}|\mathrm{n}\rangle\langle\mathrm{n}|, \tag{68}$$

$$\hat{M} = (\hat{M}^2)^{1/2} = \nu\sum_\mathrm{n} \lambda_n|\mathrm{n}\rangle\langle\mathrm{n}|, \tag{69}$$

$$\hat{M}^{-1} = (\hat{M}^2)^{-1/2} = \nu\sum_\mathrm{n} \frac{1}{\lambda_n}|\mathrm{n}\rangle\langle\mathrm{n}|, \tag{70}$$

$$e^{\pm i\hat{M}t} = e^{\pm it(\hat{M}^2)^{1/2}} = \sum_\mathrm{n} e^{\pm i\nu t\lambda_n}|\mathrm{n}\rangle\langle\mathrm{n}|, \tag{71}$$

where $\nu = \pm 1$. Without loss of generality we take $\nu = +1$. The above formulas can be used in calculating the matrix elements of $\hat{A}(t)$ and $\hat{A}^+(t)$.

Taking the eigenvectors $|\mathrm{n}\rangle = |n, k_x, s\rangle$ and $|\mathrm{n}'\rangle = |n', k_x', s'\rangle$ with $n' = n+1$, we calculate matrix elements $\hat{A}_{\mathrm{n,n'}}(t)$ using $\hat{A}(t)$ given in Eqs. (58) and (59). The selection rules for $\hat{A}_{\mathrm{n,n'}}(0)$ are $k_x = k_x'$, while there are no selection rules for s, s'. The matrix element of $\hat{M}^{-1}\hat{A}(0)\hat{\Omega}$ appearing in Eqs. (58) and (59) is

$$\langle\mathrm{n}|\hat{M}^{-1}\hat{A}(0)\hat{\Omega}|\mathrm{n}'\rangle = \frac{1}{\lambda_n}\hat{A}(0)_{\mathrm{n,n'}} s'\omega_{n'} = s'\hat{A}(0)_{\mathrm{n,n'}}. \tag{72}$$

In the last equality we used $\omega_{n'} = \omega_{n+1}$ and $\lambda_n = \omega\sqrt{n+1}$. Then

$$\hat{A}_1(t)_{\rm n,n'} = \frac{1}{2}e^{i(s\omega_n-\lambda_n)t}(1+s')\hat{A}(0)_{\rm n,n'}, \tag{73}$$

$$\hat{A}_2(t)_{\rm n,n'} = \frac{1}{2}e^{i(s\omega_n+\lambda_n)t}(1-s')\hat{A}(0)_{\rm n,n'}. \tag{74}$$

Thus the matrix element of $\hat{A}(t)_{\rm n,n'} = \hat{A}_1(t)_{\rm n,n'} + \hat{A}_2(t)_{\rm n,n'}$ is a sum of two terms, of which the first is nonzero for $s' = +1$, while the second is nonzero for $s' = -1$. One can show that the matrix elements given in Eqs. (73) and (74) are equal to the matrix elements of the Heisenberg operator $\hat{A}(t)_{\rm n,n'} = \langle {\rm n}|e^{i\Omega t}\hat{A}(0)e^{-i\Omega t}|{\rm n'}\rangle$. For $\hat{A}^+(t)_{\rm n',n} = \hat{A}_1^+(t)_{\rm n',n} + \hat{A}_2^+(t)_{\rm n',n}$ one obtains in a similar way

$$\hat{A}_1^+(t)_{\rm n',n} = \frac{1}{2}e^{i(+\lambda_n-s\omega_n)t}(1+s')\hat{A}^+(0)_{\rm n',n}, \tag{75}$$

$$\hat{A}_2^+(t)_{\rm n',n} = \frac{1}{2}e^{i(-\lambda_n-s\omega_n)t}(1-s')\hat{A}^+(0)_{\rm n',n}. \tag{76}$$

Formulas (73)-(76) describe the time evolution of the matrix elements of $\hat{A}(t)$ and $\hat{A}^+(t)$ calculated between two eigenstates of $\hat{\Omega}$. The frequencies appearing in the exponents are of the form $\pm\omega_{n+1} \pm \omega_n$. The intraband terms characterized by $\omega_n^c = \omega_{n+1} - \omega_n$ correspond to the cyclotron motion, while the interband terms characterized by $\omega_n^Z = \omega_{n+1} + \omega_n$ describe ZB. Different values of s, s' in the matrix elements of $\hat{A}_1(t)_{\rm n,n'}, \hat{A}_2(t)_{\rm n',n}, \hat{A}_1^+(t)_{\rm n',n}, \hat{A}_2^+(t)_{\rm n',n}$ give contributions either to the cyclotron or to the ZB motion. The exact compact results given in Eqs. (73)-(76) indicate that our choice of $\hat{A}(t)$ and $\hat{A}^+(t)$ operators for the description of electrons in graphene in a magnetic field was appropriate.
To complete the operator considerations of ZB we calculate the matrix elements of $\hat{A}(t)$. We take $|{\rm n}\rangle = |n, k_x, 1\rangle$ and $|{\rm n'}\rangle = |n+1, k_x, s\rangle$. The matrix elements of $\hat{A}(t)_{\rm n,n'}$ for the cyclotron and ZB components are

$$\hat{A}(t)_{\rm n,n'} = (\sqrt{n+1} \pm \sqrt{n}\,)e^{i\omega(\sqrt{n} \mp \sqrt{n+1}\,)t}, \tag{77}$$

where the upper signs correspond to the cyclotron and the lower ones to the ZB motion, respectively. For $\hat{A}^+(t)_{\rm n,n'}$ there is

$$\hat{A}^+(t)_{\rm n,n'} = (\sqrt{n+1} \pm \sqrt{n}\,)e^{i\omega(\sqrt{n+1} \mp \sqrt{n}\,)t}. \tag{78}$$

The above results describe exact dynamics of electrons in monolayer graphene in the presence of an external magnetic field. This description is equivalent to the one given in the text but it is more general since it is not related to a specific electron wave packet.

7. Appendix B

We briefly discuss here the classical electron velocity and mass for a linear energy band of monolayer graphene, as they are often subjects of misunderstandings. Let us consider the conduction band and take $p \geq 0$, where the pseudo-momentum is $\boldsymbol{p} = \hbar\boldsymbol{k}$. Then the band dispersion is $E = up$ and the classical velocity is

$$v_i = \frac{\partial E}{\partial p_i} = \frac{dE}{dp}\frac{\partial p}{\partial p_i} = \frac{dE}{dp}\frac{p_i}{p} = \frac{dE}{dp}\frac{1}{p}\delta_{ij}p_j, \tag{79}$$

where δ_{ij} is the Kronecker delta function and we use the sum convention over the repeated index $j = 1, 2$. The electron mass tensor $\hat{m}$ relating the velocity to pseudo-momentum is *defined* by $\hat{m}\boldsymbol{v} = \boldsymbol{p}$. Then the inverse mass tensor $(1\hat{/}m)$ is defined by

$$v_i = \left(\frac{1}{m}\right)_{ij} p_j. \tag{80}$$

Equating Eq. (79) with Eq. (80) we obtain

$$\left(\frac{1}{m}\right)_{ij} = \frac{dE}{dp}\frac{1}{p}\delta_{ij}. \tag{81}$$

Thus the inverse mass tensor is a scalar: $1/m = (dE/dp)(1/p)$. Using the initial band dispersion one has $dE/dp = u$, so that $m = p/u = E/u^2$. This equality can be seen in two ways. First, it gives

$$E = mu^2, \tag{82}$$

which is analogous to the Einstein relation between the particle energy and mass. Second, the formula

$$m = \frac{E}{u^2} \tag{83}$$

states that the mass vanishes at $E = 0$ (or $p = 0$), but is nonzero for $E > 0$ (or $p > 0$). These relations hold also for a more general "semi-relativistic" case of narrow-gap semiconductors, see Zawadzki (1997; 2006).

If a force $\boldsymbol{F}$ is applied to the electron, the relation

$$\frac{d\boldsymbol{p}}{dt} = \boldsymbol{F} \tag{84}$$

is always valid. In our case the pseudo-momentum $p = mu$, where u is the constant velocity. Thus the force does not accelerate the electron, it only changes its mass.

One should add that, if one defined the mass by the relation of the force to acceleration: $\hat{M}\boldsymbol{a} = \boldsymbol{F}$, the inverse mass would be given by the *second derivative* of the energy with respect to pseudo-momentum. For the linear band of graphene: $E = up$, the second derivative of E vanishes, so that such a mass would be infinitely large for all energies. Thus, it is not a useful quantity.

8. References

Ajiki, H. & Ando, T. (1993). Magnetic Properties of Carbon Nanotubes. *J. Phys. Soc. Jpn* Vol. 62, 2470–2480

Barut, A. O. & Bracken A. J. (1981). Zitterbewegung and the internal geometry of the electron. *Phys. Rev. D* Vol. 23, No. 10, 15 May 1981, 2454–2463

Bjorken, J. D. & Drell S. D. (1964). *Relativistic Quantum Mechanics*, McGraw-Hill, ISBN 9780072320022, New York

Bohm, D. (1952). *Quantum Theory*, Prentice-Hall, ISBN 0486659690, New York

Cannata, F.; Ferrari, L.; & Russo, G. (1990). Dirac-like behaviour of a non-relativistic tight binding Hamiltonian in one dimension. *Sol. St. Comun.* Vol. 74, Issue 4, April 1990, 309–312

Castro Neto, A. H.; Guinea, F.; Peres, N. M. R.; Novoselov, K. S. & Geim, A. K. (2009). The electronic properties of graphene. *Rev. Mod. Phys.* Vol. 81, No. 1, January–March 2009 2009, 109–162

Cserti, J. & David, G. (2006). Unified description of Zitterbewegung for spintronic, graphene, and superconducting systems. *Phys. Rev. B* Vol. 74, No. 17, 15 November 2006, 172305 [4 pages] (arXiv:0604526)

Dreisow, F.; Heinrich, M.; Keil, R.; Tunnermann, A.; Nolte, S.; Longhi, S. & Szameit, A. (2010). Classical Simulation of Relativistic Zitterbewegung in Photonic Lattices. *Phys. Rev. Lett.* Vol. 105, No. 14, 28 September 2010, 143902 [4 pages]

Englman, P. & Vertesi, T. (2008). Large Berry phases in layered graphene. *Phys. Rev. B* Vol. 78, No. 20, 15 November 2008, 205311 [7 pages] (arXiv:0806.3726)

Feynmann, R. (1972). *Statistical Mechanics; a set of lectures*, W. A. Benjamin, ISBN 0201360764, Massachusetts

Gerritsma, R.; Kirchmair, G.; Zahringer, F.; Solano, E.; Blatt, R. & Roos, C. F. (2010). Quantum simulation of the Dirac equation. *Nature* Vol. 463, 7 January 2010, 68–71 (arXiv:0909.0674)

Greiner, W. (1994). *Relativistic Quantum Mechanics*, Springer, ISBN 3540674578, Berlin

Jackson, J. D. (1975). *Classical Electrodynamics*, John Wiley &. Sons, ISBN 9780471309321, New York

Jaynes, E. T. & Cummings, F. W. (1963). Comparison of quantum and semiclassical radiation theories with application to the beam maser. *Proc IEEE* Vol. 51, No. 1, January 1963, 89–109

Johanning, M.; Varron, A. F. & Wunderlich, C (2009). Quantum simulations with cold trapped ions. *J. Phys. B* Vol. 42, No. 15, 14 August 2009, 154009 [27 pages] (arXiv:0905.0118)

Katsnelson, M. I. (2006). Zitterbewegung, chirality, and minimal conductivity in graphene. *Europ. Phys. J. B* Vol. 51, No. 2, May 2006, 157–160 (arXiv:0512337)

Kronig, R. L. & Penney, W. (1931). Quantum Mechanics of Electrons in Crystal Lattices. *Proc. Roy. Soc. London* Vol. 130, No. 814, 3 February 1931, 499–513

Krueckl, V. & Kramer, T. (2009). Revivals of quantum wave packets in graphene. *New J. Phys.* Vol. 11, No. 9, September 2009, 093010 [22 pages]

Lamata, L.; Leon J.; Schatz, T. & Solano, E. (2007). Dirac Equation and Quantum Relativistic Effects in a Single Trapped Ion. *Phys. Rev. Lett.* Vol. 98, No. 25, 22 June 2007, 253005 [4 pages] (arXiv:0701208)

Leibfried, D.; Blatt, R.; Monroe, C. & Wineland, D. (2003). Quantum dynamics of single trapped ions. *Rev. Mod. Phys.* Vol. 75, No. 1, January–March 2003, 281–324

Lock, J. A. (1979). The Zitterbewegung of a free localized Dirac particle. *Am. J. Phys.* Vol. 47, No. 9, September 1979, 797–802

Lurie, D. & Cremer, S. (1970). Zitterbewegung of quasiparticles in superconductors. *Physica* Vol. 50, No. 2, 30 November 1970, 224–240

Maksimova, G. M, Demikhovskii, V Y & Frolova, E V (2008). Wave packet dynamics in a monolayer graphene. *Phys. Rev. B* Vol. 78, No. 23, 15 December 2008, 235321 [7 pages] (arXiv:0809.0367)

Martinez, J. C.; Jalil, M. B. A. & Tan, S. G. (2010). Klein tunneling and zitterbewegung and the formation of a polarized p-n junction in graphene. *Appl. Phys. Lett.* Vol. 97, No. 6, 9 Aug 2010, 062111 [3 pages](arXiv:0806.0222)

McCann, E. & Fal'ko V. I. (2006). Landau-Level Degeneracy and Quantum Hall Effect in a Graphite Bilayer. *Phys. Rev. Lett.* Vol. 96, No. 8, 3 March 2006, 086805 [4 pages] (arXiv:0510237)

Novoselov, K. S.; Geim.; A. K.; Morozov, S. V.; Jiang, D.; Zhang, Y.; Dubonos, S. V.; Grigorieva, I. V. & Firsov, A. A. (2004). Electric Field Effect in Atomically Thin Carbon Films. *Science* Vol. 306, No. 5696, 22 October 2004, 666–669 (arXiv:0410631)

Romera, E. & de los Santos, F. (2009). Revivals, classical periodicity, and zitterbewegung of electron currents in monolayer graphene. *Phys. Rev.* B Vol. 80, No. 16, 15 October 2009, 165416 [5 pages] (arXiv:0911.4367)

Rusin, T. M. & Zawadzki, W. (2007a). Zitterbewegung of nearly-free and tightly-bound electrons in semiconductors. *J. Phys. Cond. Matter* Vol. 19, No. 13, 4 April 2007, 136219 [18 pages] (arXiv:0605384)

Rusin, T. M. & Zawadzki, W. (2007b). Transient Zitterbewegung of charge carriers in mono- and bilayer graphene, and carbon nanotubes. *Phys. Rev. B* Vol. 76, No. 19, 15 November 2007, 195439 [7 pages] (arXiv:0702425)

Rusin, T. M. & Zawadzki, W. (2008). Zitterbewegung of electrons in graphene in a magnetic field. *Phys. Rev. B* Vol. 78, No. 12, 15 September 2008, 125419 [9 pages] (arXiv:0712.3590)

Rusin, T. M. & Zawadzki, W. (2009). Theory of electron Zitterbewegung in graphene probed by femtosecond laser pulses. *Phys. Rev. B* Vol. 80, No. 4, 15 July 2009, 045416 [9 pages] (arXiv:0812.4773)

Sakurai, J. J. (1987). *Modern Quantum Mechanics*, Addison-Wesley, ISBN 9780201539295, New York

Schliemann, J.; Loss, D. & Westervelt, R. M. (2005). Zitterbewegung of Electronic Wave Packets in III-V Zinc-Blende Semiconductor Quantum Wells. *Phys. Rev. Lett.* Vol. 94, No. 20, 27 May 2005, 206801 [4 pages] (arXiv:0410321)

Schliemann, J. (2008). Cyclotron motion in graphene. *New J. Phys.* Vol. 10, April 2008, 034024 [9 pages] (arXiv:0802.1396)

Schrodinger, E. (1930). Ueber die kraeftefreie Bewegung in der relativistischen Quantenmechanik. *Sitzungsber. Preuss. Akad. Wiss. Phys. Math.* Kl. 24, 418–428. Schrodinger's derivation is reproduced in Barut & Bracken (1981)

Semenoff, G. W. (1984). Condensed-Matter Simulation of a Three-Dimensional Anomaly. *Phys. Rev. Lett.* Vol. 53, No. 26, 24 December 1984, 2449–2452

Slonczewski, J. C. & Weiss, P. R. (1958). Band Structure of Graphite. *Phys. Rev.* Vol. 109, No. 2, January 1958, 272–279

Smith, R. A. (1961). *Wave Mechanics of Crystalline Solids*, Chapman & Hall, ISBN 0412093707, London

Vonsovskii, S. V.; Svirskii, M. S. & Svirskaya, L. M. (1990). Zitterbewegung of a band electron. *Teor. Matem. Fizika* Vol. 85, No. 2, November, 1990, 211–221; *Theor. and Math. Physics*, Vol. 85, No. 2, 1159–1167

Wallace, P. R. (1947). The Band Theory of Graphite. *Phys. Rev.* Vol. 71, No. 9, May 1947, 622–634

Wang, Y. X.; Yang, Z. & Xiong, S. J. (2010). Study of Zitterbewegung in graphene bilayer with perpendicular magnetic field. *Europhys. Lett.* Vol. 89, No. 1, January 2010, 17007 [6 pages]

Winkler, R.; Zulicke, U. & Bolte, J. (2007). Oscillatory multiband dynamics of free particles: The ubiquity of zitterbewegung effects. *Phys. Rev. B* Vol. 75, No. 20, 15 May 2007, 205314 [10 pages] (arXiv:0609005)

Zawadzki, W. (1997). Semirelativity in semiconductors. *High Magnetic Fields in the Physics of Semiconductors II*, Landwehr, G. & Ossau. W, 755–768 World Scientific, ISBN 981-02-3990-4, Singapore

Zawadzki, W. (2005). Zitterbewegung and its effects on electrons in semiconductors. *Phys. Rev. B* Vol. 72, No. 8, 15 August 2005, 085217 [4 pages] (arXiv:0411488)

Zawadzki, W. (2006). One-dimensional semirelativity for electrons in carbon nanotubes. *Phys. Rev. B* Vol. 74, No. 20, 15 November 2006, 205439 [4 pages] (arXiv:0510184)

Zawadzki, W. & Rusin, T. M. (2008). Zitterbewegung (trembling motion) of electrons in narrow-gap semiconductors. *J. Phys. Cond. Matter* Vol. 20, No. 45, 12 November 2008, 454208 [5 pages] (arXiv:0805.0478)

Zawadzki, W. & Rusin, T. M. (2010). Nature of electron Zitterbewegung in crystalline solids. *Phys. Lett. A* Vol. 374, No. 34, 26 July 2010, 3533–3537 (arXiv:0909.0463)

Zhang, X. & Liu, Z. (2008). Extremal Transmission and Beating Effect of Acoustic Waves in Two-Dimensional Sonic Crystals. *Phys. Rev. Lett.* Vol. 101, No. 26, 31 December 2008, 264303 [4 pages] (arXiv:0804.1978)

16

Graphene and Cousin Systems

L.B Drissi[1], E.H Saidi[1,2] and M. Bousmina[1]
[1]MAScIR-Inanotech, Institute for Nanomaterials and Nanotechnology, Rabat
[2]LPHE- Modelisation et Simulation, Faculté des Sciences Rabat
Morocco

1. Introduction

Besides its simple molecular structure, the magic of *2D* graphene, a sheet of carbon graphite, is essentially due to two fundamental electronic properties: First for its peculiar band structure where valence and conducting bands intersect at two points K_+ and K_- of the reciprocal space of the *2D* honeycomb making of graphene a zero gap semi-conductor. Second, for the ultra relativistic behavior of the charge carriers near the Fermi level where the energy dispersion relation $E = E(p)$ behaves as a linear function in momenta; $E(p) = v_f p + O(p^2)$. This typical property, which is valid for particles with velocity comparable to the speed of light, was completely unexpected in material science and was never suspected before *2004*; the year where a sheet of *2D* graphene has been experimentally isolated (Geim & Novoselov, 2007; Novoselov et al., 2004). From this viewpoint, graphene is then a new material with exotic properties that could play a basic role in the engineering of electronic devices with high performances; it also offers a unique opportunity to explore the interface between condensed interface between condensed matter physics and relativistic Dirac theory where basic properties like chirality can be tested; and where some specific features, such as numerical simulation methods, can be mapped to *4D* lattice gauge theory like lattice QCD (Boriçi, 2008; Capitani et al., 2010; Creutz, 2008). Although looking an unrealistic matter system, interest into the physical properties of graphene has been manifested several decades ago. The first model to analyze the band structure of graphite in absence of external fields was developed by Wallace in *1947* (Wallace, 1947); see also (Slonczewski & Weiss, 1958). Since then, several theoretical studies have been performed on graphene in the presence of a magnetic field (Haldane, 1983)-(Goerbig et al., 2006). The link between the electronic properties of graphene and $(2+1)$-dimensional Dirac theory was also considered in many occasions; in particular by Semenoff, Fradkin and Haldane during the *80*-th of the last century (Castro-Neto et al., 2009; Haldane, 1988; Semenoff, 1984); see also (Jackiw & Pi, 2007; 2008) and refs therein.

In this book chapter, we use the tight binding model as well as the $SU(3)$ hidden symmetry of *2D* honeycomb to study some physical aspects of *2D* graphene with a special focus on the electronic properties. We also develop new tools to study some of graphene's cousin systems such as the *1D*- poly-acetylene chain, cumulene, poly-yne, Kekulé cycles, the *3D* diamond and the *4D* hyperdiamond models. As another application of the physics in higher dimension, we also develop the relation between the so called four dimensional graphene first studied in (Bedaque et al., 2008; Creutz, 2008; Drissi et al., 2011 a); and *4D* lattice quantum chromodynamics (QCD) model considered recently in the lattice quantum field theory (QFT)

literature to deal with QCD numerical simulations (Capitani et al., 2009 a;b).
The presentation is as follows: *In section 2*, we review the main lines of the electronic properties of *2D* graphene and show, amongst others, that they are mainly captured by the $SU(3)$ symmetry of the *2D* honeycomb. *In section 3*, we study higher dimensional graphene type systems by using the power of the hidden symmetries of the underlying lattices. *In section 4*, we give four examples of graphene's derivatives namely the *1D*- poly-acetylene chain, having a $SU(2)$ invariance, as well as Kekulé cycles thought of as a particular *1D*- system. We also study the *3D* diamond model which exhibits a $SU(4)$ symmetry; the corresponding *2D* model, with $SU(3)$ invariance, is precisely the graphene considered in section 2. *In section 5*, we develop the four dimensional graphene model living on the *4D* hyperdiamond lattice with a SU(5) symmetry. *In section 6*, we study an application of this method in the framework of *4D* lattice QCD. Last section is devoted to conclusion and comments.

2. Two dimensional graphene

First, we give a brief review on the tight binding modeling the physics of *2D* graphene; then we study its electronic properties by using hidden symmetries. We show amongst others that the *2D* honeycomb is precisely the weight lattice of $SU(3)$ (Drissi et al., 2010); and the two Dirac points are given by the roots of $SU(3)$. This study may be also viewed as a first step towards building graphene type systems in diverse dimensions.

2.1 Tight binding model

Graphene is a two dimensional matter system of carbon atoms in the sp^2 hybridization forming a *2D* honeycomb lattice. This is a planar system made of two triangular sublattices $\mathcal{A}_2$ and $\mathcal{B}_2$; and constitutes the building block of the layered *3D* carbon graphite. Since its experimental evidence in *2004*, the study of the electronic properties of graphene with and without external fields has been a big subject of interest; some of its main physical aspects were reviewed in (Castro-Neto et al., 2009) and refs therein. This big attention paid to the *2D* graphene, its derivatives and its homologues is because they offer a real alternative for silicon based technology and bring together issues from condensed matter and high energy physics (Giuliani et al., 2010)-Chakrabarti et al. (2009) allowing a better understanding of the electronic band structure as well as their special properties.
In this section, we focus on a less explored issue of *2D* graphene by studying the link between specific electronic properties and a class of hidden symmetries of the *2D* honeycomb. These symmetries allow to get more insight into the transport property of the electronic wave modes and may be used to approach the defects and the boundaries introduced in the graphene monolayer (Cortijo & Vozmediano, 2009). The existence of these hidden symmetries; in particular the remarkable hidden $SU(3)$ invariance considered in this study, may be motivated from several views. For instance from the structure of the first nearest carbon neighbors like for the typical $\langle A_0\text{-}B_1\rangle$, $\langle A_0\text{-}B_2\rangle$, $\langle A_0\text{-}B_3\rangle$ as depicted in triangle of fig(1). These doublets $A_0\text{-}B_1$, $A_0\text{-}B_2$, $A_0\text{-}B_3$ are basic patterns generating the three $SU(2)$ symmetries contained in the hidden $SU(3)$ invariance of honeycomb. The A-B patterns transform in the isospin $\frac{1}{2}$ representations of $SU(2)$ and describe the electronic wave doublets $\phi_{\pm\frac{1}{2}} = [a(\mathbf{r}), b(\mathbf{r})]$ interpreted as quasi-relativistic *2D* spinors in the nearby of the Dirac points (Castro-Neto et al., 2009). The $SU(3)$ hidden symmetry of honeycomb is also encoded in the second nearest neighbors $\langle\langle A_0\text{-}A_i\rangle\rangle$ and $\langle\langle B_0\text{-}B_i\rangle\rangle$, $i = 1, ..., 6$ which capture data on its adjoint representation where the six $\langle\langle A_0\text{-}A_i\rangle\rangle$ (and similarly for $\langle\langle B_0\text{-}B_i\rangle\rangle$) are precisely associated

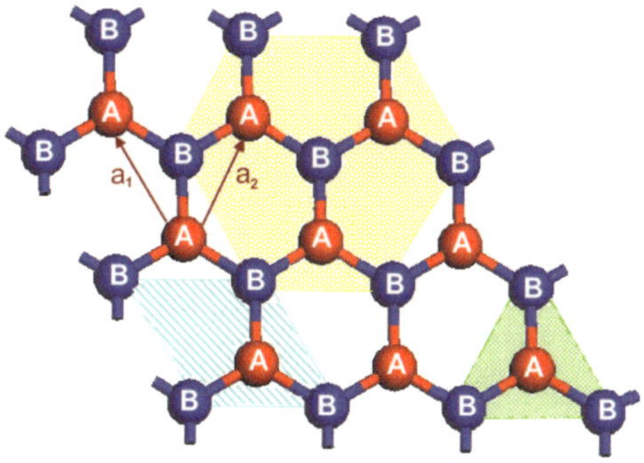

Fig. 1. Sublattices A and B of the honeycomb with unit cell given by dashed area. A-type carbons are given by red balls and B-type atoms by blue ones. Each carbon has three first nearest neighbors as shown by the triangle; and six second nearest ones.

with the six roots of $SU(3)$ namely $\pm\boldsymbol{\alpha}_1, \pm\boldsymbol{\alpha}_2, \pm\boldsymbol{\alpha}_3$; see below. In addition to above mentioned properties, hidden symmetries of graphene are also present in the framework of the tight binding model with hamiltonian,

$$H = -t\sum_{\mathbf{r}_i}\sum_{n=1}^{3} a_{\mathbf{r}_i} b^{\dagger}_{\mathbf{r}_i+\mathbf{v}_n} - t' \sum_{\mathbf{r}_i,\mathbf{r}_j}\left(a_{\mathbf{r}_i} a^{\dagger}_{\mathbf{r}_j} + b_{\mathbf{r}_i} b^{\dagger}_{\mathbf{r}_j}\right) + hc\ , \tag{2.1}$$

where $t \simeq 2.8eV$ is the hopping energy; and where the fermionic creation and annihilation operators a, b, $a^{\dagger}$, $b^{\dagger}$ are respectively associated to the pi-electrons of each atom of the sublattices $\mathcal{A}_2$ and $\mathcal{B}_2$. The three *relative* vectors $\mathbf{v}_1$, $\mathbf{v}_2$, $\mathbf{v}_3$ define the first nearest neighbors, see fig(2) for illustration. These 2D vectors are globally defined on the honeycomb and obey the remarkable constraint equation

$$\mathbf{v}_1 + \mathbf{v}_2 + \mathbf{v}_3 = \mathbf{0}, \tag{2.2}$$

which, a priori, encodes also information on the electronic properties of graphene. Throughout this study, we show amongst others, that the three above mentioned $SU(2)$'s are intimately related with these $\mathbf{v}_n$'s which, as we will see, are nothing but the weight vectors $\boldsymbol{\lambda}_n$ of the $SU(3)$ symmetry; i.e $\mathbf{v}_n = a\frac{\boldsymbol{\lambda}_n}{\|\boldsymbol{\lambda}_n\|}$. The wave functions $\phi_{\boldsymbol{\lambda}_n}(\mathbf{r})$ of the delocalized electrons are organized into a complex $SU(3)$ triplet of waves as given below

$$\begin{pmatrix} |\boldsymbol{\lambda}_1\rangle \\ |\boldsymbol{\lambda}_2\rangle \\ |\boldsymbol{\lambda}_3\rangle \end{pmatrix} \equiv \underline{\mathbf{3}}\ , \qquad \boldsymbol{\lambda}_1 + \boldsymbol{\lambda}_2 + \boldsymbol{\lambda}_3 = \mathbf{0}. \tag{2.3}$$

The symbol $\underline{\mathbf{3}}$ refers to the 3-dimensional representation of $SU(3)$; say with dominant weight $\boldsymbol{\lambda}_1$. We also show that the mapping of the condition $\sum_{n=1}^{3}\boldsymbol{\lambda}_n = \mathbf{0}$ to the momentum space can be interpreted as a condition on the conservation of total momenta at each site of honeycomb. This connection with $SU(3)$ representations opens a window for more insight into the study of the electronic correlations in *2D* graphene and its cousin systems by using symmetries.
The organization of this section is as follows: *In subsection 2*, we exhibit the $SU(3)$ symmetry of graphene. We also give a field theoretic interpretation of the geometric constraint equation

$\mathbf{v}_1 + \mathbf{v}_2 + \mathbf{v}_3 = \mathbf{0}$ both in real and reciprocal honeycomb. We also use the simple roots and the fundamental weights of hidden $SU(3)$ symmetry to study aspects of the electronic properties of *2D* graphene. *In subsection 3*, we develop the relation between the energy dispersion relation $E(k_x, k_y)$ and the hidden $SU(3)$ symmetry. Comments regarding the link between graphene bilayers and symmetries are also given.

2.2 Symmetries and electronic properties

2.2.1 Hidden symmetries of graphene

In dealing with pristine *2D* graphene, one immediately notices the existence of a hidden $SU(3)$ group symmetry underlying the crystallographic structure of the honeycomb lattice and governing the hopping of the pi-electrons between the closed neighboring carbons. To exhibit this hidden $SU(3)$ symmetry, let us start by examining some remarkable features on the graphene lattice and show how they are closely related to $SU(3)$. Refereing to the two sublattices of the graphene monolayer by the usual letters $\mathcal{A}_2$ and $\mathcal{B}_2$ generated by the vectors $\mathbf{a}_1 = d(\sqrt{3}, 0)$, $\mathbf{a}_2 = \frac{d}{2}(-\sqrt{3}, 3)$; together with the three relative $\mathbf{v}_1 = \frac{d}{2}(\sqrt{3}, 1)$, $\mathbf{v}_2 = \frac{d}{2}(-\sqrt{3}, 1)$, $\mathbf{v}_3 = -\mathbf{v}_1 - \mathbf{v}_2$ with carbon-carbon distance $d \simeq 1.42\ A$; and denoting by $\phi_A(\mathbf{r}_i)$ and $\phi_B(\mathbf{r}_j)$ the wave functions of the corresponding pi-electrons, one notes that the interactions between the first nearest atoms involve two kinds of trivalent vertices capturing data on $SU(3)$ symmetry, see fig(2) for illustration. This hidden $SU(3)$ invariance can be

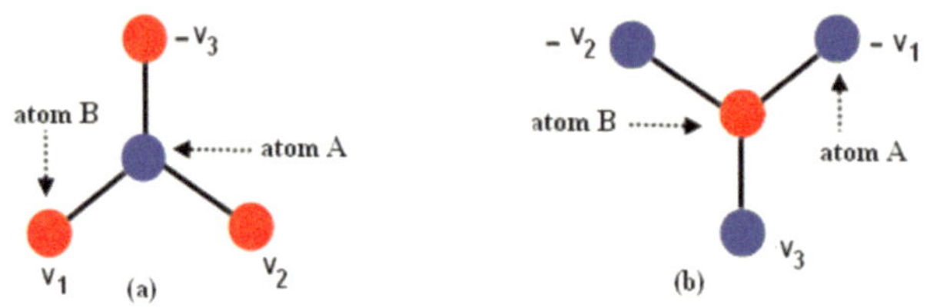

Fig. 2. (a) Nearest neighbors of a A- type atom. (b) Nearest neighbors of a B-type atom. These two configurations are precisely the representations **3** and **3*** of $SU(3)$.

made more explicit by remarking that the relative vectors $\mathbf{v}_1$, $\mathbf{v}_2$, $\mathbf{v}_3$ describing the three first closed neighbors to a $\mathcal{A}$- type carbon at site $\mathbf{r}_i$ of the honeycomb, together with their opposites $-\mathbf{v}_n$ for $\mathcal{B}$-type carbons, are precisely the weight vectors of the 3-dimensional representations of the $SU(3)$ symmetry, $\mathbf{v}_n = d\sqrt{\frac{3}{2}}\boldsymbol{\lambda}_n$. For readers not familiar with representation group theory terminology, we give here below as well in the beginning of section 3.a summary on the $SU(N)$ symmetry.

some useful tools on $SU(3)$

Roughly, the $SU(3)$ symmetry is the simplest extension of the $SU(2)$ symmetry group behind the spin of the electron. The basic relation $\mathbf{v}_1 + \mathbf{v}_2 + \mathbf{v}_3 = \mathbf{0}$ of the honeycomb, which upon setting $\mathbf{v}_n = d\sqrt{\frac{3}{2}}\boldsymbol{\lambda}_n$, reads also as $\boldsymbol{\lambda}_1 + \boldsymbol{\lambda}_2 + \boldsymbol{\lambda}_3 = \mathbf{0}$. This constraint relation has an interpretation in $SU(3)$ representation theory; it should be put in one to one correspondence with the well known $SU(2)$ relation $(\frac{1}{2} - \frac{1}{2}) = 0$ of the spin $\frac{1}{2}$ representation;

$$\left(\begin{array}{c} \left|\frac{+1}{2}\right\rangle \\ \left|\frac{-1}{2}\right\rangle \end{array}\right) \equiv \underline{\mathbf{2}}\,, \qquad \frac{1}{2} - \frac{1}{2} = 0, \tag{2.4}$$

see also eq(3.3) for details. The basic properties of the $SU(3)$ symmetry are encoded in the so called Cartan matrix K_{ij} and its inverse K_{ij}^{-1} which read as

$$K_{ij} = \begin{pmatrix} 2 & -1 \\ -1 & 2 \end{pmatrix}, \qquad K_{ij}^{-1} = \begin{pmatrix} \frac{2}{3} & \frac{1}{3} \\ \frac{1}{3} & \frac{2}{3} \end{pmatrix}. \tag{2.5}$$

These matrices can be also written as the intersection of 2D- vectors as $K_{ij} = \boldsymbol{\alpha}_i \cdot \boldsymbol{\alpha}_j$, $K_{ij}^{-1} = \boldsymbol{\omega}_i \cdot \boldsymbol{\omega}_j$ where $\boldsymbol{\alpha}_1$ and $\boldsymbol{\alpha}_2$ are the two simple roots of $SU(3)$ and where ω_1 and ω_2 are the corresponding two fundamental weights which related to the simple roots by the following duality relation

$$\boldsymbol{\alpha}_i \cdot \boldsymbol{\omega}_j = \delta_{ij}, \qquad \boldsymbol{\alpha}_i = K_{ij}\boldsymbol{\omega}_j. \tag{2.6}$$

Using these tools, the honeycomb relation $\boldsymbol{\lambda}_1 + \boldsymbol{\lambda}_2 + \boldsymbol{\lambda}_3 = \mathbf{0}$ is naturally solved in terms of the fundamental weights as follows

$$\boldsymbol{\lambda}_1 = \omega_1,\ \boldsymbol{\lambda}_2 = \omega_2 - \omega_1,\ \boldsymbol{\lambda}_3 = \omega_2 . \tag{2.7}$$

We also have the following relations between the $\mathbf{a}_i$ vectors and the $\mathbf{v}_i$ ones: $\mathbf{a}_1 = (\mathbf{v}_1 - \mathbf{v}_2)$, $\mathbf{a}_2 = \mathbf{v}_2 - \mathbf{v}_3$ and $\mathbf{a}_3 = \mathbf{v}_3 - \mathbf{v}_1$. Notice that the vectors $\pm\mathbf{a}_1$, $\pm\mathbf{a}_2$, $\pm\mathbf{a}_3$ are, up to the scale factor $d\sqrt{\frac{3}{2}}$, precisely the six roots of the $SU(3)$ symmetry

$$\mathbf{a}_1 = d\sqrt{\tfrac{3}{2}}\boldsymbol{\alpha}_1,\ \mathbf{a}_2 = d\sqrt{\tfrac{3}{2}}\boldsymbol{\alpha}_2,\ \mathbf{a}_3 = -d\sqrt{\tfrac{3}{2}}\left(\boldsymbol{\alpha}_1 + \boldsymbol{\alpha}_2\right), \tag{2.8}$$

where we have also used the remarkable relation between roots and weights that follow from eqs (2.5-2.6).

$$\boldsymbol{\alpha}_1 = 2\omega_1 - \omega_2,\ \boldsymbol{\alpha}_2 = 2\omega_2 - \omega_1 \tag{2.9}$$

2.2.2 Electronic properties

Quantum mechanically, there are two approaches to deal with the geometrical constraint relation (2.2). The first one is to work in real space and think about it as the conservation law of total space-time probability current densities at each site $\mathbf{r}_i$ of the honeycomb. The second approach relies on moving to the reciprocal space where this constraint relation and the induced electronic properties get a remarkable interpretation in terms of $SU(3)$ representations.

1) conservation of total current density

In the real space, the way we interpret eq(2.2) is in terms of the relation between the time variation of the probability density $\rho(t,\mathbf{r}_i) = |\phi(t,\mathbf{r}_i)|^2$ of the electron at site $\mathbf{r}_i$ and the sum $\sum_{n=1}^{3} \mathbf{J}_{\mathbf{v}_n}(t,\mathbf{r}_i) = \mathbf{J}(t,\mathbf{r}_i)$ of incoming and outgoing probability current densities along the $\mathbf{v}_n$-directions. On one hand, because of the *equiprobability* in hopping from the carbon at $\mathbf{r}_i$ to each one of the three nearest carbons at $\mathbf{r}_i + \mathbf{v}_n$, the norm of the $\mathbf{J}_{\mathbf{v}_n}$- vector current densities should be equal and so they should have the form

$$\mathbf{J}_{\mathbf{v}_n}(t,\mathbf{r}_i) = j(t,\mathbf{r}_i)\,\mathbf{e}_n\ ,\ n = 1,2,3\ . \tag{2.10}$$

These probability current densities together with the unit vectors $\mathbf{e}_n = \frac{\mathbf{v}_n}{d}$ pointing in the different $\mathbf{v}_n$- direction; but have the same non zero norm: $\|\mathbf{J}_{\mathbf{v}_1}\| = \|\mathbf{J}_{\mathbf{v}_2}\| = \|\mathbf{J}_{\mathbf{v}_3}\| = |j|$. Substituting in the above relation, the total probability current density $\mathbf{J}(t,\mathbf{r}_i)$ at the site $\mathbf{r}$ and time t takes then the factorized form

$$\mathbf{J}(t,\mathbf{r}) = \frac{j(t,r)}{d}\left(\sum_n \lambda_n\right). \tag{2.11}$$

On the other hand, by using the Schrodinger equation $i\hbar\frac{\partial\phi}{\partial t} = \left(-\frac{\hbar^2}{2m}\nabla^2 + V\right)\phi$ describing the interacting dynamics of the electronic wave at $\mathbf{r}$, we have the usual conservation equation,

$$\frac{\partial\rho(t,\mathbf{r})}{\partial t} + \operatorname{div}\mathbf{J}(t,\mathbf{r}) = 0 \ , \tag{2.12}$$

with probability density $\rho(t,\mathbf{r})$ as before and $J = \frac{i\hbar}{2m}(\phi\nabla\phi^* - \phi^*\nabla\phi)$ with m the mass of the electron and $\phi = \phi(t,\mathbf{r})$ its wave. Moreover, assuming $\frac{\partial\rho}{\partial t} = 0$ corresponding to stationary electronic waves $\phi(t,\mathbf{r}) = e^{i\omega t}\phi(\mathbf{r})$, it follows that the space divergence of the total current density vanishes identically; $\operatorname{div}\mathbf{J} = 0$. This constraint equation shows that generally $\mathbf{J}$ should be a curl vector; but physical consideration indicates that we must have $\mathbf{J}(t,\mathbf{r}) = 0$, in agreement with Gauss-Stokes theorem $\int_{\mathcal{V}}\operatorname{div}\mathbf{J}\, d\mathcal{V} = \int_{\partial\mathcal{V}}\mathbf{J}.d\sigma$ leading to the same conclusion. Combining the property $\mathbf{J}(t,\mathbf{r}) = 0$ with its factorized expression $\frac{j}{d}(\sum_n \mathbf{v}_n)$ given by eq(2.11) together with $j \neq 0$, we end with the constraint relation $\sum_n \mathbf{v}_n = 0$.

2) conservation of total phase

In the dual space of the electronic wave of graphene, the constraint relation (2.2) may be interpreted in two different, but equivalent, ways; first in terms of the conservation of the total relative phase $\Delta\varphi_{tot} = \sum \mathbf{k}.\Delta\mathbf{r}$ of the electronic waves induced by the hopping to the nearest neighbors. The second way is in terms of the conservation of the total momenta at each site of the honeycomb.

Decomposing the wave function $\phi(\mathbf{r})$, associated with a A-type carbon at site $\mathbf{r}$, in Fourier modes as $\sum_{\mathbf{k}} e^{i2\pi\mathbf{k}\cdot\mathbf{r}}\,\tilde{\phi}(\mathbf{k})$; and similarly for the B-type neighboring ones $\phi(\mathbf{r}+\mathbf{v}_n) = \sum_{\mathbf{k}} e^{i2\pi\mathbf{k}\cdot\mathbf{r}}\,\tilde{\phi}_n(\mathbf{k})$ with $\mathbf{k} = (k_x, k_y)$, we see that $\tilde{\phi}(k)$ and the three $\tilde{\phi}_n(k)$ are related as

$$\tilde{\phi}_n(k) = e^{i2\pi\theta_n}\tilde{\phi}(k) \ , \ n = 1,2,3 \ , \tag{2.13}$$

with relative phases $\theta_n = \mathbf{k}\cdot\mathbf{v}_n$. These electronic waves have the same module, $|\tilde{\phi}_n(k)|^2 = |\tilde{\phi}(k)|^2$; but in general non zero phases; $\theta_1 \neq \theta_2 \neq \theta_3$. This means that in the hop of an electron with momentum $\mathbf{p} = \hbar\mathbf{k}$ from a site $\mathbf{r}_i$ to the nearest one at $\mathbf{r}_i + \mathbf{v}_n$, the electronic wave acquires an extra phase of an amount θ_n; but the probability density at each site is invariant. Demanding the total relative phase to obey the natural condition,

$$\theta_1 + \theta_2 + \theta_3 = 0 \ , \quad \operatorname{mod}(2\pi), \tag{2.14}$$

one ends with the constraint eq(2.2). Let us study two remarkable consequences of this special conservation law on the θ_n phases by help of the hidden $SU(3)$ symmetry of graphene. Using eq(2.7), which identifies the relatives $\mathbf{v}_n$ vectors with the weight vectors ω_n, as well as the duality relation $\boldsymbol{\alpha}_i \cdot \boldsymbol{\omega}_j = \delta_{ij}$ (2.6), we can invert the three equations $\theta_n = \mathbf{k}\cdot\mathbf{v}_n$ to get the momenta $\mathbf{p}_n = \hbar\mathbf{k}_n$ of the electronic waves along the $\mathbf{v}_n$-directions. For the two first θ_n's, that is $n = 1,2$, the inverted relations are nicely obtained by decomposing the 2D wave vector $\mathbf{k}$ along the $\boldsymbol{\alpha}_1$ and $\boldsymbol{\alpha}_2$ directions; that is $\mathbf{k} = k_1\boldsymbol{\alpha}_1 + k_2\boldsymbol{\alpha}_2$; and end with the following particular solution,

$$\theta_1 = k_1 d,\, \theta_2 = (k_2 - k_1)\, d,\, \theta_3 = -k_2 d\ . \tag{2.15}$$

2.3 Band structure

We first study the case of graphene monolayer; then we extend the result to the case of graphene bilayers by using the corresponding hidden symmetries.

2.3.1 Graphene monolayer

By considering a graphene sheet and restricting the tight binding hamiltonian (2.1) to the first nearest neighbor interactions namely,

$$H = -t \sum_{\mathbf{r}_i} \sum_{n=1}^{2} a_{\mathbf{r}_i} b^{\dagger}_{\mathbf{r}_i + \mathbf{v}_n} + hc, \tag{2.16}$$

we can determine the energy dispersion relation and the delocalized electrons by using the $SU(3)$ symmetry of the $2D$ honeycomb. Indeed performing the Fourier transform of the various wave functions, we end with the following expression of the hamiltonian in the reciprocal space

$$H = -t \sum_{\mathbf{k}} \left(a^{+}_{\mathbf{k}}, b^{+}_{\mathbf{k}}\right) \begin{pmatrix} 0 & \bar{\varepsilon}_{\mathbf{k}} \\ \varepsilon_{\mathbf{k}} & 0 \end{pmatrix} \begin{pmatrix} a_{\mathbf{k}} \\ b_{\mathbf{k}} \end{pmatrix} . \tag{2.17}$$

The diagonalization of this hamiltonian leads to the two eigenvalue $E_{\pm} = \pm t \left|\varepsilon_{\mathbf{k}}\right|$ giving the energy of the valence and conducting bands. In these relations, the complex number $\varepsilon_{\mathbf{k}}$ is an oscillating wave vector dependent function given by $\varepsilon_{\mathbf{k}} = e^{idQ_1} + e^{idQ_2} + e^{-id(Q_1+Q_2)}$ where we have set $Q_l = \mathbf{k}.\lambda_l$. This relation, which is symmetric under permutation of the three Q_i, can be also rewritten by using the fundamental weights as follows,

$$\varepsilon_{\mathbf{k}} = \left(e^{id[\mathbf{k}.\omega_1]} + e^{id[\mathbf{k}.(\omega_2 - \omega_1)]} + e^{-id[\mathbf{k}.\omega_2]}\right) . \tag{2.18}$$

Up on expanding the wave vector as $\mathbf{k} = k_1 \boldsymbol{\alpha}_1 + k_2 \boldsymbol{\alpha}_2$, this relation reads also as $\varepsilon_{\mathbf{k}} = e^{ik_1 d} + e^{i(k_2 - k_1)d} + e^{-ik_2 d}$. Notice that from (2.18), we learn that $\varepsilon_{\mathbf{k}}$ is invariant under the translations $\mathbf{k} \to \mathbf{k} + \frac{2\pi}{d}\left(N_1 \boldsymbol{\alpha}_1 + N_2 \boldsymbol{\alpha}_2\right)$ with N_1, N_2 arbitrary integers; thanks to the duality relation $\boldsymbol{\alpha}_i.\boldsymbol{\omega}_j = \delta_{ij}$.

Notice also that near the origin $\mathbf{k} = \mathbf{0}$, we have $\varepsilon_{\mathbf{k}} = 3 + O\left(\mathbf{k}^2\right)$, in agreement with non relativistic quantum mechanics. The three terms which are linear terms in $\mathbf{k}$ cancel each others due to the $SU(3)$ symmetry. Notice moreover that the Hamiltonian (2.16) has Dirac zeros located, up to lattice translations, at the following wave vectors

$$(k_1, k_2) = \begin{cases} \frac{2\pi}{3d}(1,0)\ ,\ -\frac{2\pi}{3d}(1,0) \\ \frac{2\pi}{3d}(0,1)\ ,\ -\frac{2\pi}{3d}(0,1) \\ \frac{2\pi}{3d}(1,1)\ ,\ -\frac{2\pi}{3d}(1,1) \end{cases} \tag{2.19}$$

Notice that these six zero modes, which read also as

$$\mathbf{K}_1^{\pm} = \pm\tfrac{2\pi}{3d}\left(2\omega_1 - \omega_2\right), \mathbf{K}_2^{\pm} = \pm\tfrac{2\pi}{3d}\left(-\omega_1 + 2\omega_2\right), \mathbf{K}_3^{\pm} = \pm\tfrac{2\pi}{3d}\left(\omega_1 + \omega_2\right)\ , \tag{2.20}$$

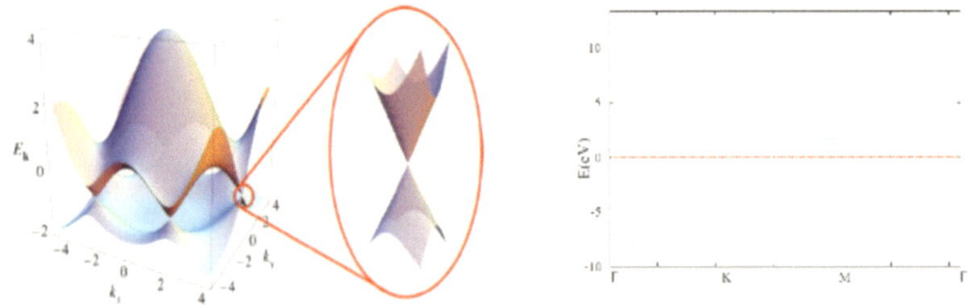

Fig. 3. On left the band structure of the graphene monolayer where one recognizes the Dirac points. On right, it is shown the relativistic behavior near a Dirac point where conducting and valence bands touch. On right the band structure in GGA approximation using QE code.

are not completely independent; some of them are related under lattice translations. For instance, the three $\mathbf{K}_i^+$ are related to each others as follows

$$\mathbf{K}_1^+ + \tfrac{2\pi}{d}\omega_2 = \mathbf{K}_2^+ + \tfrac{2\pi}{d}\omega_1 = \mathbf{K}_3^+ \ . \tag{2.21}$$

The same property is valid for the other three $\mathbf{K}_i^-$'s; so one is left with the usual $\mathbf{K}_\pm$ Dirac zeros of the first Brillouin zone,

$$\mathbf{K}_\pm = \pm\frac{2\pi}{3d}\left(\omega_1 + \omega_2\right) = \pm\frac{2\pi}{3d}\left(\alpha_1 + \alpha_2\right). \tag{2.22}$$

These two zeros are not related by lattice translations; but are related by a $\mathbb{Z}_2$ symmetry mapping the fundamental weights and the simple roots to their opposites.
We end this section by noting that the group theoretical approach developed in this study may be also used to deal with graphene multi-layers and cousin systems. Below, we describe briefly the bilayers; the cousin systems are studied in next sections.

2.3.2 Bilayer graphene

Bilayer graphene was studied for the first time in McCann & Falko (2006). It was modeled as two coupled hexagonal lattices including inequivalent sites in the two different layers that are ranged in the Bernal stacking (*the stacking fashion of graphite where the upper layer has its B sublattice on top of sublattice A of the underlying layer*) as showed in the figure (4). This leads

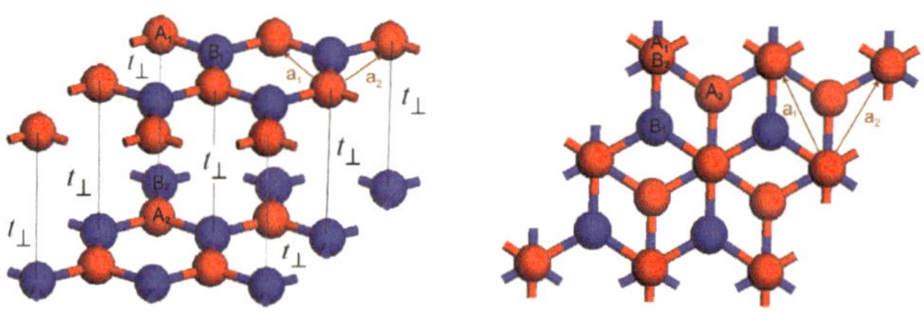

Fig. 4. On right bilayer graphene in the AB stacking allotrope with hop energy $t_\perp$ between the layers. On right bilayer graphene projection in the x-y plane.

to a break of the D_{6h} Bravais symmetry of the lattice with respect to the c axis. Comparing bilayer graphene to monolayer one, we notice that its unit cell contains four atoms. There exist other arrangements such as the AA stacking, where the two lattices are directly above each other and bonds form between the same sublattices. The AB stacking arrangement was experimentally verified in epitaxial graphene by Ohta et al. (Ohta et al., 2007) . The tight-binding model describing bilayer graphene is an extension of the one corresponding to the monolayer (2.1), by adding interlayer hopping elements $H = H_1 + H_2 + H_\perp$ where H_i are as in (2.1) and where

$$H_\perp = -t_\perp \sum_{\mathbf{r}_i} \sum_{n=1}^{3} a_1(\mathbf{r}_i)\, b_2^\dagger(\mathbf{r}_i) + a_2(\mathbf{r}_i)\, b_1^\dagger(\mathbf{r}_i) + hc\ , \tag{2.23}$$

with $t_\perp$ is the hop energy of the pi-electrons between layers calculated to be $t_\perp \sim \frac{t}{10}$ (Charlier et al., 1991). From the view of hidden symmetries, the bilayer graphene has a symmetry type $SU(3) \times SU(2) \times SU(3)$; each $SU(3)$ factor is associated with a graphene sheet; while the $SU(2)$ corresponds to the transitions between the two layers and is associated with propagation along the z-direction of the $3D$-space.
Applying Fourier transform, the above hamiltonian can be rewritten in the following form:

$$H = -t \sum_{\mathbf{k}} \left(a_{1\mathbf{k}}^+, b_{1\mathbf{k}}^+, a_{2\mathbf{k}}^+, b_{2\mathbf{k}}^+\right) \begin{pmatrix} 0 & \varepsilon_k & 0 & \frac{t_\perp}{t} \\ \varepsilon_k^* & 0 & \frac{t_\perp}{t} & 0 \\ 0 & \frac{t_\perp}{t} & 0 & \varepsilon_k \\ \frac{t_\perp}{t} & 0 & \varepsilon_k^* & 0 \end{pmatrix} \begin{pmatrix} a_{1\mathbf{k}} \\ b_{1\mathbf{k}} \\ a_{2\mathbf{k}} \\ b_{2\mathbf{k}} \end{pmatrix}, \tag{2.24}$$

with $\varepsilon_\mathbf{k}$ is as in eq(2.18). The diagonalization of this hamiltonian leads to the following energy dispersion relations,

$$E_\mathbf{k}^\pm = \pm\frac{1}{t}\sqrt{(t_\perp - t\varepsilon_\mathbf{k}^*)(t_\perp - t\varepsilon_\mathbf{k})},\ E_k^{\pm\prime} = \pm\frac{1}{t}\sqrt{(t_\perp + t\varepsilon_\mathbf{k}^*)(t_\perp + t\varepsilon_\mathbf{k})}\ , \tag{2.25}$$

The corresponding band structure has two additional bands, π and π^* states having lower energy bands, that is consequence of the number of atoms per unit cell. Neutral bilayer graphene is gapless McCann & Falko (2006) and exhibits a variety of second-order effects. The studies on bilayer graphene show that it has many common physical properties

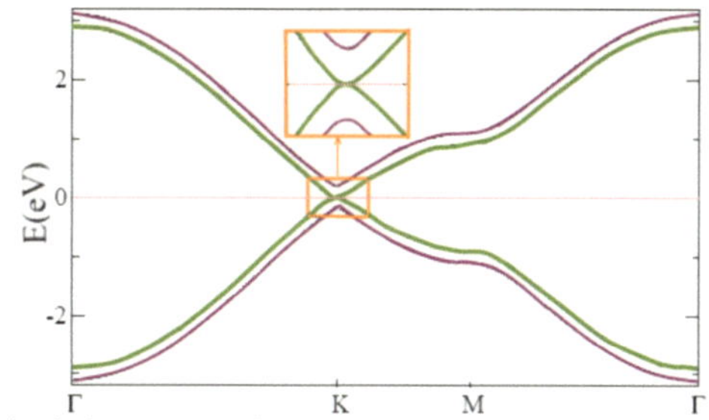

Fig. 5. Band structure of the bilayer graphene

with the monolayer, such as the exceptionally high electron mobility and high mechanical stability (Ohta et al., 2007)-Novoselov et al. (2006). The synthesis of bilayer graphene thin films was realized by deposition on a silicon carbide (*SiC*) substrate (Ohta et al., 2006). The measurements of their electronic band structure, using angle-resolved photo-emission spectroscopy (*ARPES*), suggest the control of the gap at the K Point by applying Coulomb potential between the two layers. This tuning of the band gap changed the biased bilayer from a conductor to a semiconductor.

3. Higher dimensional graphene systems

Motivated by the connection between *2D* graphene and $SU(3)$ symmetry, we study in this section the extension of the physics of *2D* graphene in diverse dimensions; that is *1D, 2D, 3D, 4D*, and so on; the *2D* case is obviously given by *2D* graphene and its multi-layers considered in previous section. The precited dimensions are not all of them realizable in condensed matter physics; but their understanding may help to get more insight on the specific properties of *2D* graphene since the $SU(3)$ is the second element of the $SU(N)$ symmetries series.
First we develop our proposal regarding higher dimensional graphene systems that are based on $SU(N)$ symmetry including the particular *1D* poly-acetylene chain which corresponds to $SU(2)$ symmetry. Then, we compute the energy dispersion relation of these kinds of lattice quantum field theory (QFT). Explicit examples of such lattice fermionic models will be studied in the next sections.

3.1 The $SU(N)$ model

Higher dimensional graphene systems are abstract extensions of *2D* graphene; the analogue of the *2D* honeycomb is given by a real N-dimensional lattice $\mathcal{L}_{su(N)}$. The quantum hamiltonian describing these systems is a generalization of (2.1) and reads as follows,

$$H_N = -t\sum_{\mathbf{r}_i}\left(\sum_{n=1}^{N} a_{\mathbf{r}_i} b^{\dagger}_{\mathbf{r}_i+\mathbf{v}_n}\right) + hc - t'\sum_{\mathbf{r}_i}\sum_{n<m=1}^{N}\left(a_{\mathbf{r}_i} a^{\dagger}_{\mathbf{r}_i+\mathbf{V}_{nm}} + b_{\mathbf{r}_i} b^{\dagger}_{\mathbf{r}_i+\mathbf{V}_{nm}}\right), \tag{3.1}$$

where $a_{\mathbf{r}_i}, b_{\mathbf{r}_i+\mathbf{v}_n}, a^{\dagger}_{\mathbf{r}_i}, b^{\dagger}_{\mathbf{r}_i+\mathbf{v}_n}$ are fermionic annihilation and creation operators living on $\mathcal{L}_{su(N)}$. Moreover the vectors $\mathbf{v}_1, ..., \mathbf{v}_N$ are, up to a global scale factor, the fundamental weights of the N-dimensional representation of the $SU(N)$ symmetry constrained by the typical property

$$\mathbf{v}_1 + ... + \mathbf{v}_N = 0. \tag{3.2}$$

The vectors $\mathbf{V}_{nm} = (\mathbf{v}_n - \mathbf{v}_m)$ are, up to a scale factor, precisely the $N(N-1)$ roots of $SU(N)$; they obey as well the group property $\sum \mathbf{V}_{nm} = 0$.
These particular features of H_N let understand that its physical properties are expected to be completely encoded by the hidden $SU(N)$ symmetry of the model. Below, we show that this is indeed the case; but for simplicity we will focus on the first term of H_N; i.e working in the limit $t' \to 0$.

3.1.1 Useful tools on $SU(N)$ symmetry

Since one of our objectives in this paper is to use the $SU(N)$ symmetry of the crystals to study higher dimensional graphene systems; and seen that readers might not be familiar with these tools; we propose to give in this subsection some basic tools on $SU(N)$ by using explicit examples.
a) cases $SU(2)$ and $SU(3)$
The $SU(2)$ symmetry is very familiar in quantum mechanics; it is the symmetry that describes the spin of the electrons and the quantum angular momentum states.
Roughly speaking, the $SU(2)$ symmetry is a 3-dimensional space generated by three matrices which can be thought of as the usual traceless Pauli matrices

$$\sigma^0 = \begin{pmatrix} \frac{1}{2} & 0 \\ 0 & -\frac{1}{2} \end{pmatrix}, \sigma^- = \begin{pmatrix} 0 & 0 \\ 1 & 0 \end{pmatrix}, \sigma^+ = \begin{pmatrix} 0 & 1 \\ 0 & 0 \end{pmatrix}, \tag{3.3}$$

involving one diagonal matrix σ^0, giving the charge operator, and two nilpotent matrices $\sigma^\pm$ interpreted as the step operators or equivalently the creation and annihilation operators in the language of quantum mechanics. These three matrices obey commutation relations $[\sigma^0, \sigma^\pm] = \pm 2\sigma^\pm$ that define the $su(2)$ algebra. Observe also the traceless property of the charge operator $Tr\sigma^0 = \frac{1}{2} - \frac{1}{2} = 0$, which should be related to the constraint relation (3.2) with $N = 2$.
The $SU(3)$ symmetry group is 8-dimensional space generated by *8* matrices which can be denoted as

$$h_1, h_2, e^{\pm\alpha_1}, e^{\pm\alpha_2}, e^{\pm(\alpha_1+\alpha_2)}, \tag{3.4}$$

with h_1, h_2 two diagonal matrices defining the charge operators and six step operators $e^{\pm\alpha_1}$, $e^{\pm\alpha_2}$, $e^{\pm\alpha_3}$ playing the role of creation and annihilation operators. The $e^{\pm\alpha_i}$'s are nilpotent and are related as $(e^{-\alpha_i})^+ = e^{+\alpha_i}$. An example of these matrices is given by the following 3×3 matrices

$$h_1 = \begin{pmatrix} \mu_1 & 0 & 0 \\ 0 & \mu_2 & 0 \\ 0 & 0 & \mu_3 \end{pmatrix}, \quad h_2 = \begin{pmatrix} \mu_1' & 0 & 0 \\ 0 & \mu_2' & 0 \\ 0 & 0 & \mu_3' \end{pmatrix},$$

$$e^{+\alpha_1} = \begin{pmatrix} 0\,1\,0 \\ 0\,0\,0 \\ 0\,0\,0 \end{pmatrix}, \quad e^{+\alpha_2} = \begin{pmatrix} 0\,0\,0 \\ 0\,0\,1 \\ 0\,0\,0 \end{pmatrix}, \quad e^{+\alpha_3} = \begin{pmatrix} 0\,0\,1 \\ 0\,0\,0 \\ 0\,0\,0 \end{pmatrix} \tag{3.5}$$

with the traceless property of the charge operators which reads as follows

$$\boldsymbol{\lambda}_1 + \boldsymbol{\lambda}_2 + \boldsymbol{\lambda}_3 = 0 \, , \, \boldsymbol{\lambda}_i = \begin{pmatrix} \mu_i \\ \mu_i' \end{pmatrix}, \tag{3.6}$$

and which should be compared with the case $N = 3$ in (3.2).
The vectors $\boldsymbol{\alpha}_1$ and $\boldsymbol{\alpha}_2$ are the simple roots encountered in the previous section; their scalar product $\boldsymbol{\alpha}_i.\boldsymbol{\alpha}_j$ gives precisely the Cartan matrix $\mathbf{K}_{ij}$ of eq(2.5).
b) case $SU(N)$
In the general case $N \geq 2$, the corresponding $SU(N)$ symmetry is $(N^2 - 1)$-dimensional space generated by $(N^2 - 1)$ matrices; $N - 1$ of them are diagonal

$$h_1, \ldots h_{N-1}, \tag{3.7}$$

and are interpreted as the charge operators; and $N(N-1)$ step operators giving the creation and annihilation operators $e^{+\alpha}$, $e^{-\alpha}$ with $\boldsymbol{\alpha}$ standing for generic roots containing the two following:
(**a**) *N-1* simple ones namely $\boldsymbol{\alpha}_1, \boldsymbol{\alpha}_2, \ldots, \boldsymbol{\alpha}_{N-1}$ (together with their opposites) whose scalar products $\boldsymbol{\alpha}_i.\boldsymbol{\alpha}_j$ give precisely the following $(N-1) \times (N-1)$ Cartan matrix

$$\mathbf{K} = \begin{pmatrix} 2 & -1 & 0 & \cdots & 0 & 0 \\ -1 & 2 & -1 & & 0 & 0 \\ 0 & -1 & 2 & & 0 & 0 \\ \vdots & & & \ddots & & \vdots \\ 0 & 0 & 0 & & 2 & -1 \\ 0 & 0 & 0 & \cdots & -1 & 2 \end{pmatrix}, \tag{3.8}$$

(b) non simple roots given by linear (positive and negative) combinations of the simple ones; these roots are given by $\pm\beta_{ij} = \pm\left(\alpha_i + ... + \alpha_j\right)$ with $1 \leq i < j \leq N-1$.
Notice that the above Cartan matrix **K** and its inverse

$$\mathbf{K}^{-1} = \begin{pmatrix} \frac{N}{N+1} & \frac{N-1}{N+1} & \frac{N-2}{N+1} & \cdots & \frac{2}{N+1} & \frac{1}{N+1} \\ \frac{N-1}{N+1} & \frac{2(N-1)}{N+1} & \frac{2(N-2)}{N+1} & \cdots & \frac{4}{N+1} & \frac{2}{N+1} \\ \frac{N-2}{N+1} & \frac{2(N-2)}{N+1} & \frac{3(N-2)}{N+1} & \cdots & \frac{6}{N+1} & \frac{3}{N+1} \\ \vdots & \vdots & \vdots & \ddots & \vdots & \vdots \\ \frac{2}{N+1} & \frac{4}{N+1} & \frac{6}{N+1} & \cdots & \frac{2(N-1)}{N+1} & \frac{N-1}{N+1} \\ \frac{1}{N+1} & \frac{2}{N+1} & \frac{3}{N+1} & \cdots & \frac{N-1}{N+1} & \frac{N}{N+1} \end{pmatrix} \tag{3.9}$$

capture many data on the $SU(N)$ symmetry; they give in particular the expression of the simple roots $\alpha_1, \alpha_2, ..., \alpha_{N-1}$ in terms of the fundamental weights $\omega_1, ..., \omega_{N-1}$ and vice versa; that is $\alpha_i = \sum_j K_{ij}\omega_j$ and $\omega_i = \sum_j K_{ij}^{-1}\alpha_j$. Recall that simple roots and fundamental weights obey the duality property $\alpha_i.\omega_j = \delta_{ij}$; we also have $\omega_i.\omega_j = K_{ij}^{-1}$.

3.1.2 The lattice $\mathcal{L}_{su(N)}$

The lattice $\mathcal{L}_{su(N)}$ is a real $(N-1)$- dimensional crystal with two superposed integral sublattices $\mathcal{A}_N$ and $\mathcal{B}_N$; each site $\mathbf{r_m}$ of these sublattices is generated by the $SU(N)$ simple roots $\alpha_1, ..., \alpha_{N-1}$;

$$\mathbf{r_m} = m_1\alpha_1 + m_2\alpha_2 + ...m_{N-1}\alpha_{N-1} , \tag{3.10}$$

with m_i integers; for illustration see the schema (a), (b), (c) of the figure (6) corresponding respectively to $N = 2, 3, 4$; and which may be put in one to one with the sp^1, sp^2 and sp^3 hybridization of the carbon atom orbital $2s$ and $2p$.

On each lattice site $\mathbf{r}_m$ of $\mathcal{L}_{su(N)}$; say of A-type, lives a quantum state $A_{\mathbf{r}_m}$ coupled to the

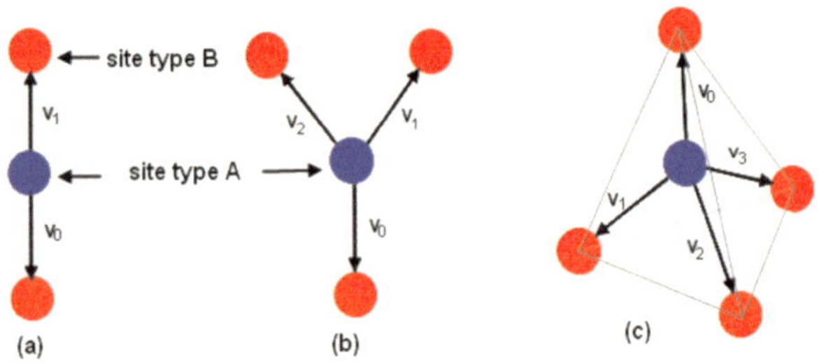

Fig. 6. (a) 1A+2B lattice sites of $\mathcal{L}_{su(2)}$; A-type in blue and B-type in red; the 2B form a $su(2)$ doublet. (b) 1A+3B sites of $\mathcal{L}_{su(3)}$; the 3B form a $su(3)$ triplet. (c) 1A+4B sites of $\mathcal{L}_{su(4)}$ with 4B sites forming a regular tetrahedron.

nearest neighbor states; in particular the first nearest states $B_{\mathbf{r}_m+\mathbf{v}_i}$ and the second nearest ones $A_{\mathbf{r}_m+\mathbf{V}_{ij}}$.
Generally, generic sites in $\mathcal{L}_{su(N)}$ have the following properties:
(1) N first nearest neighbors with relative position vectors $\mathbf{v}_i$ constrained as

$$\mathbf{v}_1 + \ldots + \mathbf{v}_N = 0 . \tag{3.11}$$

These constraint relations are solved in terms of the $SU(N)$ weight vectors $\boldsymbol{\lambda}_i$ (resp. $-\boldsymbol{\lambda}_i$) of the fundamental (anti-fundamental) representation as follows

$$\mathbf{v}_i = a\boldsymbol{\lambda}_i \equiv d\frac{\boldsymbol{\lambda}_i}{\|\boldsymbol{\lambda}_i\|}\ , \tag{3.12}$$

where d is the relative distance between the closest $\mathcal{L}_{su(N)}$ sites. The $\boldsymbol{\lambda}_i$'s which satisfy $\boldsymbol{\lambda}_1 + \ldots + \boldsymbol{\lambda}_N = 0$ can be nicely expressed in terms of the fundamental weights $\omega_1, ..., \omega_{N-1}$ as follows

$$\boldsymbol{\lambda}_1 = \omega_1,\ \boldsymbol{\lambda}_i = \omega_i - \omega_{i-1},\ \boldsymbol{\lambda}_N = -\omega_{N-1}\ . \tag{3.13}$$

From the QFT view, this means that the quantum states at $\mathbf{r}_m + \mathbf{v}_i$ sites are labeled by the $\boldsymbol{\lambda}_i$ weights as $B_{\mathbf{r}_m+\mathbf{v}_i} \equiv B_{\boldsymbol{\lambda}_i}(\mathbf{r}_m)$ and so the multiplet

$$\begin{pmatrix} |\boldsymbol{\lambda}_1 > \\ \vdots \\ |\boldsymbol{\lambda}_N > \end{pmatrix} \equiv \underline{\mathbf{N}}, \qquad \boldsymbol{\lambda}_1 + \ldots + \boldsymbol{\lambda}_N = 0, \tag{3.14}$$

transform in the fundamental representation of $SU(N)$.
(2) $N(N-1)$ second nearest neighbors of A-type with relative position vectors $\mathbf{V}_{ij}$ given by $\mathbf{v}_i - \mathbf{v}_j$ and obeying the constraint relation $\sum_{i,j}\mathbf{V}_{ij} = 0$. This condition is naturally solved by (3.11) and (3.12) showing that the relative vectors between second nearest neighbors are proportional to $SU(N)$ roots $\boldsymbol{\beta}_{ij}$ like

$$\mathbf{V}_{ij} = a\boldsymbol{\beta}_{ij}\ ,\ \boldsymbol{\beta}_{ij} = \boldsymbol{\lambda}_i - \boldsymbol{\lambda}_j, \tag{3.15}$$

and so the condition $\sum \mathbf{V}_{ij} = 0$ turns to a $SU(N)$ property on its adjoint representation labeled by the roots.

3.2 Energy dispersion relation

Restricting the analysis to the first nearest neighbors described by eq(3.1) in the limit $t' \to 0$, the hamiltonian H_N on $\mathcal{L}_{su(N)}$ reduces to

$$H_N = -t\sum_{\mathbf{r}_i}\left(\sum_{n=1}^{N} a_{\mathbf{r}_i} b^{\dagger}_{\mathbf{r}_i+\mathbf{v}_n}\right) + hc\ , \tag{3.16}$$

where now $\mathbf{r}_i$ and $\mathbf{v}_n$ are $(N-1)$- dimensional vectors. By using the Fourier transform of the field operators $a_{\mathbf{r}_i}$ and $B^{\pm}_{\mathbf{r}_m+\mathbf{v}_i}$ namely,

$$a_{\mathbf{r}_i} \sim \sum_{\mathbf{k}} e^{i\mathbf{k}.\mathbf{r}_m} a^{\pm}_{\mathbf{k}}\ ,\ b_{\mathbf{r}_m+\mathbf{v}_i} \sim \sum_{\mathbf{k}} e^{i\mathbf{k}.(\mathbf{r}_m+\mathbf{v}_i)} b_{\mathbf{k}} \tag{3.17}$$

we can put the hamiltonian H_N as a sum over the wave vectors $\mathbf{k}$ in the following way;

$$H_N = \sum_{\mathbf{k}} \left(a^{\dagger}_{\mathbf{k}}, b^{\dagger}_{\mathbf{k}}\right) \begin{pmatrix} 0 & \varepsilon_{\mathbf{k}} \\ \varepsilon^{*}_{\mathbf{k}} & 0 \end{pmatrix} \begin{pmatrix} a_{\mathbf{k}} \\ b_{\mathbf{k}} \end{pmatrix}, \tag{3.18}$$

with $\varepsilon_{\mathbf{k}} = t\sum_i e^{ia\mathbf{k}.\boldsymbol{\lambda}_i}$. This complex number can be also written as $t\sum_i e^{ia\mathbf{k}.(\omega_i - \omega_{i-1})}$ with $\omega_{-1} = 0 = \omega_N$. The energy dispersion relation of the "valence" and "conducting" bands are obtained by diagonalizing the hamiltonian H_N; they are given by $\pm|\varepsilon_{\mathbf{k}}|$ with,

$$|\varepsilon_{\mathbf{k}}| = t\sqrt{N+2\sum_{i<j=1}^{N}\cos\left[a\mathbf{k}.\left(\boldsymbol{\lambda}_i-\boldsymbol{\lambda}_j\right)\right]} \ . \tag{3.19}$$

Notice that $|\varepsilon_{\mathbf{k}}|$ depends remarkably in the difference of the weights $\boldsymbol{\lambda}_i-\boldsymbol{\lambda}_j$; which by help of eq(3.13), can be completely expressed in terms of the fundamental weights.
To get the Fermi wave vectors $\mathbf{k}_F$ for which the oscillating multi-variable function $\varepsilon_{\mathbf{k}} = t\sum_l e^{ia\mathbf{k}.\lambda_l}$ vanish, we will proceed as follows: First, we work out an explicit example; then we give the general result. To that purpose, we expand the wave vector $\mathbf{k}$ in the weight vector basis as follows,

$$\mathbf{k} = \sum_{i=1}^{N-1} Q_i\boldsymbol{\omega}_i,\ (Q_1,...,Q_N) \in \mathbb{R}^N \ , \tag{3.20}$$

and focus on working out the solution for the particular case where all the Q_i's are equal, i.e: $Q_1 = Q_2 = ... = Q_{N-1} = Q$. General solutions are obtained from this particular case by performing lattice translations along the $\boldsymbol{\omega}_i$-directions; this leads to the new values $Q_l = Q + \frac{2\pi n_l}{N}$ with n_l integers. Obviously, one may also expand the wave vector $\mathbf{k}$ like

$$\mathbf{k} = \sum_{i=1}^{N-1} k_i\boldsymbol{\alpha}_i \ ,\ (k_1,...,k_N) \in \mathbb{R}^N \ . \tag{3.21}$$

But this is equivalent to (3.20); the relation between the Q_l's and the k_l's is obtained by substituting $\boldsymbol{\alpha}_i = \sum_j K_{ij}\boldsymbol{\omega}_j$ into (3.21) and identifying it with (3.20). To compute the factors $e^{ia\mathbf{k}.\lambda_l}$, we express the vectors $\boldsymbol{\lambda}_l$ in terms of the simple roots as follows

$$\boldsymbol{\lambda}_1 = \boldsymbol{\omega}_1,\ \boldsymbol{\lambda}_2 = \boldsymbol{\omega}_1 - \boldsymbol{\alpha}_1,\ \ldots\ \boldsymbol{\lambda}_N = \boldsymbol{\omega}_1 - \boldsymbol{\alpha}_1 - ... - \boldsymbol{\alpha}_{N-1}, \tag{3.22}$$

then we use the root/weight duality relation $\boldsymbol{\omega}_i.\boldsymbol{\alpha}_j = \delta_{ij}$ as well as the simple choice $Q_l = Q$ to put the scalar product $\mathbf{k}.\boldsymbol{\lambda}_l$ into the following form $(\mathbf{k}.\boldsymbol{\lambda}_l) = (\mathbf{k}.\boldsymbol{\omega}_1) - lQ$, $l = 1,...,N-1$. Putting this expression back into $\varepsilon_{\mathbf{k}}$ and setting $\xi = e^{iaQ}$, we obtain $\varepsilon_{\mathbf{k}} = e^{ia\mathbf{k}.\omega_1}\left[1+\xi+...+\xi^{N-1}\right] = 0$ which is exactly solved by the N-*th* roots of unity namely

$$Q = \pm\tfrac{2s\pi}{aN}\ ,\ s = 1,...,\left[\tfrac{N}{2}\right] . \tag{3.23}$$

Therefore the Dirac points are, up to lattice translations, located at the wave vectors $\mathbf{k}_s = \pm\frac{2s\pi}{aN}\sum_i\boldsymbol{\omega}_i = \pm\frac{2s\pi}{aN}\sum_i\boldsymbol{\alpha}_i$.

4. Leading models

In this section, we study the cases $N = 2, 4$ as $N = 3$ corresponds precisely to the *2D* graphene before. The case $N = 5$ will be studied in the next section seen its remarkable relation with *4D* lattice QCD.

4.1 The $su\,(2)$ model

In this case, the lattice $\mathcal{L}_{su(2)}$, which is depicted in the figure (7), is a one dimensional chain with coordinate positions $\mathbf{r}_m = ma$ where a is the site spacing and m an arbitrary integer.

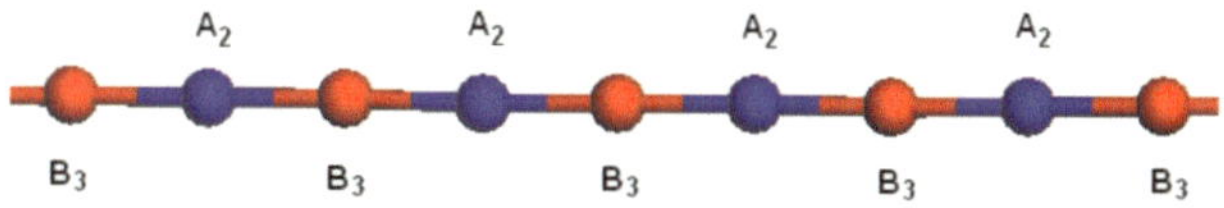

Fig. 7. the lattice $\mathcal{L}_{su(2)}$ given by the superposition of two sublattices $\mathcal{A}_{su(2)}$ (in blue) and $\mathcal{B}_{su(2)}$ (in red). The atoms may be thought of as carbons in the sp^1 hybridization state.

Examples of carbon chains with delocalized electrons are given by one of the three following molecules

chain	molecule	delocalized electrons
polyacetylene	$... - CH = CH - CH = CH - CH - ...$	1
cumulene	$... = C = C = C = C = C = ...$	2
poly-yne	$... - C \equiv C - C \equiv C - C \equiv C - ...$	2

(4.1)

These molecules can be taken as the graphene bridge ultimately narrowed down to a few-carbon atoms or a single-atom width (Giritet et al., 2009; Jun, 2008; Koskinen et al., 2008). Each site of $\mathcal{L}_{su(2)}$ has two first nearest neighbors forming an $su(2)$ doublet; and two second nearest ones that are associated with the two roots $\pm\alpha$ of $su(2)$ in agreement with the generic result summarized in the table,

nearest neighbors	$SU(N)$	$SU(2)$	$SU(3)$	$SU(4)$	$SU(5)$
first	N	2	3	4	5
second	$N(N-1)$	2	6	12	20

(4.2)

In the $SU(2)$ lattice model, eqs(3.2) read as

$$\begin{aligned} \mathbf{v}_0 + \mathbf{v}_1 &= 0 \quad , (a) \\ \mathbf{V}_{01} &= \mathbf{v}_0 - \mathbf{v}_1 \; , (b) \end{aligned} \tag{4.3}$$

and are solved by the fundamental weights $\lambda_1 = +\frac{1}{2}$, $\lambda_2 = -\frac{1}{2}$ of the $SU(2)$ fundamental representation; i.e the isodoublet.

1) polyacetylene

The hamiltonian of the polyacetylene, where each carbon has one delocalized electron, is given by

$$H_t = -t \sum_m \left(a_{\mathbf{r}_m} b^+_{\mathbf{r}_m + a} + a_{\mathbf{r}_m} b^+_{\mathbf{r}_m - a} \right) + hc \; . \tag{4.4}$$

Substituting $N = 2$ in (3.19), we get the following energy dispersion relation

$$|\varepsilon_k| = t\sqrt{2 + 2\cos(2ak)} \tag{4.5}$$

which is also equal to $2t\cos(ka)$ in agreement with the expression $\varepsilon_k = t\left(e^{iak} + e^{-iak}\right)$; see also figure (8). Moreover, the vanishing condition $\varepsilon_2(k) = 0$ is solved by the wave vectors $k_\pm = \pm\frac{\pi}{2a} \bmod \left(\frac{2\pi}{a}\right)$.

2) cumulene and poly-yne

In the case of cumulene and poly-yne, the two delocalized electrons are described by two wave functions $\phi^1_{\mathbf{r}_m}, \phi^2_{\mathbf{r}_m}$. The tight binding hamiltonian modeling the hopping of these electrons is a

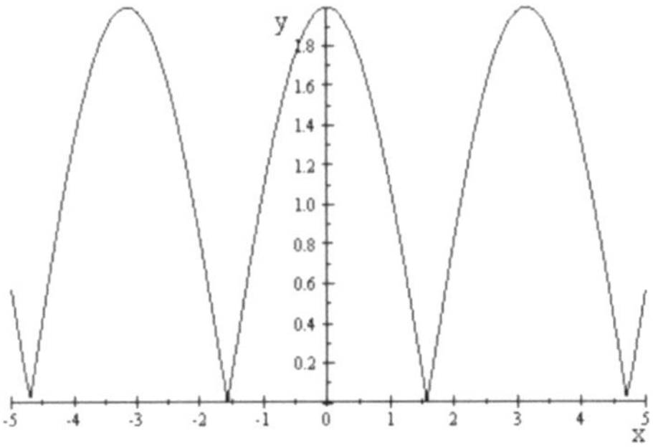

Fig. 8. Energy dispersion relation of 1D poly-acetylene chain.

generalization of H_t. Let $a^{\alpha}_{\mathbf{r}_m}, a^{\alpha+}_{\mathbf{r}_m}, \alpha = 1,2$ (resp. $b^{\alpha}_{\mathbf{r}_m \pm a}, b^{\alpha+}_{\mathbf{r}_m \pm a}$) be the annihilation and creation operators at the site $\mathbf{r}_m$ (resp. $\mathbf{r}_m \pm a$), the hamiltonian reads as follows

$$H_{t,t'} = -\sum_m \sum_{\alpha,\beta=1}^{2} \left(a^{\alpha}_{\mathbf{r}_m} t_{\alpha\beta} b^{\beta+}_{\mathbf{r}_m+a} + a^{\alpha}_{\mathbf{r}_m} t'_{\alpha\beta} b^{\beta+}_{\mathbf{r}_m-a} \right) + hc \; , \tag{4.6}$$

where $t_{\alpha\beta}$ and $t'_{\alpha\beta}$ are hop energy matrices which are identical for cumulene ($t_{\alpha\beta} = t'_{\alpha\beta}$), but different for poly-yne ($t_{\alpha\beta} \neq t'_{\alpha\beta}$). Mapping this hamiltonian to the reciprocal space, we get

$$H_{t,t'} = -2\sum_m \sum_{\alpha,\beta=1}^{2} \left(a^1_k, b^1_k, a^2_k, b^2_k\right) \begin{pmatrix} 0 & A & 0 & B \\ A^* & 0 & C^* & 0 \\ 0 & C & 0 & D \\ B^* & 0 & D^* & 0 \end{pmatrix} \begin{pmatrix} a^{1+}_k \\ b^{1+}_k \\ a^{2+}_k \\ b^{2+}_k \end{pmatrix} , \tag{4.7}$$

with

$$\begin{aligned} A(k) &= t_{11} e^{ika} + t'_{11} e^{-ika} \\ B(k) &= t_{12} e^{ika} + t'_{12} e^{-ika} \\ C(k) &= t_{21} e^{ika} + t'_{21} e^{-ika} \\ D(k) &= t_{22} e^{ika} + t'_{22} e^{-ika} \end{aligned} \tag{4.8}$$

Now, using the fact that the two delocalized electrons are indistinguishable, it is natural to assume the following relations on the hop energies $t_{11} = t_{22}$, $t_{12} = t_{21}$ and the same thing for the $t'_{\alpha\beta}$ matrix. This leads to the relations $A = D$, $B = C$ and so the above hamiltonian simplifies. In this case, the four energy eigenvalues are given by

$$\begin{aligned} E_{\pm} &= \pm\sqrt{(A^* + B^*)(A + B)} \, , \\ E'_{\pm} &= \pm\sqrt{(A^* - B^*)(A - B)} \, , \end{aligned} \tag{4.9}$$

and the zeros modes are given by $e^{2ika} = -\frac{t'_{11}}{t_{11}} = -\frac{t'_{12}}{t_{12}}$. Since in the case of cumulene we have $t_{\alpha\beta} = t'_{\alpha\beta}$, it follows that the zero modes are located as $k = \pm\frac{\pi}{2a} \bmod \frac{2\pi}{a}$.

3) nanoruban

We end this paragraph noting that such analysis may be also extended to the particular case of the periodic chain made by the junction of hexagonal cycles as depicted in the figure (9). This chain, which can be also interpreted as the smallest graphene nanoruban, is very particular

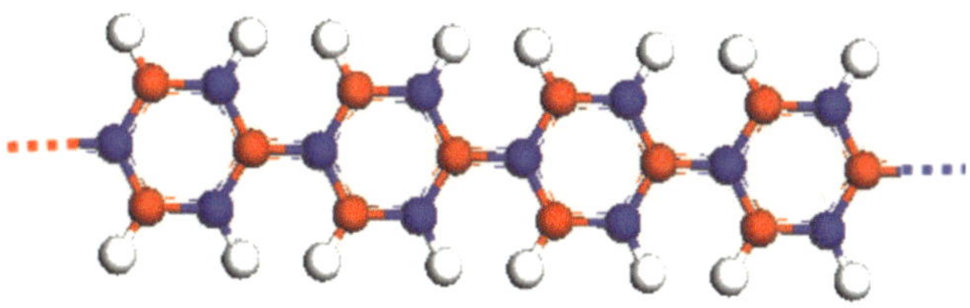

Fig. 9. a periodic chain in 3D space with unit cells given by hexagonal cycles. Each cycle has six delocalized electrons.

from several issues; first its unit cells can be taken as given by the hexagonal cycles; second amongst the *6* carbons of the unit cycle, *4* of them have two first nearest neighbors and the *2* others have three first nearest ones. The third particularity is that the tight binding description of this chain is somehow more complicated with respect to the previous examples. Below we focus on the electronic properties of a given cycle by using the same approach we have been considering in this study.

4.2 Kekulé cycles

Kekulé cycles are organic molecules named in honor to the German chemist Friedrich Kekulé known for his works on tetravalent structure of carbon and the cyclic structure of benzene C_6H_6. These molecules; in particular the family C_nH_n with $n \geq 3$, may be thought of as one dimensional cycles living in the *3D* space; they involve carbon atoms (eventually other atoms such as Nitrogen) arranged in a cyclic lattice with both σ- and π-bonds. All these carbon atoms are in the sp^2 hybridization; they have *3n* covalent σ-bonds defining a *quasi-planar* skeleton; and *n* delocalized π-bonds with Pi electron orbital expanding in the normal direction as shown in the examples of fig(10). Our interest into Kekulé molecules, in particular to the $C_{2N}H_{2N}$

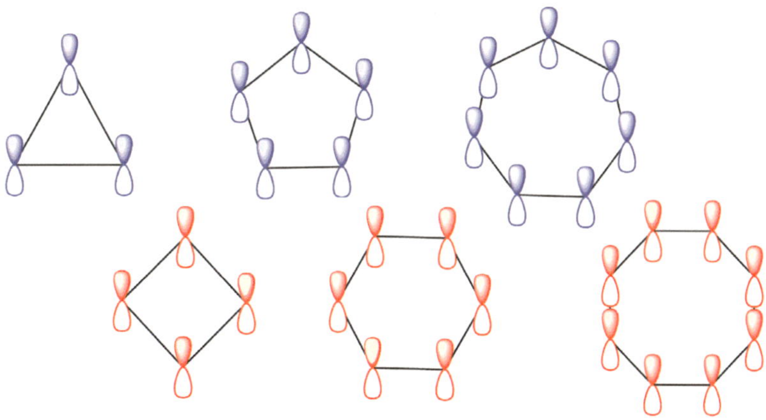

Fig. 10. Six examples of Kekulé cycles type C_n H_n with $n = 3, 4, 5, 6, 7, 8$. The cations C^+ of these molecules form a heavy skeleton represented by n-polygons. The orbitals in the normal direction are associated with the delocalized Pi-electrons.

family, comes from the fact that they can be viewed as the *1D* analogue of the *2D* graphene monolayer; they may be also obtained from the poly-acetylene chain by gluing the ends. It is then interesting to explore the electronic properties of this special class of systems by using the tight binding model and symmetries. To illustrate the method, we focus on the benzene C_6H_6 thought of as the superposition of two C_3H_3 sub-molecules as depicted in figure (11). From group theory view, the positions $\mathbf{v}_1, \mathbf{v}_2, \mathbf{v}_3, \mathbf{v}_4, \mathbf{v}_5, \mathbf{v}_6$ of the carbon atoms are given by

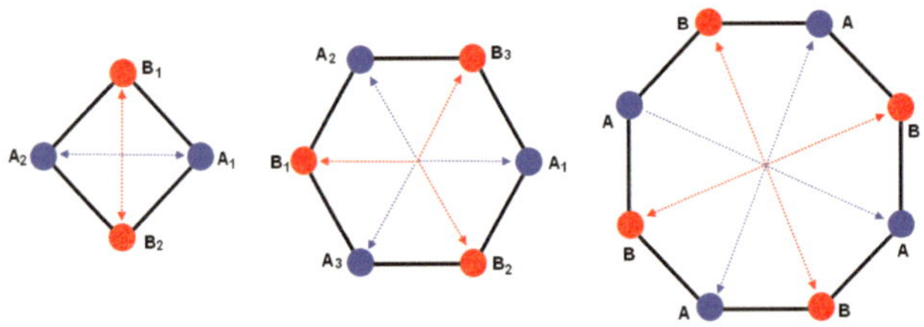

Fig. 11. Kekulé molecules as the superposition of two sublattices. Sublattice A in blue and sublattice $\mathcal{B}$ in red. Except the benzene, these molecules are generally are non planar.

the six roots of the $SU(3)$ symmetry.; that is

$$\mathbf{v}_i = a\boldsymbol{\alpha}_i \ , \ \mathbf{v}_{3+i} = -a\boldsymbol{\alpha}_i \ , \tag{4.10}$$

where $a = 1.39$ Å and where the three $\boldsymbol{\alpha}_i$'s are as in section 2.
tight binding description
The electronic properties of the C_6H_6 are captured by the pi-electrons of the carbons. Denoting by $\mathbf{a}^{\dagger}_{\mathbf{r}_i}, \mathbf{a}_{\mathbf{r}_i}$ (resp. $\mathbf{b}^{\dagger}_{\mathbf{r}_i}, \mathbf{b}_{\mathbf{r}_i}$) the usual electronic creation and annihilation operators associated with the A_i (B_j) atoms in the sublattice $\mathcal{A}_{benz}$ ($\mathcal{B}_{benz}$), the tight binding hamiltonian of the benzene, restricted to first nearest neighbors, reads as follows,

$$H_{benz} = -t \sum_{m=-\infty}^{\infty} \sum_{l=1}^{3} \left(\sum_{j=1}^{2} \mathbf{a}_{\mathbf{r}_{l,m}} \mathbf{b}^{\dagger}_{\mathbf{r}_{l,m}+\mathbf{v}_{l,j}} \right) + hc \ . \tag{4.11}$$

In this relation, the position vectors $\mathbf{r}_{lm}$ have two indices; l and m. The first one takes the values $l = 1, 2, 3$; it indexes the three atoms in $\mathcal{A}_{benz}$; and the three ones in $\mathcal{B}_{benz}$. These positions are as follows,

$$\mathbf{r}^A_{2l-1,m} = \mathbf{r}^A_{1m}, \mathbf{r}^A_{3m}, \mathbf{r}^A_{5m} \ , \ \mathbf{r}^B_{2l,m} = \mathbf{r}^B_{2m}, \mathbf{r}^B_{4m}, \mathbf{r}^B_{6m} \ . \tag{4.12}$$

The second integer is an arbitrary number ($m \in \mathbb{Z}$); it captures the periodicity of the cycle and encodes in some sense the rotational invariance with respect to the axis of the planar molecule. To fix the ideas, think about $\mathbf{r}_{lm}$ as the *l-th* electron in the sublattice $\mathcal{A}_{benz}$; that is $\mathbf{r}_{lm} \equiv \mathbf{r}^A_{2l-1,m}$. After a hop of this electron to the two first nearest carbons in $\mathcal{B}_{benz}$, the new position is

$$\mathbf{r}^B_{2l,m} = \mathbf{r}_{lm} + \mathbf{v}_{lj} \ , \ j = \pm \ . \tag{4.13}$$

where the $\mathbf{v}_{lj}$s are the relative positions of the first nearest neighbors.
Taking the Fourier transform of the creation and annihilation operators, $c^{\pm}_{\mathbf{r}_n} = \sum_k e^{\pm i\mathbf{k}.\mathbf{r}_n} c^{\pm}_{\mathbf{k}}$ with $c^{\pm}_{\mathbf{k}}$ standing for $a^{\pm}_{\mathbf{k}}, b^{\pm}_{\mathbf{k}}$, we get an expression involving the product of three sums $\sum_{\mathbf{k}} \sum_{\mathbf{k}'} \sum_m$. Then, using the discrete rotational invariance with respect to the axis of the molecule, we can eliminate the sum $\sum_m$ in terms of a Dirac delta function $\delta_2(\mathbf{k} - \mathbf{k}')$ and end, after

integration with respect $\mathbf{k}'$, with the following result,

$$\mathcal{H}_{benz} = \sum_{\mathbf{k}} \left(a_{\mathbf{k}}, b_{\mathbf{k}}\right) \begin{pmatrix} 0 & \varepsilon_{\mathbf{k}} \\ \varepsilon_{\mathbf{k}}^* & 0 \end{pmatrix} \begin{pmatrix} a_{\mathbf{k}}^\dagger \\ b_{\mathbf{k}}^\dagger \end{pmatrix} . \tag{4.14}$$

with

$$\varepsilon_{\mathbf{k}} = -t \sum_{l=1}^{3} \left(e^{i\mathbf{k}.\mathbf{v}_{l-}} + e^{i\mathbf{k}.\mathbf{v}_{l+}} \right) . \tag{4.15}$$

Notice that like in graphene, the above hamiltonian has two eigenvalues $\pm |\varepsilon_{\mathbf{k}}|$. Moreover, substituting the $\mathbf{v}_{l\pm}$'s by their explicit expressions in terms of the $SU(3)$ roots $\boldsymbol{\alpha}_l$, we obtain the following dispersion relation together with a constraint relation capturing the planarity property of the molecule

$$\varepsilon_{\mathbf{k}} = -2t \sum_{l=0}^{2} \cos\left(\tfrac{a\sqrt{2}}{2}\mathbf{k}.\boldsymbol{\alpha}_l\right), \; \eta_{\mathbf{k}} = \sum_{l=0}^{2} \sin\left(\tfrac{a\sqrt{2}}{2}\mathbf{k}.\boldsymbol{\alpha}_l\right) = 0 . \tag{4.16}$$

Notice that the constraint equation $\eta_{\mathbf{k}} = 0$ allows us to express the k_2 component of the wave vector in terms of k_1 and vice versa as depicted in fig(12). This relation plays a crucial role in the determination of the wave vectors at the Fermi level.

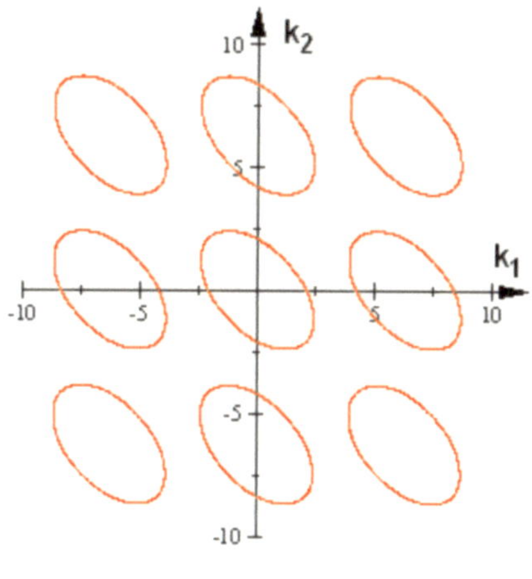

Fig. 12. the plot of the energy dispersion relation ε_{k_1,k_2} and the constraint relation $\eta_{k_1,k_2} = \sin k_1 + \sin k_2 - \sin(k_1 + k_2) = 0$ in the reciprocal space.

4.3 The diamond model

The diamond model lives on the lattice $\mathcal{L}_{su(4)}$; this is a 3-dimensional crystal given by the superposition of two isomorphic sublattices $\mathcal{A}_4$ and $\mathcal{B}_4$ along the same logic as in the case of the *2D* honeycomb. Each site $\mathbf{r}_m$ in $\mathcal{L}_{su(4)}$ has *4* first nearest neighbors at $(\mathbf{r}_m + \mathbf{v}_i)$ forming the vertices of a regular tetrahedron. A way to parameterize the relative positions $\mathbf{v}_i$ with respect to the central position at $\mathbf{r}_m$ is to embed the tetrahedron inside a cube; in this case we have:

$$\begin{aligned} \mathbf{v}_1 &= \tfrac{d}{\sqrt{3}}(-1,-1,+1) , \; \mathbf{v}_2 = \tfrac{d}{\sqrt{3}}(-1,+1,-1) \\ \mathbf{v}_3 &= \tfrac{d}{\sqrt{3}}(+1,-1,-1) , \; \mathbf{v}_0 = \tfrac{d}{\sqrt{3}}(+1,+1,+1) \end{aligned} \tag{4.17}$$

Clearly these vectors satisfy the constraint relation $\mathbf{v}_0 + \mathbf{v}_1 + \mathbf{v}_2 + \mathbf{v}_3 = 0$. Having these expressions, we can also build the explicit positions of the *12* second nearest neighbors; these are given by $\mathbf{V}_{ij} = \mathbf{v}_i - \mathbf{v}_j$; but are completely generated by the following basis vectors

$$\mathbf{R}_1 = \tfrac{d}{\sqrt{3}}\left(2,2,0\right),\ \mathbf{R}_2 = \tfrac{d}{\sqrt{3}}\left(0,-2,2\right),\ \mathbf{R}_3 = \tfrac{d}{\sqrt{3}}\left(-2,2,0\right) \tag{4.18}$$

that are related to $\mathbf{V}_{ij}$ as $\mathbf{R}_i = \mathbf{V}_{(i-1)i}$. We also have:

- the intersection matrix of the $\mathbf{R}_i$ vectors

$$\mathbf{R}_i.\mathbf{R}_j = \frac{4d^2}{3}\mathbf{K}_{ij} \tag{4.19}$$

with

$$\mathbf{K}_{ij} = \begin{pmatrix} 2 & -1 & 0 \\ -1 & 2 & -1 \\ 0 & -1 & 2 \end{pmatrix},\ \mathbf{K}_{ij}^{-1} = \begin{pmatrix} \frac{3}{4} & \frac{2}{4} & \frac{1}{4} \\ \frac{2}{4} & \frac{4}{4} & \frac{2}{4} \\ \frac{1}{4} & \frac{2}{4} & \frac{3}{4} \end{pmatrix} \tag{4.20}$$

- the special relation linking the $\mathbf{R}_i$'s and $\mathbf{v}_0$,

$$\tfrac{3}{4}\mathbf{R}_1 + \tfrac{2}{4}\mathbf{R}_2 + \tfrac{1}{4}\mathbf{R}_3 = \mathbf{v}_0\ . \tag{4.21}$$

Concerning the vector positions of the remaining *9* second neighbors, *3* of them are given by $-R_1, -R_2, -R_3$ and the other *6* by the linear combinations $R_4 = V_{02}$, $R_5 = V_{13}$, $R_6 = V_{03}$ with

$$V_{02} = R_1 + R_2,\ V_{13} = R_2 + R_3,\ V_{03} = R_1 + R_2 + R_3. \tag{4.22}$$

From this construction, it follows that generic positions $\mathbf{r}_{\mathbf{m}}^A \equiv \mathbf{r}_{\mathbf{m}}$ and $\mathbf{r}_{\mathbf{m}}^B$ in the $\mathcal{A}_4$ and $\mathcal{B}_4$ sublattices are given by

$$\begin{aligned} \mathcal{A}_4 &: \mathbf{r}_{\mathbf{m}} = m_1\mathbf{R}_1 + m_2\mathbf{R}_2 + m_3\mathbf{R}_3\ , \\ \mathcal{B}_4 &: \mathbf{r}_{\mathbf{m}}^B = \mathbf{r}_{\mathbf{m}} + \mathbf{v}\ , \end{aligned} \tag{4.23}$$

where $\mathbf{m} = (m_1,m_2,m_3)$ is an integer vector and where the shift vector $\mathbf{v} = \mathbf{r}_{\mathbf{m}}^B - \mathbf{r}_{\mathbf{m}}^A$ is one of $\mathbf{v}_i$'s in (4.17).

1) Energy dispersion relation

First notice that as far as the electronic properties are concerned, the figures (a), (b), (c) of (6) are respectively associated with the sp^1, sp^2 and sp^3 hybridizations of the atom orbitals; i.e:

figures	hybridization	example of molecules
(6-a)	sp^1	acetylene
(6-b)	sp^2	graphene
(6-c)	sp^3	diamond

(4.24)

In (6-a) and (6-b), the atoms have delocalized pi- electrons that capture the electronic properties of the lattice atoms and have the following dispersion relation,

$$\left|\varepsilon_{su(N)}\left(\mathbf{k}\right)\right| = t_1\sqrt{N + 2\sum_{i<j=0}^{N-1}\cos\left[a\mathbf{k}.\left(\boldsymbol{\lambda}_i - \boldsymbol{\lambda}_j\right)\right]} \tag{4.25}$$

with $N = 2,3$. However, in the case of sp^3, the atoms have no delocalized pi-electrons;

they only have strongly correlated sigma- electrons which make the electronic properties of systems based on $\mathcal{L}_{su(4)}$ different from those based on $\mathcal{L}_{su(3)}$ and $\mathcal{L}_{su(2)}$. Nevertheless, as far as tight binding model idea is concerned, one may consider other applications; one of which concerns the following toy model describing a system based on the lattice $\mathcal{L}_{su(4)}$ with dynamical vacancy sites.

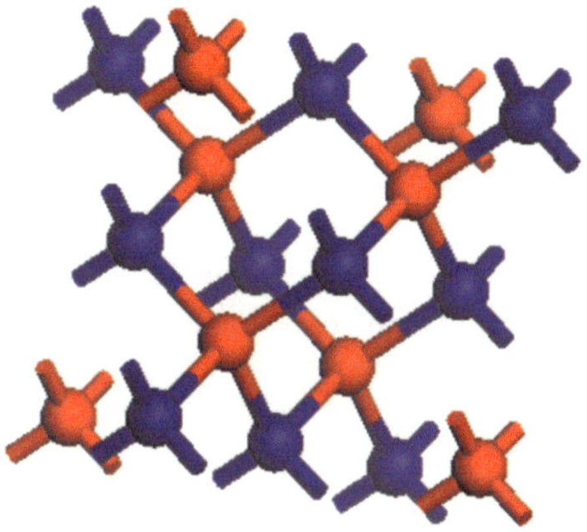

Fig. 13. the lattice $\mathcal{L}_{su(4)}$ with sublattices $\mathcal{A}_{su(4)}$ (in blue) and $\mathcal{B}_{su(4)}$ (in red). Each atom has *4* first nearest neighbors, forming a tetrahedron, and *12* second nearest ones.

2) Toy model

This is a lattice QFT on the $\mathcal{L}_{su(4)}$ with dynamical particles and vacancies. The initial state of the system correspond to the configuration where the sites of the sublattice $\mathcal{A}_4$ are occupied by particles and those of the sublattice $\mathcal{B}_4$ are unoccupied.

sublattice	initial configuration	quantum state
$\mathcal{A}_4$	particles at $\mathbf{r}_m$	$\mathbf{A}_{\mathbf{r}_m}$
$\mathcal{B}_4$	vacancy at $\mathbf{r}_m + \mathbf{v}$	$\mathbf{B}_{\mathbf{r}_m+\mathbf{v}}$

(4.26)

Then, the particles (vacancies) start to move towards the neighboring sites with movement modeled by hops to first nearest neighbors. Let $A_{\mathbf{r}_m}$ and $B_{\mathbf{r}_m+\mathbf{v}_i}$ be the quantum states describing the particle at $\mathbf{r}_m$ and the vacancy at $\mathbf{r}_m + \mathbf{v}_i$ respectively. Let also $A^{\pm}_{\mathbf{r}_m}$ and $B^{\pm}_{\mathbf{r}_m+\mathbf{v}_i}$ be the corresponding creation and annihilation operators. The hamiltonian describing the hops of the vacancy/particle to the first nearest neighbors is given by

$$H_4 = -t\left(\sum_{i=0}^{3} A^{-}_{\mathbf{r}_m} B^{+}_{\mathbf{r}_m+v_i} + hc\right). \tag{4.27}$$

By performing the Fourier transform of the wave functions $A^{\pm}_{\mathbf{r}_m}$, $B^{\pm}_{\mathbf{r}_m+v_i}$, we end with the dispersion energy $\pm t\,|\varepsilon_{\mathbf{k}}|$ where

$$\varepsilon_{\mathbf{k}} = \sqrt{4 + 2\sum_{i<j}\cos\left(\mathbf{k}.\mathbf{V}_{ij}\right)}\ , \tag{4.28}$$

and $\mathbf{V}_{ij}$ are as in (4.21-4.22). The Dirac points are located at $\mathbf{k}_s = \pm\frac{s\pi}{2a}\sum_{i=1}^{3}\omega_i$ with $s = 1, 2$.

5. Four dimensional graphene

The so called four dimensional graphene is a QFT model that lives on the *4D* hyperdiamond; it has links with lattice quantum chromodynamics (QCD) to be discussed in next section. In this section, we first study the *4D* hyperdiamond; then we use the results of previous section to give some physical properties of *4D* graphene.

5.1 Four dimensional hyperdiamond

Like in the case of *2D* honeycomb, the *4D* hyperdiamond may be defined by the superposition of two sublattices $\mathcal{A}_4$ and $\mathcal{B}_4$ with the following properties:

- sites in $\mathcal{A}_4$ and $\mathcal{B}_4$ are parameterized by the typical *4d*- vectors $\mathbf{r_n}$ with $\mathbf{n} = (n_1, n_2, n_3, n_4)$ and n_i's arbitrary integers. These lattice vectors are expanded as follows

$$\mathcal{A}_4:\ \mathbf{r_n} = n_1\,\mathbf{a}_1 + n_2\,\mathbf{a}_2 + n_3\,\mathbf{a}_3 + n_4\,\mathbf{a}_4\ ,\ \mathcal{B}_4:\ \mathbf{r}'_\mathbf{n} = \mathbf{r_n} + \mathbf{e}_5\ , \tag{5.1}$$

  where $\mathbf{a}_1$, $\mathbf{a}_2$, $\mathbf{a}_3$, $\mathbf{a}_4$ are primitive vectors generating these sublattices; and $\mathbf{e}_5$ is a shift vector which we describe below.

- the vector $\mathbf{e}_5$ is a global vector taking the same value $\forall\ \mathbf{n}$; it is a shift vector giving the relative positions of the $\mathcal{B}_4$ sites with respect to the $\mathcal{A}_4$ ones, $\mathbf{e}_5 = \mathbf{r}'_\mathbf{n} - \mathbf{r_n}$, $\forall\ \mathbf{n}$.

The $\mathbf{a}_l$'s and $\mathbf{e}_5$ vectors can be chosen as

$$\mathbf{a}_1 = \mathbf{e}_1 - \mathbf{e}_5,\ \mathbf{a}_2 = \mathbf{e}_2 - \mathbf{e}_5,\ \mathbf{a}_3 = \mathbf{e}_3 - \mathbf{e}_5,\ \mathbf{a}_4 = \mathbf{e}_4 - \mathbf{e}_5 \tag{5.2}$$

with

$$\begin{aligned}
\mathbf{e}_1^\mu &= \tfrac{1}{4}\left(+\sqrt{5}, +\sqrt{5}, +\sqrt{5}, +1\right),\\
\mathbf{e}_2^\mu &= \tfrac{1}{4}\left(+\sqrt{5}, -\sqrt{5}, -\sqrt{5}, +1\right),\\
\mathbf{e}_3^\mu &= \tfrac{1}{4}\left(-\sqrt{5}, -\sqrt{5}, +\sqrt{5}, +1\right),\\
\mathbf{e}_4^\mu &= \tfrac{1}{4}\left(-\sqrt{5}, +\sqrt{5}, -\sqrt{5}, +1\right),
\end{aligned} \tag{5.3}$$

and $\mathbf{e}_5 = -\sum_{i=1}^{4}\mathbf{e}_i$. Notice also that the 5 vectors $\mathbf{e}_1$, $\mathbf{e}_2$, $\mathbf{e}_3$, $\mathbf{e}_4$, $\mathbf{e}_5$ define the first nearest neighbors to $(0,0,0,0)$ and satisfy the constraint relations,

$$\mathbf{e}_i.\mathbf{e}_i = 1,\ \mathbf{e}_i.\mathbf{e}_j = \cos\vartheta_{ij} = -\tfrac{1}{4},\quad i \neq j\ , \tag{5.4}$$

showing that the $\mathbf{e}_i$'s are distributed in a symmetric way since all the angles satisfy $\cos\vartheta_{ij} = \frac{-1}{4}$; see also figure (14) for illustration.

some specific properties

From the figure (14) representing the first nearest neighbors in the *4D* hyperdiamond and their analog in *2D* graphene, we learn that each $\mathcal{A}_4$- type node at $\mathbf{r_n}$, with some attached wave function $A_{\mathbf{r_n}}$, has the following closed neighbors:

Fig. 14. On left the 5 first nearest neighbors in the pristine *4D* hyperdiamond with the properties $\|\mathbf{e}_i\| = 1$ and $e_1 + e_2 + e_3 + \mathbf{e}_4 + \mathbf{e}_5 = 0$. On right, the 3 first nearest in pristine 2D graphene with $\|\mathbf{e}_i\| = 1$ and $e_1 + \mathbf{e}_2 + \mathbf{e}_3 = 0$.

- 5 first nearest neighbors belonging to $\mathcal{B}_4$ with wave functions $B_{\mathbf{r_n}+d\mathbf{e}_i}$; they are given by:

lattice position	*attached wave*
$\mathbf{r_n} + d\mathbf{e}_1$	$B_{\mathbf{r_n}+d\mathbf{e}_1}$
$\mathbf{r_n} + d\mathbf{e}_2$	$B_{\mathbf{r_n}+d\mathbf{e}_2}$
$\mathbf{r_n} + d\mathbf{e}_3$	$B_{\mathbf{r_n}+d\mathbf{e}_3}$
$\mathbf{r_n} + d\mathbf{e}_4$	$B_{\mathbf{r_n}+d\mathbf{e}_4}$
$\mathbf{r_n} + d\mathbf{e}_5$	$B_{\mathbf{r_n}+d\mathbf{e}_5}$

(5.5)

Using this configuration, the typical tight binding hamiltonian describing the couplings between the first nearest neighbors reads as

$$-t\sum_{\mathbf{r_n}}\sum_{i=1}^{5} A_{\mathbf{r_n}} B^{+}_{\mathbf{r_n}+d\mathbf{e}_i} + hc\ . \tag{5.6}$$

where t is the hop energy and where d is the lattice parameter.
Notice that in the case where the wave functions at $\mathbf{r_n}$ and $\mathbf{r_n} + d\mathbf{e}_i$ are rather given by two component Weyl spinors

$$A^{a}_{\mathbf{r_n}} = \begin{pmatrix} A^{1}_{\mathbf{r_n}} \\ A^{2}_{\mathbf{r_n}} \end{pmatrix}\ ,\ \bar{B}^{\dot{a}}_{\mathbf{r_n}+d\mathbf{e}_i} = \begin{pmatrix} \bar{B}^{\dot{1}}_{\mathbf{r_n}+d\mathbf{e}_i} \\ \bar{B}^{\dot{2}}_{\mathbf{r_n}+d\mathbf{e}_i} \end{pmatrix}, \tag{5.7}$$

together with their adjoints $\bar{A}^{\dot{a}}_{\mathbf{r_n}}$ and $\bar{B}^{a}_{\mathbf{r_n}+d\mathbf{e}_i}$, as in the example of *4D* lattice QCD to be described in next section, the corresponding tight binding model would be,

$$-t\sum_{\mathbf{r_n}}\sum_{i=1}^{5}\left[\sum_{\mu=1}^{4} \mathbf{e}^{\mu}_i \left(A^{a}_{\mathbf{r_n}} \sigma^{\mu}_{a\dot{a}} \bar{B}^{\dot{a}}_{\mathbf{r_n}+d\mathbf{e}_i}\right)\right] + hc\ . \tag{5.8}$$

where the $\mathbf{e}^{\mu}_i$'s are as in (5.3); and where $\sigma^1, \sigma^2, \sigma^3$ are the Pauli matrices and $\sigma^4 = I_{2\times 2}$. Notice moreover that the term $\sum_{i=1}^{5} \mathbf{e}^{\mu}_i \left(A^{a}_{\mathbf{r_n}} \sigma^{\mu}_{a\dot{a}} \bar{B}^{\dot{a}}_{\mathbf{r_n}}\right)$ vanishes identically due to $\sum_{i=1}^{5} \mathbf{e}^{\mu}_i = 0$.

- *20* second nearest neighbors belonging to the same $\mathcal{A}_4$ with the wave functions $A_{\mathbf{r_n}+d(\mathbf{e}_i-\mathbf{e}_j)}$; they read as

$$\begin{array}{l} r_{\mathbf{n}} \pm d\left(\mathbf{e}_1 - \mathbf{e}_2\right), r_{\mathbf{n}} \pm d\left(\mathbf{e}_1 - \mathbf{e}_3\right), r_{\mathbf{n}} \pm d\left(\mathbf{e}_1 - \mathbf{e}_4\right), \\ r_{\mathbf{n}} \pm d\left(\mathbf{e}_1 - \mathbf{e}_5\right), r_{\mathbf{n}} \pm d\left(\mathbf{e}_2 - \mathbf{e}_3\right), r_{\mathbf{n}} \pm d\left(\mathbf{e}_2 - \mathbf{e}_4\right), \\ r_{\mathbf{n}} \pm d\left(\mathbf{e}_2 - \mathbf{e}_5\right), r_{\mathbf{n}} \pm d\left(\mathbf{e}_3 - \mathbf{e}_4\right), r_{\mathbf{n}} \pm d\left(\mathbf{e}_3 - \mathbf{e}_5\right), \\ r_{\mathbf{n}} \pm d\left(\mathbf{e}_4 - \mathbf{e}_5\right). \end{array} \tag{5.9}$$

The *5* vectors $\mathbf{e}_1, \mathbf{e}_2, \mathbf{e}_3, \mathbf{e}_4, \mathbf{e}_5$ are, up to a normalization factor namely $\frac{\sqrt{5}}{2}$, precisely the weight vectors λ_0, λ_1, λ_2, λ_3, λ_4 of the *5*-dimensional representation of $SU(5)$; and the *20* vectors $\left(\mathbf{e}_i - \mathbf{e}_j\right)$ are, up to a scale factor $\frac{\sqrt{5}}{2}$, their roots $\beta_{ij} = \left(\lambda_i - \lambda_j\right)$. We show as well that the particular property $\mathbf{e}_i.\mathbf{e}_j = -\frac{1}{4}$, which is constant $\forall\ \mathbf{e}_i$ and $\mathbf{e}_j$, has a natural interpretation in terms of the Cartan matrix of $SU(5)$.

2D/4D Correspondence

First notice that a generic bond vector $\mathbf{e}_i$ in the hyperdiamond links two sites in the same unit cell of the *4D* lattice as shown on the typical coupling term $A_{\mathbf{r_n}} B^+_{\mathbf{r_n}+d\mathbf{e}_i}$. This property is quite similar to the action of the usual γ^μ matrices on *4D* (Euclidean) space time spinors which links the components of spinors.

Mimicking the tight binding model of *2D* graphene, it has been proposed in (Bedaque et al., 2008) a graphene inspired model for *4D* lattice QCD. There, the construction relies on the use of the following:

- the naive correspondence between the bond vectors $\mathbf{e}_i$ and the γ^i matrices

$$\mathbf{e}_i \longleftrightarrow \gamma_i\ ,\ i = 1, ..., 4\ , \tag{5.10}$$

together with

$$\begin{array}{l} -\mathbf{e}_5 = \mathbf{e}_1 + \mathbf{e}_2 + \mathbf{e}_3 + \mathbf{e}_4\ , \\ -\Gamma_5 = \gamma_1 + \gamma_2 + \gamma_3 + \gamma_4\ . \end{array} \tag{5.11}$$

- as in the case of *2D* graphene, $\mathcal{A}_4$-type sites are occupied by left $\phi^a_{\mathbf{r}}$ and right $\bar{\phi}^{\dot{a}}_{\mathbf{r}}$ 2-component Weyl spinors. $\mathcal{B}_4$-type sites are occupied by right $\bar{\chi}^{\dot{a}}_{\mathbf{r}+d\mathbf{e}_i}$ and left $\chi^a_{\mathbf{r}+d\mathbf{e}_i}$ Weyl spinors.

lattice	2D graphene	4D hyperdiamond
$\mathcal{A}_4$-sites at $\mathbf{r}_n$	$A_{\mathbf{r}}$	$\phi^a_{\mathbf{r}}$, $\bar{\phi}^{\dot{a}}_{\mathbf{r}}$
$\mathcal{B}_4$-sites at $\mathbf{r}_n + d\mathbf{e}_i$	$B^+_{\mathbf{r}+de_i}$	$\bar{\chi}^{\dot{a}}_{\mathbf{r}+d\mathbf{e}_i}$, $\chi^a_{\mathbf{r}+d\mathbf{e}_i}$
couplings	$A_{\mathbf{r}} B^+_{\mathbf{r}+de_i}$ $B_{\mathbf{r}+de_i} A^+_{\mathbf{r}}$	$\mathbf{e}^\mu_i \left(\phi^a_{\mathbf{r}} \sigma^\mu_{a\dot{a}} \bar{\chi}^{\dot{a}}_{\mathbf{r}+d\mathbf{e}_i}\right)$ $\mathbf{e}^\mu_i \left(\chi^a_{\mathbf{r}+d\mathbf{e}_i} \bar{\sigma}^\mu_{a\dot{a}} \bar{\phi}^{\dot{a}}_{\mathbf{r}}\right)$

(5.12)

where the indices $a = 1, 2$ and $\dot{a} = \dot{1}, \dot{2}$; and where summation over μ is in the Euclidean sense.

For later use, it is interesting to notice the two following:

(a) in *2D* graphene, the wave functions $A_{\mathbf{r}}$ and $B_{\mathbf{r}+de_i}$ describe polarized electrons in first nearest sites of the *2D* honeycomb. As the spin up and spin down components of the electrons contribute equally, the effect of spin couplings in *2D* graphene is ignored.

(b) in the *4D* hyperdiamond, we have *4+4* wave functions at each $\mathcal{A}_4$-type site or $\mathcal{B}_4$-type one. These wave functions are given by:

(i) $\phi^a = \left(\phi^1_{\mathbf{r}_n}, \phi^2_{\mathbf{r}_n}\right)$ and $\bar{\phi}^{\dot{a}}_{\mathbf{r}} = \left(\bar{\phi}^{\dot{1}}_{\mathbf{r}_n}, \bar{\phi}^{\dot{2}}_{\mathbf{r}_n}\right)$ having respectively positive and negative γ^5 chirality,

(ii) $\bar{\chi}^{\dot{a}}_{\mathbf{r}+d\mathbf{e}_i} = \left(\bar{\chi}^{\dot{1}}_{\mathbf{r}+d\mathbf{e}_i}, \bar{\chi}^{\dot{2}}_{\mathbf{r}+d\mathbf{e}_i}\right)$ and $\chi^a_{\mathbf{r}+d\mathbf{e}_i} = \left(\chi^1_{\mathbf{r}+d\mathbf{e}_i}, \chi^2_{\mathbf{r}+d\mathbf{e}_i}\right)$ having respectively negative and positive γ^5 chirality.

By mimicking the *2D* graphene study, we have the couplings

$$\begin{array}{l} \mathbf{e}^\mu_i \sigma^\mu_{1\dot{1}} \left(\phi^1_{\mathbf{r}} \, \bar{\chi}^{\dot{1}}_{\mathbf{r}+d\mathbf{e}_i}\right) , \mathbf{e}^\mu_i \sigma^\mu_{2\dot{2}} \left(\phi^2_{\mathbf{r}} \, \bar{\chi}^{\dot{2}}_{\mathbf{r}+d\mathbf{e}_i}\right) \\ \mathbf{e}^\mu_i \bar{\sigma}^\mu_{1\dot{1}} \left(\chi^1_{\mathbf{r}+d\mathbf{e}_i} \, \bar{\phi}^{\dot{1}}_{\mathbf{r}}\right) , \mathbf{e}^\mu_i \bar{\sigma}^\mu_{2\dot{2}} \left(\chi^2_{\mathbf{r}+d\mathbf{e}_i} \, \bar{\phi}^{\dot{2}}_{\mathbf{r}}\right) \end{array} \tag{5.13}$$

building the hamiltonian

To describe 4D lattice fermions, one considers *4D* space time Dirac spinors together with the following γ^μ matrices realizations,

$$\begin{array}{l} \gamma^1 = \tau^1 \otimes \sigma^1 \ , \gamma^2 = \tau^1 \otimes \sigma^2 \ , \gamma^3 = \tau^1 \otimes \sigma^3 \ , \\ \gamma^4 = \tau^2 \otimes I_2 \ \ , \gamma^5 = \tau^3 \otimes I_2 \ \ , \end{array} \tag{5.14}$$

where the τ^i's are the Pauli matrices acting on the sublattice structure of the hyperdiamond lattice,

$$\tau^1 = \begin{pmatrix} 0 & 1 \\ 1 & 0 \end{pmatrix}, \tau^2 = \begin{pmatrix} 0 & -i \\ i & 0 \end{pmatrix}, \tau^3 = \begin{pmatrix} 1 & 0 \\ 0 & -1 \end{pmatrix}. \tag{5.15}$$

The 2×2 matrices σ^i satisfy as well the Clifford algebra $\sigma^i\sigma^j + \sigma^j\sigma^i = 2\delta^{ij} I_2$ and act through the coupling of left/right Weyl spinors at neighboring sites

$$\phi^a_{\mathbf{r}} \sigma^\mu_{a\dot{a}} \, \bar{\chi}^{\dot{a}}_{\mathbf{r}+d\frac{\sqrt{5}}{2}\lambda_i} - \chi^a_{\mathbf{r}} \bar{\sigma}^\mu_{a\dot{a}} \, \bar{\phi}^{\dot{a}}_{\mathbf{r}-d\frac{\sqrt{5}}{2}\lambda_i} = \left(\phi_{\mathbf{r}} \sigma^\mu \bar{\chi}_{\mathbf{r}+d\frac{\sqrt{5}}{2}\lambda_i} - \chi_{\mathbf{r}} \bar{\sigma}^\mu \bar{\phi}_{\mathbf{r}-d\frac{\sqrt{5}}{2}\lambda_i}\right) \tag{5.16}$$

where $\sigma^\mu = \left(\sigma^1, \sigma^2, \sigma^3, +iI_2\right)$ and $\bar{\sigma}^\mu = \left(\sigma^1, \sigma^2, \sigma^3, -iI_2\right)$. For later use, it is interesting to set

$$\begin{array}{l} \sigma^\mu.e^\mu_1 = \frac{\sqrt{5}}{4}\sigma^1 + \frac{\sqrt{5}}{4}\sigma^2 + \frac{\sqrt{5}}{4}\sigma^3 + \frac{i}{4} I_2, \\ \bar{\sigma}^\mu.e^\mu_1 = \frac{\sqrt{5}}{4}\sigma^1 + \frac{\sqrt{5}}{4}\sigma^2 + \frac{\sqrt{5}}{4}\sigma^3 - \frac{i}{4} I_2, \end{array} \tag{5.17}$$

and similar relations for the other $\sigma.e_i$ and $\bar{\sigma}.\mathbf{e}_i$.

Now extending the tight binding model of 2D graphene to the 4D hyperdiamond; and using the weight vectors λ_i instead of e_i, we can build a free fermion action on the 4D lattice by attaching a two-component left-handed spinor $\phi^a(\mathbf{r})$ and right-handed spinor $\bar{\phi}^{\dot{a}}_{\mathbf{r}}$ to each A_4-node r, and a right-handed spinor $\bar{\chi}^{\dot{a}}_{\mathbf{r}+d\frac{\sqrt{5}}{2}\lambda_i}$ and left-handed spinor $\chi^a_{\mathbf{r}+d\frac{\sqrt{5}}{2}\lambda_i}$ to every B_4-node at $r + d\frac{\sqrt{5}}{2}\lambda_i$.

The hamiltonian, describing hopping to first nearest-neighbor sites with equal probabilities in all five directions λ_i, reads as follows:.

$$H_4 = \sum_{\mathbf{r}} \sum_{i=0}^{4} \left(\phi_{\mathbf{r}} \sigma^{\mu} \bar{\chi}_{\mathbf{r}+d\frac{\sqrt{5}}{2}\lambda_i} - \chi_{\mathbf{r}} \bar{\sigma}^{\mu} \bar{\phi}_{\mathbf{r}-d\frac{\sqrt{5}}{2}\lambda_i} \right) \lambda_i^{\mu} \,. \tag{5.18}$$

Expanding the various spinorial fields $\xi_{\mathbf{r}\pm\mathbf{v}}$ in Fourier sums as $\int \frac{d^4k}{(2\pi)^4} e^{-i\mathbf{k}.\mathbf{r}} \left(e^{\mp i\mathbf{k}.\mathbf{v}} \xi_{\mathbf{k}} \right)$ with $\mathbf{k}$ standing for a generic wave vector in the reciprocal lattice, we can put the field action H_4 into the form

$$H_4 = i \sum_{\mathbf{k}} (\bar{\phi}_{\mathbf{k}}, \bar{\chi}_{\mathbf{k}}) \begin{pmatrix} 0 & -iD \\ i\bar{D} & 0 \end{pmatrix} \begin{pmatrix} \phi_{\mathbf{k}} \\ \chi_{\mathbf{k}} \end{pmatrix} \tag{5.19}$$

where we have set

$$D = \sum_{l=0}^{4} D_l e^{id\frac{\sqrt{5}}{2}\mathbf{k}.\lambda_l} = \sum_{\mu=1}^{4} \sigma^{\mu} \left(\sum_{l=0}^{4} \lambda_l^{\mu} e^{id\frac{\sqrt{5}}{2}\mathbf{k}.\lambda_l} \right) , \tag{5.20}$$

with

$$D_l = \sum_{\mu=1}^{4} \sigma^{\mu} \lambda_l^{\mu} = \begin{pmatrix} \lambda_l^3 + i\lambda_l^4 & \lambda_l^1 - i\lambda_l^2 \\ \lambda_l^1 + i\lambda_l^2 & \lambda_l^3 - i\lambda_l^4 \end{pmatrix} , \tag{5.21}$$

and $p_l = \mathbf{k}.\lambda_l = \sum_{\mu} k_{\mu} \lambda_l^{\mu}$. Similarly we have

$$\bar{D} = \sum_{l=0}^{4} \bar{D}_l e^{-id\frac{\sqrt{5}}{2}\mathbf{k}.\lambda_l} = \sum_{\mu=1}^{4} \bar{\sigma}^{\mu} \left(\sum_{l=0}^{4} \lambda_l^{\mu} e^{-id\frac{\sqrt{5}}{2}\mathbf{k}.\lambda_l} \right) . \tag{5.22}$$

5.2 Energy dispersion and zero modes

To get the dispersion energy relations of the 4 waves components $\phi_{\mathbf{k}}^1$, $\phi_{\mathbf{k}}^2$, $\chi_{\mathbf{k}}^1$, $\chi_{\mathbf{k}}^2$ and their corresponding 4 holes, one has to solve the eigenvalues of the Dirac operator (5.19). To that purpose, we first write the 4-dimensional wave equation as follows,

$$\begin{pmatrix} 0 & -iD \\ i\bar{D} & 0 \end{pmatrix} \begin{pmatrix} \phi_{\mathbf{k}} \\ \chi_{\mathbf{k}} \end{pmatrix} = E \begin{pmatrix} \phi_{\mathbf{k}} \\ \chi_{\mathbf{k}} \end{pmatrix} , \tag{5.23}$$

where $\phi_{\mathbf{k}} = (\phi_{\mathbf{k}}^1, \phi_{\mathbf{k}}^2)$, $\chi_{\mathbf{k}} = (\chi_{\mathbf{k}}^1, \chi_{\mathbf{k}}^2)$ are Weyl spinors and where the 2×2 matrices D, $\bar{D}$ are as in eqs(5.20,5.22). Then determine the eigenstates and eigenvalues of the 2×2 Dirac operator matrix by solving the following characteristic equation,

$$\det \begin{pmatrix} -E & 0 & D_{11} & D_{12} \\ 0 & -E & D_{21} & D_{22} \\ \bar{D}_{11} & \bar{D}_{21} & -E & 0 \\ \bar{D}_{12} & \bar{D}_{22} & 0 & -E \end{pmatrix} = 0 \tag{5.24}$$

from which one can learn the four dispersion energy eigenvalues $E_1(\mathbf{k})$, $E_2(\mathbf{k})$, $E_3(\mathbf{k})$, $E_4(\mathbf{k})$ and therefore their zeros.

1) computing the energy dispersion

An interesting way to do these calculations is to act on (5.23) once more by the Dirac operator to bring it to the following diagonal form

$$\begin{pmatrix} D\bar{D} & 0 \\ 0 & D\bar{D} \end{pmatrix} \begin{pmatrix} \phi_{\mathbf{k}} \\ \chi_{\mathbf{k}} \end{pmatrix} = E^2 \begin{pmatrix} \phi_{\mathbf{k}} \\ \chi_{\mathbf{k}} \end{pmatrix} . \tag{5.25}$$

Then solve separately the eigenvalues problem of the 2-dimensional equations $D\bar{D}\phi_{\mathbf{k}} = E^2\phi_{\mathbf{k}}$ and $\bar{D}D\chi_{\mathbf{k}} = E^2\chi_{\mathbf{k}}$. To do so, it is useful to set

$$u(\mathbf{k}) = \vartheta^1 + i\vartheta^2 \ , \ v(\mathbf{k}) = \vartheta^3 + i\vartheta^4 \tag{5.26}$$

with $\vartheta^\mu = \sum_l \lambda_l^\mu \exp(id\frac{\sqrt{5}}{2}k.\lambda_l)$. Notice that in the continuous limit, we have $\vartheta^\mu \to id\frac{\sqrt{5}}{2}\mathbf{k}^\mu$,

$$u(\mathbf{k}) \to id\frac{\sqrt{5}}{2}\left(\mathbf{k}^1 + i\mathbf{k}^2\right), \ v(\mathbf{k}) \to id\frac{\sqrt{5}}{2}\left(\mathbf{k}^3 + i\mathbf{k}^4\right) . \tag{5.27}$$

Substituting (5.26) back into (5.20) and (5.22), we obtain the following expressions,

$$D\bar{D} = \begin{pmatrix} |u|^2 + |v|^2 & 2\bar{u}v \\ 2u\bar{v} & |u|^2 + |v|^2 \end{pmatrix} , \ \bar{D}D = \begin{pmatrix} |u|^2 + |v|^2 & 2\bar{u}\bar{v} \\ 2uv & |u|^2 + |v|^2 \end{pmatrix} \tag{5.28}$$

By solving the characteristic equations of these 2×2 matrix operators, we get the eigenstates $\phi_{\mathbf{k}}^{a\prime}, \chi_{\mathbf{k}}^{a\prime}$ with their corresponding eigenvalues $E_{\pm}^2$,

eigenstates		eigenvalues
$\phi_{\mathbf{k}}^{1\prime} =$	$\sqrt{\frac{v\bar{u}}{2\|u\|\|v\|}}\phi_{\mathbf{k}}^1 + \sqrt{\frac{u\bar{v}}{2\|u\|\|v\|}}\phi_{\mathbf{k}}^2$	$E_+^2 = \|u\|^2 + \|v\|^2 + 2\|u\|\|v\|$
$\phi_{\mathbf{k}}^{2\prime} =$	$-\sqrt{\frac{v\bar{u}}{2\|u\|\|v\|}}\phi_{\mathbf{k}}^1 + \sqrt{\frac{u\bar{v}}{2\|u\|\|v\|}}\phi_{\mathbf{k}}^2$	$E_-^2 = \|u\|^2 + \|v\|^2 - 2\|u\|\|v\|$
$\chi_{\mathbf{k}}^{1\prime} =$	$\sqrt{\frac{\bar{u}\bar{v}}{2\|u\|\|v\|}}\chi_{\mathbf{k}}^1 + \sqrt{\frac{uv}{2\|u\|\|v\|}}\chi_{\mathbf{k}}^2$	$E_+^2 = \|u\|^2 + \|v\|^2 + 2\|u\|\|v\|$
$\chi_{\mathbf{k}}^{2\prime} =$	$-\sqrt{\frac{\bar{u}\bar{v}}{2\|u\|\|v\|}}\chi_{\mathbf{k}}^1 + \sqrt{\frac{uv}{2\|u\|\|v\|}}\chi_{\mathbf{k}}^2$	$E_-^2 = \|u\|^2 + \|v\|^2 - 2\|u\|\|v\|$

(5.29)

By taking square roots of $E_{\pm}^2$, we obtain 2 positive and 2 negative dispersion energies; these are

$$E_{\pm} = +\sqrt{(|u| \pm |v|)^2}, \ E_{\pm}^* = -\sqrt{(|u| \pm |v|)^2} \tag{5.30}$$

which correspond respectively to particles and the associated holes.

2) determining the zeros of $E_{\pm}$ and $E_{\pm}^*$

From the above energy dispersion relations, one sees that the zero modes are of two kinds: $E_+^2 = 0$, $E_-^2 = 0$; and $E_-^2 = 0$ *but* $E_+^2 = E_{+\min}^2 \neq 0$. Let us consider the case $E_+^2 = E_-^2 = 0$; in this situation the zero modes are given by those wave vectors $\mathbf{K}_F$ solving the constraint relations $u(\mathbf{K}_F) = v(\mathbf{K}_F) = 0$. These constraints can be also put in the form

$$\begin{aligned} &\lambda_0^\mu e^{id\frac{\sqrt{5}}{2}\mathbf{K}_F.\lambda_0} + \lambda_1^\mu e^{id\frac{\sqrt{5}}{2}\mathbf{K}_F.\lambda_1} + \\ &+\lambda_2^\mu e^{id\frac{\sqrt{5}}{2}\mathbf{K}_F.\lambda_2} + \lambda_3^\mu e^{id\frac{\sqrt{5}}{2}\mathbf{K}_F.\lambda_3} + \lambda_4^\mu e^{id\frac{\sqrt{5}}{2}\mathbf{K}_F.\lambda_4} = 0 \end{aligned} \tag{5.31}$$

for all values of $\mu = 1, 2, 3, 4$, or equivalently like $d\frac{\sqrt{5}}{2}K_F.\lambda_l = \frac{2\pi}{5}N + 2\pi N_l$. Notice that setting $\mathbf{k} = \mathbf{K}_F + \mathbf{q}$ with small $q = \|\mathbf{q}\|$ and expanding D and $\bar{D}$, eq(5.23) gets reduced to the following familiar wave equation in Dirac theory

$$\frac{d\sqrt{5}}{2}\sum_{\mu=1}^{4}\mathbf{q}_{\mu}\begin{pmatrix}0 & \sigma^{\mu}\\ \bar{\sigma}^{\mu} & 0\end{pmatrix}\begin{pmatrix}\phi_{\mathbf{k}}\\ \bar{\chi}_{\mathbf{k}}\end{pmatrix}=E\begin{pmatrix}\phi_{\mathbf{k}}\\ \bar{\chi}_{\mathbf{k}}\end{pmatrix}. \tag{5.32}$$

6. Graphene and lattice QCD

In this section, we would like to deepen the connection between *2D* graphene and *4D* lattice QCD. This connection has been first noticed by M.Creutz (Creutz, 2008) and has been developed by several authors seen its convenience for numerical simulations in QCD .

6.1 More on link graphene/lattice QCD

2D graphene has some remarkable properties that can be used to simulate *4D* lattice QCD. Besides chirality, one of the interesting properties is the existence of two Dirac points that can be interpreted as the light quarks up and down. This follows from the study of the zero modes of the 2×2 Dirac operator which corresponds also to solve the vanishing of the following energy dispersion relation

$$\sum_{l=1}^{l}\cos ak_l + i\sum_{l=1}^{l}\sin ak_l = 0\ , \tag{6.1}$$

which has two zeros as given by (2.22).
To make contact with lattice QCD, we start by recalling the usual *4D* hamiltonian density of a free Dirac fermion $\Psi=\left(\psi^1,\psi^2,\bar{\chi}_{\dot{1}},\bar{\chi}_{\dot{2}}\right)$ living in a euclidian space time,

$$H=\frac{1}{2}\int d^4x\left(\sum_{\mu=1}^{4}\bar{\Psi}(x)\,\gamma^{\mu}\frac{\partial\Psi(x)}{\partial x^{\mu}}+hc\right), \tag{6.2}$$

where γ^{μ} are the usual 4×4 Dirac matrices given by (5.14). Then, we discretize this energy density H by thinking about the spinorial waves $\Psi\left(x^1,...,x^4\right)$ as $\Psi_{\mathbf{r}_n}$ living at the $\mathbf{r}_n$-nodes of a four dimensional lattice $\mathbb{L}_4$ and its space time gradient $\frac{\partial\Psi(x)}{\partial x^{\mu}}$ like $\frac{1}{a}\left(\Psi_{\mathbf{r}_n+a\boldsymbol{\mu}}-\Psi_{\mathbf{r}_n}\right)$. The field $\Psi_{\mathbf{r}_n+a\boldsymbol{\mu}}$ is the value of the Dirac spinor at the lattice position $\mathbf{r}_n+a\boldsymbol{\mu}$ with the unit vectors $\boldsymbol{\mu}$ giving the four relative positions of the first nearest neighbors of $\mathbf{r}_n$. Putting this discretization back into (6.2), we end with the free fermion model

$$H=\frac{1}{2a}\sum_{\mathbf{r}_n}\left(\sum_{\mu=1}^{4}\left[\bar{\Psi}_{\mathbf{r}_n}\gamma^{\mu}\Psi_{\mathbf{r}_n+a\boldsymbol{\mu}}-\bar{\Psi}_{\mathbf{r}_n+a\boldsymbol{\mu}}\gamma^{\mu}\Psi_{\mathbf{r}_n}\right]\right). \tag{6.3}$$

The extra two term $\bar{\Psi}_{\mathbf{r}_n}\Gamma\Psi_{\mathbf{r}_n}$ and $\left(\bar{\Psi}_{\mathbf{r}_n}\Gamma\Psi_{\mathbf{r}_n}\right)^{+}$ with $\Gamma=\frac{1}{2}\sum_{\mu}\gamma^{\mu}$ cancel each other because of antisymmetry of the spinors. Clearly, this hamiltonian looks like the tight binding hamiltonian describing the electronic properties of the *2D* graphene; so one expects several similarities for the two systems.
Mapping the hamiltonian (6.3) to the Fourier space, we get $H=\sum_{\mathbf{k}}\left(\bar{\Psi}_{\mathbf{k}}\mathcal{D}\Psi_{\mathbf{k}}\right)$ with Dirac operator $\mathcal{D}=\frac{i}{a}\sum_{\mu=1}^{4}\gamma^{\mu}\sin\left(ak_{\mu}\right)$,where we have set $k_{\mu}=(\mathbf{k}.\boldsymbol{\mu})$; giving the wave vector component along the $\boldsymbol{\mu}$-direction. The $\mathcal{D}$- operator is a 4×4 matrix that depends on the wave vector components (k_1,k_2,k_3,k_4) and has 2^4 zeros located as

$$k_1=0,\ \tfrac{\pi}{a}\ ;\ k_2=0,\ \tfrac{\pi}{a}\ ;\ k_3=0,\ \tfrac{\pi}{a}\ ;\ k_4=0,\ \tfrac{\pi}{a}\ . \tag{6.4}$$

However, to apply these formalism to 4D lattice QCD, the number of the zero modes of the Dirac operator should be two in order to interpret them as the light quarks up and down. Following (Creutz, 2008), this objective can be achieved by modifying (6.3) so that the Dirac operator takes the form

$$\mathcal{D} = \frac{i}{a}\sum_{\mu=1}^{4}\gamma^{\mu}\sin\left(ak_{\mu}\right) + \frac{i}{a}\sum_{\mu=1}^{4}\gamma'^{\mu}\cos\left(ak_{\mu}\right) \tag{6.5}$$

where γ'^{μ} is some 4×4 matrix that is introduced in next subsection.

6.2 *Boriçi-Creutz* fermions

Following (Capitani et al., 2009 a;b) and using the 4-component Dirac spinors $\Psi_{\mathbf{r}} = \left(\phi_{\mathbf{r}}^{a}, \bar{\chi}_{\mathbf{r}}^{\dot{a}}\right)$, the *Boriçi-Creutz (BC)* lattice action of *free* fermions reads in the position space, by dropping mass term m_0, as follows:

$$H_{BC} \sim \tfrac{1}{2a}\sum_{\mathbf{r}}\left(\sum_{\mu=1}^{4}\bar{\Psi}_{\mathbf{r}}\Upsilon^{\mu}\Psi_{\mathbf{r}+a\mu} - \sum_{\mu=1}^{4}\bar{\Psi}_{\mathbf{r}+\mathbf{a}\mu}\bar{\Upsilon}^{\mu}\Psi_{\mathbf{r}}\right) - \tfrac{2i}{a}\sum_{\mathbf{r}}\bar{\Psi}_{\mathbf{r}}\Gamma\Psi_{\mathbf{r}} \tag{6.6}$$

where, for simplicity, we have dropped out gauge interactions; and where $\Upsilon^{\mu} = \gamma^{\mu} + i\gamma'^{\mu}$; which is a kind of complexification of the Dirac matrices.
Moreover, the matrix Γ appearing in the last term is a 4×4 matrix linked to γ^{μ}, γ'^{μ} as follows:

$$\gamma'^{\mu} = \Gamma - \gamma^{\mu},\ 2\Gamma = \sum_{\mu=1}^{4}\gamma^{\mu},\ \gamma^{\mu} + i\gamma'^{\mu} = \Upsilon^{\mu},\ \gamma^{\mu}\gamma^{\nu} + \gamma^{\nu}\gamma = 2\delta^{\mu\nu\mu}\ , \tag{6.7}$$

Mapping (6.6) to the reciprocal space, we have

$$H_{BC} \sim \sum_{\mathbf{k}}\bar{\Psi}_{\mathbf{k}}\mathcal{D}_{BC}\Psi_{\mathbf{k}} \tag{6.8}$$

where the massless Dirac operator $\mathcal{D}_{BC}$ is given by

$$\begin{aligned}\mathcal{D}_{BC} = &+\tfrac{1}{2a}\left(\Upsilon_{\mu} - \bar{\Upsilon}_{\mu}\right)\cos\left(ak_{\mu}\right)\\ &+\tfrac{i}{2a}\left(\Upsilon_{\mu} + \bar{\Upsilon}_{\mu}\right)\sin\left(ak_{\mu}\right) - \tfrac{2i}{a}\Gamma\ .\end{aligned} \tag{6.9}$$

Upon using $\Upsilon_{\mu} + \bar{\Upsilon}_{\mu} = 2\gamma_{\mu}$ and $\Upsilon_{\mu} - \bar{\Upsilon}_{\mu} = 2i\gamma'_{\mu}$, we can put $\mathcal{D}_{BC}$ in the form

$$\mathcal{D}_{BC} = D_{\mathbf{k}} + \bar{D}_{\mathbf{k}} - \frac{2i}{a}\Gamma \tag{6.10}$$

with

$$D_{\mathbf{k}} = \tfrac{i}{a}\left(\sum_{\mu=1}^{4}\gamma^{\mu}\sin ak_{\mu}\right),\ \bar{D}_{\mathbf{k}} = \tfrac{i}{a}\left(\sum_{\mu=1}^{4}\gamma'^{\mu}\cos ak_{\mu}\right), \tag{6.11}$$

where $k_{\mu} = \mathbf{k}.\boldsymbol{\mu}$. In the next subsection, we will derive the explicit expression of these k_{μ}'s in terms of the weight vectors $\boldsymbol{\lambda}_l$ of the 5-dimensional representation of the $SU(5)$ symmetry as well as useful relations.
The zero modes of $\mathcal{D}_{BC}$ are points in the reciprocal space; they are obtained by solving $\mathcal{D}_{BC} =$

0; which leads to the following condition

$$\sum_{\mu=1}^{4}\gamma^{\mu}\left(\sin aK_{\mu}-\cos aK_{\mu}\right)-\Gamma\left(2-\sum_{\mu=1}^{4}\cos aK_{\mu}\right)=0\ . \tag{6.12}$$

This condition is a constraint relation on the wave vector components K_{μ}; it is solved by the two following wave vectors:

$$\begin{array}{l}\text{point } K_{BC}: K_1=K_2=K_3=K_4=0\ \ ,\\ \text{point } K'_{BC}: K'_1=K'_2=K'_3=K'_4=\frac{\pi}{2a}\ ,\end{array} \tag{6.13}$$

that are interpreted in lattice QCD as associated with the light quarks up and down.
Notice that if giving up the γ'_{μ}- terms in eqs(6.6-6.8); i.e $\gamma'_{\mu}\rightarrow 0$, the remaining terms in D_{BC} namely $D_{\mathbf{k}}\sim\gamma^{\mu}\sin aK_{\mu}$ have 16 zero modes given by the wave components $K_{\mu}=0,\pi$. By switching on the γ'_{μ}-terms, *14* zeros are removed.

6.3 Hyperdiamond model

The hamiltonian H_{BC} is somehow very particular; it let suspecting to hide a more fundamental property which can be explicitly exhibited by using hidden symmetries. To that purpose, notice that the price to pay for getting a Dirac operator with two zero modes is the involvement of the complexified Dirac matrices Υ^{μ}, $\bar{\Upsilon}^{\mu}$ as well as the particular matrix Γ. Despite that it violates explicitly the $SO(4)$ Lorentz symmetry since it can be written as

$$\Gamma=\frac{1}{2}\sum_{\mu=1}^{4}\gamma^{\mu}v_{\mu},\ v_{\mu}=\begin{pmatrix}1\\1\\1\\1\end{pmatrix}, \tag{6.14}$$

the matrix Γ plays an important role in studying the zero modes. The expression of the matrix Γ (6.7) should be thought of as associated precisely with the solution of the constraint relation $2\Gamma-\sum_{\mu=1}^{4}\gamma^{\mu}=0$ that is required by a hidden symmetry of the *BC* model. This invariance is precisely the $SU(5)$ symmetry of the *4D* hyperdiamond to be identified below. Moreover, the *BC* hamiltonian H_{BC} lives on a *4D* lattice $\mathbb{L}_4^{BC}$ generated by $\boldsymbol{\mu}\equiv\mathbf{v}_{\mu}$; i.e the vectors

$$\mathbf{v}_1=\begin{pmatrix}\mathbf{v}_1^x\\ \mathbf{v}_1^y\\ \mathbf{v}_1^z\\ \mathbf{v}_1^t\end{pmatrix},\ \mathbf{v}_2=\begin{pmatrix}\mathbf{v}_2^x\\ \mathbf{v}_2^y\\ \mathbf{v}_2^z\\ \mathbf{v}_2^t\end{pmatrix},\ \mathbf{v}_3=\begin{pmatrix}\mathbf{v}_3^x\\ \mathbf{v}_3^y\\ \mathbf{v}_3^z\\ \mathbf{v}_3^t\end{pmatrix},\ \mathbf{v}_4=\begin{pmatrix}\mathbf{v}_4^x\\ \mathbf{v}_4^y\\ \mathbf{v}_4^z\\ \mathbf{v}_4^t\end{pmatrix} \tag{6.15}$$

These $\boldsymbol{\mu}$-vectors look somehow ambiguous to be interpreted both by using the analogy with *4D* graphene prototype; and also from the $SU(5)$ symmetry view. Indeed, to each site $\mathbf{r}\in\mathbb{L}_4^{BC}$ there should be *5* first nearest neighbors that are rotated by $SU(5)$ symmetry. But from the *BC* hamiltonian we learn that the first nearest neighbors to each site $\mathbf{r}$ are:

$$\mathbf{r}\rightarrow\left\{\begin{array}{l}\mathbf{r}+a\mathbf{v}_1\\ \mathbf{r}+a\mathbf{v}_2\\ \mathbf{r}+a\mathbf{v}_3\\ \mathbf{r}+a\mathbf{v}_4\end{array}\right. . \tag{6.16}$$

The fifth missing one, namely $\mathbf{r} \to \mathbf{r}+a\mathbf{v}_5$ may be interpreted in the BC fermions as associated with the extra term involving the matrix Γ. To take into account the five nearest neighbors, we have to use the rigorous correspondence $\Gamma^{\mu} \to \mathbf{v}_{\mu}$ and $\Gamma^{5} \to \mathbf{v}_5$ which can be also written in a combined form as follows $\Gamma^{M} \to \mathbf{v}_M$ with $\Gamma^{M} = (\Gamma^{\mu}, \Gamma^{5})$ and $\mathbf{v}_M = (\mathbf{v}_{\mu}, \mathbf{v}_5)$. Because of the $SU(5)$ symmetry properties, we also have to require the condition $\mathbf{v}_1 + \mathbf{v}_2 + \mathbf{v}_3 + \mathbf{v}_4 + \mathbf{v}_5 = 0$ characterizing the 5 first nearest neighbors. To determine the explicit expressions of the matrices Γ_M in terms of the usual Dirac ones, we modify the BC model (6.6) as follows

$$H'_{BC} \sim \frac{1}{2a} \sum_{\mathbf{r}} \left(\sum_{M=1}^{5} \bar{\Psi}_{\mathbf{r}} \Gamma^{M} \Psi_{\mathbf{r}+a\mathbf{v}_M} - \sum_{M=1}^{5} \bar{\Psi}_{\mathbf{r}+a\mathbf{v}_M} \Gamma^{M} \Psi_{\mathbf{r}} \right), \tag{6.17}$$

exhibiting both $SO(4)$ and $SU(5)$ symmetries and leading to the following free Dirac operator

$$\begin{aligned} \mathcal{D} = & \frac{i}{2a} \sum_{\mu=1}^{4} \left(\Gamma_{\mu} + \bar{\Gamma}_{\mu}\right) \sin\left(ak_{\mu}\right) + \frac{i}{2a} \left(\Gamma_5 + \bar{\Gamma}_5\right) \sin\left(ak_5\right) \\ & \frac{1}{2a} \sum_{\mu=1}^{4} \left(\Gamma_{\mu} - \bar{\Gamma}_{\mu}\right) \cos\left(ak_{\mu}\right) + \frac{1}{2a} \left(\Gamma_5 - \bar{\Gamma}_5\right) \cos\left(ak_5\right) \end{aligned} \tag{6.18}$$

where $k_M = \mathbf{k}.\mathbf{v}_M$ and where $\Pi_{M=1}^{5} \left(e^{iak_M}\right) = 1$, $\sum_{M=1}^{5} k_M = 0$ expressing the conservation of total momenta at each lattice site. Equating with (6.9-6.10-6.11), we get the identities

$$\Upsilon_{\mu} + \bar{\Upsilon}_{\mu} = \Gamma_{\mu} + \bar{\Gamma}_{\mu}, \; \Upsilon_{\mu} - \bar{\Upsilon}_{\mu} = \Gamma_{\mu} - \bar{\Gamma}_{\mu}, \tag{6.19}$$

and

$$\frac{i}{2a} \left(\Gamma_5 + \bar{\Gamma}_5\right) \sin\left(ak_5\right) + \frac{1}{2a} \left(\Gamma_5 - \bar{\Gamma}_5\right) \cos\left(ak_5\right) = -\frac{4i}{2a} \Gamma. \tag{6.20}$$

Eqs(6.19) are solved by $\Gamma_{\mu} = \Upsilon_{\mu}$; that is $\Gamma_{\mu} = \gamma^{\mu} + i\left(\Gamma - \gamma^{\mu}\right)$ while

$$\begin{aligned} \Gamma_5 &= -2i\Gamma \;\text{ for } \sin\left(ak_5\right) = 0 \,, \\ \Gamma_5 &= -2\Gamma \;\text{ for } \sin\left(ak_5\right) = 1 \,. \end{aligned} \tag{6.21}$$

where $k_5 = -\left(k_1 + k_2 + k_3 + k_4\right)$. In this 5-dimensional approach, the ambiguity in dealing with the μ-vectors is overcome; and the underlying $SO(4)$ and $SU(5)$ symmetries of the model in reciprocal space are explicitly exhibited.

7. Conclusion and comments

Being a simple lattice-carbon based structure with delocalized electrons, graphene has been shown to exhibit several exotic physical properties and chemical reactions leading to the synthesis of graphene type derivatives such as graphAne and graphOne. In this book chapter, we have shown that graphene has also very remarkable hidden symmetries that capture basic physical properties; one of these symmetries is the well known $SU(2)$ invariance of the unit cells that plays a crucial role in the study of the electronic properties using first principle calculations. Another remarkable hidden invariance, which has been developed in this work, is the $SU(3)$ symmetry that captures both crystallographic and physical properties of the graphene. For instance, first nearest neighbors form 3-dimensional representations of $SU(3)$; and the second nearest neighbor ones transform in its adjoint. Moreover, basic constraint relations like $v_1 + v_2 + v_3 = 0$ is precisely a $SU(3)$ group property; and its solutions are

exactly given by group theory. Furthermore, the location of the Dirac zero modes of graphene is also captured by $SU(3)$ seen that these points are given by $\pm\frac{2\pi}{3d}\boldsymbol{\alpha}_1$, $\pm\frac{2\pi}{3d}\boldsymbol{\alpha}_2$, $\pm\frac{2\pi}{3d}\boldsymbol{\alpha}_3$ where the $\boldsymbol{\alpha}_i$'s are the $SU(3)$ roots that generate the reciprocal space.
On the other hand, from $SU(3)$ group theory's point of view, graphene has cousin systems with generic $SU(N)$ symmetries where the integer N takes the values $2, 3, 4, ...$ The leading graphene cousin systems are linear molecules with hidden SU(2) invariance; this is precisely the case of poly-acetylene, cumulene and poly-yne studied in section 4. The graphene cousin systems with hidden SU(4) and SU(5) symmetries are given by *3D* diamond; and *4D* hyperdiamond which has an application in *4D*- lattice QCD.
Finally, it is worth to mention that the peculiar and unique properties of graphene are expected to open new areas of applications due to its important electronic, spintronic, mechanical and optical properties. The challenge is find low-cost-processes for producing graphene and graphene-based structures and to tune its properties to the targeted applications such as the replacement of silicon in the field of new-type of semiconductors and new electronics, new data-storage devices, new materials with exceptional mechanical properties and so on.
Various attempts are also made to incorporate other atoms within the structure of graphene or combine the graphene-based structures with other materials in sandwich type structure or in chemical way by binding it to various molecules with divers topologies and functionalities.

8. References

P. F. Bedaque, M. I. Buchoff, B. C. Tiburzi, A. Walker-Loud, Broken Symmetries from Minimally Doubled Fermions, *Phys. Lett. B,* 662 (2008), 449–457,

A. Boriçi, Creutz fermions on an orthogonal lattice, *Phys. Rev. D,* 78 (October 2008) 074504-074506,

S. Capitani, J. Weber, H. Wittig, Minimally doubled fermions at one loop, *Phys. Lett. B* 681, (2009), 105-112,

S. Capitani, J. Weber, H. Wittig, Minimally doubled fermions at one-loop level, *Proceedings of Science, The XXVII International Symposium on Lattice Field Theory - LAT2009,* 1-7, (July 26-31 2009), Peking University, Beijing, China,

S. Capitani, M. Creutz, J. Weber, H.Wittig, *J. High Energy Phys,* Renormalization of minimally doubled fermions, Volume 2010, Number 9, ,2010, 27,

A.H. Castro-Neto, F. Guinea, N. M. R. Peres, K. S. Novoselov and A. K. Geim, The electronic properties of graphene, *Rev. Mod. Phys.* 81, (January 2009), 109–162,

D. Chakrabarti, S.Hands, A.Rago, Topological Aspects of Fermions on a Honeycomb Lattice, *J. High Energy Phys,* Volume 2009, Issue 06 (June 2009) 060,

J. C. Charlier, X. Gonze, J. P. Michenaud, *Phys. Rev. B.* 43, (1991), 4579–4589,

A. Cortijo, M A. H. Vozmediano, *Nucl.Phys.B* 807, (2009), 659-660,

M. Creutz, Four-dimensional graphene and chiral fermions, *J.High Energy Phys,* Volume 2008, Issue 04 (April 2008) 017,

M. Di Pierro, An Algorithmic Approach to Quantum Field Theory, *Int.J.Mod.Phys.A,* 21, (2006), 405-448,

L.B Drissi, E.H Saidi, M.Bousmina, Electronic properties and hidden symmetries of graphene, *Nucl Phys B,* Vol 829, (April 2010), 523-533,

L.B Drissi, E.H Saidi, M.Bousmina, Four dimensional graphene

L.B Drissi, E.H Saidi, M.Bousmina, Graphene, Lattice QFT and Symmetries, *J. Math. Phys.* 52, (February 2011), 022306-022319,

L. B Drissi, E. H Saidi, On Dirac Zero Modes in Hyperdiamond Model, ISSN arXiv:1103.1316,

D. D. Ferrante, G. S. Guralnik, Mollifying Quantum Field Theory or Lattice QFT in Minkowski Spacetime and Symmetry Breaking, ISSN arXiv:hep-lat/0602013,
A. K. Geim, K. S. Novoselov, The rise of graphene, *Nature Materials*, 6, 183, (2007), 183-191,
A. Giuliani, V. Mastropietro, M. Porta, Lattice gauge theory model for graphene, *Phys.Rev.*B, 82, (September 2010), 121418-121421(R),
C. O. Giritet, J C. Meyer, R. Erni, M D. Rossell, C. Kisielowski, L. Yang, C-H Park, M. F. Crommie, M L. Cohen, S G. Louie, A. Zettl, Graphene at the Edge: Stability and Dynamics, *Science*, 23, (March 2009), 1705-1708,
M.O. Goerbig, R. Moessner, B. Doucot, Electron interactions in graphene in a strong magnetic field, *Phys. Rev.* B, 74, (October 2006), 161407-161410,
F. D. M. Haldane, Fractional Quantization of the Hall Effect: A Hierarchy of Incompressible Quantum Fluid States, *Phys. Rev. Lett.* 51, (August 1983), 605-608,
F. D. M. Haldane, Model for a Quantum Hall Effect without Landau Levels: Condensed-Matter Realization of the "Parity Anomaly", *Phys. Rev. Lett.* 61, (October 1988), 2015-2018,
R. Jackiw, S.-Y. Pi, Chiral Gauge Theory for Graphene, *Phys. Rev. Lett.* 98 (June 2007), 266402-266405,
R. Jackiw, S.-Y. Pi, Persistence of zero modes in a gauged Dirac model for bilayer graphene, *Phys. Rev. B*, 78, (October 2008), 132104-132106,
S. Jun, Density-functional study of edge stress in graphene, *Phys. Rev. B*, 78, (August 2008), 073405-073408,
T. Kimura, T. Misumi, Characters of Lattice Fermions Based on the Hyperdiamond Lattice, *Prog. Theor. Phys*, 124 (2010), 415-432,
P. Koskinen, S. Malola, H. Hakkinen, Self-Passivating Edge Reconstructions of Graphene, *Phys.Rev. Lett.*, 101, (September 2008), 115502-115505,
E. McCann, V.I. Falko, Landau-Level Degeneracy and Quantum Hall Effect in a Graphite Bilayer, *Phys. Rev. Lett.* 96 (June 2006), 086805-086808,
K. Nomura, A. H. Mac Donald, Quantum Hall Ferromagnetism in Graphene, *Phys. Rev. Lett.* 96, (June 2006), 256602-256605,
K. S. Novoselov, A. K. Geim, S. V. Morozov, D. Jiang, Y. Zhang, S. V. Dubonos, I. V. Grigorieva, A. A. Firsov, (2004). Electric Field Effect in Atomically Thin Carbon Films, *Science*, 306, 5696, (October 2004), 666-669,
K. S. Novoselov, E. McCann, S. V. Morozov, V. I. Fal'ko, M. I. Katsnelson, U. Zeitler, D. Jiang, F. Schedin, A. K. Geim, Unconventional quantum Hall effect and Berry's phase of 2pi in bilayer graphene, *Nat. Phys.* 2 (2006),177-180,
T. Ohta, A. Bostwick, T. Seyller, K. Horn, E. Rotenberg, Controlling the Electronic Structure of Bilayer Graphene, *Science*, 313, 5789 (August 2006), 951-954,
T. Ohta, A. Bostwick, J. L. McChesney, T. Seyller, K. Horn, E. Rotenberg, Interlayer Interaction and Electronic Screening in Multilayer Graphene Investigated with Angle-Resolved Photoemission Spectroscopy, *Phys. Rev. Lett.* 98 (May 2007), 206802-206805,
P.R Wallace, The Band Theory of Graphite, *Phys Rev*, 71, (May 1947), 622–634,
J. C. Slonczewski and P. R. Weiss, Band Structure of Graphite, *Phys. Rev.* 109, (January 1958), 272–279,
G.V. Semenoff, Condensed-Matter Simulation of a Three-Dimensional Anomaly, *Phys. Rev. Lett.* 53, (December 1984), 2449–2452,
Y. Zheng and T. Ando, Hall conductivity of a two-dimensional graphite system, *Phys. Rev. B* 65, (June 2002), 245420-245430,

Y. Zhang, Y-Wen Tan, H L. Stormer, P. Kim, Experimental observation of the quantum Hall effect and Berry's phase in graphene, *Nature* 438, (November 2005), 201-204,

Y. Zhang, Z. Jiang, J. P. Small, M. S. Purewal, Y.-W. Tan, M. Fazlollahi, J. D. Chudow, J. A. Jaszczak, H. L. Stormer, and P. Kim, Landau-Level Splitting in Graphene in High Magnetic Fields, *Phys. Rev. Lett.* 96, (April 2006), 136806-136809.

17

Single-Particle States and Elementary Excitations in Graphene Bi-Wires: Minding the Substrate

Cesar E.P. Villegas and Marcos R.S. Tavares
Universidade Federal do ABC,
Centro de Ciências Naturais e Humanas, Santo André, São Paulo
Brazil

1. Introduction

Graphene has become a subject of intense theoretical and experimental interest since it was realized in laboratory in 2004 (Novoselov et al., 2004). As being considered as a true two-dimensional (2D) material, researchers worldwide have focused their work on studying its electronic properties and the promising technological applications in designing nanoelectronic devices based on it (Castro Neto et al., 2009; Castro Neto, 2010; Peres, 2010). In fact, such studies revealed that graphene is a zero-gap 2D semiconductor. Furthermore, novel and intriguing effects concerning ballistic transport (Miao et al., 2007), the absence of Anderson localization (Beenakker, 2008), the anomalous electron-electron (e-e) interaction (Das Sarma et al., 2010), and so on, have also attracted much attention. Most of these effects arise as a consequence of the graphene band structure, in which the energy presents pretty linear dispersion relation around the Dirac point. As a direct consequence, the low-lying excitations can be described, in the first place, by the 2D Dirac-Weyl Hamiltonian, and the quasiparticles in graphene being described as massless fermions (Semenoff, 1984).

Also, it turned out to be very common to tune the carrier densities in graphene by manipulating local gate voltages (Novoselov et al., 2004; Williams et al., 2007). This led to the creation of tunable graphene-based microstructures. By the same token, it was not surprising to see that, for a perpendicular incidence of flowing quasiparticles across the electrostatic barrier, the absence of backscattering might led to a perfect transmission (Katsnelson et al., 2006) through the barrier. Indeed, the localization of electron (or hole) states by electrostatic quantum confinement in graphene turned out to be a challenging task by virtue of Klein tunneling. But, recent experimental results (Williams et al., 2011) have shown the real possibility of creating electronic waveguides by using electrostatic gates in graphene with pretty much high efficiency. The waveguide is realized by controlling the charge density in the sample through an external gate, which forms a *p-n-p* junction. In this quasi-one-dimensional (Q1D) channel, the Fermi-energy mismatching across the junction serves as a refraction index to quantify the guidance efficiency loss. Although the conductance along the wire is treated as ballistic in this case, it has been mentioned that effects such as: (i) junction roughness (Williams et al., 2011); (ii) the interaction between the graphene sheet and the underlying substrate (Chiu et al., 2010) ; and (iii) the charge-charge correlation might affect the guidance efficiency (Zhang & Fogler, 2008).

As a matter of fact, the effects induced by the electron-electron (e-e) correlation in these doped Q1D systems have been subject of intense theoretical study. (Zhang & Fogler, 2008) The most fundamental quantity to study e-e correlation is the dielectric function, which should provide us with the dynamical screening properties of the Fermionic system embedded in the sample. In fact, such a quantity has been defined, within the random phase approximation (RPA), at zero and non-zero temperature, and for doped and undoped samples with the presence of the spin-orbit interaction and strain (Das Sarma et al., 2010)(Hwang & Das sarma, 2007)(Pellegrino et al., 2010)(Brey & Fertig, 2007).
Effects induced by the interaction between graphene sheet and the underlying substrate are also an important issue which is being very addressed recently. There have been much experimental concern about substrate induced effects on the graphene sheets. Anisotropic effects have been observed (Yang et al., 2010)(Villegas et al., 2010) on the conductivity and claimed to be induced by the interaction between the graphene sheet and the underlying substrate.
This chapter focuses on the study of the electronic properties of confined electrons in bi-wires based on graphene. We theoretically propose a model to describe two coupled waveguides (quantum wires) based on graphene which has eventually been deposited over some sort of substrate. The graphene-substrate interaction is considered here through a phenomenological parameter which is taken into account in an *ad-hoc* manner. (Villegas et al., 2010) The first section is devoted to review the main points on how to confine electrons in graphene through electrostatic gate potentials. The second section presents our model and shows the single particle spectra of graphene bi-wires, taking into account the graphene-substrate interaction phenomenologically in the Dirac hamiltonian, whose solution is given by a pseudo-spinor. Such an effect can be induced in the sample by considering the interaction between the graphene sheet and the substrate such as boron nitrate (BN) (Giovannetti et al., 2007). Furthermore, in order to control the pseudo-spinor components in the barrier region, we theoretically explore and manage the interaction between the substrate and the graphene sheet, considering it in order to keep a reasonable guiding efficiency. We then study, in Sec. III, the dynamical screening properties of the Fermion gas which is eventually laid in the structure. We calculate the dielectric response function within the RPA, which turned out to be a good approximation describing e-e correlations in graphene even for small charge densities (Hwang & Das sarma, 2007). The roots of this function provide us with the collective excitations in the system. Our results show that the graphene-substrate interaction induced effects play an important role in obtaining the bare Coulomb potential and the dielectric function of the system. Such a role manifests itself in the dispersion relation of the optical and acoustical plasmon modes. We found that the graphene-substrate interaction might eventually increase the damping effects on these modes.

2. Electrostatic confinement of electrons in graphene

Nano-engineering serves to yield structures that are not found *in natura* anywhere. Indeed, most of such structures have been realized as an essential feature of quantum mechanics, i.e., they are a result of quantum confinement, which constitutes the basic building block of electronic devices. Although in graphene the quantum confinement by electrostatic potentials seems to be a difficult job by virtue of the absence of backscattering, which is a direct consequence of pseudo-chiral behavior of quasiparticles, many theoretical (Pereira et al.,

2007) and experimental (Williams et al., 2011) works have attemptedly proposed different approaches to study and overcome this difficulty. In the following, we discuss the main points regarding these proposals.

2.1 Etched structures

Etching the graphene leads to the formation of 1D quantum wires. These structures are commonly mentioned in the literature as graphene nanoribbons (GNRs) (Brey & Fertig, 2006). It is well known that the semiconducting gap in GNRs is strongly dependent on the ribbon width, and also on the kind of the ribbon edge (Son & Cohen, 2006). The ribbon depends on the kind of the edge termination, leading to different boundary conditions. Since the graphene is a hexagonal bound of carbon atoms, two types of termination can be etched: the armchair and the zig-zag terminations. It has been shown that the narrower the ribbon, the bigger the semiconducting gap. Moreover, very narrow GNRs evidenced the existence of surface states and miniband formation in the long wavelength limit $\mathbf{k} \rightarrow \mathbf{0}$. (Brey & Fertig, 2006) As a matter of fact, the combination between ribbon width and edge termination turned out to be a crucial point in determining if the sample is metallic or semiconductor. Therefore, etching these structures requires very sensible controlling of the edge termination, and do not assure a standard technique in obtaining identical materials. This may be now-days the main disadvantage in etching graphene structure to produce 1D quantum structures producing the same kind of material.

2.2 Infinite mass boundary conditions

The infinite mass boundary condition approach was proposed by Berry and Mondragon (Berry & Mondragon, 1987) in order to overcome the difficulty in confining neutrino relativistic billiards. They considered the possibility of taking into account the neutrino quasi-mass induced effects in theoretically treating the electrostatic gate potential $V(x)$. The neutrino mass term $M(x)$ was considered to be x-coordinate dependent, so that the effective gate potential might be written as $V^{eff}(x) \rightarrow V(x) + M(x)c^2\sigma_z$, where c is the light velocity and σ_z the z-direction Pauli matrix. The same approach can be used to study the gate potential providing charge confinement in graphene (Peres et al., 2009). In principle, such a gate potential should form a long stripe based on graphene. We schematically show $V^{eff}(x)$ in Fig.1 and simply picture the different regions regarding the mass term $M(x)$. To graphene, one assumes the light velocity $c \rightarrow v_F = 1.1x10^6 m/s$. In region II the mass term $M(x)$ can be considered zero by virtue of vanishing graphene-substrate interaction, while in regions I and III it can assume finite values. Notice that in the case of neutrinos, the $M(x)$ was considered to be infinite (hard wall approach), since they theoretically sought for vanishing the Fermi velocity outside the well. In this sense, it was claimed that this kind of technique might lead to the confinement of charges to region II, because the effective potential should be strong enough to enable wavefunction reflection at the interface of the potential (at $x = \pm L/2$). The difference between this technique and the one presented above concerns to the ribbon (or stripe) edge. Here, the edge termination does not play the same role as in the previous case, and the boundary conditions are the same as in the standard quantum well (or finite mass) confinement (Lin et al., 2008). We finally recall that the gate potential naturally imposes the continuity of the wavefunction at the interfaces $x = \pm L/2$, but not of its derivative.

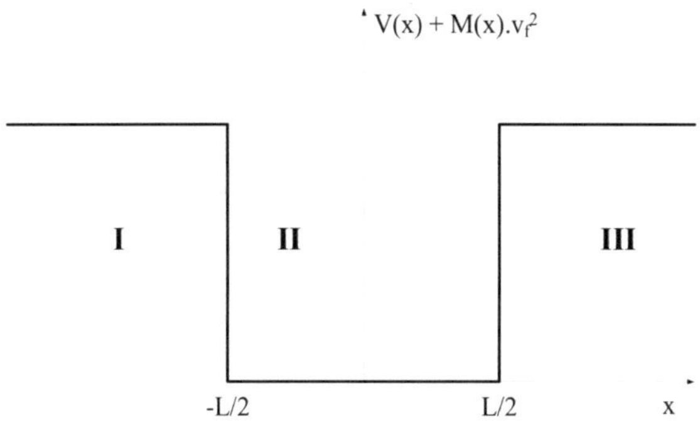

Fig. 1. Schematic representation of the infinite mass confinement potential profile. The mass M is zero in region II and infinite elsewhere.

2.3 Graphene-substrate interaction

Another usual technique improving confinement of electrons in graphene is to help opening up a semiconducting gap in the energy spectrum. In particular, it was shown that the most effective way to do that is to have the graphene, and its eventual substrate, embedded in boron nitrate material (Giovannetti et al., 2007), provided it is also a hexagonal lattice. The interaction between the different atoms in the whole sample causes then an anisotropy associated to the different number of electrons in the sublattices A and B that form the unit cell of the graphene honeycomb (Semenoff, 1984). That is responsible to open up the gap. Experiments have shown that this induced gap can vary from 50 up to 260 meV (Lu et al., 2007; Zhou et al., 2007). We mention that the effects of such asymmetry in the sublattices can be theoretically incorporated into the Dirac Hamiltonian by considering a sort of phenomenological (or *quasi-mass*) parameter $\Delta\sigma_z$ in an ad-hoc manner. It is worth to mention that this parameter $\Delta\sigma_z$ has the same nature as those representing the Rashba spin-orbit interaction in graphene (Kane & Mele, 2005; Min et al., 2006).

2.4 Tuning the Fermi level in p-n junctions

A recent theoretical proposal to improve and study the confinement of electrons in graphene is based on the so-called Goos-Hanchen (GH) effect (Goos & Hanchen, 1947). The GH effect has been shown for the first time back in the 40s and accounts for a universal shift that a light beam, which is reflected from a given medium, can suffer. Such a shift can be attributed to a sort of residual transmission into the medium which has the lower refractive index, causing the evanescence of the electromagnetic wave. The electronic analogy of the GH effect can also manifest itself in graphene *p-n-p* junctions (Beenakker et al., 2006). As it was mentioned, electrostatic gates actually control the charge-type doping in the graphene and eventually create a *p-n* junction. In such structures, the Fermi energy plays the role of a refraction index, since it assumes different values as it crosses the junction. The main advantage of this technique is that it enables one to easily control the carrier density in the sample, providing the designing of *electronic* waveguide structures. As a mater of fact, recent experiments have shown the possibility of creating waveguides in graphene *p-n-p* junctions, leading to a variety of promising applications (Cheianov et al., 2007; Williams et al.,

2011). Such a work also dealt with the direction-dependent transmission induced effects in graphene (Pereira et al., 2007). In any case, it was shown that, by tuning the Fermi energy in given *p-n-p* junctions, it was certainly possible to form electronic waveguides with reasonable efficiency in graphene, leading to the promising fabrication of nanodevices based on electronic confinement in graphene.

3. Single particle spectra in graphene quantum wires

In response to the points discussed above and motivated by the promising applications of quantum confinement in graphene, we study below the single particle spectrum of quasi-confined electrons in graphene *p-n-p* junctions. We consider an electrostatic gate potential, whose some sort of arrangement can eventually be responsible for forming two quantum wires (double *p-n-p* junctions) coupled by a barrier of width L_b and height V_b. Our model also deals with a more generic situation where the interaction between the substrate and graphene sheet is properly considered in both doped regions .
The low-energy band structure for a single valley[1] (K point) in graphene can be described by the 2D Dirac-like Hamiltonian (Semenoff, 1984) in the form

$$H_{2D} = \gamma \begin{pmatrix} 0 & k_x - ik_y \\ k_x + ik_y & 0 \end{pmatrix} + \beta\sigma_z, \tag{1}$$

where σ_z is the Pauli matrix, $\gamma = \sqrt{3}\hbar a_0 t/2$ is a parameter associated to the lattice constant $a_0 = 0.246$ nm and to the nearest neighbor hopping energy $t \approx 2.8$ eV in graphene. As mentioned before, the parameter β accounts for an asymmetry associated to different number of electrons in sublattices A and B, that form the unit cell of the honeycomb. The values for β might be strongly dependent on the substrate over which the graphene sheet is embedded. The stronger is the interaction between graphene sheet and the substrate, the larger is the β parameter which is responsible for opening an energy gap in the Dirac cone spectrum. This is the parameter which will quantify here, in an *ad-hoc* manner, the graphene-substrate interaction strength. In this way, one should be able to investigate such effects theoretically by considering the parameter $\beta \neq 0$ in the Eq. (1). Therefore, the effects associated to the mentioned parameter will be carefully studied here.

3.1 Triple barrier (Double well)

In order to study the electronic guidance in double coupled waveguides based on graphene, one should be able to seek the effects of imposing a confinement potential $U(x)$, as pictured in Fig.(2), onto the Hamiltonian H_{2D}. This potential $U(x)$, as generated by electrostatic gates, leads to two transversal quantum wires along the y-direction with widths L_1 and L_2 separated by a barrier of length L_b. The tunable barrier between the wires, the region III on Fig.(2), is assumed to be responsible for controlling the coupling between the two *p-n* junctions. As mentioned above, the issue of weather the epitaxially-growth graphene can or cannot be gated is not important here, because one can ad-hocly assigned the parameter β to other phenomenological effects as well, such as the spin-orbit interaction, the impurities or imperfections in the sample, and so forth (Ziegler, 1996). Therefore, the eigenstates of

[1] It should be pointed out that the two lattice points K and K′ decouple from each other in the graphene Hamiltonian (1) only when the disorder generated by the potential is much smaller than the lattice parameter (Ando & Nakanishi, 1998).

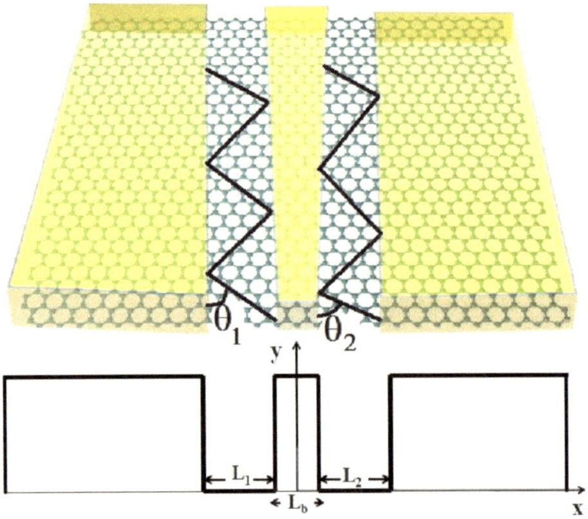

Fig. 2. Schematic representation of the potential profile forming a biwaveguide based on graphene. It shows two coupled channels of widths L_1 and L_2, separated by a barrier of width L_b and height U_0. Two different reflection angles θ_1 and θ_2 can be considered in the regions II and IV, respectively, by virtue of different graphene-substrate interaction strength.

the Hamiltonian $H = H_{2D} + U(x)$ can be written as a two-component spinor in the form (Villegas & Tavares, 2010; Villegas et al., 2010)

$$|\psi(x,y)> \simeq e^{ik_y y} \begin{bmatrix} \chi_A(x) \\ \chi_B(x) \end{bmatrix}, \quad (2)$$

where $\chi_A(x)$ and $\chi_B(x)$ are the pseudo-spin wavefunction components. Here we choose expressions for $\chi_A(x)$ and $\chi_B(x)$ that naturally satisfies the secular equation $H|\psi(x,y)> = E|\psi(x,y)>$ and represent electronic states that are confined by the potential $U(x)$. It should be recalled at this point that the Fermi level in the sample should play the role of a refraction index and, as consequence, the quasiparticles might undergo to internal reflections as they eventually cross the potential interfaces, giving rise to the phenomena analogous to the GH shift discussed above. Thus, the wavefunctions for the component $\chi_A(x)$ in different regions, can be written in terms of stationary functions within the well regions and evanescent functions within the barrier region, i.e.,

$$\chi_A(x) \sim \begin{cases} C_I e^{\kappa_I x} & region\ I \\ C_{II} \sin q_{II} x + D_{II} \cos q_{II} x & region\ II \\ C_{III} e^{\kappa_{III}} x + D_{III} e^{-\kappa_{III} x} & region\ III \\ C_{IV} \sin q_{IV} x + D_{IV} \cos q_{IV} x & region\ IV \\ D_V\ e^{-\kappa_V x} & region\ V \end{cases} \quad (3)$$

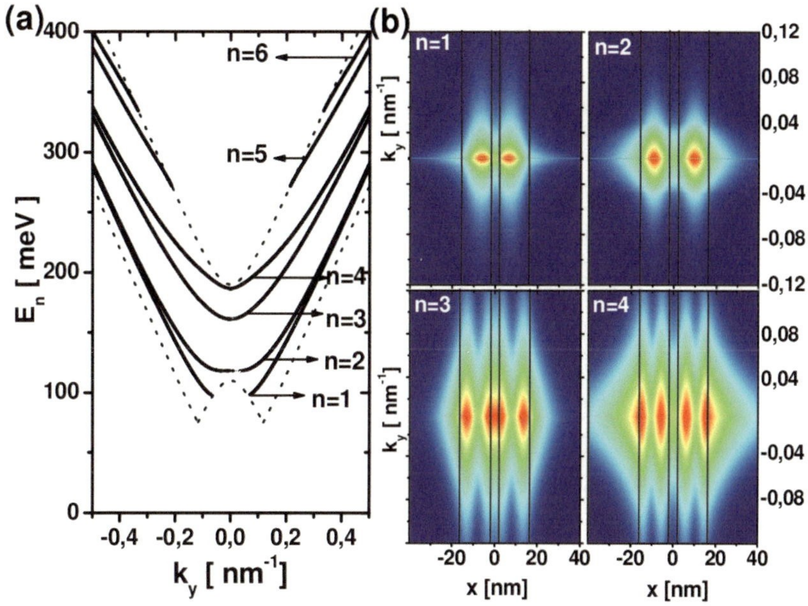

Fig. 3. (Color online) (a) Dispersion relations, $E_n(k_y)$, for a symmetric structure with $L_1 = L_2 = 14$ nm, $L_b = 4$ nm, and $U_0 = 150$ meV. Here, $\beta = 40$ meV over all layers in the structure. The dashed lines represent the edges of the regions where the confined state branches lay down. (b) Probability density function $\rho_{nk_y}(x)$ for $n = 1, ..., 4$, plotted in arbitrary units, showing the spreading of states in the graphene wires. Thin vertical lines indicate the position of the potential edges.

Here, the wavevector along the x-direction

$$\kappa_i = \gamma^{-1}\sqrt{(\gamma k_y)^2 + (\beta_i)^2 - \left[E(k_y) - U_0\right]^2} \tag{4}$$

for $i = I, III, V$ and

$$q_j = \gamma^{-1}\sqrt{E(k_y)^2 - (\gamma k_y)^2 - \left(\beta_j\right)^2} \tag{5}$$

for $j = II, IV$. Notice that the parameter β may assume a different value in each region and this is a form to simulate different substrate interaction strengths or dopings. Due to the perfect similarity between spin-orbit and Dirac Hamiltonians, such a parameter can be linked, in an *ad-hoc* manner, to a spin-orbit interaction occurring in the samples without gate. Finally, the set of coefficients $\{C_I, C_{II}, C_{III}, C_{IV}\}$ and $\{D_{II}, D_{III}, D_{IV}, D_V\}$ indicate the intensity of propagating and anti-propagating waves, respectively within each doped region. This type of solution for $\chi_A(x)$ in the Eq. (3) was considered before (Pereira et al., 2007) where the component χ_B is also written straightforwardly, since $H|\psi> = E|\psi>$ leads to two coupled differential equations for the components $\chi_A(x)$ and $\chi_B(x)$ which are solved through the same *anzatz*(3). The energy dispersions $E_n(k_y)$ in Eqs. (4) and (5), are obtained by imposing continuity conditions for both spinor components $\chi_A(x)$ and $\chi_B(x)$ at the edges of the confining potential $U(x)$. From these continuity conditions, one is able to write down a set of equations in the matrix form

$$\hat{M}C^T = 0, \tag{6}$$

where $\hat{M}(E_n, k_y)$ is the 8×8 matrix shown in the Appendix A and C is a 1×8 matrix formed by the amplitude of propagating and anti-propagating waves in the sequence: $(C_I, C_{II}, D_{II}, C_{III}, D_{III}, C_{IV}, D_{IV}, D_V)$. Notice that the dispersions $E_n(k_y)$ are obtained through the non-trivial solutions of the Eq. (6) and implying that

$$\det\left[\hat{M}(E_n, k_y)\right] = 0. \tag{7}$$

We use optimized numerical routines to find the n roots of Eq. (7) for a given value of the wavevector k_y.
At this point, we should define

$$\theta_n^j = \arctan(k_y / q_n^j) \tag{8}$$

as the reflection angles for the regions j = II and IV. These angles are schematically indicated in Fig.(2) and represent the n possible angles of reflection in the regions j = II and IV through which the guidance occurs. One can really read out from Eq. (5) that, in considering $\beta_{II} \neq \beta_{IV}$, a spatial-anisotropy induced effect should take place, as the effective depths of each potential well in the regions j = II and IV are different from each other.

3.1.1 Extended states

Let us first assume an unique value for the ad-hoc parameter $\beta_{i=j} \equiv \beta = 40$ meV, for all layers shown in Fig.(2). Then we solve Eq. (7) for a given value of k_y and determine a number n of different roots. Figure 3(a) shows the first six solutions as a function of k_y and these energy dispersions, $E_n(k_y)$, map out the confined modes in the graphene double wire system described by the lateral confining potential, $U(x)$. Due to interwire tunneling, the pairs of *bound* (symmetric) and *antibound* (antisymmetric) states with energy eigenvalues E_1 and E_2, E_3 and E_4, etc, show larger (smaller) energy separation for smaller (larger) values of wavevector k_y. These energy branches were found in a region limited by the dotted lines, which are roughly given by the expressions

$$[(\gamma k_y)^2 + \beta^2]^{1/2} \tag{9}$$

and

$$\pm [(\gamma k_y)^2 + \beta^2]^{1/2} + U_0. \tag{10}$$

We stress that the inverse of the lifetime of any subband displayed in Fig. 3(a) is assumed as being of the order of the numerical precision required to solve the Eq. (7) (Nguyen et al., 2009). Although this may seem an approximation to the numerical solution, the high precision required in our calculation produces very reliable results.
The effects associated to wide states can be better addressed when the probability density function given by

$$\rho_{nk_y}(x) = |\chi_A(x)|^2 + |\chi_B(x)|^2, \tag{11}$$

is drawn for each subband n as a function of the wavevector k_y. Figure 3(b) shows $\rho_{nk_y}(x)$, plotted in arbitrary units, for the same lowest four states of Fig. 3(a), showing regions of maximum and minimum intensities for localized and quasi-extended states in our potential model. Clearly, the larger (smaller) is the wavefunction amplitude in regions I and V, the weaker (stronger) is the evidence for localization effects on the structure. The states n = 1, 2, and 3 show stronger confinement to the well regions while state $n = 4$ display the quasi-extended character with a large probability amplitude over the whole structure and a

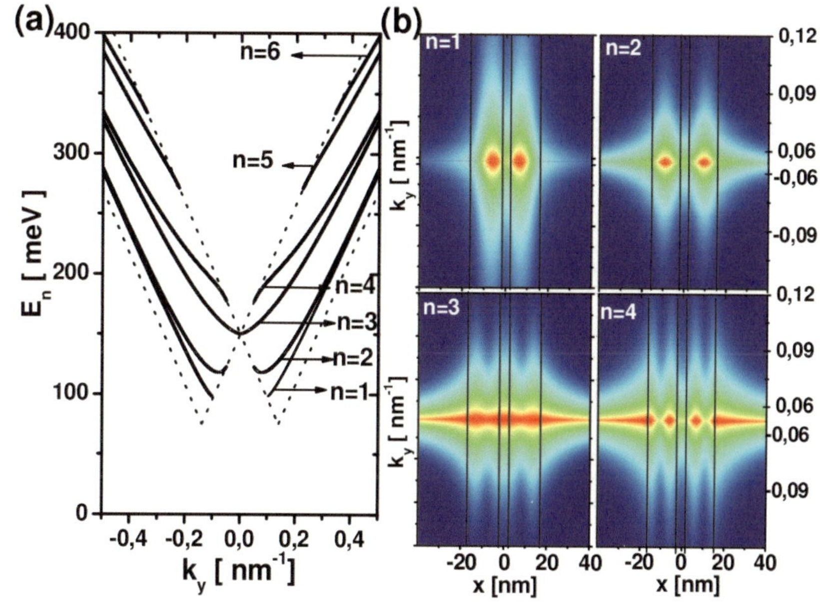

Fig. 4. (Color online) - The same energy dispersions and probability densities as in Fig. 2, but without the **substrate-graphene coupling** term ($\beta = 0$ in all layers) in the pseudo-relativistic Hamiltonian.

long-range exponential decay for small values of k_y in the regions I and V. As the wavevector $|k_y|$ increases, these extended states become more and more localized inside the well regions. In Fig. 4 we are showing the same results as in Fig. 3, but with the parameter β being zero in all regions. As noticed, the state $n = 1$ shows only a slightly larger spreading along the wires for increasing values of k_y and the state $n = 2$ increases the spreading along the wires and into the lateral barrier regions. However, the amplitude probability for states 3 and 4 show very large extended character in all regions for small values of k_y. In other words, the addition of a substrate-graphene coupling term to the Dirac-like Hamiltonian increases the lateral localization character of the extended states.

3.1.2 Anisotropic values of β : effects on the conductance

With the choice of different values for the parameter β in each region, the system should reveal interesting effects. In order to analyze these effects, we define auxiliary quantities such as $\Delta = \beta_I - \beta_{III}$, $\beta_{II} = \beta_{IV} = 0$, $\beta_I = \beta_V$ and $\beta_{III} = 10$ meV. When $\Delta \neq 0$, the interaction between graphene sheet and substrate in regions I and III have different strength values. Furthermore, one can also read out from Eq. (4) that the wavevector $\kappa_I \neq \kappa_{III}$ when $\Delta \neq 0$. This situation mimics a spatial anisotropy as if the effective barrier heights in regions I and III were different from each other. We predict that these spatial anisotropy induced by finite Δ can be verified experimentally through a measure of the electrical conductance along the y-direction. Very recently, this conductance, as a function of well width, has been investigated theoretically in one dimensional channel formed by a single graphene p-n junction without the interaction parameter β (Beenakker et al., 2006). It was shown that a minimum in the energy dispersion $E_n(k_y)$ would generate a plateau in the conductance along the channel. In addition, each minimum in the ground-state energy $E_1(k_y)$ would contribute independently with an

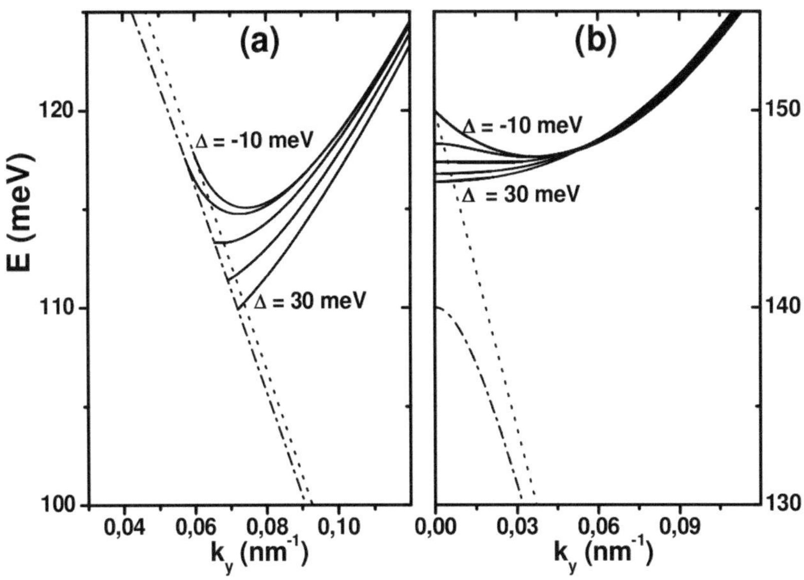

Fig. 5. (Color online) - Effects of **substrate-graphene** coupling strength on the shape of the energy dispersion in a double wire system based on graphene sheet grown over a substrate. Here, the double wire structure is symmetric with widths: $L_1 = L_2 = 9$ nm, $L_b = 4$ nm and $U_0 = 150$ meV. The **strength values**, β_i, are taken different in the layers according to: $\Delta = \beta_I - \beta_{III}$, $\beta_{II} = \beta_{IV} = 0$, $\beta_I = \beta_V$ and $\beta_{III} = 10$ meV.

amount of e^2/h per spin and per valley degrees of freedom to the conductance. Therefore, two minima in our dispersion relation $E_1(k_y)$ should be responsible for a first step in the quantum conductance at $8e^2/h$. It is claimed by (Beenakker et al., 2006) that this result can identify a novel pseudo-spin dependent scattering mechanism in graphene thus, it should manifest itself as a $8e^2/h$ conductance step in a bipolar *p-n* junction. We are sure that in coupled double wire graphene systems this conductance contribution would be simply enhanced by a factor two with a plateau occurring near $16e^2/h$ if the coupling between the wires is weak, which is not the case here.

Nevertheless, these conclusions have motivated us to study the role of the spatial anisotropy inducing modifications to the minimum found in the graphene double wire dispersion relations. We have chosen to calculate the eigenvalues $E_n(k_y)$ for narrower well layer systems where the quantum mechanical regime becomes fully achieved. In Fig. 5 we are showing only one side of the energy dispersion $E_n(k_y)$ for states $n = 1$ (panel (a)) and $n = 2$ (panel (b)) in a symmetric structure where $L_1 = L_2 = 9$ nm, $L_b = 4$ nm and $U_0 = 150$ meV. At this point we should recall that our theory can also be applied to graphene nanoribbons, whose widths are typically of the order of 5 nm (Ritter & Lyding, 2009). As the parameter Δ increases, from -10 to 30 meV in steps of 10 meV, we note minima in both energy branches being systematically destroyed. Therefore, it is possible to claim that the physical mechanisms leading to a plateau in the conductance near $16e^2/h$ will be destroyed for a finite value of Δ. Furthermore, the dotted and dash-dotted lines limit the confined states region at given critical k_y value associated to an asymmetry induced by Δ. Both dotted lines are given by Eq. (10) with the minus sign. These results evidence that, for a more general situation regarding the

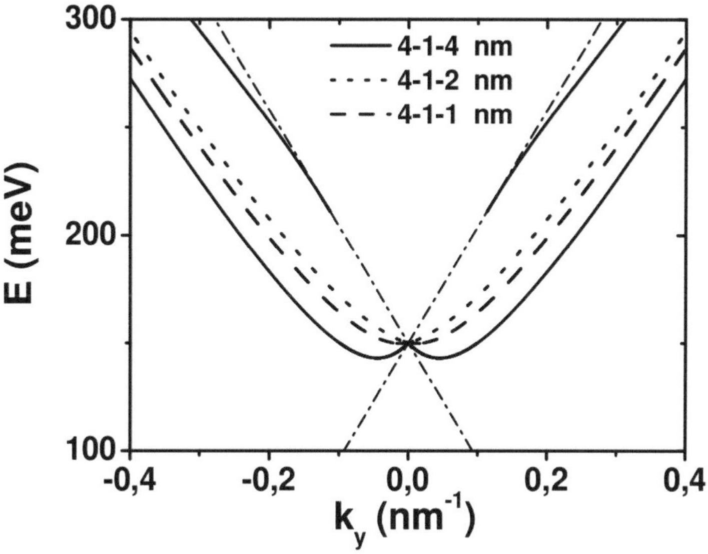

Fig. 6. (Color online) - Effects of increasing spatial anisotropy on the $E_1(k_y)$ minima for $\beta = 0$. Different indicate different values for the set of parameters $L_1 - L_b - L_2$ nm, and where $U_0 = 150$ meV.

graphene sheet and substrate interaction strength, the parameter β in the Eqs. (9) and (10) turns out to be equal to $\max(\beta_{II}, \beta_{IV})$ and $\min(\beta_I, \beta_{III})$, respectively. This informs us that the dotted energy lines limiting the confined state region can be determined by an appropriated choice of the parameter strength β. As a final test to this graphene coupled double wire model, we are showing in Fig. 6 the effects of increasing the spatial asymmetry on the ground-state double minima occurring for the structure $L_1 = 4$ nm, $L_b = 1$ nm, $L_2 = 4$ nm (referred as $4 - 1 - 4$) when we neglect any interaction between graphene sheet and the substrate in all layers of Fig. 2. As can be noted, when L_2 decreases (dashed lines), the two degenerate minima near $k_y = \pm 0.30 nm^{-1}$ merge into a single one minimum at $k_y = 0$. Therefore, in this lowest channel mode the total amount added to the conductance plateau of $16e^2/h$ for a symmetric structure $4 - 1 - 4$ will change back to the total amount $8e^2/h$ in the asymmetric structures shown in this figure.

3.1.3 Guidance efficiency

We show in Fig. 7(a) the dispersion relation E_n of the confined (guided) states (modes) in a symmetrical structure. An spatial anisotropy has been induced here by assuming $\beta_{II} = 0$ and $\beta_{IV} = 40$ meV. The branches E_n indicate the n different roots of the Eq. (7) for any given wavevector k_y. These branches were found numerically in a region limited by the long- and short-dashed lines, which are roughly given by the expressions $E = \gamma k_y$ and $E = \pm\gamma k_y + U_0$, respectively. Outside this region, the quasiparticle can be assumed to be free from the effects of $U(x)$. The intersections of these dashed lines with each branch n indicate the points (k_y, E_n) which determine, through the Eq. (8), the critical angles $\theta_c(E_n)$. The guidance through the y-direction in the whole structure is only allowed for those angles greater than $\theta_c(E_n)$. The two lowest energy branches, E_0 and E_1, get coupled strongly as the wavevector k_y increases.

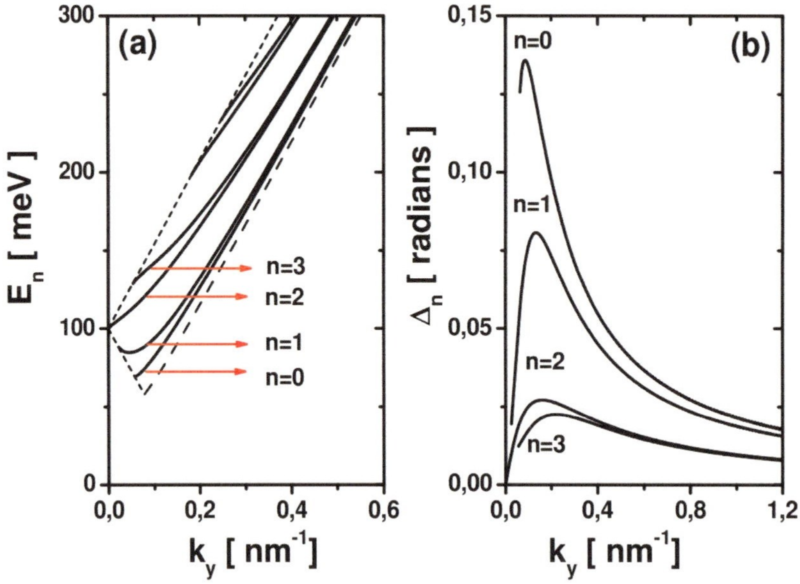

Fig. 7. (Color online) (a) The energy dispersion relation $E_n(k_y)$ for an structure whose parameters are $L_1 = L_2 = 20$ nm, $L_b = 10$ nm and $U_0 = 100$ meV. The six lowest subbands (or modes) are shown. (b) The parameter Δ_n as a function of the wavevector k_y for the 4 lowest energy modes.

The same effect occurs with the two subsequent branches E_2 and E_3. Such sort of effect is not present in the single waveguide case. Notice that, with such a spectrum, we thoroughly map out the allowed coupled guided modes in our biwire system. Furthermore, it is interesting to notice that the guided mode of the second order, $n = 2$, is the only one which is present in the long wavelength limit $k_y \to 0$, with the energy $E_{n=2}(k_y \to 0) \simeq U_0$.

It is worth to mention that, for a given value of the wavevector k_y, the reflection angles θ_n^{II} and θ_n^{IV} assume different values when $\beta_{II} \neq \beta_{IV}$.Indeed, the spatial anisotropy effect induced by considering different graphene-substrate interaction strength in each waveguide moved us to define and analyze the quantity

$$\Delta_n = \left| \theta_n^{j=II} - \theta_n^{j=IV} \right| \tag{12}$$

as a function of the wavevector k_y. It serves as an auxiliary quantity showing how the reflections in the two waveguides are actually taking place. When Δ_n is identically equal to zero, the reflections in the two waveguides are occurring strictly in phase. We plot Δ_n in Fig. 7(b) for $n = 0, 1, 2$ and 3. All parameters are the same as in Fig. 7(a). We found maxima in the Δ_n as a function of k_y for all the considered subbands n. The corresponding wavevector values of these maxima indicate the parallel momenta (or energy) for which the out-of-phase reflections in the two wires lead to the most destructive effect.

We stress that all guided modes found in Fig. 7(a), i.e., the different n branches, can be considered as of infinite lifetime since we are considering purely real energies only (Nguyen et al., 2009).

3.2 Double barrier (One well)

For the sake of completeness, we now show results for an single waveguide, of width 50 mn, in the absence of any substrate interaction, i.e., $\beta_j \equiv \beta = 0$ for all regions. To produce a single waveguide, we take the barrier width $L_b \to 0$ and the well width $L_1 = L_2 = 25$ nm. It is not difficult to realize that such parameters simply lead to the wavevector $k_y \to \gamma^{-1} E_n \sin\theta$, where θ is the reflection angle within the single waveguide defined in the Eq. (8). Therefore, we solve Eq. (7) in order to find the n different roots E_n for a given angle θ. In this way, we plot in Fig. 8(a) the dispersion relation E_n of the confined (guided) states (modes) in this single waveguide as a function of the reflection angle θ. As one can notice, there is much evidence of the existence of guided modes up to the seventh order. These modes were found in a region limited by dotted lines, which are given by $\gamma k_y + U_0$ and $U_0 - \gamma k_y$. The electrons whose energies are located outside this region are not allowed to be guided with any value for θ whatsoever. Intersections of the branches E_n with the dotted lines indicate the critical angle for each mode n. They are marked with circles in Fig. 8(a). Beyond these angles, the guidance should take place in the structure. We finally plot in Fig.8(b) the probability density function $\rho_{n\theta}(x)$ in terms of θ (instead of k_y) for the four lowest guided modes. By showing them, we completely mapped out the values of the reflection angle θ for which the guidance takes place with the most probability.

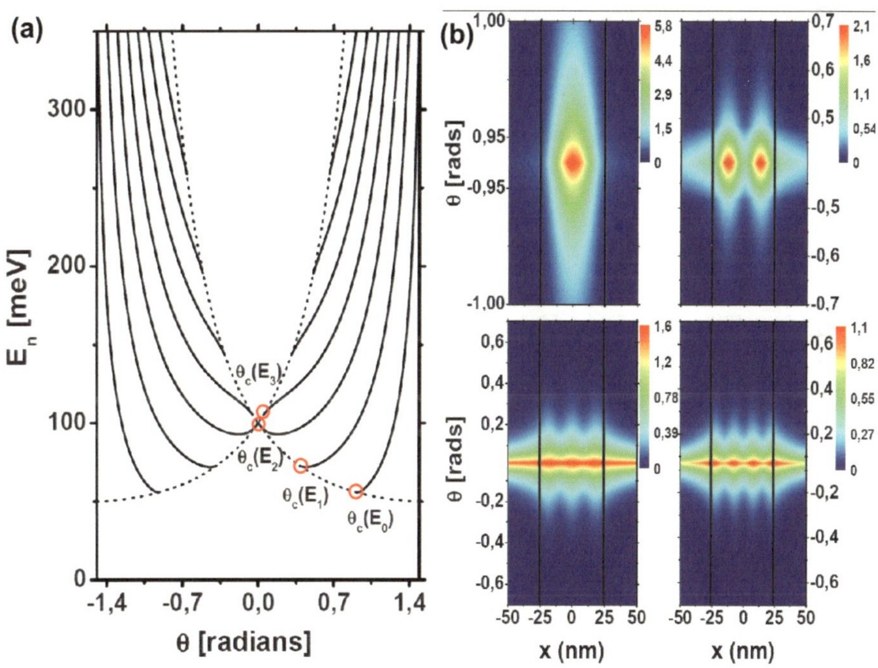

Fig. 8. (Color online) (a) The energy dispersion E_n as a function of the reflection angle θ for a single waveguide of width 50 *nm*. Here, $U_0 = 100$ meV. The red circles indicate the critical angles: $\theta_c(E_0) = 0.92$ rad, $\theta_c(E_1) = 0.40$ rad, $\theta_c(E_2) = 0.0$ rad and $\theta_c(E_3) = 0.05$ rad. (b) The probability density function $\rho_{n\theta}(x)$ corresponding to the 4 lowest energy branches showed in part (a).

4. Collective excitations in graphene quantum wires

Collective excitations (plasmon modes) are the most elementary excitations due to Coulomb interaction. These modes should be seen experimentally through the Raman inelastic scattering spectra (Bostwick et al., 2007). Within the linear response theory, plasmon modes can be theoretically modeled by finding out the poles of the dielectric function, which will be written here within the RPA. Since we are dealing with two coupled graphene stripes, it shall be worthy to develop below the fundamental expressions of interest in order to properly get the dielectric function (Ehrenreich & Cohen, 1959).

4.1 Triple barrier in weak tunneling regime

For the sake of simplicity, we set the solutions for the pseudo-spinor $\phi_{nk}(x)$ by assuming an extremely weak tunneling regime between the wires. In such a regime, the system can be considered as two independent waveguides. This procedure simplifies the model and represents a very feasible experimental situation in which the gate voltage, as shown in Fig.2, is increased to values up to 1 eV. The results shown in (Williams et al., 2011) supports such an approach. The graphene based waveguides turn out to be of a high efficiency, which suggests that the Klein tunneling through the barriers turn out be a less important effect. Indeed, armchair edged GNRs are considered to be no longer a gapless semiconductor. In such a regime, the pseudo-spinor components are considered to be identically zero outside both wires, including the barrier region.

Thus, we neglect the coupling between higher subbands and assume charges (electrons) occupying the lowest subband only, so that subband index n is treated here as waveguide index. The proposed analytical solutions for the pseudo-spinor can assume the following form (Villegas & Tavares, 2011):

$$\phi_{nk}(x) = \begin{pmatrix} A_n \sin k_n^0 \left[x + (-1)^{n+1} L_b/2\right] \\ B_n k_n^0 \cos k_n^0 \left[x + (-1)^{n+1} L_b/2\right] - k \sin k_n^0 \left[x + (-1)^{n+1} L_b/2\right] \end{pmatrix}. \tag{13}$$

Here, the coefficients $A_n = \sqrt{2/L_n}$ and $B_n = -i\gamma A_n / \left[E_n + \beta\right]$, with $E_{n=1,2}(k) = \gamma\sqrt{k^2 + (k_n^0)^2 + (\beta\gamma^{-1})^2}$ being the hyperbolic eigenvalues of H. Notice that $k_n^0 = \pi/L_n$ is the quantized wave-vector corresponding to the fundamental quantum state in both waveguides

4.2 Linear response approach

We then start writing out the usual screened Coulomb potential for the Fermions embedded in our structure Jackson (1998),

$$V^s(r) = \frac{1}{4\pi\epsilon_0} \int dr\prime \frac{\rho(r\prime)}{|r - r\prime|}, \tag{14}$$

where ϵ_0 is the static background dielectric constant which depends on the substrate over which the graphene sheet lies (Ando, 2006). Here, the RPA density-density correlation function

$$\rho(r,\omega) = \frac{e^2}{L_y} \sum_{m,n,k,q} \psi^*_{mk}(x,y)\, \psi_{n,k+q}(x,y) \Pi_{mn}(k,q,\omega)\, V_{mn}(k,q) \tag{15}$$

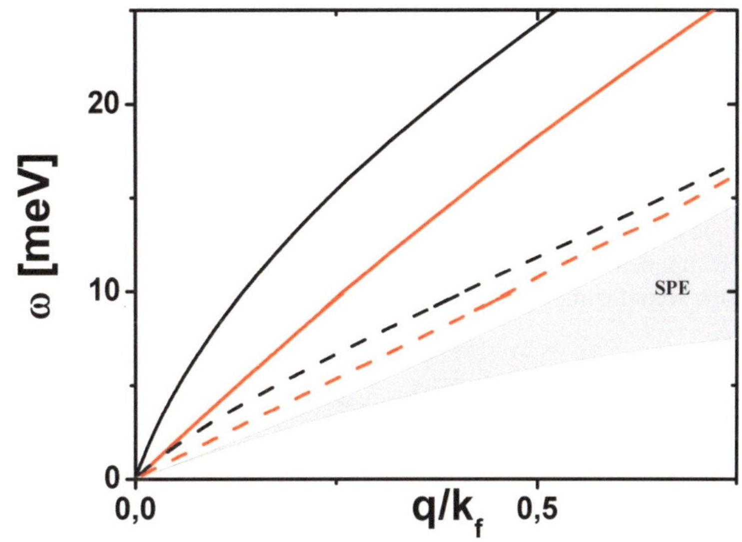

Fig. 9. The optical ω_+ (black branches) and acoustical ω_- (red branches) plasmons for the background values: (i) $\epsilon_0 = 1$ (solid-line branches) and (ii) $\epsilon_0 = 3.9$ (dashed-line branches). The Fermi wavevector $k_f = 0.7x10^6\ cm^{-1}$ and the sample parameters are $L_1 = L_2 = 20$ nm. We keep $\beta = 0$.

is written in terms of both the noninteracting polarizability function,

$$\Pi_{mn}(k,q,\omega) = g_v g_s \frac{f\left[E_m(k)\right] - f\left[E_n(k+q)\right]}{E_m(k) - E_n(k+q) - \hbar\omega - i\delta\hbar}, \tag{16}$$

with g_v and g_s being the valley and spin degeneracy, respectively, and the total Coulomb potential $V \equiv V^{ext} + V^s$, where V^{ext} is the external potential. The wavevector q is the usual transferred momentum due to Coulomb interaction and ω is the external frequency. Since the total potential $V_{mn}(k,q)$ in Eq. (15) is a function of two wave-vectors, one should Fourier transform the potential $V^s(r)$, and take advantage from Eq. (15), to write

$$V^s_{mn}(k,q,\omega) = \frac{1}{L_y} \sum_{m',n',k\prime,q'} \Pi_{m'n'}(k',q',\omega) \times$$

$$v_{mnm'n'}(k\prime,k,q) \times V_{m'n'}(k\prime,q'), \tag{17}$$

where

$$v_{mnm'n'}(k,k',q) = \frac{2e^2}{\epsilon_0} \int dx' \int dx\, \phi^*_{m,k}(x)\phi_{n,k+q}(x) \times$$

$$K_0(q\left|x-x'\right|)\phi^*_{m',k'+q}(x')\phi_{n',k'}(x'). \tag{18}$$

Here, K_0 is the zero-order modified Bessel function of the second kind.

The 1D spatial confinement used here couples the motion in the x-direction to the motion in the y-direction through the wavevector k, leading $v_{mnm'n'}$ to be dependent also on the wavevectors k and k'. We mention that this turns out to be a difficulty, since it

makes the dielectric function depends not only on the wave-vector q, but also on the wavevectors k and k'(Brey & Fertig, 2007). As a matter of fact, for 1D semiconductor systems (Wendler & Grigoryan, 1996; Xia & Hai, 2002) in the presence of perpendicular magnetic field, this difficulty has been overcame by treating the k and the spatial dependencies separated from each other, and expanding the wavefunction in a perturbation series in such a way that the dielectric function could be projected into a finite submatrix, whose dimension is the same as the perturbation order. In this work, we also overcome this difficulty by using the same technique and further taking the advantage from the natural relation between the pseudo-spinors components.
Notice first that the spinor component

$$\phi^B_{nk}(x) = -\frac{i\gamma}{[E_n(k)+\beta]}\left[\frac{d\phi^A_{nk}}{dx} - k\phi^A_{nk}\right] \tag{19}$$

can be written in terms of its counterpart $\phi^A_{nk}(x)$ by virtue of the symmetry of the Eq. 1 only. This leads the Coulomb potential form factor to be written as

$$v_{mnm'n'}(k,k',q) = -\gamma^2\eta A^{(0)}_{mnm'n'} + A^{(1)}_{mnm'n'}k' + A^{(2)}_{mnm'n'}k'^2 + A^{(3)}_{mnm'n'}, \tag{20}$$

where

$$\eta(k',q) = [(E_m(k')+\beta)(E(k'+q)+\beta)]^{-1},$$

and

$$A^{\mu}_{mnm'n'} = a^{(0,\mu)}_{mnm'n'} + a^{(1,\mu)}_{mnm'n'}k + a^{(2,\mu)}_{mnm'n'}k^2 \text{ with } (\mu = 0,1,2, \text{ and } 3).$$

Secondly, notice that the k'-dependence in Eq. (20) has been explicitly taken out. The coefficients $a^{(0,\mu)}_{ijmn}(k,q)$, $a^{(1,\mu)}_{ijmn}(k,q)$, and $a^{(2,\mu)}_{ijmn}(k,q)$ for $\mu = 0,1,2,$ and 3 are shown in the Appendix B and involves the spatial integrals shown in Eq. (18). The summation of the terms on k' appearing in the Eq. (20) does not arise as a consequence of any perturbative approach on $\psi_{nk}(x,y)$.It rather shows up directly due to the nature of the Dirac equation, leading to the Eq. (19). At this point, and in order to get the exact dielectric constant as a function of q and ω only, we use the same technique as in Refs. (Li & Das Sarma, 1991; Rodríguez & Tejedor, 1994) and define

$$X_{mn}(q,\omega) = \frac{1}{L_y}\sum_k V^s_{mn}(k,q,\omega), \tag{21}$$

so that the dynamically dielectric function come out straightforwardly after the standard procedure in obtaining the relation $\varepsilon(q,\omega) = V^s/V$.
Because of the weak tunneling condition and provided the system we are considering in this paper is symmetric ($L_1 = L_2$ in Fig. 1), there are only two different elements of v, i.e., $v_{1111}(q,k,k') = v_{2222}(q,k,k') = V_A$ and $v_{1122}(q,k,k') = v_{2211}(q,k,k') = V_C$. (Tavares & Hai, 2000) Furthermore, in this work we consider the same 1D charge density $N_{1D} = 2k_F/\pi$ (with $k_F = 0.7\times10^6 cm^{-1}$) for both waveguides, so that the polarizability $\Pi_{11} = \Pi_{22} = \Pi_0(k,q,\omega)$. The collective excitation spectra is then obtained through the condition of vanishing determinant of the matrix $\varepsilon(q,\omega)$. This condition leads to the equation $\varepsilon_+\varepsilon_- = 0$, where

$$\varepsilon_{\pm}(q,\omega) = \left[1 - \sum_{k,k'}(V_A \pm V_C)\Pi_0\right]. \tag{22}$$

The roots of $\varepsilon_+ = 0$ and $\varepsilon_- = 0$ provide the in-phase optical and out-phase acoustic plasmon modes, respectively. These modes can be observed via inelastic light scattering spectroscopy experiments. The acoustic mode represents densities in the two wires fluctuating out of phase, whereas the optical mode represents densities in two wires fluctuating in phase. In the following, we analyze these modes carefully.

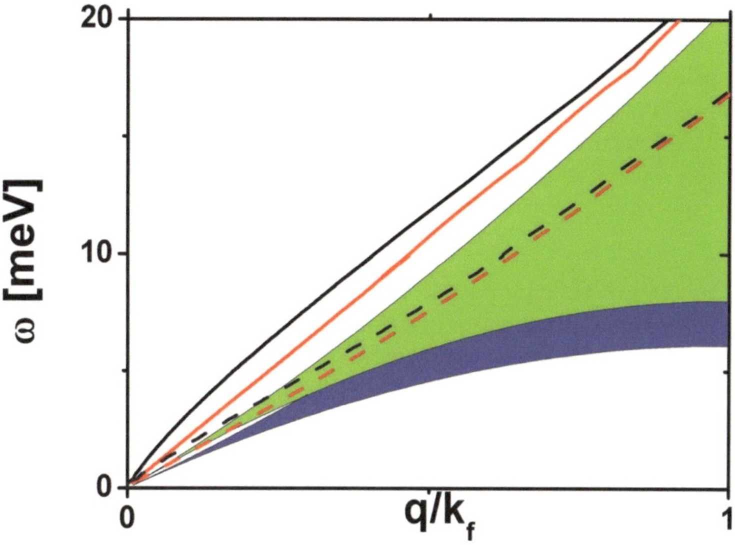

Fig. 10. The optical ω_+ (black branches) and acoustical ω_- (red branches) plasmons for different values of the parameter β : (i) $\beta = 0$ (solid-line branches) and (ii) $\beta = 75meV$ (dashed-line branches). The Fermi wavevector and the sample parameters are the same as in the previous Figure. Here we keep $\epsilon_0 = 3.9$. The green [blue] continuum is the SPE region for the case (i) [(ii)].

4.3 Results and discussions

Firstly, let us discuss the interesting effects that different choices for the *background* dielectric constant ϵ_0 have on the response functions ε_+ and ε_- and, as consequence, on the plasmon dispersion relation. This dielectric constant accounts for the kind of material that serves as a substrate over which the graphene sheet is eventually deposited. Here, we take two values for it: (i) $\epsilon_0 = 1$ and (ii) $\epsilon_0 = 3,9$. The first one represents the most elementar case where the graphene sheet is suspended, while the case (ii) represents SiO_2 as a substrate over which the graphene sheet is deposited. The goal here is to show the effects of continuously changing ϵ_0 on the plasmon modes and also on the single-particle excitations (SPE). For such a purpose, we show in Fig.9 the dispersion relation of the acoustic $\omega_-(q)$ (red branches) and optical $\omega_+(q)$ (black branches) plasmon modes, as a function of the transferred wavevector q, for the two values of ϵ_0 discussed above. The solid lines represent the results for $\epsilon_0 = 1$, while results for $\epsilon_0 = 3,9$ are shown in dashed lines. The shadow area is the continua where SPE occurs.

These excitations are defined by a continuum region in the ω-q space where $Im[\Pi_0(q,\omega)] \neq 0$ and are responsable for Landau damping the plasmon modes as they enter into (or approach to) the shaded continuum. As ϵ_0 increases, the plasmons get closer and closer to the SPE continuum. These results then suggest that these modes should be experimentally seen more easily for materials with less values of ϵ_0, since they are far from the Landau damping region. At this point we should briefly mention that, in the presence of effective tunneling between the wells, the acoustic mode $\omega_-(q)$ in Fig.9 should present a non-zero value in the long wavelength limit $q \to 0$. The same behavior certainly occurs if $\epsilon_0 = 11,68$, which corresponds to the background value for GaAs. Despite of it, the results presented here should not be much different when an effective, but still weak, tunneling is considered. Moreover, these results turn out be the first step in theoretically approaching tunneling effects, which is certainly responsible for losing guidance efficiency in the ballistical waveguides. In Fig. 10, we choose to show the same results as in Fig. 9, but now varying the parameter β and keep the dielectric constant $\epsilon_0 = 3.9$. As the parameter β increases, the SPE continuum changes accordingly. Such an effect on the SPE mimics an re-scaling of the Fermi surface as the parameter β increases.

5. Conclusion

In summary, we have studied single-particle properties of Dirac Fermions laterally confined to coupled double wire graphene structures which are being embedded over different substrates, and explored for 2D and 1D transport recently. We have added an ad-hoc term to the pseudo-relativistic Dirac equation which controls the interaction between graphene sheets and the substrate. Both the strength of this interaction as well as the spatial asymmetry on the double wire structures can affect the extended and localized character of states and the shape of the minima in the energy dispersion of the lowest subbands. We may claim that these effects will manifest themselves on the quantum conductance of the system and, eventually, changing the conductance plateau from $16e^2/h$ to $8e^2/h$. We also have shown the existence of guided modes up to the seventh order in single and double waveguides based on graphene. Our results also show that different choices for β in both waveguides induces spatial anisotropy effects on the guided modes, reflecting anomalous effects on their probability density functions. We believe that these results are timely and may stimulate further theoretical and experimental investigations on these graphene based structures.

We also have theoretically studied the acoustical and optical plasmon modes in coupled graphene quantum wires in the extremely weak tunneling regime. In particular, attention was devoted to the effects induced by the interaction between the graphene sheet and the substrate. This interaction has been considered through an *ad-hoc* parameter β in the 2D Dirac-like Hamiltonian modeling a more general diatomic system in which the graphene lattice sites A and B might have different number of electrons. We have calculated the Coulomb potential elements for this massless Fermion gas. We have shown that the parameter β might be serving to screen the Coulomb interaction in the system. Such an effect manifests itself in the dispersion relation of the optical and acoustical plasmon modes. We found that the parameter β should eventually increase the damping effects on these modes.

6. Acknowledgments

The authors are indebted to Gilmar E. Marques for helpfull discussions. The authors also thank to FAPESP (Brazilian agency) and UFABC for financial support.

7. Appendix A

The 8×8 matrix $\hat{M}$ is given by

$$\begin{bmatrix} \frac{f_-(E+\beta_{II})L}{\gamma e^{-\kappa_I\theta}} & m_{12} & m_{13} & 0 & 0 & 0 & 0 & 0 \\ 0 & m_{22} & m_{23} & m_{24} & m_{25} & 0 & 0 & 0 \\ 0 & 0 & 0 & m_{34} & m_{35} & m_{36} & m_{37} & 0 \\ 0 & 0 & 0 & 0 & 0 & m_{46} & m_{47} & \frac{f_+(E+\beta_{IV})L}{\gamma e^{-\kappa_I\delta}} \\ e^{-\kappa_I\theta} & \sin q_{II}\theta & -\cos q_{II}\theta & 0 & 0 & 0 & 0 & 0 \\ 0 & \sin q_{II}\sigma & -\cos q_{II}\sigma & e^{-\kappa_{III}\sigma} & e^{\kappa_{III}\sigma} & 0 & 0 & 0 \\ 0 & 0 & 0 & e^{\kappa_{III}\sigma} & e^{-\kappa_{III}\sigma} & -\sin q_{IV}\sigma & -\cos q_{IV}\sigma & 0 \\ 0 & 0 & 0 & 0 & 0 & -\sin q_{IV}\delta & -\cos q_{IV}\delta & e^{-\kappa_I\delta} \end{bmatrix}, \quad (23)$$

where

$$\begin{aligned} m_{12} &= L(k_y \sin q_{II}\theta + q_{II}\cos q_{II}\theta), \quad & m_{13} &= L(q_{II}\sin q_{II}\theta - k_y \cos q_{II}\theta), \\ m_{22} &= L(k_y \sin q_{II}\sigma + q_{II}\cos q_{II}\sigma), \quad & m_{23} &= L(q_{II}\sin q_{II}\sigma - k_y \cos q_{II}\sigma), \\ m_{24} &= L\gamma^{-1}g_-(E+\beta_{II})e^{-\kappa_{III}\sigma}, \quad & m_{25} &= L\gamma^{-1}g_+(E+\beta_{II})e^{\kappa_{III}\sigma}, \\ m_{34} &= L\gamma^{-1}g_-(E+\beta_{IV})e^{-\kappa_{III}\sigma}, \quad & m_{35} &= L\gamma^{-1}g_+(E+\beta_{IV})e^{\kappa_{III}\sigma}, \\ m_{36} &= L(q_{IV}\cos q_{IV}\sigma - k_y \sin q_{IV}\sigma), \quad & m_{36} &= -L(k_y \cos q_{IV}\sigma + q_{IV}\sin q_{IV}\sigma), \\ m_{46} &= L(q_{IV}\cos q_{IV}\delta - k_y \sin q_{IV}\delta), \quad & m_{36} &= -L(k_y \cos q_{IV}\delta + q_{IV}\sin q_{IV}\delta). \end{aligned}$$

Here, we also define $f_\pm = \gamma(k_y \pm \kappa_I) \,/\, (E - U + \beta_I), g_\pm = \gamma(k_y \pm \kappa_{III}) \,/\, (E - U + \beta_{III}), \theta = (2L_1 + L_b)/2, \sigma = L_b/2$, and $\delta = (2L_2 + L_b)/2$.

8. Appendix B

The coefficients $a^{(0,\mu)}_{ijmn}(k,q)$, $a^{(1,\mu)}_{ijmn}(k,q)$ and $a^{(2,\mu)}_{ijmn}(k,q)$ for $\mu = 0,...,3$ are related to the expansion of the Coulomb form factor and have the following form:

$$\begin{aligned} a^{(0,0)}_{ijmn}(k,q) = \frac{2e^2}{\epsilon_0}\int dx' \int dx[\phi_{A,i}(x)\phi_{A,j}(x) + \gamma^2\eta(\phi'_{A,i}(x)\phi'_{A,j}(x) + \\ q\phi'_{A,i}(x)\phi_{A,j}(x))]K_0(q\,|x - x'|)[\phi'_{A,m}(x')\phi'_{A,n}(x') + q\phi'_{A,m}(x')\phi_{A,n}(x')]. \end{aligned}$$

9. References

Ando, T.; & Nakanishi, T. (1998). Impurity Scattering in Carbon Nanotubes - Absence of Back Scattering. *J. Phys. Soc. Jpn.* Vol. 67, (1998) 1704-1713.

Ando, T.; (2006). Screening Effect and Impurity Scattering in Monolayer Graphene. *J. Phys. Soc. Jpn.* Vol. 75, (2006) 074716.

Beenakker, C. W. J. (2008). Colloquium: Andreev reflection and Klein tunneling in graphene. *Rev. Mod. Phys.* Vol. 80, (2008) 1337–1354

Beenakker, C.W. J.; Sepkhanov, R. A.; Akhmerov, A. R.; & Tworzydło, J. (2009). Quantum Goos-Hanchen effect in graphene. *Phys. Rev. Lett.* Vol. 102, (2009) 146804-146808

Berry, M. V.; & Mondragon, R. J. (1987). Neutrino billiards: time-reversal symmetry-breaking without magnetics fields. *Proc. R. Soc. Lond. A.* Vol. 412, (1987) 53-74

Bostwick, A.; Ohta, T.; Seyller, T.; Horn, K.; & Rotenberg, E. (2007). Quasiparticle dynamics in graphene. *Nature Physics.* Vol. 3, (2007) 36 - 40

Brey, L.; Fertig, H. A. (2006). Electronic states of graphene nanoribbons studied with the Dirac equation. *Phys. Rev. B,* Vol. 73, (2006) 235411-235416

Brey, L.; & Fertig, H. A.; (2007). Elementary electronic excitations in graphene nanoribbons. *Phys. Rev. B* Vol. 75, (2007) 125434-125440

Castro Neto, A. H.; Guinea, F.; Peres, N. M. R.; Novoselov, K. S.; & Geim, A. K. (2009). The electronic properties of graphene. *Rev. Mod. Phys,* Vol. 81, (2009) 109–162

Castro Neto, A. H. (2010). The carbon new age. *Materialstoday,* Vol. 13, (2010) 12-17

Cheianov, V. V.; Fal'ko, V.; & Altshuler, B. L. (2007). The Focusing of Electron Flow and a Veselago Lens in Graphene p-n Junctions. *Science,* Vol. 315, (2007) 1252-1255

Chiu, H-Y.; Perebeinos, V.; Lin, Y-M.; & Avouris, P. (2010). Controllable p-n Junction Formation in Monolayer Graphene Using Electrostatic Substrate Engineering. *Nano Letters,* Vol. 10, (2010), 4634–4639

Hwang, E. H.; & Das sarma, S. (2007). Dielectric function, screening, and plasmons in two-dimensional graphene. *Phys. Rev. B,* Vol. 75, (2007) 205418-205424

Das Sarma, S.; Adam, S.; Hwang, E. H.; Rossi, E. (2010). Electronic transport in two dimensional graphene. *arXiv:1003.4731v2*

Ehrenreich, H.; & Cohen, M. H. (1959). Self-Consistent Field Approach to the Many-Electron Problem. *Phys. Rev.* Vol. 115, (1959) 786-790

Giovannetti, G.; Khomyakov, P. A.; Brocks, B.; Kelly, P. J.; & van den Brink, J. (2007). Substrate-induced band gap in graphene on hexagonal boron nitride: Ab initio density functional calculations. *Phys. Rev. B,* Vol. 76, (2007) 073103-073107

Goos, F.; & Hanchen, H. (1947). Ein neuer und fundamentaler versuch zur totalreflexion. *Ann. Phys.* Vol. 1, (1947) 333-346

Jackson, J. D. (1998). Classical Electrodynamics. *New York: John Wiley & Sons,* (1998).

Kane, C. L.; & Mele, E. J. (2005). Quantum spin Hall effect in graphene. *Phys. Rev. Lett,* Vol. 95, (2005) 226801-226805

Katsnelson, M. I.; Novoselov, K. S.; & Geim, A. K. (2006). Chiral tunnelling and the Klein paradox in graphene. *Nature Phys.* Vol. 2, (2006) 620-625

Li, Q. P.; & Das Sarma, S. (1991). Elementary excitation spectrum of one-dimensional electron systems in confined semiconductor structures: Finite magnetic field. *Phys. Rev. B,* Vol. 44, (1991) 6277–6283

Lin, Y.-M.; Perebeinos, V.; Chen, Z.; & Avouris P. (2008). Electrical observation of subband formation in graphene nanoribbons. *Phys. Rev. B* Vol. 78, (2008) 161409-161413

Lu, Y. H.; He, P. M.; & Feng, Y. P. (2007). Graphene on metal surface: gap opening and n-doping. *arXiv:0712.4008v1*

Min, H.; Hill, J. E.; Sinitsyn, N. A.; Sahu, B. R.; Kleinman, L.; & MacDonald, A. H. (2006). Intrinsic and Rashba spin-orbit interactions in graphene sheets. *Phys. Rev. B,* Vol. 74, (2006) 165310-165315

Miao, F.; Wijeratne, S.; Zhang, Y.; Coskun, U. C.; Bao, W.; & Lau, C. N. (2007). Phase-coherent transport in graphene quantum billiards. *Science,* Vol. 317, (2007) 1530-1533

Nguyen, H.C.; Hoang, M.T.; & Nguyen, V.L. (2009). Quasi-bound states induced by one-dimensional potentials in graphene. *Phys. Rev. B.* Vol. 79 (2009) 035411-035417

Novoselov, K. S.; Geim, A. K.; Morozov, S. V. ; Jiang, D.; Zhang, Y.; Dubonos, S. V.; Grigorieva, I. V.; & Firsov, A. A. (2004). Electric field effect in atomically thin carbon films. *Science,* Vol. 306, (2004) 666-669.

Pereira, J. M.; Mlinar, V.; Peeters, F. M.; & Vasilopoulos, P. (2007). Confined states and direction-dependent transmission in graphene quantum wells. *Phys. Rev. B,* Vol. 74 (2007) 045424-045428

Pereira, J. M.; Mlinar, V.; Peeters, F. M.; & Vasilopoulos, P. (2007). Confined states and direction-dependent transmission in graphene quantum wells. *Phys. Rev. B,* Vol. 74 (2007) 045424-045428

Peres, N. M. R.; Rodrigues, J.N.B.; Stauber, T.; & Lopes dos Santos, J. M. B. (2009). Dirac electrons in graphene-based quantum wires and quantum dots. *J. Phys.: Condens. Matter,* Vol. 21, (2009) 344202-344220

Peres N. M. R. (2010). Colloquium: The transport properties of graphene: An introduction. *Rev. Mod. Phys,* Vol. 82, (2010) 2673-2700

Ritter, K. A.; & Lyding, J. W. (2009). The influence of edge structure on the electronic properties of graphene quantum dots and nanoribbons. *Nature Materials* Vol. 8, (2009) 235 - 242

Rodríguez, F. J.; & Tejedor, C. (1994). Nonlocal interaction and Fermi-edge singularities in quasi-one-dimensional systems with a transverse magnetic field. *Phys. Rev. B* Vol. 49, (1994) 16781-16784

Semenoff, Gordon W. (1984). Condensed-Matter simulation of a three-dimensional anomaly. *Phys. Rev. Lett.* Vol. 53, (1984) 2449–2452.

Son, Y-W; Cohen, M. L.; & Louie, S. G. (2006). Energy gaps in graphene nanoribbons. *Phys. Rev. Lett.* Vol. 97, (2006) 216803-216807

Tavares, M.R.S. & Hai, G.-Q. (2000). Inelastic Coulomb scattering rates due to acoustic and optical plasmon modes in coupled quantum wires. *Phys. Rev. B* Vol. 61, (2000) 7564-7570

Villegas, C. E. P.; & Tavares, M. R. S. (2010). Strongly coupled modes in bi-waveguides based on graphene. *Solid state communications,* Vol 150 (2010) 1350-1354

Villegas, C. E. P.; Tavares, M. R. S.; & Marques, G. E. (2010). Anisotropy induced localization of pseudo-relativistic spin states in graphene double quantum wire structures. *Nanotechnology,* Vol 21, (2010) 365401-365407

Villegas, C. E. P.; & Tavares, M. R. S. (2010). Substrate interaction effects on optical and acoustical plasmons in bi-waveguides based on graphene. *Diamond and related Materials,* Vol 20, (2011) 170-173

Wendler, L.; & Grigoryan, V. G. (1996). Collective and single-particle excitations of the quasi-one-dimensional electron gas in the presence of a magnetic field. *Phys. Rev. B,* Vol. 54, (1996) 8652-8675

Williams, J. R.; DiCarlo, L.; & Marcus, C. M. (2007) Quantum Hall effect in a gate-controlled p-n junction of graphene. *Science,* Vol. 317, (2007) 638-641.

Williams, J. R.; Low, T.; Lundstrom, M. S.; & Marcus, C. M. (2011) Gate-controlled guiding of electrons in graphene. *Nature Nanotechnology,* Vol. 6, (2011) 222–225

Xia, J.-B.; & Hai, G.-Q. (2002). Collective and single-particle excitations in coupled quantum wires in magnetic fields. *Phys. Rev. B* Vol. 65, (2002) 245326-245331

Yakes, M.K. *et al.* (2010). Conductance Anisotropy in Epitaxial Graphene Sheets Generated by Substrate Interactions. NanoLetters Vol 10, (2010) 1559-1562

Zhang, L. M.; & Fogler, M. M. (2008). Nonlinear Screening and Ballistic Transport in a Graphene p-n Junction. *Phys. Rev. Lett.* Vol. 100, (2008) 116804-116808.

Zhou, S. Y.; Gweon, G.-H.; Fedorov, A. V.; First, P. N.; De Heer, W. A.; Lee, D.-H.; Guinea, F.; Castro Neto, A. H.; & Lanzara, A. (2007). Substrate-induced bandgap opening in epitaxial graphene. *Nature Mater.* Vol. 6, (2007) 770-775.

Ziegler, K. (1996). Dirac fermions with disorder in two dimensions: Exact results. *Phys Rev. B.* Vol. 53, (1996) 9653-9657.